Polymer Surfaces

Polymer Surfaces

From Physics to Technology

Fabio Garbassi
Istituto Guido Donegani, EniChem S.p.A., Novara, Italy

Marco Morra
Nobil Bio Ricerche, Villafranca D'Asti, Italy

Ernesto Occhiello
Istituto Guido Donegani, EniChem S.p.A., Novara, Italy

JOHN WILEY & SONS

Chichester · New York · Brisbane · Toronto · Singapore

Other Wiley Editorial Offices

John Wiley & Sons, Inc., 605 Third Avenue,
New York, NY 10158-0012, USA

Jacaranda Wiley Ltd, 33 Park Road, Milton
Queensland 4064, Australia

John Wiley & Sons (Canada) Ltd, 22 Worcester Road,
Rexdale, Ontario M9W 1L1, Canada

John Wiley & Sons (SEA) Pte Ltd, 37 Jalan Pemimpin #05–04,
Block B, Union Industrial Building, Singapore 2057

British Library Cataloguing in Publication Data

A catalogue record for this book is available from the British Library

ISBN 0 471 93817 3

Typeset in 10/12pt Times by Thomson Press (India) Ltd., New Delhi
Printed and bound in Great Britain by Biddles Ltd, Guildford, Surrey

To our wives

Solo dopo aver conosciuto la superficie delle cose ci si puo' spingere a cercare quello che c'e' sotto.
Ma la superficie delle cose e' inesauribile.

Italo Calvino, *Palomar* (Einaudi, Torino)

Only after knowing the surface of things can one throw himself into the search for what is underneath.
But the surface of things is endless.

Contents

Preface   xi

PART I: INTRODUCTORY REMARKS   1

Chapter 1   The Origin of Surface Properties   3
   1.1   Van der Waals' forces   6
   1.2   Electrostatic forces   14
   1.3   The DLVO theory   27
   1.4   Structural interactions   32
   1.5   Short-range interactions (hydrogen, acid–base, covalent)   43
   References   45

Chapter 2   Dynamics of Polymer Surfaces   49
   2.1   Fundamental aspects of polymer surface dynamics   50
   2.2   Experimental evidence of polymer surface dynamics   57
   References   65

PART II: CHARACTERIZATION METHODS   67

Chapter 3   Spectroscopic Methods   69
   3.1   Ion scattering spectroscopy   69
   3.2   Secondary ion mass spectroscopy   75
   3.3   X-ray photoelectron spectroscopy   92
   3.4   Internal reflection spectroscopy   113
   3.5   Diffuse reflectance spectroscopy   122
   3.6   Photoacoustic spectroscopy   128
   3.7   Other vibrational techniques   133
   3.8   Transmission spectroscopies   143
   3.9   Nuclear magnetic resonance spectroscopy   144
   3.10   Comparison of spectroscopic methods   148
   References   150

Chapter 4   Surface Energetics and Contact Angle   161
   4.1   Contact angles and the Young equation   164

4.2   Experimental measurement of contact angles   166
4.3   Contact angle hysteresis   171
4.4   From contact angle to surface tension   182
4.5   Surface thermodynamic parameters by inverse gas
      chromatography   193
References   195

Chapter 5   New and Emerging Methods   200
5.1   Surface force measurements   200
5.2   The evaluation of electrostatic interactions   207
5.3   Reflectivity of neutrons   216
References   217

PART III: MODIFICATION TECHNIQUES   221

Chapter 6   Physical Modifications   223
6.1   Flame treatments   223
6.2   Corona treatments   225
6.3   'Cold' plasma treatments   226
6.4   'Hot' plasma treatments   229
6.5   UV treatments   230
6.6   Laser treatments   231
6.7   X-ray and γ-ray treatments   232
6.8   Electron beam treatments   232
6.9   Ion beam treatments   233
6.10 Metallization   234
6.11 Sputtering   235
References   237

Chapter 7   Chemical Modifications   242
7.1   Wet treatments   242
7.2   Surface grafting   257
References   269

Chapter 8   Bulk Modifications   274
8.1   Polymer blend surfaces   275
8.2   Block copolymer surfaces   285
References   296

PART IV: APPLICATIONS   299

Chapter 9   Wettability   301
9.1   Hydrophilic surfaces   305
9.2   Hydrophobic surfaces   319
References   329

Chapter 10   Adhesion   332
    10.1   Theories of adhesion   332
    10.2   Measurement of adhesion   340
    10.3   Methods for modifying adhesion   348
    References   373

Chapter 11   Barrier Properties   379
    11.1   Coating   379
    11.2   Sulphonation and fluorination   382
    11.3   Evaporation   385
    11.4   PECVD and sputtering   389
    References   390

Chapter 12   Biomedical Materials   395
    12.1   Blood contacting devices   396
    12.2   Contact lenses   416
    References   430

Chapter 13   Friction and Wear   436
    13.1   Compounding   437
    13.2   Chemical coating   442
    13.3   High energy density technologies   445
    References   446

Index   455

Preface

The idea to write a book on polymer surfaces came from the rapid growth of the subject and from the fact that the numerous books published on it, even if in some cases dealing with fundamentals, were more a collection of contributions from different authors on their own research work than an organic and exhaustive treatise on the subject. Moreover, the recent prolific scientific literature is spread over a large number of journals concerning different fields, and so is only available with difficulty to the single scientist. We conceived a content that starts from the physical principles and proceeds to the more important application aspects. We intended also to focus on the latter aspect, considering mostly the scientific literature as well as the patent sources.

This book is divided into 13 chapters, grouped in four parts. The first part concerns some fundamentals, treating the origin of superficial properties of polymers and their dynamic aspects. The second part discusses the methods that characterize the polymer surfaces, highlighting the spectroscopic methods and those connected with surface energetics. The third part deals with techniques able to modify the properties of polymer surfaces. Finally, the fourth part is on applicative aspects, such as wettability, adhesion, biocompatibility, etc., all important both from the technological and the scientific points of view.

We have attempted to review the above subjects giving the state of the art for all of them; however we lay no claim to completeness, owing to the great amount of published literature.

Novara, March 1993

Fabio Garbassi
Marco Morra
Ernesto Occhiello

Part I

INTRODUCTORY REMARKS

Chapter 1

The Origin of Surface Properties

Surface forces are at the origin of surface properties exhibited by all kinds of materials [1–4]. They are caused by the forces acting between the basic component of matter, and result from a summation of all the interatomic forces acting between all of the atoms of the material involved, those of the two bodies plus any intervening medium. It must be noted that, because of many-body effects, the summation of interatomic forces is not simply a pairwise addition. However, some of the correct trends in the interaction behaviour can be understood by considering that as a first approximation. When a material interacts with another, or with the surrounding environment, it is the nature of the surface field of forces that determines the kind of interaction. Its knowledge allows, in principle, the understanding and control of its behaviour, to design or modify it in order to perform at best a given function. The key problem of this chapter, and, in general, of the whole book, is portrayed in Figure 1.1, where the interaction free energy between two macroscopic bodies as a function of separation D is shown in a qualitative way (we shall adopt the symbols W and D for the free energy of interaction and separation of macroscopic bodies, and w and r for the same quantities when interactions involve only molecules).

A number of issues raised by Figure 1.1 must be addressed, in order to understand the interaction between bodies:

(1) The functional dependence of W on D. It determines the range of the interaction, that is the distance at which one body feels the presence of the other.

(2) The depth of the potential well, and the distance at which the minimum is found. These two parameters control the strength of the interaction and the distance between the two bodies at equilibrium. Very often, especially in colloid science, the former term is expressed in units of the thermal energy kT,

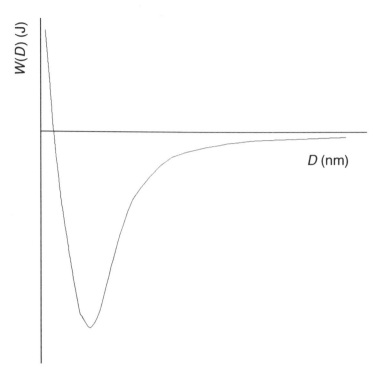

Figure 1.1   A generic interaction free energy versus distance curve.

k being the Boltzmann constant and T the absolute temperature. The rationale for this choice is to use the Boltzmann factor kT as a gauge for the strength of the interaction, since it represents the disorganizing effect of thermal motion. Thus, if the strength of the interaction exceeds kT, bodies or molecules will organize (i.e. coalesce, coagulate, condense, bond, etc.), while if thermal energy wins out, the unorganized form will prevail.

(3) The form of $W(D)$. As discussed later in this chapter, when two forces of opposite sign and different dependence on D are operating, interaction potentials such as the one shown in Figure 1.2(a) are not uncommon. If the energy barrier is far higher than thermal energy kT, the two bodies will repel each other, even if the stable equilibrium state is the aggregated form. Particles whose interaction free energy is shown in Figure 1.2(b) will adhere, but in the secondary minimum, and their adhesion energy will be much lower than expected on purely thermodynamic grounds. It is clear that, in such circumstances, the knowledge of only the free energy minimum is of little help in understanding the nature of the interaction.

The last part of this century has witnessed a great deal of development in the

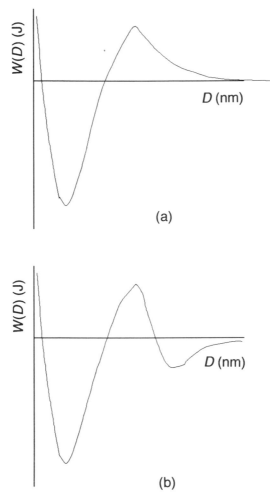

Figure 1.2 Possible forms of the interaction free energy versus distance curve.

theoretical description and the understanding of surface forces [5,6]. On the experimental side, the development of the surface force apparatus [7] (see Chapter 5) was a major breakthrough, since it allows the measurement of the forces that arise as two bodies approach.

The importance of surface forces is best appreciated in the field of colloid science, where the very high surface to volume ratio magnifies the role of surface forces. Many, if not all, of the theories describing surface forces were first developed for colloidal systems.

The aim of this chapter is to present these theories, to describe how the basic properties of matter give rise to interaction functions between bodies such as those shown in the previous figures. Of course, in a field where knowledge is rapidly increasing, the approach will be rather descriptive, and our goal will be to give the basis for the understanding of the (surface) structure–properties relationships that will be discussed in the rest of the book. The reader will be referred to specific textbooks for in depth discussion on a given topic.

1.1   VAN DER WAALS' FORCES

Van der Waals' forces is the collective name given to a set of forces characterized by the same power dependence on distance. The important parameters in this kind of interaction are the dipole moment, which arises from the uneven charge distribution of molecule, and the atomic polarizability, which basically indicates the tendency to charge redistribution when the molecule is subjected to an electrical field. Owing to this process of charge redistribution, a molecule becomes a dipole and gives rise to an electrical field. Finally, an important contribution to the van der Waals' field of forces is given by the interaction between the instantaneous dipole moment arising from the instantaneous position of electrons with respect to the nucleus.

The fundamental importance of van der Waals' forces stems from the fact that they are always involved in the interaction between bodies, unlike other kinds of forces that require a particular feature (such as an electric charge or a layer of adsorbed macromolecules, as discussed below).

In order to obtain the analytical form of the van der Waals interaction between bodies, let us recall some expressions that belong to the field of the interaction between isolated molecules. From basic electrostatics, the potential between two interacting dipoles in vacuum, is given by:

$$w(r,\Theta_1,\Theta_2,\phi) = -\frac{u_1 u_2}{4\pi\varepsilon_0 r^3}[2\cos\Theta_1\cos\Theta_2, -\sin\Theta_1\sin\Theta_2\cos\phi] \quad (1.1)$$

where u_n is the dipole moment of molecule n, ε_0 is the permittivity of the free space, r is the distance between the centre of dipoles and the angles as defined in Figure 1.3. When the thermal energy is greater than the energy that tries to align molecules (equation (1.1)), dipoles will be freely rotating. In this case the interaction free energy can be obtained by subjecting equation (1.1) to a Boltzmann average over all the space, which, while leaving the dipoles free to rotate, will give more weight to the more energetically favourable orientations [8]:

$$e^{-w(r)/kT} = \int e^{-w(r)/kT}\, d\Omega / \int d\Omega \quad (1.2)$$

where $d\Omega = \sin\Theta\, d\Theta\, d\phi$ (Figure 1.3) and the integration is over all the space. From equations (1.1) and (1.2) it can be shown that the dipole–dipole interaction

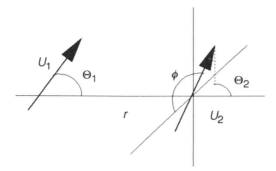

Figure 1.3 Coordinate system for interacting dipoles.

free energy of rotating, permanent dipoles in vacuum is given by:

$$w(r) = -\frac{u_1^2 u_2^2}{3(4\pi\varepsilon_0)^2 kTr^6} = -C_k/r^6 \tag{1.3}$$

which is the formula developed by Keesom in 1921 [9, 10]. This kind of interaction is usually called the Keesom or orientation interaction.

In a similar way it is possible to obtain the free energy of the Debye or induction interaction. In this case the starting point is the dipole moment that is induced in a given molecule by the electrical field E arising from a nearby dipolar molecule, that is:

$$u_{\text{ind}} = \alpha E \tag{1.4}$$

where α is the atomic polarizability. It is possible to show that the free energy of the dipole–induced dipole interaction, between molecules 1 and 2 is given by:

$$w(r) = -\frac{u_1^2 \alpha_2 + u_2^2 \alpha_1}{(4\pi\varepsilon_0)^2 r^6} = -C_D/r^6 \tag{1.5}$$

as shown by Debye in 1921 [11, 12].

The last contribution to the overall van der Waals' force is the London or dispersion interaction. Since this kind of interaction does not require any particular molecular feature (such as a permanent dipole in, at least, one of the two partners, as in the previous cases) it acts between all atoms and molecules. In fact, it originates from the interaction between instantaneous dipole moment given by the instantaneous distribution of electrons with respect to the positively charged nucleus. The electric field of the instantaneous dipole polarizes any nearby atom or molecule, inducing in this way a dipole moment in it. The calculation of the free energy of interaction requires knowledge of the electrons distribution, that is a quantum mechanical treatment, and was first successfully performed by London [13], who proposed the following equation, for the interaction between two dissimilar hydrogen-like atoms:

$$w(r) = -\frac{3\alpha_1\alpha_2 h v_1 v_2}{2(4\pi\varepsilon_0)^2 r^6(v_1 + v_2)} = -C_{\mathrm{L}}/r^6 \tag{1.6}$$

where h is Planck's constant and v_n is the frequency of the electron in the first orbit of a Bohr atom. Its role in equation (1.6) can be intuitively understood by considering that it is linearly proportional to the Bohr radius a_0, and hence to the instantaneous dipole moment $u = a_0 e$, where e is the electronic charge. Experimentally, the value of v_n is obtained from the dispersion of the refractive index, hence the name dispersion forces.

The interaction free energies described by equations (1.3) to (1.6) share the common feature of an inverse dependence on the sixth power of the distance. Thus they can be written in a single equation, which describes the total van der Waals' contribution to the free energy of interaction:

$$w_{\mathrm{vdw}} = -[C_{\mathrm{D}} + C_{\mathrm{K}} + C_{\mathrm{L}}]/r^6 = -C_{\mathrm{vdw}}/r^6 \tag{1.7}$$

Equation (1.7) together with a repulsive, very short-ranged, potential arising from the overlap between the electron clouds of atoms (Born repulsion), describes the interactions between isolated molecules in vacuum, while our goal is to express the free energy of interaction of macroscopic bodies in a given medium. In order to accomplish the first task, let us consider a system comprising a molecule and a flat solid composed of like molecules [14]. If we assume that interactions are simply additive, i.e. not complicated by many-body effects, it is clear that the overall free energy of interaction can be obtained by integrating the molecule–molecule interaction potential over all the solid, as indicated in Figure 1.4. If ρ is the number density of molecules in the solid, then the number of molecules in the ring indicated in Figure 1.4 is $2\pi\rho x\,dx\,dz$. Thus:

$$w(D) = -2\pi C\rho \int_{z=D}^{z=\infty} dz \int_{x=0}^{x=\infty} \frac{x\,dx}{(z^2+x^2)^{6/2}} = -\frac{2\pi C\rho}{4}\int_D^\infty \frac{dz}{z^4} = -\frac{\pi C\rho}{6D^3} \tag{1.8}$$

When a similar integration is performed for bodies of different geometries, the results shown in Figure 1.5 are obtained. It can be seen that, in every case, the free energy or interaction is described by a relationship of the kind:

$$W(D) \propto U A/D^n \tag{1.9}$$

where U is a factor which contains numerical constants and relevant dimensions of the bodies involved (safe, of course, when the free energy interaction is expressed per unit area), and $A = \pi^2 C\rho_1\rho_2$ is called the Hamaker constant. Equation (1.9) shows all the important features of van der Waals interactions between macroscopic bodies, that is:

- The free energy of interactions involving macroscopic bodies depends on the dimension of the bodies (through the factor U), as is expected from the fact that free energy is an extensive property.

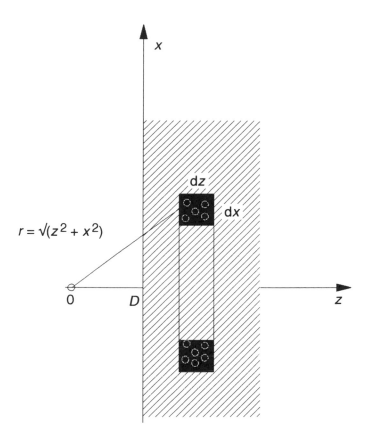

Figure 1.4   Method of integrating the interaction free energies of a molecule near a wall.

- Van der Waals' forces between macroscopic bodies are more long-ranged than between molecules, and the exact power law they obey depends on the shape of the bodies. In this respect, a point to note which is not contained in the above expressions, is that, at separations above about 5 nm, van der Waals' forces start to decay more rapidly, owing to the so called retardation effect. This effect arises because it takes a finite time for the electric field propagating from one instantaneous dipole to reach a second atom, induce a dipole and come back to the first. If this time is long enough, that is if the separation between the atoms involved is greater than about 5 nm, the first dipole will have rotated and will no longer be correlated with the induced one.
- The effect of the chemical and physical nature of the materials constituting the bodies involved in the interaction on the interaction itself, is described by the Hamaker constant, which contains the relevant atomic or molecular parameters in the constant C.

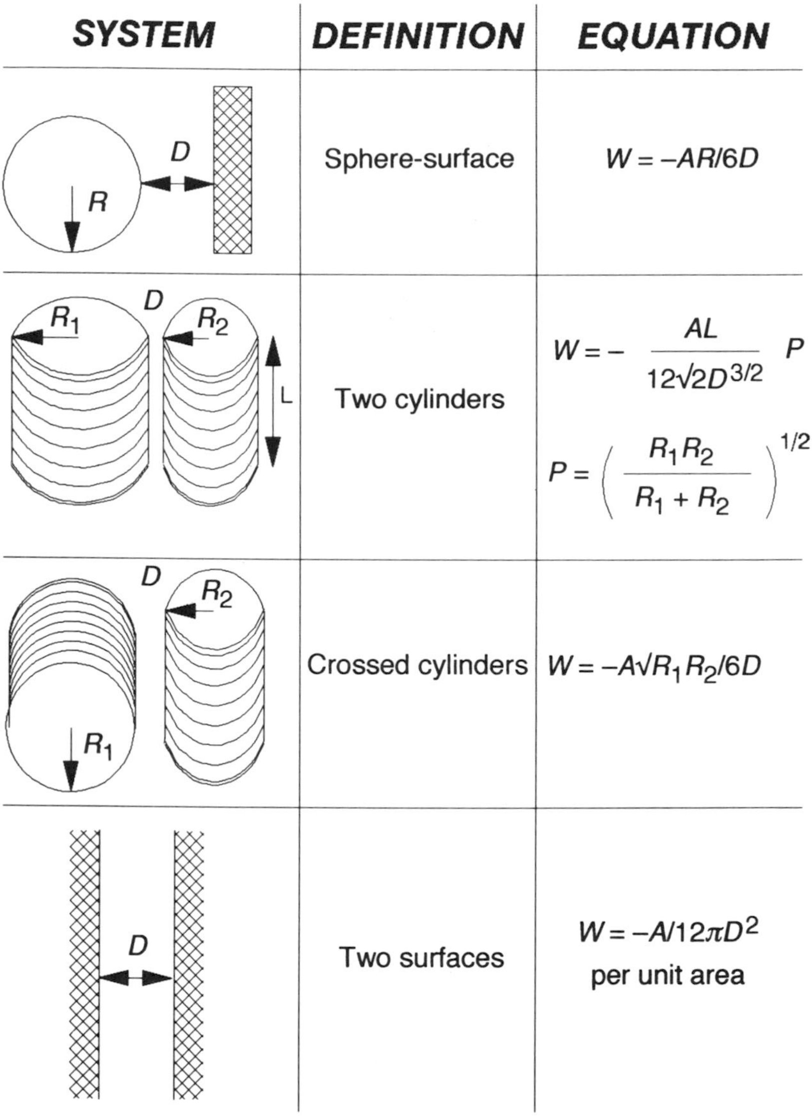

SYSTEM	DEFINITION	EQUATION
	Sphere-surface	$W = -AR/6D$
	Two cylinders	$W = -\dfrac{AL}{12\sqrt{2}D^{3/2}}\,P$ $P = \left(\dfrac{R_1 R_2}{R_1 + R_2}\right)^{1/2}$
	Crossed cylinders	$W = -A\sqrt{R_1 R_2}/6D$
	Two surfaces	$W = -A/12\pi D^2$ per unit area

Figure 1.5   Van der Waals' interaction free energies between bodies of different geometries.

A further point to note in Figure 1.5, is that the free energy of interacting spheres can be calculated from the expression for flat surfaces and:

$$F(D)_{\text{sph.}} = 2\pi\left(\frac{R_1 R_2}{R_1 + R_2}\right) W(D)_{\text{flat}} \tag{1.10}$$

followed by an integration, since:

$$W(D) = - \int F(D)\, dD \qquad (1.11)$$

Equation (1.10), which is called the Derjaguin approximation [15], is often extremely useful since, in general, it is easier to derive expressions for the interaction potentials of flat rather than curved surfaces. A rigorous derivation of equation (1.10) would show that it is applicable so long as the range of interaction and separation D is much less than the radii of spheres [16].

Coming back to equation (1.8), the integration we performed implies the assumption of simple pairwise additivity, which means that we considered the properties of interacting atoms unaffected by the presence of neighbouring atoms. Actually, the atomic polarizability can and usually does change when surrounded by other atoms. Moreover, the whole interaction is complicated by many-body effects, since the instantaneous dipole field of an atom (A) induces dipole moments not only in a second atom (B), but also in any neighbouring atoms, whose induced dipole will, in turn, affect atom B. In other words, the field from atom A reaches atom B both directly and through other atoms, which means that the assumption of pairwise additivity is not completely sound, especially in condensed phases, where the mean distance between atoms is small and many-body effects cannot be ignored.

This problem can be solved by a completely different approach, proposed by Lifshitz in 1955 [17–19]. Ignoring the atomic structure of materials, the Lifshitz theory treats bodies as continuous media. Basically, the Lifshitz or macroscopic approach (while the former treatment is usually called microscopic, since it is based on atomic or molecular parameters) considers the interactions between electromagnetic waves emanating from macroscopic bodies. Since these electromagnetic fields are completely described by the complex dielectric spectrum [20], this is the only material property that can be used in order to specify the force between macroscopic bodies. The detailed original treatment is very complicated [21], and requires sophisticated mathematics, but several more accessible accounts have subsequently been published [22, 23]. Of course, the results given in Figure 1.5 and the general equation (1.9) remain valid, the only thing which has changed is the way in which the Hamaker constant is calculated. The Lifshitz approach has the great merit of automatically incorporating many body effects and of being readily applicable to interactions in a third medium.

From a practical point of view it allows the calculation of the Hamaker constant of interacting bodies simply on the basis of bulk properties such as the dielectric constant and the refractive index. The following is a useful approximate expression for the dispersion contribution to the Hamaker constant of two different bodies (1 and 2) interacting across a medium (3), none of them being a conductor (i.e. a metal) [24]:

$$A_L = \frac{3h v_e (n_1^2 - n_3^2)(n_2^2 - n_3^2)}{8\sqrt{2}(n_1^2 + n_3^2)^{1/2}(n_2^2 + n_3^2)^{1/2}[(n_1^2 + n_3^2)^{1/2} + (n_2^2 + n_3^2)^{1/2}]} \qquad (1.12)$$

where v_e is the main electronic adsorption frequency in the UV (assumed to be the same for the three bodies). Equation (1.11) gives the London contribution to the total van der Waals' interaction. At the same level of approximation, the contribution from Keesom and Debye forces can be calculated by:

$$A_{KD} = -kT \frac{\varepsilon_1 - \varepsilon_3}{\varepsilon_1 + \varepsilon_3} \frac{\varepsilon_2 - \varepsilon_3}{\varepsilon_2 + \varepsilon_3} \qquad (1.13)$$

where ε_i is the static dielectric constant of phase i. The total Hamaker constant is given, of course, by the sum:

$$A = A_L + A_{KD} \qquad (1.14)$$

Even if more complex analytic formulae can be found in literature [25, 26], equations (1.12) to (1.14) give good agreement with more rigorously computed or experimentally determined values. As an example, Table 1.1 shows a comparison between the rigorously calculated literature values of the Hamaker constant (where available) and those obtained from the previous equations, together with the input values we used, for the interaction between two identical media (three polymers) in air and in water. From equations (1.13) and (1.14) it is clear that in air, the Hamaker constant is in every case completely dominated by the London contribution. In water, the overall strength of the interaction is reduced, since the lowering of the London term due to the decreased mismatch between the refractive indices is not completely recovered by increasing the Keesom and Debye contribution, brought on by the high value of the static dielectric constant of water.

Some general features of the van der Waals' interactions can be deduced from

Table 1.1 Hamaker constant for two identical polymers interacting across air ($\varepsilon_3 = 1, n_3 = 1$) and water ($\varepsilon_3 = 80, n_3 = 1.333$), calculated by equations (1.12) to (1.14) and by exact solutions ([1] and references therein)

Polymer	ε	n	V_e $(10^{15} s^{-1})$	Medium	Hamaker constant A Eqn. (1.14) $(10^{-20}$ J$)$	Rigorous
Polystyrene	2.55	1.557	2.3	Air	6.5	6.6, 7.9
Polyvinyl chloride	3.2	1.527	2.9	Air	7.5	7.8
PTFE	2.1	1.359	2.9	Air	3.8	3.8
Polystyrene	2.55	1.557	2.7	H_2O	1.4	0.95, 1.3
PTFE	2.1	1.359	2.9	H_2O	0.29	0.33

equations (1.12) and (1.13); first of all, the van der Waals force between two bodies in vacuum is always attractive, regardless of whether these bodies are similar or dissimilar. On the other hand, in a third medium, the van der Waals' forces are always attractive for two identical bodies, while they can be either attractive or repulsive for two different bodies. Repulsive van der Waals' forces in water have been discussed by Neumann and coworkers [27].

The development of surface force apparatus has directly allowed the testing of theoretical speculations. All the general features have been confirmed, including the retardation effects previously discussed [28–30]. As an example, Figure 1.6 [31], shows the attractive van der Waals' force between two curved mica surfaces. At distances above 5 nm, the experimental points cease to follow the inverse square dependence on distance, which is the correct dependence in a non-retarded regimen.

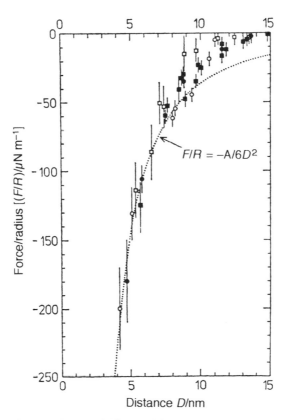

Figure 1.6 Attractive van der Waals' force (*F*) between mica surfaces in water, showing retardation effects at distances above 5 nm. (Reproduced by permission of the Royal Society of Chemistry from ref. [31].)

As already discussed at the beginning of this section, van der Waals' forces play a pivotal role in the interaction between bodies, since they are always present. However, in the great majority of cases they co-operate with other kinds of forces, that will be discussed in the rest of this chapter, starting with interactions arising from electrical charges.

1.2   ELECTROSTATIC FORCES

Surfaces of materials can assume a net charge by several different mechanisms. Among them, rubbing or contact electrification in air, and ionization of suitable groups in water or high dielectric constant solvents, are probably the most common. Whatever the charging mechanism, the outcome is that a further interaction plays a role in the global field of forces arising from bodies. Its effect is clearly observed when dry powders are handled in a dry atmosphere, or by the annoying adhesive tendency exhibited by common plastic films. On the other hand, the stability of suspensions of similar colloidal particles, which, according to the results of the previous section, should experience attractive forces, and hence should coagulate, is often a consequence of electrostatic forces in water.

The control of electrostatic interactions between bodies in air, as can be easily understood, is the subject of extensive on going research. Also, in the case of electrostatic interactions in water the list of technological fields where it plays a role is impressive, ranging from traditional colloid applications to biomaterials (Chapter 12).

The force between charged bodies in air follows the well-known inverse square relationship described by Coulomb's law. This means that it is a long-range force and, using Coulomb's with typical charge-distance values, it is clear that it is also very strong. The last point is easily confirmed by the previously noted behaviour of dry powders, which shows that, when charging occurs, it is a major determinant of the behaviour of the system.

The basic tenets of the electrostatic interaction in water of high dielectric constant solvents, a traditional pillar of colloid chemistry [1, 2, 32], will be briefly reviewed in the following section.

1.2.1   Electrostatic Interaction in Liquid Medium

Surfaces of polymers immersed in water, or in a high dielectric constant solvent, can acquire a charge basically through two different mechanisms, that is ionization of suitable groups and specific adsorption of ions [33]. The first mechanism requires that dissociable groups such as carboxyl, sulphate, amino etc. are present on the surface, where they can be either intentionally introduced or inadvertently produced, for instance due to surface oxidation that frequently occurs at the molten polymer– mould interface, during the preparation stage [34]. Adsorption of charged molecules leads to charged surfaces when it occurs in a specific form, that is when ions of the same sign are preferentially adsorbed (of course, non-specific adsorption leads to the accumulation of both negative and positive ions, hence no net charge can accumulate). The best example of molecules that can induce a net surface charge by specific adsorption is given by natural and synthetic ionic surfactants.

Whichever the charging mechanism, the most fundamental electrostatic parameters that describe the interaction of a charged interface with the surrounding environment, are the charge and the potential. It is customary to express the former as surface density (often the word 'density' is omitted), indicated by the symbol σ or σ_0 and expressed in C/m^2. It is important to realize that this definition automatically implies the assumption of a homogeneous charge distribution, a picture that, thinking at what a polymer surface can look like, appears the exception rather than the rule. It is, however, true that the effect of an uneven charge distribution on the surface–surrounding environment interaction is very much a matter of scale: when the distance between charged sites on the surface is very short as compared to the typical dimension of interacting bodies (such as in the interaction between charged macroscopic plates), heterogeneity can safely be neglected. On the other hand, when studying, for instance, the adsorption of proteins on block copolymers surfaces, the soundness of the assumption of homogeneous surface charge should be carefully evaluated.

The description of the interaction free energy between charged bodies in water or high dielectric constant solvents, can start, as an intuitive level, with Figure 1.7(a). It can easily be assumed that an isolated charged plate will induce a given distribution in surrounding ions, that is a dependence of the ion density on the distance from the plate. At this stage, we can disregard the precise analytical formula of this function (it will be discussed later in this section), all we need is the intuitive feeling that the concentration of ions whose charge is opposite to that of the surface is increased with respect to the bulk value close to the surface (giving rise to what is called an electric double layer) and decreases, as the distance from the surface increases. If two such charged surfaces are brought close together, the situation described in Figure 1.7(b) will arise, where it is shown that the ion concentration between the surfaces is always higher than ρ_∞. This means that an osmotic pressure will operate between surfaces, which, together with the electric field effect, constitute the force opposing the approaching of the surfaces. As Langmuir suggested [35], at equilibrium, the total force will be the same for every x between 0 and D, while the relative weight of the two contributions will change. Now, at $x = D/2$, the electric field is zero, for symmetry, and the total force per unit area is given by the osmotic pressure contribution, that is (for a dilute solution):

$$P = kT(\rho_m - \rho_\infty) \qquad (1.15)$$

where ρ_m is of course the ion density at midplane.

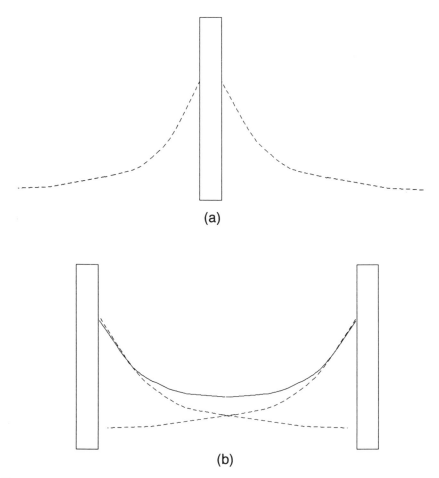

Figure 1.7 Counter-ion distribution around an isolated charged surface (a) and two interacting surfaces (b).

It follows that, in order to calculate the force per unit area, and hence, by integration, the free energy of interaction of two planar surfaces and, applying the Derjaguin approximation, the free energy of interaction of spheres, it is necessary to calculate the ion concentration at midplane or, in general, the ion concentration profile induced by a charged plate.

In order to do that, we return to the isolated surface of Figure 1.7(a). The ion density must now be expressed more precisely, and must be related to the relevant physical parameters: the ion distribution is, of course, a function of the potential ϕ, but the overall interaction is not purely electrostatic, since ions are subjected to thermal motion. Thus, the density of ith ion at distance x away from the surface can be described by the Boltzmann distribution:

$$\rho_{xi} = \rho_{\infty i} \exp(-ze\phi_x/kT) \qquad (1.16)$$

where $\rho_{\infty i}$ is the number density at $x = \infty$, that is the bulk density.

The assumption of an ion density controlled by the combined action of electrostatic interaction and thermal motion is the basis of the Gouy–Chapman theory of the diffuse double layer, which we will discuss at some length [36–38]. The point is, of course, to give an analytical expression to $\phi = \phi(x)$, and this requires some mathematics. First, let us combine equation (1.16) with the well-known Poisson's equation, which describes the net charge density at a given x:

$$-\varepsilon\varepsilon_0 \left(\frac{d^2\phi}{dx^2}\right) = \sum_i z_i e \rho_i \qquad (1.17)$$

so that:

$$-\varepsilon\varepsilon_0 \left(\frac{d^2\phi}{dx^2}\right) = \sum_i z_i e \rho_{\infty i} e^{-ze\phi/kT} \qquad (1.18)$$

where, of course, the assumption of a planar surface, allows the problem to be treated in only one dimension. Equation (1.18) is called the Poisson–Boltzmann equation.

Now let us consider the value of the first derivative of ϕ with respect to x in selected points. At $x = \infty$ the field must vanish, that is:

$$\left(\frac{d\phi}{dx}\right)_{x=\infty} = 0 \qquad (1.19)$$

while the value at $x = 0$, that is at the surface, can be calculated as follows: if σ is the surface charge (density), electroneutrality requires that the following relation holds true:

$$\sigma = -\int_\infty^0 \sum_i z_i e \rho_i \, dx \qquad (1.20)$$

or, combining with equation (1.17) and condition (1.19):

$$\sigma = \varepsilon\varepsilon_0 \int_\infty^0 \left(\frac{d^2\phi}{dx^2}\right) dx = \varepsilon\varepsilon_0 \left(\frac{d\phi}{dx}\right)_0 \qquad (1.21)$$

and finally:

$$\left(\frac{d\phi}{dx}\right)_{x=0} = \frac{\sigma}{\varepsilon\varepsilon_0} \qquad (1.22)$$

The previous equations give a useful relation, which is usually called the contact value theorem [39]. If we take the derivative of the ion density with respect to the distance from the surface, we obtain:

$$\frac{d\sum_i \rho_{xi}}{x} = -\sum_i \rho_{\infty i} z_i \frac{e}{kT} e^{-ze\phi/kT} \frac{d\phi}{dx} \qquad (1.23a)$$

(using equation (1.18))

$$= \frac{\varepsilon\varepsilon_0}{kT}\left(\frac{d^2\phi}{dx^2}\right)\frac{d\phi}{dx} = \frac{\varepsilon\varepsilon_0}{2kT}\frac{d}{dx}\left(\frac{d\phi}{dx}\right)^2 \qquad (1.23b)$$

If the derivative (1.23a, b) is integrated between infinity and x, it yields:

$$\sum_i \rho_{xi} = \sum_i \rho_{\infty i} + \frac{\varepsilon\varepsilon_0}{2kT}\left(\frac{d\phi}{dx}\right)^2 \qquad (1.24)$$

which, in the case of the surface ($x = 0$) becomes (equation (1.22)):

$$\sum_i \rho_{0i} = \sum_i \rho_{\infty i} + \frac{\sigma^2}{2kT\varepsilon\varepsilon_0} \qquad (1.25)$$

which is the contact value theorem, and gives the relation between the bulk and the surface ion density in terms of the surface charge.

We will use the previous results to discuss, for simplicity, the case of a charged surface immersed in a solution of a 1:1 electrolyte, such as Na^+Cl^-. The relationship between surface charge and surface potential (ϕ_0) can be obtained by substituting equation (1.16) in (1.25), that is:

$$[Na^+]_\infty e^{-e\phi_0/kT} = [Cl^-]_\infty e^{e\phi_0/kT} = [Na^+]_\infty + [Cl^-]_\infty + \frac{\sigma^2}{2kT\varepsilon\varepsilon_0} \qquad (1.26)$$

where the number of ions is expressed in the practical unit of molar concentration, that is the number of moles in a dm^{-3} (which can be obtained from the number density in m^{-3} by dividing for the product of the Avogadro number N_a times the dm^{-3}/m^{-3} ratio, that is 6.023×10^{26}). Since $[Na^+]_\infty = [Cl^-]_\infty$, equation (1.26) can be written:

$$(e^{e\phi_0/kT} + e^{-e\phi_0/kT} - 2) = \frac{\sigma^2}{[Na^+]_\infty 2kT\varepsilon\varepsilon_0} \qquad (1.27a)$$

or, since $(e^x + e^{-x} - 2) = (e^{x/2} - e^{-x/2})^2$:

$$(e^{e\phi_0/kT} - e^{-e\phi_0/kT})^2 = \frac{\sigma^2}{[Na^+]_\infty 2kT\varepsilon\varepsilon_0} \qquad (1.27b)$$

Equation (1.27) is generally called the Grahame equation [40], and allows the calculation of the surface potential induced by a given surface charge density and ion concentration or vice versa. The calculation of σ from a given ϕ_0 is straightforward. The reverse is easily done, when it is recognized that equation (1.27) is a quadratic in ϕ_0/kT, and yields two symmetrical roots:

$$\phi_0 = \frac{kT}{e} \ln\left(\frac{y \pm (y-4)^{1/2}}{2}\right) \tag{1.28}$$

where $y = 2 + \sigma^2/[\text{Na}^+]_\infty 2kT\varepsilon\varepsilon_0$.

Equation (1.28) can be used to obtain a quantitative feeling about surface potentials. The area occupied by polar head groups of surfactant molecules in water is less than $1\,\text{nm}^2$, say $0.6\,\text{nm}^2$. Within this approximation, then, a fully ionized surface of carboxylic groups will bear a charge of $-0.27\,\text{C/m}^2$, which is a typical surface charge density value. Table 1.2 shows the surface potential, calculated from equation (1.28), of a surface bearing a $-0.27\,\text{C/m}^2$ charge, immersed in a solution of 1:1 electrolyte at $25\,°\text{C}$, as a function of the electrolyte concentration. As the concentration increases by six orders of magnitude, the surface potential falls from some hundreds to some tenths of mV.

We will return to the interplay between charge, potential and concentration at $x = 0$ below. We now consider the variation of the potential away from the surface, this can be done combining equations (1.6) and (1.24). Taking care of the earlier observations, it is possible to write:

$$(e^{e\phi_0/2kT} - e^{-e\phi/2kT})^2 = \frac{\varepsilon\varepsilon_0}{[\text{Na}^+]_\infty 2kT}\left(\frac{d\phi}{dx}\right) \tag{1.29}$$

or:

$$\sinh^2\left(\frac{e\phi}{2kT}\right) = \frac{\varepsilon\varepsilon_0}{8[\text{Na}^+]_\infty kT}\left(\frac{d\phi}{dx}\right)^2 \tag{1.30}$$

which finally yields:

$$\frac{d\phi}{dx} = (8[\text{Na}^+]_\infty kT/\varepsilon\varepsilon_0)^{1/2} \text{Sinh}\left(\frac{e\phi}{2kT}\right) \tag{1.31}$$

Table 1.2 Surface potential calculated from equation (1.28) as a function of 1:1 electrolyte concentration (surface charge $= -0.27\,\text{C/m}^2$)

Electrolyte Concentration (M)	Surface potential (mV)
1×10^{-6}	-434
1×10^{-3}	-256
1	-81

Integration of equation (1.31) gives the expression for the potential as a function or x, that is:

$$\phi_x = \frac{2kT}{e}\ln\left(\frac{1 + ue^{-kx}}{1 - ue^{-kx}}\right) \approx \frac{4kT}{e}ue^{-kx} \tag{1.32}$$

where:

$$u = \tanh\left(e\phi_0/4kT\right) \tag{1.33}$$

and:

$$\mathbf{k} = (e^2\sum_i \rho_{\infty i}z_i^2/kT\varepsilon\varepsilon_0)^{1/2} \tag{1.34}$$

($\mathbf{k}$ is given in bold in order to avoid confusion with the Boltzmann constant). Equation (1.34) gives a very important and useful parameter, whose meaning can be fully appreciated, once it is observed that at low potential, below about 25 mV, equation (1.32) simplifies to:

$$\phi_x = \phi_0\exp\left(-\mathbf{k}x\right) = \phi_0\exp\left(-x/\mathbf{k}^{-1}\right) \tag{1.35}$$

Equation (1.35) is obtained following the so-called Debye–Huckel approximation, which works when $z_ie\phi < kT$. In this case, the exponential (1.16) can be expanded as a power series, with only the first order term retained, which greatly simplifies the following calculation. At the same level of approximation, equation (1.27) becomes (Figure 1.8):

$$\sigma = \varepsilon\varepsilon_0\mathbf{k}\phi_0 \tag{1.36}$$

Coming back to equation (1.35), it is clear that $\mathbf{k}^{-1}$, whose units are length, can be identified as the distance over which the potential has decayed to $1/e$ of the value at $x = 0$. In other words $\mathbf{k}^{-1}$, which is called the Debye length, gives a measure of the thickness of the double layer, which, by the very assumption made in writing equation (1.16) (that is the ions are subjected simultaneously to electrostatic interaction and thermal motion) is diffuse and cannot be characterized by a discrete thickness value. Thus, $\mathbf{k}^{-1}$ is usually called the 'diffuse double layer thickness', even if it must be always remembered that, strictly speaking, the thickness of a diffuse double layer is infinite.

Now let us consider briefly the effect of the various parameters we discussed on the distribution of ions around a charged interface. Table 1.3 gives the calculated Debye lengths for several different electrolytes at different concentrations. The presence of multivalent ions greatly reduces $\mathbf{k}^{-1}$, as can be deduced from equation (1.34), and from Figure 1.7(b) it can be seen that this will reduce the effective range of interaction between charged surfaces.

Figure 1.9 shows the concentration profiles (ratioed to the bulk concentration) at constant surface charge as a function of the electrolyte (1:1) concentration. From the previous discussion we may expect the Debye length $\mathbf{k}^{-1}$ to increase as the concentration decreases (equations (1.34)). This means that, as the electrolyte

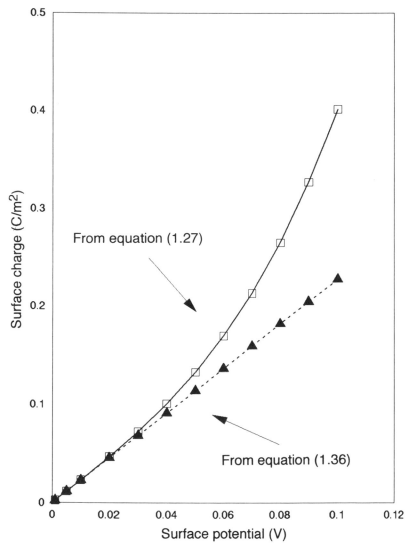

Figure 1.8 Calculation of surface charge from surface potential using equation (1.27) and the simplified form (1.36).

concentration is lowered, the distance from the surface increases where the ion concentration is higher than the bulk concentration.

At constant surface charge, the surface potential is a decreasing function of the ion concentration (equations (1.27) and (1.36)). This is the reason why the lower the concentration, the higher the surface 'crowding'.

Table 1.3 Calculated Debye length $(\mathbf{k}^{-1})$ for several concentrations of different electrolytes

Electrolyte	Concentration (m)	Debye length (nm)
1:1	1	0.3
1:1	1×10^{-2}	3.0
1:1	1×10^{-4}	30.4
2:1	1	0.2
2:1	1×10^{-2}	1.7
2:1	1×10^{-4}	17.6
2:2	1	0.1
2:2	1×10^{-2}	1.5
2:2	1×10^{-4}	15.2

The effect of increasing surface charge at constant electrolyte concentration is reported in Figure 1.10. Here the Debye length is constant and, as the surface charge is increased, the surface potential and the number of ions confined close to the surface increases.

The last observation underlines the basic limit of the outlined Gouy–Chapman theory: no upper limit exists to the crowding of ions at the surface, and through equation (1.27), it is possible to calculate surface charge densities that, if discussed in terms of surface area occupied by the charged species, will yield unrealistically low figures. The point is that in the previous theory no care was taken of ion dimensions, and actually one assumption of the Gouy–Chapman theory is the treatment of ions as dimensionless point charge. A further point to note is that equation (1.16) contains no mention of chemical interactions, that is to the specific adsorption of charged species that was described as one of the mechanisms responsible for the build-up of charges at surfaces.

Both these problems are addressed in the Stern modification of what is known as the Gouy–Stern double layer [41]. The basic idea is to limit the correction to the first ionic layer facing the surface, while the outer part is as previously described.

Following the Stern theory, the first ionic layer is located at a distance δ from the charged surface, where δ is the radius of (hydrated) ions (Figure 1.11). The plane passing at $x = \delta$, parallel to the surface, is called the Stern plane. The values of $x < \delta$ are inaccessible to ions, due to their finite dimensions and in this way the ion size in the direction perpendicular to the surface is taken into account. The plane parallel to it, the interaction of this first ionic layer with the surface, is described by a Langmuir isotherm or:

$$\frac{n_{ads}}{N} = \frac{Kn_{\infty}}{1 + Kn_{\infty}} \tag{1.37}$$

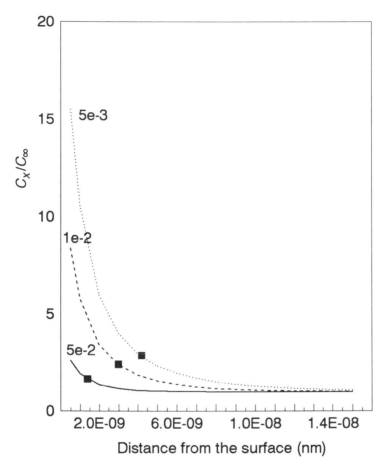

Figure 1.9 Concentration profiles of counter-ions at constant surface charge as a function of 1:1 electrolyte concentration (indicated in the figure). Markers show the Debye length for a given concentration.

where n_{ads} is the number of adsorbed ions, N the total number of adsorption surface sites, n_∞ is the ion concentration in bulk solution and K is a constant.

The finite number of adsorption surface sites automatically takes into account the finite size of ions. The specific or chemical adsorption is described by the affinity constant K, that is:

$$K = \exp(ze\phi_\delta + \Phi) \tag{1.38}$$

where Φ is the specific chemical energy associated with the adsorption and ϕ_δ is the potential at the Stern surface. To calculate this, it must be noted that the

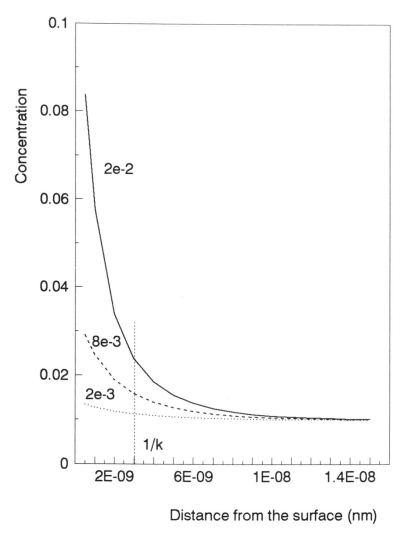

Figure 1.10   The effect of increasing surface charge on the counter-ion profile (0.1 [M] 1:1 electrolyte. Surface charge is indicated in the figure). Dotted line shows the Debye length.

charged surface and the adsorbed ionic layer resemble a parallel plate capacitor, whose electrical field drop is given by [42]:

$$\left(\frac{d\phi}{dx}\right) = \frac{\sigma}{\varepsilon\varepsilon_0}$$

(1.39)

If the derivative in equation (1.39) is approximated by the ratio of increments,

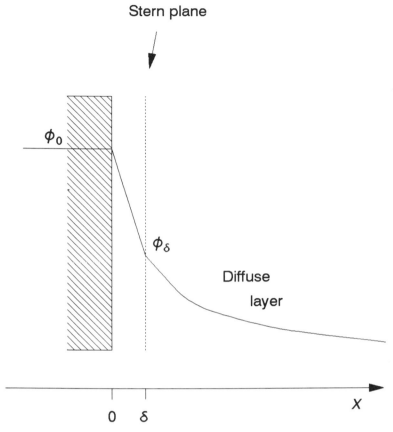

Figure 1.11 The Stern modification of the Gouy–Chapman double layer (Gouy–Stern layer).

then:

$$\frac{\phi_0 - \phi_\delta}{\delta} = \frac{\sigma_0}{\varepsilon_s \varepsilon_0} \frac{Kn_\infty}{1 + Kn_\infty} \tag{1.40}$$

where the charge of the Stern plane has been obtained by the product of the charge at saturation coverage (σ_0) times the fraction of occupied sites. The subscript in the permittivity of the solvent reminds us that surface-induced orientation of solvent molecules affects the local dielectric constant. This part of the double layer is called the inner or compact double layer. At the Stern plane the potential starts to follow the Gouy–Chapman theory, the only modifications being that ϕ_0 must be substituted by ϕ_δ and x must be measured starting from δ.

The Stern theory has several consequences: if specific adsorption of counter-ions occurs, ϕ_δ can be much lower than expected on a purely electrostatic approach. In cases of marked chemical affinity for ions whose charge is opposite to that of the surface, specific ion adsorption can lead to what is called superequivalent adsorption, or a reversal of the charge sign at the Stern layer. These simple examples show that the details of the affinity of charge-bearing species towards a given surface play a major role in interbody interactions.

Several other double-layer models have been introduced: since the transition from the compact to the diffuse part seems a little abrupt (Stern corrections involve only the first layer from the surface), all these models share the common feature of introducing a further step between the compact and the diffuse region of the double layer. As an example, in a model first proposed by Grahame, a distinction is made between an inner Helmholtz plane (iHp), where specific adsorption occurs, and an outer Helmholtz plane (oHp), where the diffuse layer starts (the name arises from the fact that Helmholtz introduced the model of a 'molecular' parallel plate condenser to describe electrokinetic phenomena).

It must be noted that the Stern treatment introduces some parameters (the distance δ, the specific chemical energy for adsorption Φ) that are not only very difficult to measure but also specific for a given ion. From one side, it can appear disturbing that the most captivating feature of the Gouy–Chapman theory, that is its generality, is lost, but on the other hand it is very important to appreciate that general, macroscopic treatments are not completely satisfactory when it comes to the details of interactions, where the description of interacting species at the molecular level must be taken into account. This will become increasingly evident in the rest of this chapter. At present, since our goal is to describe the interaction free energy between macroscopic bodies, it is the outer part of the double layer that concerns us (Figure 1.7). Thus, we will use the equation previously derived following the Gouy–Chapman theory, and the only way that the break-down of its basic assumptions will affect us, is that we should use, instead of the surface potential ϕ_0, the potential of the plane at which the diffuse double layer begins (the Stern plane or the oHp). To avoid confusion, however, we will follow the use of most textbooks and continue to use the symbol ϕ_0 for the potential at the inner limit of the diffuse double layer. While the previous discussion made it clear that it is very difficult to calculate it, in Chapter 5 we will see that conventional 'surface potential' measurements do actually measure the potential at a location very close to the outer limit of the compact double-layer region.

It is now time to return to Figure 1.7 and equation (1.15), which now can be written (for a 1:1 electrolyte) as:

$$P = kT\rho_\infty(e^{e\phi_m/kT} + e^{-e\phi_m/kT} - 2) \approx e^2\phi^{m2}\rho_\infty/kT \qquad (1.41)$$

where ϕ_m is, of course, the potential at midplane and the last term arises from the power series expansion of the exponential term which, with the assumption of a small midplane potential, can be truncated after the leading term. With the further assumption that the total midplane potential is the sum of the potential of each surface at $x = D/2$, then, using equation (1.32):

$$P = 64 \, kT \rho_\infty u^2 e^{-kD} \tag{1.42}$$

so that finally, the interaction free energy for unit area of two flat surfaces can be written as:

$$W(D) = -(64 \, kT \rho_\infty u^2/k) e^{-kD} \tag{1.43}$$

whence, using the Derjaguin approximation, the free energy of interaction between two spheres of radius R can be calculated, that is:

$$W(D) = -(64 \, \pi kTR \rho_\infty u^2/k^2) e^{1kD} \tag{1.44}$$

It is clear that situations in which electrostatic forces alone determine the total interaction between bodies do not occur in nature, since with them, there will always be the contribution of van der Waals' forces. The combination of van der Waals' and electrostatic effects in water (or high dielectric constant solvent) is the subject of the next section.

1.3   THE DLVO THEORY

Macroscopic bodies in water or high dielectric constant solvents, always interact by means of van der Waals' forces. Moreover, as we have discussed, if their surfaces are charged either by intrinsic characteristics (i.e. the presence of ionizable groups) or by the specific adsorption of ions from the solution, they will also interact by electrostatic force. The combined action of van der Waals' and electrostatic forces in water is described by the so-called DLVO (from the initials of Derjaguin, Landau, Vervey and Overbeek) theory of colloid stability [43, 44]. This theory was developed in the late 1940s by the Russian (DL) and the Dutch school (VO) and since then has represented a basic pillar of colloid science [45]. Its first merit was to correctly describe the effect of the electrolyte concentration on colloid stability, and its soundness in the prediction of the interactions between macroscopic bodies was later confirmed by the direct measurement of surface forces, which we will discuss in Chapter 5.

In the DLVO theory, the total free energy of interaction is expressed by the sum of the van der Waals' and the electrostatic term, thus, for the unit area of similar, interacting surfaces (Figure 1.5, equation (1.43)):

$$W(D) = -(64 kT \rho_\infty u^2/k) e^{-kD} - (A)/(12\pi D) \tag{1.45}$$

Equation (1.42) allows some consideration. First, the functional form of the two contributions shows that the van der Waals' attraction always prevails at small distances, since it is a power law term which tends to infinity as separation tends to zero, while the electrostatic term remains finite. Then, the van der Waals' term is insensitive to variations in the electrolyte concentration or to modification of the pH, which, in turn, greatly affect the electrostatic term. Thus for a given system, the former term can be considered as a constant that depends solely on the nature of interacting species, while the latter can be affected by environmental conditions. This is reflected in Figure 1.12, which is probably the most often represented figure in colloid science and that, in the present case, was calculated from equation (1.45), assuming a surface charge of $-0.08\,C/m^2$ and a Hamaker

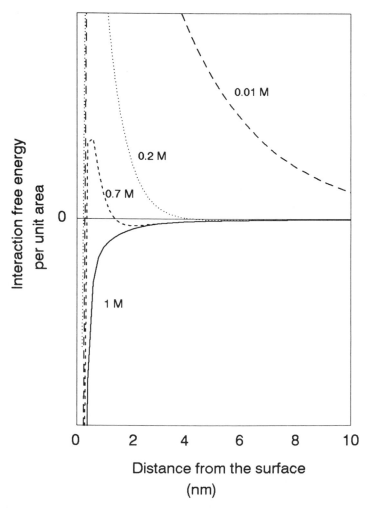

Figure 1.12   Interaction free energy between surfaces interacting by van der Waals' and electrostatic forces. Values used in the calculation: Surface charge $= -0.08\,C/m^2$. Hamaker constant 8×10^{-21}. The concentration of the 1:1 electrolyte is written on the Figure.

constant of 8×10^{-21} J. As can be seen, at low electrolyte concentration, the interaction is long-ranged and peaks close to the surface. The high energy barrier prevent the surfaces coming into contact. As the electrolyte concentration increases the reciprocal Debye length decreases, and so does the distance at which the electrostatic repulsion begins. At a given concentration it is possible to observe a secondary minimum, usually not deep enough to permit the bodies to coagulate. At high electrolyte concentration, the range of electrostatic repulsion is so reduced that the overall interaction is controlled by the attractive van der Waals' forces: in this case surfaces coalesce.

As an example of the interplay of van der Waals' and electrostatic forces in systems involving polymer surfaces in water, let us consider a subject adapted from a paper of Norde and Lyklema [46], that is the study of the initial events of protein and bacterial adhesion to solid surfaces following a colloid chemical approach. Even if protein–surface and bacteria–surface adhesion are highly specialized events, involving biological interactions [47], the colloidal chemical approach, that is the description of interaction in the framework of the DLVO theory, can give a useful insight into the early stages of the process. In later stages, the response and the adaptation of biological particles to the surrounding environment shifts the core of the problem to a more specific approach, beyond the DLVO regime.

Following this approach, both protein and bacteria are considered as colloidal particles, i.e. they are characterized by a radius R, a surface potential ϕ, and a Hamaker constant A. Together with the surface potential and Hamaker constant of the surface they have to adhere to, and the electrolyte concentration in the liquid medium, these parameters provide the information we need in order to compute the free energy of interaction:

$$W(D) = W_A(D) + W_E(D) \tag{1.46}$$

where (Figure 1.5):

$$W_A(D) = (A_{123} R_2)/(6D) \tag{1.47}$$

and [46]:

$$W_E(D) = \pi \varepsilon \varepsilon_0 R_2 (\phi_{13}^2 + \phi_{23}^2) \left[\frac{2\phi_{13}\phi_{23}}{\phi_{13}^2 + \phi_{23}^2} \ln \left(\frac{1 + e^{-kD}}{1 - e^{-kD}} \right) + \ln(1 - e^{-2kD}) \right] \tag{1.48}$$

where subscripts 1, 2 and 3 refer to the surface, the particle and the medium respectively.

Figure 1.13 shows the calculated curves, while the set of data used in the calculation, reflecting typical practical conditions, are given in Table 1.4. Some features are immediately evident, that is:

• The bacterium–surface interaction is, in general, stronger than the protein–surface interaction. This is, of course, a consequence of the greater

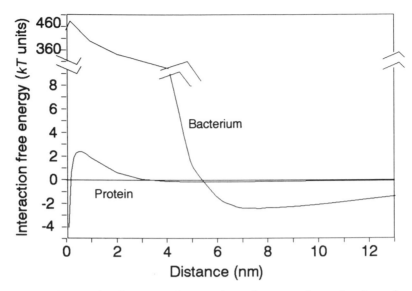

Figure 1.13 Interaction free energy between bacteria or proteins and a charged surface (calculated from equation (1.48)). Input data are shown in Table 1.4.

Table 1.4 Set of values used for the calculation of the bacteria–surface and protein–surface interaction (equation 1.48) [46]

Surface potentials: $-30\,mV$
1:1 Electrolyte concentration: 1×10^{-1} [M]
Temperature: $25\,°C$

Bacteria	Protein molecules
Radius: 5×10^{-7} m	Radius: 5×10^{-9} m
$A_{132} = 1 \times 10^{-21}$ J	$A_{132} = 5 \times 10^{-21}$ J

dimension of bacteria as compared to protein, that is, of the greater number of atoms involved in the interaction, as previously discussed.

- Bacteria cannot reach the primary minimum, since the free energy barrier at $D = 1$ nm is far higher than thermal energy kT. However, the curve shows a secondary minimum at $D = 8$ nm, whose depth is of the order of some kT unit. Bacteria approaching the surface are captured by this secondary minimum, giving rise, at this stage, to an unspecific, weak and reversible adhesion.
- In the protein–surface interaction curve, the maximum amounts to a few kT units and there is no sizeable secondary minimum (this is, because in this case, the van der Waals contribution is too low to win over the electrostatic energy at a distance greater than 1 nm). This means that a fraction of protein

molecules arriving at the surface can pass the energy barrier and fall in the primary minimum, where specific, non-DLVO, short-range interactions determine the final outcome of the adhesion process.

Experimental findings are reported in good agreement with the outlined theory [46].

This example clearly shows the interdisciplinary and wider-ranging applicability of the general concepts that form the basis of the DLVO theory. However, as

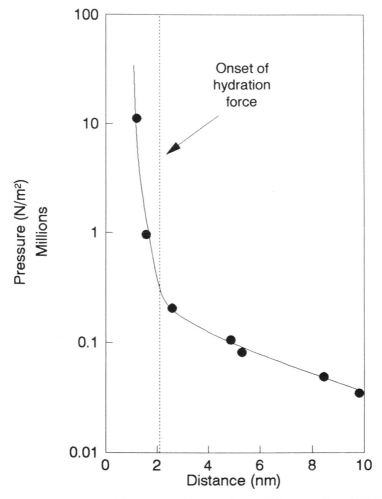

Figure 1.14 Measured repulsive pressure between charged bilayer surfaces (90% lecithin, 10% phosphatidylglycerol) in water, showing the onset of hydration forces below 2 nm separation. (Reprinted with permission from Cowley *et al.*, *Biochemistry*, **17**, 3163. Copyright (1978) American Chemical Society.)

already noted, treatments based on general, macroscopic properties, where the media are treated as structureless continua, may be expected to fail when the distance between bodies become of the order of molecular dimension or when the nature of interacting species gives rise to specific interactions. This is clearly seen in Figure 1.14, taken from ref. [48], where the repulsive pressure between charged bilayer surfaces in water is shown. While at $D > 2$ nm the overall interaction is well illustrated by a DLVO curve, at smaller separations an additional repulsion is observed. This is due to the so-called 'hydration force', arising from the extra energy needed to dehydrate the very hydrophilic head groups. This points to the need for a more specific approach to interacting bodies, that will constitute the rest of this chapter and will involve forces more directly related to the size and shape (steric forces) and to the chemical nature (acid-base and covalent forces) of interacting molecules and particles.

1.4 STRUCTURAL INTERACTIONS

In the present section we will proceed following the way highlighted by the last part of the discussion on the electrical double layer and by Figure 1.14: interacting solids surfaces, and all species participating in the interaction (such as the intervening medium) are not structureless solids or zero size points and theories based on such approximations are expected to break down when the separation between interacting bodies approaches molecular dimensions. This can happen at very small separations (of the order of some nm at most) with ordinary molecules or at much longer distances (some tens of nm) when interaction involves macromolecules. As discussed by Horn [3], there is no general consensus on the best terminology for forces arising from (quoting Adamson [49]) the 'graininess' of the matter, partly because of the different meaning given to words such as 'structure' and 'solvation' by chemists and physics. We will collectively call them 'structural' since they arise from the structure of the species or of the medium between the surfaces. In particular, we will shortly discuss forces arising from the constraints imposed on a 'hard spheres' liquid by approaching surfaces. Then those situations will be described where there is some kind of interaction between the liquid and the surface, so that the ordering of the liquid molecules close to the surface is not only controlled by 'geometrical' constraints. In this case, the term 'solvation forces' will be used, following Horn [3] (Israelachvili, in ref. [1] also includes the former interaction under the term 'solvation' forces). Finally, we will describe forces between polymer-covered surfaces which, following the general usage, as discussed below, will be called 'steric'.

A smooth rigid surface has a strong effect on the packing of molecules of the liquid medium. Since the second part of 1970s, theoretical investigations have shown that molecules prefer to order in layers parallel to the surface, since this is the more efficient way of packing [50–55]. When two surfaces come close together, preferred situations are those where the distance is roughly an integer multiple of the molecular diameter, since this allows all molecules to be arranged between the surfaces in rows parallel to the surface (Figure 1.15(a)). When the separation is intermediate between two successive integer multiples, molecules cannot dispose in the preferred packing mode (Figure 1.15(b)). Accordingly, their free energy will be higher than that of properly packed molecules. This means that, owing to the constraints imposed by rigid boundaries, one must expect an oscillating force as surface separation varies, with a distance between maxima close to the molecular diameter. Of course, the amplitude of this oscillating force should decay with separation, following the decrease of the weight of the layering effect of the surfaces on the overall interaction. The continuing improvement of the surface force apparatus in 1981 allowed the experimental verification of oscillatory forces between surfaces [56]. Figure 1.16 is reprinted from that paper, and shows the recorded force between two curved mica surfaces in octamethyl-cyclotetrasiloxane (OMCTS), an inert liquid, with a nearly spherical, non-polar molecule with a diameter of about 0.9 nm. Similar forces were later measured with other liquids [57, 58].

It is clear that the oscillating behaviour arises, both theoretically and experimentally, because of the molecularly smooth boundary surfaces. Thus, it can be

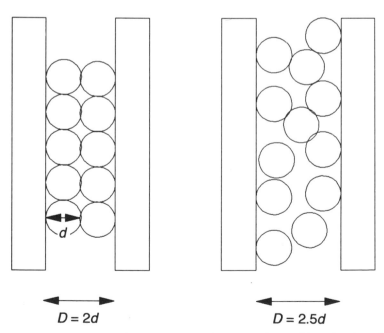

$$D = 2d \qquad D = 2.5d$$

Figure 1.15 Effect of separation between surfaces on the packing density of liquid molecules.

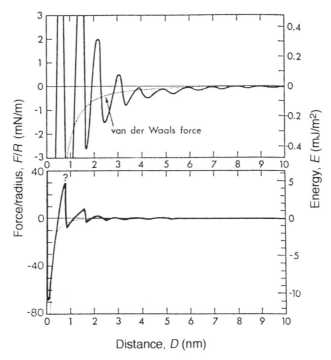

Figure 1.16 Measured oscillatory forces between two curved mica surfaces in octamethylcyclotetrasiloxane (OCMTS). The distance between the maxima is similar to the molecular diameter of OCMTS. (Reproduced by permission of the American Institute of Physics from ref. [56].)

expected that the oscillations will be smoothed out if surfaces are rough, as compared to molecular dimensions. Even if a complete theoretical argument is still lacking, it has been experimentally shown that oscillatory solvation forces are reduced in range, and ultimately vanish in a van der Waals' curve, when surfaces are rough or permeable to liquids [59].

A further class of structural forces arises from ordered boundary layers built by specific surface–liquid interactions (solvation forces, according to Horn [31]). As two surfaces approach, solvation layers overlap, and forces arise because the need to rearrange is constrained by the surface–liquid interaction molecule. As Horn points out [3], our knowledge of solvation forces is based almost entirely on experimental evidence, and theoretical work is still in its infancy [60–64]. Water and its solutions are probably the most extensively studied subjects, since their unusual properties [65] give rise to strong solvation (hydration) effects. A classical example was shown in Figure 1.14. Another intriguing result, highlighting the effect of solute, comes from extensive experiments by Pashley and

Israelachvili [66–68]. They found that, while the interaction between molecularly smooth mica surfaces in dilute electrolyte solution follows the DLVO theory, repulsive hydration forces arise at higher salt concentration, specific to each electrolyte. Since the strength and the range of repulsive forces increase with the hydration number of cations, they are most likely caused by the energy needed to dehydrate bound cations, that bind to the negatively charged mica surfaces retaining some water of hydration [69].

This interaction is far more complicated than the simple molecular packing effects previously described. The latter are surely taking place also in this case, and Israelachvili and Pashley measured oscillatory forces between mica surfaces, below 1.5 nm separation, with a mean periodicity close to the diameter of a water molecule [70]. However, other poorly understood effects are concurrently operating, such as image interaction [64] and decaying, cooperative hydrogen bonding interaction [60].

When two hydrophobic surfaces approach in water, they sense a strong attraction, and jump into adhesive contact. This 'hydrophobic force' can be considered a solvation force, since it probably arises from the liberation of water layers from the constraint of staying close to a hydrophobic surface [71, 72]. As in the well-known hydrophobic effect, water molecules close to a hydrophobic particle must assume a highly ordered structure, so as to keep the hydrogen bonding interaction with other water molecules and minimize the disturbing effect of non-hydrogen bonding molecules. The minimization of the hydrophobic surface–water interface allows a reduction in the number of water molecules in an entropically unfavourable condition, hence the attractive force. Even if the previous explanation seems the most likely, it is still difficult to understand the extremely long range of this force, which can be measured at surface separation to be as large as 70 nm [3]. To date, no completely satisfactory theory exists.

Finally, we consider the forces arising between polymer-covered surfaces. Also in this case, the effect of adsorbed macromolecules in the interaction between macroscopic bodies is best appreciated in the field of colloidal stabilization. Since the beginning of his history, man has exploited naturally occurring macromolecules to stabilize colloidal suspensions such as inks and paints. In the 1950s, this kind of interaction was called 'steric interaction' [73], a rather unfortunate choice, since the term 'steric' had already a precise identity and a completely different meaning in the language of organic chemistry. However, even if some suggestion of a different terminology has been made [74], the name 'steric' has gained widespread and indisputed acceptance, and we will follow the general usage. We will shortly discuss the origin and the effects of steric interaction by a quick description of some of the results obtained in the field of the polymeric stabilization of colloids. For a complete survey of theories and literature up to 1983, the reader is referred to a book by Napper [75].

The general problem of the interaction between surfaces covered by macromolecules (either adsorbed, or grafted, or constituted by mobile branches of a well-engineered block copolymer) can, with a certain degree of arbitrariness, be described as in Figure 1.17. At separation $D \geqslant 2L_s$ (Figure 1.17(a)), where L_s is the span of surface-covering macromolecules, no interaction occurs (even if, strictly speaking, one should take care of possible effects of polymer chain-induced structuring of solvent molecules). As the separation decreases, the so-called

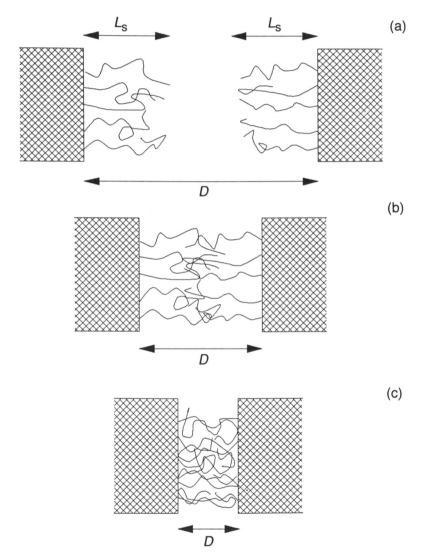

Figure 1.17 Schematic representation of the three domains of approach for polymeric chain bearing surfaces: (a) non-interpenetrational domain; (b) interpenetrational domain; (c) interpenetrational plus compressional domain.

interpenetrational domain begins, defined by $L_s \leqslant D \leqslant 2L_s$ and depicted by Figure 1.17(b). It is clear that, from a thermodynamic point of view, this step looks much like the process of the de-mixing of macromolecular chains from a solvent, thus the free energy variation can be described by classical polymer solution theories, as we will see below. This contribution is usually called the 'mixing' free energy term, and one can expect that its sign depends on the nature of the chain–solvent interaction. As the separation between the two surfaces decreases below the chains' span, that is $D < L_s$ (Figure 1.17(c)), the motion of macromolecules is severely constrained, and the number of possible chain configurations is greatly reduced. From a thermodynamic point of view, this means that, in order to account for the overall interaction, one must consider, beside the mixing contribution, also the loss of configurational entropy caused by the reduction of the number of allowable configurations. This term is usually called the 'elastic' contribution to the free energy, and is always repulsive, since it arises from an entropy loss.

We will proceed following this division, even if more recent theories try to avoid the somehow artificial separation of the interaction free energy into elastic and mixing contribution [76]. Such criticism is probably sound, since it is apparent that the two contributions are not independent, in the sense that the elastic term depends on the macromolecule configuration and this is obviously affected by the mixing term [77–79]. For an in-depth discussion on this topic, as well as an extremely complete review of the subject of the interaction between polymer-coated particles the reader is referred to the book by Napper [75].

From the previous discussion it can be seen that the description of the steric interaction is very much the realm of the chain configurations modelling and statistical description of segments of chain densities. The development of theories of steric interaction, involves a continuous refining of these models.

The first theoretical paper on the interaction between sterically stabilized particles was written by Mackor [80, 81]. Even if he used a very crude model, it is very interesting to consider it in some length, since it gives both the feeling of where theories started and an easily visualizable picture of the problem. According to the Mackor model, macromolecules adsorbed on flat colloidal particles are treated as terminally anchored rigid rods projecting into the dispersion medium (Figure 1.18(a)). As a further simplification, the variation in free energy is considered only in terms of loss of configurational entropy. In this way, the separation dependent Gibbs free energy can be written as:

$$G(D) - G(\infty) = kT \ln [W(\infty)/W(D)] \qquad (1.49)$$

where $W(x)$ is the number of accessible configurations at separation x, and the Boltzmann relationship, $S = k \ln W_2$, was applied. If the number of accessible configurations is set as proportional to the area scanned by the rigid macromolecular rod (that is the surface area of the hemisphere centred on the ball-joint and of radius L), then it is clear that it decreases from $2\pi L^2$, when separation

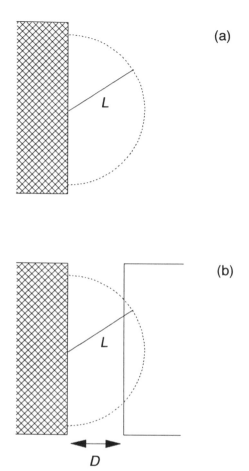

Figure 1.18   The Mackor model of steric interactions (explanations in the text).

is greater than length L, to $2\pi LD$ when $D < L$ (Figure 1.18(b)). Then, from equation (1.49):

$$G(D) - G(\infty) = -kT \ln(D/L) \tag{1.50}$$

or, if the logarithm can be expanded (D not too far from L):

$$G(D) - G(\infty) = kT(1 - D/L) \tag{1.51}$$

where only the first order term has been maintained. Equations (1.50) and (1.51) predict an increase of the free energy as the distance between the plates decreases, as well as an increase of the range of interaction with increasing macromolecular chain length. In order to account for the free energy dependence on the fractional coverage of stabilizing macromolecules, equations (1.50) and (1.51) can be multiplied by the product of the number of adsorption surface sites times the fractional coverage. It must be noted, however, that the treatment is restricted to low fractional coverage, since interaction between closely adsorbed macromolecules is not accounted for in the previous theory.

The merit of the Mackor theory is to give a quick picture of the entropy loss. However, steric interaction is not only a matter of configurational entropy, and apart from the gross oversimplification of the rigid macromolecular chain, the model completely ignores the effect of macromolecule–solvent interaction, that experimental findings indicate as a very important variable. As an example of a more refined approach, that also takes into account the chain–solvent interaction, let us consider the HVO (Hesselink, Vrij and Oveerbeck) theory [82–86]. The free energy of interaction is given, as previously discussed, by the sum of a mixing and an elastic contribution (respectively 'osmotic' and 'volume restriction' terms in the HVO notation), together with the attractive van der Waals' interaction. Thus, for the unit area of two-polymer covered surfaces:

$$W(D) = 2vkTV(D) + 2\left(\frac{2\pi}{9}\right)^{3/2} v^2 kT(\alpha^2 - 1)\langle r^2 \rangle M(D) - \frac{A}{12\pi D^2} \quad (1.52)$$

where v is the number of adsorbed chains per unit area, α and $\langle r^2 \rangle$ are respectively the conventional intramolecular expansion factor and the mean-square end-to-end length [87], and the last term is the already familiar contribution of the van der Waals' interaction between flat surfaces (Figure 1.5). The separation dependent $V(D)$ and $M(D)$ are the volume restriction and the osmotic repulsion functions, which contain all the details and the assumptions of the model. Thus let us consider briefly how they are calculated. The HVO theory assumes, as a model of polymer molecule, the so-called random flight or freely jointed chain, that is a model that describes the macromolecule as a sequence of freely articulated, connected segments. In other words, it ignores fixed valency angles and the barrier to free rotation about bonds, which means that, even if rather more refined than the previous Mackor model, the HVO macromolecular chain also suffers a certain degree of arbitrariness. Thus, the calculation goes on by evaluating the number of possible conformations for several different modes of adsorption of a freely jointed chain, and how the number of configurations is reduced when a second interface approaches, using random flight statistics on a six-choice cubic lattice. Spatial dimensions are expressed in units $(il^2)^{1/2}$, where i is the number of segments each with length l.

In order to obtain the elastic-restricted volume term, let us recall that the relative loss of configurations as a function of the separation between two surfaces is given, by definition, by the ratio between the number of possible configurations at a given separation and the total number of configurations of the 'undisturbed' adsorbed chain. The effect of the loss of configurational entropy on the Helmholtz free energy of two approaching surfaces (A and B), can be written, using the previously used Boltzmann relationship, as

$$F_{RV}(D) - F_{RV}(\infty) = -2kT \sum_i n_i \ln R_j(i, D) \tag{1.53}$$

where R is the relative loss of configuration, a quantity that depends, beside, of course, on the separation D, also on the mode of attachment j (either tail or loop) and on the number i of segments that form the adsorbed tail or loop, and the subscript RV reminds that this contribution arises from a restricted volume effect. Equation (1.53) can be rewritten as:

$$F_{RV}(D) - F_{RV}(\infty) = 2vkT V(i, D) \tag{1.54}$$

where v is the number of loops or tails adsorbed per unit area, and, in the function $V(i, d)$, the summation is replaced by an integration over i from zero to infinity. If, for simplicity, we consider the case in which macromolecules adsorb as equal loops (in which case, $n_i = v$), it is possible to demonstrate [84] that, for $D/(il^2)^{1/2} \geqslant 1$, with a good approximation:

$$V(i, D) = -2(1 - 12D^2/il^2) \exp(-6D^2/il^2) \tag{1.55}$$

As to the osmotic-mixing term, the HVO theory utilizes a result previously derived by Meier [88], that is:

$$F_M(D) - F_M(\infty) = kT V_s^2 V_1^{-1}(1/2 - \chi)U \tag{1.56}$$

where V_s and V_1 are respectively the volumes of a polymeric segment and a solvent molecule, χ is the Flory–Huggins interaction parameter and

$$U = \int_0^D (\rho_a + \rho_b)_{D^2}\, dx - \int_0^\infty (\rho_a + \rho_b)_{\infty^2}\, dx \tag{1.57}$$

where ρ_a and ρ_b are the number densities of the segments adsorbed on interface A and B per unit of volume respectively and the suffix D and ∞ indicate the distance between A and B.

Here, the quality of the solvent is taken into account by the Flory–Huggins interaction parameter, which is expressed in terms of the expansion parameter α, following the Flory's relation [89]:

$$\alpha^5 - \alpha^3 = \frac{27i^2 V_s^2 (1/2 - \chi)}{(2\pi)^{3/2} V_1 (il^2)^{3/2}} \tag{1.58}$$

The random flight dimension (il) is equated to the experimentally determined root-mean-square end-to-end distance, and the subscript zero in equation (1.58) refers to the unperturbed dimension. Thus, in general:

$$(il)^{1/2} = \langle r \rangle^{1/2} = \alpha \langle r \rangle_0^{1/2} \tag{1.59}$$

Substituting equations (1.58) and (1.59) in (1.56), leads to a mixing-osmotic term as written in equation (1.52). The separation dependent function $M(d)$, contains the integer of the segment density distributions, and in the case of equal loops and for $D/(il^2)^{1/2} \geqslant 1$ reduces, with good approximation to [84]:

$$M(i, D) = [(3\pi)^{1/2}(6D^2/(il^2) - 1)]\exp(-3D^2/il^2) \qquad (1.60)$$

As an example of the results that can be obtained from equation (1.52), Figure 1.19 shows the effect of the quality of the solvent on the interaction between the unit area of flat particles covered by homogeneous loops ($V(D)$ and $M(D)$ are calculated from equations (1.55) and (1.60)). In Figure 1.19(a), the solvent is a good one ($\alpha > 1$). The osmotic-mixing term is repulsive and is the determining one, since (Figure 1.17) it is more long-ranged than the restricted volume-elastic term. In Figure 1.91(b), surfaces are immersed in a neutral solvent ($\alpha = 1$), and this time it is the loss of entropy that prevents the surfaces coming into contact. Finally, in Figure 1.19(c), the osmotic term is attractive, since the solvent is a poor one ($\alpha < 1$) and macromolecular chains prefer to avoid the contact with solvent molecules. Also in this case it is the elastic term that, as surfaces approach, overwhelms attractive thrusts.

As explained at the beginning of this chapter, the fate of interacting species depends on the depth of the minimum: if it is of the order of thermal energy, the disorganizing effect of thermal motion will prevail. Thus, from the point of view of the steric stabilization of colloidal particles, the goal is to push the minimum of the curve far from the surface, at a point where the attractive van der Waals' force is lower than thermal energy. The minimum found in curves such as those depicted in Figure 1.19 is neither a primary minimum in the DLVO theory sense (for, in the DLVO theory, the primary minimum arises from the combination of the attractive van der Waals' force with the very short-range Born repulsion), nor a secondary minimum, since no other minimum is expected from equation (1.52). Thus, it is usually called a pseudo-secondary minimum.

The field of the interaction between surfaces bearing adsorbed polymer layers is one of those that received the greatest benefit from the development of the surface force apparatus [90–102]. As an example of an experimental measurement, Figure 1.20 shows forces recorded between mica surfaces bearing adsorbed polystyrene (PS) of molecular weights of 50 000, 100 000 and 600 000. The solvent is cyclohexane that, at the experimental temperature (24 °C) is a poor solvent for PS. The range of the interaction (some tens of nm) and the overall force involved in the process increase with PS molecular weight.

In addition to the steric stabilization of colloids, the presence of a top layer of macromolecular chains on, at least, one of the interacting macroscopic bodies, is of great importance in a large number of technological fields. For instance, the adsorption of particles from a solution is expected to reduce the number of allowable conformations of a macromolecule, much as in the crude model of Figure 1.18. Thus, a surface bearing terminally anchored macromolecular chains is expected to be less adsorptive than a rigid one. The interface between a body covered by macromolecules and a good solvent is expected to be much more

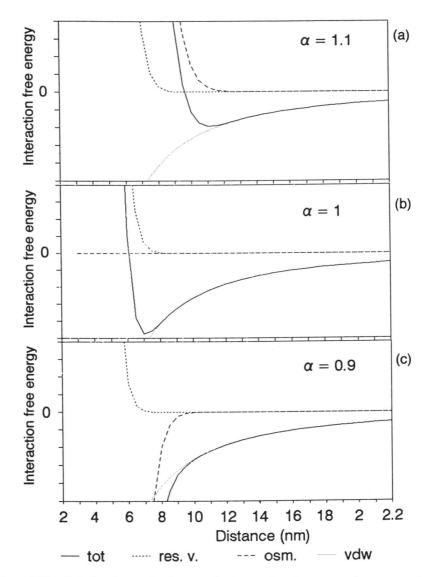

Figure 1.19 Calculated contributions to the interaction free energy between surfaces bearing adsorbed homogeneous loops (equation (1.48)). The four lines show the total restricted volume (res. v.), the osmotic (osm.), the van der Waals (vdw) contribution and their sum (tot). The Hamaker constant was taken as 1×10^{-20} J, the molecular weight of chains 6000, which means a mean square end-to-end length of 5.2 nm.

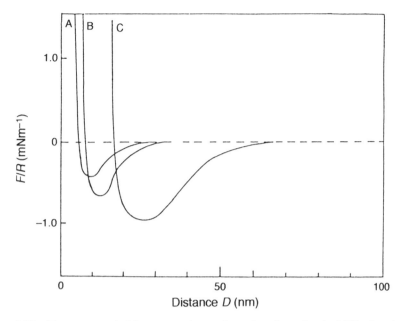

Figure 1.20 Forces recorded between mica surfaces bearing adsorbed PS of molecular weights 50 000(A), 100 000(B) and 600 000(C). At the experimental temperature of 24 °C the solvent (cyclohexane) is a poor solvent for PS. (Reprinted from ref. [102] by courtesy of Marcel Dekker, Inc.)

'open' and diffuse than the interface of the same body with a poor solvent. These different surface structures imply different answers to the surroundings. In the following chapters, we will find some examples of such effects on practical and technological applications of polymers.

1.5 SHORT-RANGE INTERACTIONS (HYDROGEN, ACID–BASE, COVALENT)

In this section we will briefly discuss forces that, due to their short-range nature, primarily determine the depth of the well shown in Figure 1.1. These forces work when suitable approaching species succeed in overcoming secondary minima and the energy barrier, and finally fall in the primary minimum built by the interplay of attractive forces and Born repulsion. In Figure 1.1, the depth of the free energy well between species that can engage covalent bonding is, on average, about 500 times deeper than when only van der Waals' forces are involved. As to the general shape of the curve, it is generally the same since the range of these forces is no more than 0.1–0.2 nm (thus, they are also called 'contact' forces [31]). It is, however, important to bear in mind that short-range forces can have indirect effects on longer-range interactions. For instance, solvation layers are created by the interplay of short-range interactions such as hydrogen bonding or chemical bonding between surface and solvent molecules. The involvement of solvation layers in the interaction between two approaching surfaces, give rise to the longer range solvation forces described in the previous section. Also, the adsorption or the grafting of macromolecules on a substrate is ultimately controlled by contact forces. The output of these very short-range events is the previously described steric interaction between macromolecule bearing surfaces, with ranges of some tens of nm.

As described in the voluminous literature on the subject [103, 104], the structural feature required to perform hydrogen bonding is an hydrogen atom covalently bonded to a highly electronegative element (such as O, N, F). The hydrogen atom, due to its tendency to become positively polarized and its small size (that means a high ratio between the polarization-induced charge and the mass of the atom) can interact strongly with nearby electronegative atoms. Thus, hydrogen bonding is expected to play a role in the interaction between polymer surfaces bearing hydroxyl, carboxyl, amino and similar groups, while polyolefins, polystyrene and related polymers cannot interact in this way.

Recently Fowkes discussed the role of acid–base interactions in contact forces [105–107]. All polymers, except saturated hydrocarbons, have acidic or basic sites in the sense of the Lewis acid–base theory [108]. Electrons of oxygens, nitrogens, sulphurs and similar elements, as well as π-electrons of polystyrene and polymers containing aromatic groups are basic sites (electron donors, following the Lewis theory). On the other hand, halogenated hydrocarbons, nitro groups and, in general, all electrophilic sites, are electron acceptor and thus Lewis acids. The acid–base interaction between polymers and their surroundings (other polymers, solvents, plasticizer, filler, etc.) control the strength of the contact interaction, whose enthalpy can be described by the four-constant Drago's equation [109, 110]:

$$-\Delta H^{ab} = C_a C_a + E_a E_b \qquad (1.61)$$

where the two constants for the base (subscript b) and two for the acid emphasize that the strength of the acid–base interaction depends not only on the ability to donate or accept electrons but also on the polarizability [111, 112]. As Fowkes explicitly points out, hydrogen bonding can be considered a subset of acid–base interactions, since its enthalpy can be accurately described by the Drago equation (1.61). We will return to this topic in the chapter on contact angles (Chapter 4), where the acid–base approach to surface energetics will be discussed.

Finally, when surface atoms and molecules come very close together, very strong forces arise from the overlap of electronic clouds and sharing of valence electrons (Table 1.1). This chemical event, where intervening atoms lose their identity and create a new species, is known as primary or covalent bonding by chemists, while physicists generally call it a very short-ranged attractive force. The details of the covalent bonding depend on the specific chemistry, that is on the electronic and geometrical configuration of the species involved, as has been unravelled by the continued effort of two centuries of modern chemical thought. While a description of covalent bonding is outside the scope of this text (it can be found in a huge number of general and advanced textbooks on chemistry and material science), it is important to note that most synthetic polymers do not bear surface functional groups suited to covalent linking. On the contrary, the chemical inertness towards the surroundings is often an appreciated property of synthetic polymers, which makes them the materials of choice where other chemically reactive materials cannot be employed. The lack of chemical reactivity poses special challenges when it comes to surface modification, where it is necessary to link the polymer (substrate) covalently with another molecule, either to impart durable specific properties to the substrate or to create a strong joint. Therefore, to introduce functional groups on polymer surfaces which can work as sites for successive covalent linking, the chemical inertness of most synthetic polymers must be overcome by some kind of 'activation step'. Activation can be carried on by treatment with very strong chemicals, by high energy density treatment, or by irradiation, and will be discussed later in this book (part III). Whatever the mechanism, it is important to appreciate that the need for often expensive and energy intensive surface treatment is the sort of cost that must be paid in order to impart some kind of chemical reactivity to surfaces of materials that often make an addition of their chemical inertness.

REFERENCES

[1] J. N. Israelachvili, *Intermolecular and Surface Forces*, Academic Press, London (1985).
[2] A. W. Adamson, *Physical Chemistry of Surfaces*, 5th edn, Wiley, New York (1990).
[3] R. G. Horn, *J. Am. Ceram. Soc.*, **73**, 1117 (1990).
[4] M. Rätzsch, H. J. Jacobash, and K. H. Freitag, *Adv. Colloid Interface Sci.*, **31**, 225 (1990).
[5] D. Tabor, *J. Colloid Interface Sci.*, **58**, 2 (1977).
[6] J. N. Israelachvili and B. W. Ninham, *J. Colloid Interface Sci.*, **58**, 14 (1977).
[7] J. N. Israelachvili and G. E. Adams, *Nature*, **262**, 77 (1976).
[8] G. S. Rushbrooke, *Trans. Faraday Soc.*, **36**, 1055 (1940).
[9] W. H. Keesom, *Physik A*, **22**, 129 (1921).
[10] W. H. Keesom, *Physik A*, **22**, 643 (1921).
[11] P. Debye, *Physik Z.*, **21**, 178 (1920).
[12] P. Debye, *Physik Z.*, **22**, 302 (1921).
[13] F. London, *Z. Phys. Chem.*, **B11**, 222 (1930).
[14] J. N. Israelachvili, *Intermolecular and Surface Forces*, Academic Press, London (1985), p.125.
[15] B. V. Derjaguin, *Kolloid Z.*, **69**, 155 (1934).
[16] J. N. Israelachvili, *Intermolecular and Surface Forces*, Academic Press, London (1985), p.130.

[17] E. M. Lifshitz, *Zh. Eksp. Teor. Fiz.*, **29**, 94 (1955).
[18] E. M. Lifshitz, *Sov. Phys.*, **2**, 73 (1956).
[19] I. E. Dzyaloshinskii, E. M. Lifshitz, and L. P. Pitaevskii, *Adv. Phys.*, **10**, 165 (1961).
[20] C. J. F. Böttcher and P. Bordewijk, *Theory of Electric Polarization*, Elsevier, Amsterdam (1967) Vol. 2.
[21] J. Mahanty and B. W. Ninham, *Dispersion Forces*, Academic Press, London (1976).
[22] D. B. Hough and L. R. White, *Adv. Coll. Interface Sci.*, **14**, 3 (1980).
[23] J. N. Israelachvili and D. Tabor, *Proc. R. Soc. London. A*, **331**, 19 (1972).
[24] J. N. Israelachvili, *Intermolecular and Surface Forces*, Academic Press, London (1985), p. 144.
[25] J. N. Israelachvili, *Proc. R. Soc. London. A*, **331**, 39 (1972).
[26] R. G. Horn and J. N. Israelachvili, *J. Chem. Phys.*, **75**, 1400 (1981).
[27] C. J. Van Oss, D. R. Absolom, and A. W. Neumann, *Colloids Surfaces*, **1**, 45 (1980).
[28] B. V. Derjaguin, I. I. Abrikossova, and E. M. Lifshitz, *Q. Rev.*, **10**, 292 (1956).
[29] B. V. Derjaguin and I. I. Abrikossova, *Phys. Chem. Solids*, **5**, 1 (1958).
[30] W. Black, J. G. V. de Jongh, J. Th. G. Overbeek, and M. J. Sparnaay, *Trans. Farad. Soc.*, **56**, 1597 (1960).
[31] J. N. Israelachvili and G. E. Adams, *J. Chem. Soc. Faraday Trans.*, **74**, 975 (1978).
[32] M. J. Sparnaay, *The Electrical Double Layer*, Pergamon Press, New York (1972).
[33] J. Lyklema, in *Surface and Interfacial Aspects of Biomedical Polymers*, J. D. Andrade (Ed.), Plenum Press, New York (1985), vol. 1, p. 293.
[34] S. Wu, *Polymer Interface and Adhesion*, Dekker, New York (1982), p. 327.
[35] I. Langmuir, *J. Chem. Phys.*, **6**, 873 (1938).
[36] G. Gouy, *J. Phys.*, **9**, 457 (1910).
[37] G. Gouy, *Ann. Phys.*, **7**, 129 (1917).
[38] D. L. Chapman, *Phil. Mag.*, **25**, 475 (1913).
[39] J. N. Israelachvili, *Intermolecular and Surface Forces*, Academic Press, London (1985), p. 165.
[40] D. C. Grahame, *Chem. Rev.*, **41**, 441 (1947).
[41] O. Stern, *Z. Elektrochem.*, **30**, 508 (1924).
[42] P. C Hiemenz, *Principles of Colloid and Surface Chemistry*, Dekker, New York (1977), p. 379.
[43] B. V. Derjaguin and L. Landau, *Acta Physicochim. Urss*, **14**, 633 (1941).
[44] E. J. W. Vervey and J. Th. G. Overbeek, *Theory of the Stability of Lyophobic Colloids*, Elsevier, Amsterdam (1948).
[45] D. J. Shaw, *Introduction to Colloid and Surface Chemistry*, Butterworth, London (1970).
[46] W. Norde, and J. Lyklema, *Colloids Surf.*, **38**, 1 (1989).
[47] *Microbial Adhesion and Aggregation*, K. C. Marshall (Ed.), Springer-Verlag, Berlin (1984).
[48] A. C. Cowley, N. L. Fuller, R. P. Rand, and V. A. Parsegian, *Biochemistry*, **17**, 3163 (1978).
[49] A. W. Adamson, *Physical Chemistry of Surfaces*, 5th edn, Wiley, New York (1990), p. 276.
[50] D. J. Mitchell, B. W. Ninhan, and B. A. Pailthorpe, *Chem. Phys. Lett.*, **51**, 257 (1977).
[51] F. F. Abraham, *J. Chem. Phys.*, **68**, 3713 (1978).
[52] J. H. Lane and T. H. Spurling, *Chem. Phys. Lett.*, **67**, 107 (1977).
[53] M. Rao, B. J. Berne, J. K. Percus, and M. H. Kalos, *J. Chem. Phys.*, **71**, 3802 (1979).
[54] D. Nicholson and N. G. Parsonage, *Computer Simulation and the Statistical Mechanics of Adsorption*, Academic Press, New York (1982), ch. 8.

[55] G. Rickayzen and P. Richmond, in *Thin Liquid Films*, I. B. Ivanov (Ed.), Dekker, New York (1985), ch. 4.
[56] R. G. Horn and J. N. Israelachvili, *J. Chem. Phys.*, **75**, 1400 (1981).
[57] H. K. Christenson, *J. Chem. Phys.*, **78**, 6906 (1983).
[58] H. K. Christenson and R. G. Horn, *Chem. Phys. Lett.*, **98**, 45 (1983).
[59] H. K. Christenson, *J. Phys. Chem.*, **90**, 4 (1986),
[60] S. Marcelja and N. Radic, *Chem. Phys. Lett.*, **42**, 129 (1976).
[61] S. Marcelja, D. J. Mitchell, B. W. Ninham, and M. J. Sculley, *J. Chem. Soc. Faraday Trans. 2*, **73**, 630 (1977).
[62] B. Jönsson, *Chem. Phys. Lett.*, **82**, 520 (1981).
[63] N. I. Christou, J. S. Whitehouse, D. Nicholson, and N. G. Parsonage, *Symp. Faraday Soc.*, **16**, 139 (1981).
[64] B. Jönsson and H. Wennerström, *J. Chem. Soc. Faraday Trans. 2*, **79**, 19 (1983).
[65] *Water: a Comprehensive Treatise*, F. Frank (Ed.), Plenum, New York (1972–1982), vols 1–7.
[66] R. M. Pashley, *J. Colloid Interface Sci.*, **80**, 153 (1981).
[67] R. M. Pashley, *J. Colloid Interface Sci.*, **83**, 531 (1981).
[68] R. M. Pashley and J. N. Israelachvili, *J. Colloid Interface Sci.*, **97**, 446 (1984).
[69] J. N. Israelachvili, *Intermolecular and Surface Forces*, Academic Press, London (1985), p. 202.
[70] J. N. Israelachvili and R. M. Pashley, *Nature*, **306**, 549 (1983).
[71] C. Y. Lee, J. A. McCammon, and P. J. Rossky, *J. Chem. Phys.*, **80**, 4448 (1984).
[72] J. C. Eriksson, S. Ljunggren, and P. M. Cleasson, *J. Chem. Soc. Faraday Trans. 2*, **85**, 163 (1989).
[73] W. Heller and T. L. Pugh, *J. Chem. Phys.*, **54**, 1778 (1954).
[74] J. M. H. Scheutjens and G. J. Fleer, *Adv. Colloid Interface Sci.*, **16**, 361 (1982).
[75] D. H. Napper, *Polymeric Stabilization of Colloidal Dispersions*, Academic Press, London (1983).
[76] P. G. deGennes, *Adv. Colloid Interface Sci.*, **27**, 189 (1987).
[77] A. K. Dolan and S. F. Edwards, *Proc. Roy. Soc. London A*, **343**, 427 (1975).
[78] D. W. J. Osmond, B. Vincent, and F. A. Waite, *Colloid Polymer Sci.*, **252**, 676 (1975).
[79] S. Levine, M. L. Thomlinson, and K. Robinson, *Faraday Disc. Chem. Soc.*, **65**, 202 (1978).
[80] E. L. Mackor, *J. Colloid Sci.*, **6**, 492 (1951).
[81] E. L. Mackor and J. H. van der Waals, *J. Colloid Sci.*, **7**, 535 (1952).
[82] F. Th. Hesselink, *J. Phys. Chem.*, **73**, 3488 (1969).
[83] F. Th. Hesselink, *J. Phys. Chem.*, **75**, 65 (1971).
[84] F. Th. Hesselink, A. Vrij, and J. Th. G. Overbeek, *J. Phys. Chem.*, **75**, 2094 (1971).
[85] F. Th. Hesselink, *J. Colloid Interface Sci.*, **50**, 606 (1975).
[86] F. Th. Hesselink, *J. Polymer Sci. Polymer Symposia*, **61**, 439 (1977).
[87] D. H. Napper, *Polymeric Stabilization of Colloidal Dispersions*, Academic Press, London (1983), p. 74.
[88] D. J. Meier, *J. Phys. Chem.*, **71**, 1861 (1967).
[89] P. J. Flory, *Principles of Polymer Chemistry*, Cornell University Press, Ithaca (1953).
[90] J. Klein and P. F. Luckham, *Macromolecules*, **17**, 1401 (1984).
[91] P. F. Luckham and J. Klein, *Macromolecules*, **18**, 721 (1985).
[92] J. Klein, *J. Chem. Soc. Faraday Trans. 1*, **79**, 99 (1983).
[93] J. N. Israelachvili, M. Tirrel, J. Klein, and Y. Almog, *Macromolecules*, **17**, 204 (1984).
[94] Y. Almog and J. Klein, *J. Colloid Interface Sci.*, **106**, 33 (1985).

[95] G. Hadziianou, S. Patel, S. Crannik, and M. Tirrel, *J. Am. Chem. Soc.*, **108**, 2869 (1986).
[96] S. Patel, M. Tirrel, and G. Hadziianou, *Colloids and Surfaces*, **31**, 157 (1988).
[97] M. A. Ansarifar and P. F. Luckham, *Polymer*, **29**, 329 (1988).
[98] H. J. Taunton, C. Toprakcioglu, L. J. Fetters, and J. Klein, *Nature*, **333**, 712 (1988).
[99] T. Afshar-Rad, A. I. Bailey, P. F. Luckham, D. Chapman, and W. MacNaughtan, *Colloids and Surfaces*, **25**, 263 (1987).
[100] T. Afshar-Rad, A. I. Bailey, P. F. Luckham, D. Chapman, and W. MacNaughtan, *Colloids and Surfaces*, **31**, 125 (1988).
[101] T. Afshar-Rad, A. I. Bailey, P. F. Luckham, D. Chapman, and W. MacNaughtan *Biochim. Biophys. Acta*, **915**, 101 (1987).
[102] P. F. Luckham, *Polym. Plast. Technol. Eng.*, **28**, 975 (1989).
[103] M. D. Joesten and L. J. Schaad, *Hydrogen Bonding*, Dekker, New York (1974).
[104] P. Schuster, G. Zundel, and C. Sandorfy, *The Hydrogen Bond*, North-Holland, Amsterdam (1976).
[105] F. M. Fowkes, in *Physicochemical Aspects of Polymer Surfaces*, K. L. Mittal (Ed.), Plenum Press, New York (1983), Vol. 2, p. 583.
[106] F. M. Fowkes, in *Surface and Interfacial Aspects of Biomedical Polymers*, J.D. Andrade (Ed.), Plenum Press, New York (1985), Vol. 1, ch. 9.
[107] F. M. Fowkes, *J. Adhesion Sci. Tech.*, **1**, 7 (1987).
[108] G. N. Lewis, *Valence and the Structure of Atoms and Molecules*, Chem. Cat. Co., New York (1923).
[109] R. S. Drago, G. C. Vogel, and T. E. Needham, *J. Am. Chem. Soc.*, **93**, 6014 (1970).
[110] R. S. Drago, L. B. Parr, and C. S. Chamberlain, *J. Am. Chem. Soc.*, **99**, 3203 (1977).
[111] G. Schwarzenbach, *Rec. Trav. Chim.*, **75**, 562 (1956).
[112] R. G. Pearson, *Hard and Soft Acids and Bases*, Dowden, Hutchinson and Ross, Stroudsburg, Pennsylvania (1973).

Chapter 2
Dynamics of Polymer Surfaces

In the previous chapter, the interaction between polymers and surroundings was discussed on the basis of the forces proceeding from the polymer surface. In this way, the emphasis was completely on the thermodynamic side of the problem, that is on the nature and the amount of the driving force for a given kind of interaction: the solid was defined by a set of properties and the resulting interaction was modelled. Discussing the steric interaction, however, it was necessary to introduce the idea of moving polymeric chains (for instance, in the Mackor model, the area scanned by the rigid macromolecular rod is involved in the calculation), that is, a dynamic effect was invoked to explain a given surface property.

In this chapter the dynamic aspects of polymer surfaces will be discussed in more length. The importance of the dynamic behaviour of polymer surfaces has only recently been stressed [1,2] although reports on dynamic properties of surface-adsorbed molecules or polymers [3–6] are much older. One of the reasons for this state of the art is that most surface physicochemical theories were classically developed for metals and ceramics, which are obviously more rigid than polymers. The typical thermal properties which are linked to the mobility of the constituents of the materials (that is the melting temperature T_m and the glass transition temperature T_g) are, in polymers, some hundredths of degree lower than the corresponding temperatures of inorganic solids. This means that, in ordinary temperature conditions, organic polymeric solids are much more mobile and enjoy a far greater freedom than their inorganic counterpart. As an extreme example, poly(dimethylsiloxane) (PDMS) is, at room temperature, about 150 °C above its T_g. One should appreciate that, in similar conditions, the overall properties of the material are deeply affected by dynamic phenomena, and, indeed, a large body of literature is devoted to the study of the dynamic properties of bulk polymers. On the other hand, probably due to its 'inorganic' heritage, the surface chemistry and physics of polymers usually disregards dynamics and assumes that surfaces are immobile and 'frozen' in a given configuration. Another compelling reason is, of course, that the experimental investigation of dynamic phenomena involving surface layers is much more difficult and elusive than the study of bulk phenomena.

As previously discussed, the suggestion of the importance of the dynamic behaviour of polymer surfaces in interfacial interactions is very recent, and originated mainly as a consequence of the observation that some polymeric systems have surface properties which are a function of the surrounding environment [1–5]. This means that polymeric surfaces can restructure in response to a change of the interfacing phase, in order to tune their surface properties with the properties of the interfacing medium. The most often quoted example, which will also be discussed in Chapter 4, is that of poly(hydroxyethylmethacrylate) (PHEMA), a polymer which exposes the hydrophobic methyl groups at the polymer–air interface and the hydrophilic hydroxyl groups when it is interfaced with water [4]. It is important to stress that the basic point to note is not that PHEMA modifies its surface structure in order to minimize the interfacial free energy (since the minimization of the free energy of the system is the direction of every natural phenomenon) but that it can do it in a time comparable with the experimental time. Thus, the emphasis is on the time taken for the rearrangement, that is on the dynamic aspect of the process. The example of PHEMA shows that, in polymers, the kinetics of motion at the polymer surface can be quick enough to affect the surface properties.

It is not easy to correlate the huge amount of work on bulk mobility of polymers to the mobility of surfaces. Intuitively, macromolecules on surfaces should enjoy a greater freedom than their bulk counterpart. On the other hand, if the interfacing phase is thermodynamically hostile, it can work as a hard wall and limit the mobility of surface chains.

In this chapter, we will review the existing work in this field. Obviously, since this discipline is still in its infancy, no general theories such as those discussed in the previous chapter and no systematic approach to the problem are available. Reflecting the present state of the art, this chapter will not review consolidated theories, rather it will collect the theoretical and experimental evidence which show that polymer surface dynamics exist and affect the surface properties of the polymer. Thus, the outline of this chapter will be very different from the previous one. First, several aspects of polymer surface dynamics will be discussed, mainly based on a recent dynamic simulation of a free polymer surface. Then, experimental evidence of the mobility of polymer surfaces will be reviewed.

2.1 FUNDAMENTAL ASPECTS OF POLYMER SURFACE DYNAMICS

To put the problem of polymer surface dynamics in a proper perspective, a few basic questions must be answered. First, which kind of motion is involved? In general, when dealing with dynamics of polymeric materials, the first point which is emphasized is that several hierarchies of motion co-exist, as shown in Figure 2.1. A first kind of motion is the rotation around the backbone and side-groups bonds. Increasing the spatial scales, the typical macromolecular movements involve large segments and the molecule as a whole. These latter mechanisms control many of the properties of interest in bulk, melt and solution of polymers. As to surfaces, outdiffusion of chains from the bulk towards the surface or reorganization of segments in the domain-like surface structure deeply affects the surface properties of polymeric blends or block copolymers, as will be discussed

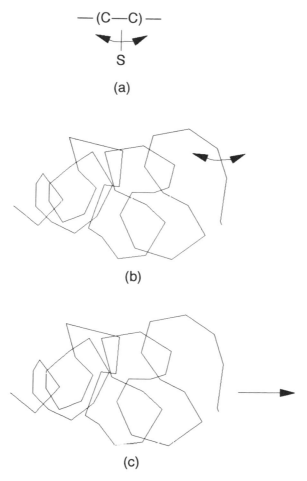

Figure 2.1 The hierarchy of motions in a randomly coiled macromolecular chain (a) Rotation about the side group (S) bond. (b) Motion on segmental scale. (c) Motion on macromolecular scale.

in Chapter 8. It is, however, important to also remember that the short-range reorientation of side-groups can have profound effects on several surface properties, as discussed later.

In Chapters 4 and 9, several aspects of wetting of polymers by liquids will be discussed. It will become clear that wettability is controlled by the outermost layers of the solid and that it is affected by the orientation of surface functional groups. Owing to the large number of important technological properties which are affected by wetting phenomena (i.e. adhesion, printing, waterproofing, etc.), it can be appreciated that the understanding of short-range motion of side-groups on the polymer surface is a central issue of polymer surface science.

From a more chemical point of view, the spatial arrangement and availability of a given functional group is known to deeply affect the yield of chemical reactions. Thus, chemical reactions involving a surface-linked functional group, as well as all the short-range, specific interactions described in the previous chapter are sensitive to the details of side-group orientation.

A further, important question is the following: is surface dynamics different from bulk dynamics? The most successful account of the dynamic behaviour of entangled chains in bulk polymers, i.e. reptation theory [7, 8], describes entangled chains as moving in a virtual tube or pipe, whose walls are made up by the locus of contacts of the polymer with non-crossable obstacles. Now, taking into account a surface chain, it may be anticipated that the nature of the wall of the virtual tube will be a function of the characteristics of the interfacing medium. Thus, in principle, a modification of the dynamic behaviour, as compared to bulk chains, is expected.

Macromolecular dynamics have been discussed in several textbooks [9–11]. It is clearly outside the scope of this text to review the existing theories on the motions of polymeric chains in bulk polymers. Rather, a discussion will be made on a recent report on a dynamic simulation of a glassy polymer surface [12] which, to the authors' knowledge, is the first to look at the free surface of an amorphous polymer in atomistic details using molecular dynamics. The system which is studied is the surface region of a thin film of glassy atactic poly(propylene) (aPP), interfaced against vacuum, at $-40\,°C$, which is about $22\,°C$ lower than the experimental T_g. To compare bulk and surface properties, the film is sufficiently thick to contain a region far from the vacuum–aPP interface. The film is modelled in a detailed microscopic manner [12, 13] and it consists of three types of interacting sites: backbone carbons, pendant hydrogens and pendant methyl groups. It is assumed that the chains are monodispersed and with a degree of polymerization of 26 ($M_w = 1110$). The basis of the molecular dynamic computer simulation techniques used are described in several textbooks [14, 15]. The simulation focuses only on the short-time dynamics of a glassy aPP free surface ($t < 400$ ps). Several interesting results arise from this study: Figure 2.2 (reprinted from ref. [12]) shows the simulated density profile of the aPP film. The film midplane lies at $z = 0\,Å$. In order to resolve spatially the dynamic characteristics

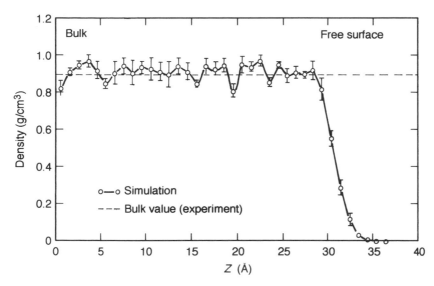

Figure 2.2 Local mass density distribution at a glassy atactic PP/vacuum interface. The experimental bulk density is indicated by a dashed line [12]. (Reprinted with permission from Mansfield and Theodorou, *Macromolecules*, **24**, 6283. Copyright (1991) American Chemical Society.)

of the film, the simulation is performed separately in each one of five different regions, whose boundary are shown in Figure 2.2 as thick marks on the upper part of the figure. It is important to note that the four inner regions, which are each 6.5 Å thick, are characterized by a density which is indistinguishable from that of the bulk polymer, while the outermost region shows a decreasing density as z increases. Local mobility is simulated within each of these regions. Details of the dynamic simulation can be found in the quoted paper. As to the results, the first dynamic aspect which is considered is the displacement of individual atoms and methyl groups on the chain. Mathematically, this kind of motion can be described by the displacement of atoms relative to the centre of mass of the chain to which they belong. The mean square displacement as a function of time in the five representative regions in shown in Figure 2.3 [12]. These data reflect only the atomic displacements, since any motion resulting from overall chain migration has been removed by subtracting from each atomic displacement the displacement of the centre of mass of the chain to which it belongs. As shown by Figure 2.3, there is a considerable enhancement of atomic mobility in the outermost region compared to the three most internal regions, a finding which confirms the intuitive feeling that the decreased constraints of macromolecules at a free surface should result in increased mobility [1]. Comparing Figures 2.2 and 2.3, it is possible to notice that the distance over which an enhanced mobility is observed

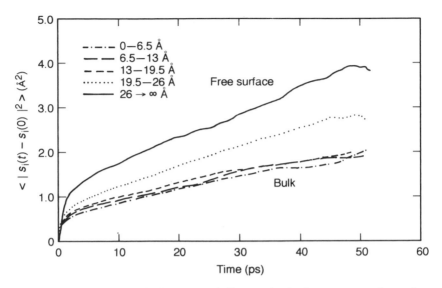

Figure 2.3 Mean-squared displacement of all atoms in the five representative regions of the model film indicated in Figure 2.2, as a function of time [12]. (Reprinted with permission from Mansfield and Theodorou, *Macromolecules*, **24**, 6283. Copyright (1991) American Chemical Society.)

is larger (by a factor of about 2) than the distance over which the density departs from its bulk value. This means that, while the enhanced surface mobility is a consequence of the decreased surface density, and hence increased free volume, the atomic level mobility does propagate for some distance into the bulk. These results are probably a consequence of the connectivity of the chain segments, which allows the chains which are exposed to the anisotropic surface region to transmit dynamical information into the underlying polymer. The interesting point is that this happens even if, as shown in Figure 2.2, the density there is indistinguishable from that of the glassy bulk.

Another dynamic property which is investigated by Mansfield and Theodorou [12] is the orientational relaxation of backbone carbon–carbon bonds. In this way it is possible to evaluate the local reorientational tendencies of the basic units of the chains. Also, in this case, it is found that surface C–C bonds lose memory of their initial orientation at a much higher rate than bulk bonds, and that the region of surface-enhanced mobility is deeper than the region of decreased density (Figure 2.2).

Finally, even in the limits of the short time-scale investigated, a look is taken at the mobility of the whole chains, as described by the mean-square displacement of the chain centre of mass. The previous findings are also confirmed in this larger scale dynamical property, and surface chains are shown to be more mobile than inner chains. Interestingly, it is found that displacements parallel to the surface are much greater than displacements normal to it. This anisotropic motion is the consequence of the anisotropic nature of the system.

In summary, the detailed molecular dynamics simulation of the free surface of a film of aPP shows a dynamic picture of the surface region which, in many respects, behaves dynamically as a polymer melt. This simulation is the first answer to the previous questions: all kinds of motion are enhanced in the surface region of the aPP film. As to the relationship between bulk and surface dynamics, within the limits of the quoted simulation, it is possible to conclude that it is incorrect to assume that the mobility of the surface chains and pendant groups is the same as bulk mobility, as described, for instance, by the bulk T_g. In the same way, also the bulk value of the characteristic temperature of so-called β relaxations (T_β), that is the rotation of a side chain about a C–C bond, should be of little help in understanding what really happens on surfaces.

In general, the quoted simulation shows that the surface region of a glassy polymer is characterized by a dynamic behaviour of its own, and that sub-surface regions (as defined by the density profile of Figure 2.2 and the results of Figure 2.3) can be, in turn, affected by the motion of the outermost layer.

The huge computational effort required by dynamic molecular simulation of polymeric chains, precludes extending the above calculation to more realistic systems, such as surfaces of polydisperse or, in general, heterogeneous systems against a given interfacing medium. In this case, several other issues must be taken into account. For instance, the inter and intrachain interactions, such as classically described by Flory's theory [16], can, in turn, affect local dynamics, since, as shown by Figure 2.3, dynamical information can be transmitted to different regions of the polymer.

Moreover, the interfacing medium can be air, water or, in general, an aqueous solution or another solid. The surface field of forces of the interfacing medium definitely affects the anisotropic nature of the system. The enhanced mobility of the outermost layer indicates that the probability of surface functional groups overturning in response to changes of the properties of the interfacing medium [1–5] is highly likely. Much experimental evidence strongly supports this view, as discussed in the following section. As to the theoretical side, a series of papers has been devoted to the thermodynamic description of the restructuring of grafted polyolefin surfaces as a function of the interfacing medium [17–20]. These polymers are made by apolar polyolefins grafted with a small amount of polar moieties (typically acrylic acid). In air, their surfaces display apolar characteristics but, as shown by Lavielle and coworkers and discussed in the next section [17,18], highly polar interfacing environments promote the restructuring of the surface so that acrylic grafts become exposed. In these papers, the decrease of the interfacial tension of grafted polyolefins against a liquid medium as a function of time, supported by the analytical evidence of several techniques, is taken as a gauge of polymer surface restructuring promoted by the environment. The problm is tackled by means of irreversible process thermodynamics, exploiting several theoretical results obtained by Sanfeld [21, 22]. An equation is first derived, which contains the contribution of orientating dipoles to the interfacial tension. A phenomenological coefficient related to the reorientation at the interface, which can be calculated from experimental values is then discussed [19]. The

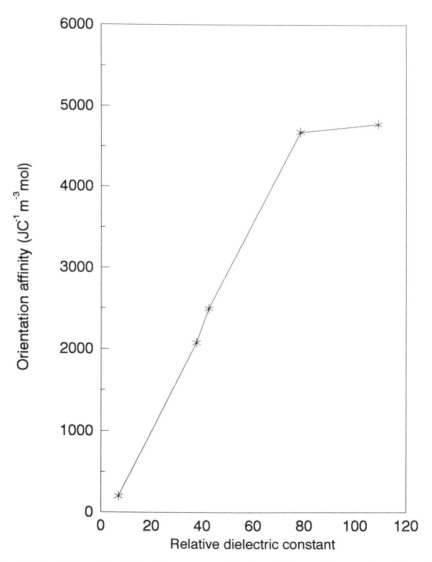

Figure 2.4   Effect of the dielectric constant of the interfacing medium on the orientation affinity of 0.5% acrylic acid-grafted polyethylene. (Plotted from data in Tables I and II of ref. [20].)

main point is, however, to evaluate the effect of the properties of the interfacing environment on the rotation of dipoles of the grafted species. As shown by the authors [20], it is possible to establish a relation between the dielectric constant of the interfacing liquid environment and the orientation affinity of the polar groups. The latter is basically the change of the interfacial free energy as a consequence of the orientation of a given dipolar moiety. The results show that the dielectric constant of the liquid medium (hence the dipole moment of the liquid molecule) has a strong effect on the orientation affinity of the polar groups on the solid polymer surface, as shown by Figure 2.4, which is plotted from part of the data of Table II of ref. [20]. This means that a polar moiety on a polymer surface feels the presence of the interfacing medium, and the higher the dielectric constant of the latter the higher the thermodynamic push towards the alignment in a proper orientation. When this thermodynamical analysis is coupled with the previous simulation of surface macromolecular dynamics, the need of attaching a dynamical dimension to the studies and the interpretation of phenomena involving polymer surfaces is strongly emphasized.

In summary, the knowledge that bulk properties of polymers are deeply affected by macromolecular dynamics naturally leads to the investigation of dynamic phenomena on polymer surfaces. From a theoretical point of view, when a closer look is taken, two main points result: surfaces are regions of enhanced mobility as compared to the bulk and the orientation of surface groups is affected by the nature of the interfacing environment. In the following section, some experimental evidence supporting these views will be discussed.

2.2 EXPERIMENTAL EVIDENCE OF POLYMER SURFACE DYNAMICS

Andrade *et al.* have discussed several experimental approaches to the evaluation of polymer surface dynamics [1]. In this section, we will review some results which strongly support the previous discussion.

The surface macromolecular dynamics simulation of Mansfield and Theodorou [12], suggests that all the kinds of motion shown in Figure 2.1 are enhanced on glassy polymer surfaces. Experimental evidence of surface-enhanced polymer segments diffusion can be found in the healing behaviour of crazes and fractures. In 1964, Kambour [23] observed the fading of interference colours on freshly produced fracture surfaces of poly(methylmethacrylate) (PMMA) at room temperature, suggesting rapid craze healing (note that the T_g of PMMA is about 100 °C). Interestingly, Kambour reports that under normal atmospheric conditions, fading can be observed in a day or two, while heat causes very rapid disappearance, confirming the thermally-activated nature of the phenomenon. Also water and, most of all, organic solvents do cause a decrease of the time taken for healing. This observation underlines another important aspect of polymer surface dynamics: the surface region of polymeric materials, can be and usually is, saturated by water (for instance, due to the moisture content of the atmosphere) or other solvents, which can work as plasticizers. Thus, the T_g of the surface chains can be further lowered with respect to the bulk chain, further increasing the mobility of surface chains with respect to bulk chains. This result, as well as others that will be discussed below, suggest that this is indeed the case.

Yang and Kramer studied the structure and coalescence of poly(styrene) (PS) craze fibril by low angle electron diffraction (LAED) [24]. The craze fibrils studied are very interesting model systems, since each chain in the fibril is no further than two molecular diameters from the fibril surface. LAED analysis shows that the fibril microstructure coarsens significantly upon ageing even at temperatures 70 °C or more below the T_g of PS. Again, the kinetics of ageing are increased greatly by increasing either ageing temperature or solvent content of the PS film. The results of the study of the kinetics of coalescence show that molecules inside tiny craze fibrils have a much higher mobility than similar molecules in the bulk, that is the segmental mobility within the small-diameter craze fibrils is much enhanced relative to the segmental mobility in the bulk polymer glass [24].

Clear evidence of these phenomena can be observed in Figure 2.5, which shows the force which is exerted when a poly (ethyleneterephthalate) (PET) fibre (*ca* 1.5 mm diameter) is immersed in a water-containing beaker [25]. As discussed in Chapter 4, this kind of measurement, the Wilhelmy plate experiment, is generally used to measure either the liquid surface tension or the contact angle of the liquid on the solid surface. While the details of the Wilhelmy plate experiment will be fully discussed in Chapter 4, at this time it is important to observe that the force–displacement curve of the as-prepared sample is very irregular, owing to the large number of cracks and crevices produced on the sample surface by the preparation routine (the large cracks, were intentionally produced by a scriber). On the other hand, after 16 hours at 35 °C, that is about 40 °C below the T_g, all crevices are healed and the force–displacement curve is now rather smooth (save, of course, for the macroscopic cracks). Thus, despite ageing being made at a temperature at which, according to the T_g, the macromolecular chains should be almost immobile, the experimental evidence shows that major macromolecular diffusion occurred. Again, this result shows that chains on polymer surfaces do have a much greater mobility than suggested by the T_g, which is, basically, a value related to bulk properties [1]. In this respect it is noteworthy to remark that the quoted work of Yang and Kramer [24] was undertaken to understand why the fibril diameter of bulk crazes [26] is lower by a factor of about three than that of air crazes. As previously described, the authors found that the discrepancy is due to the enhanced rate of coarsening of fibrils in air-exposed crazes of PS, a result which cannot be anticipated assuming that surface molecules have the same mobility as bulk ones.

A very intriguing example of experimental evidence of enhanced surface

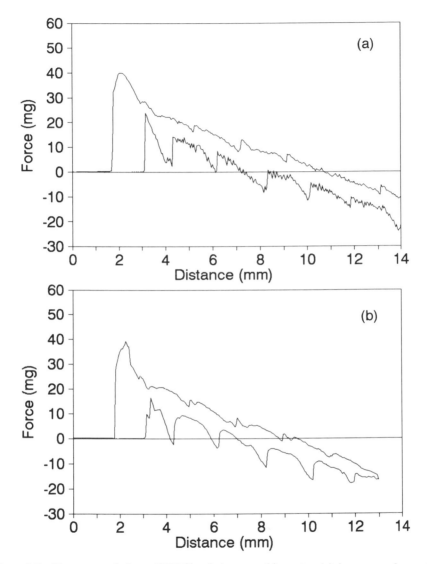

Figure 2.5 Force recorded as a PET fibre is immersed in water. (a) As prepared sample, the surface is very rough and the force line is very irregular. (b) After 16 h at 35 °C most surface cracks are healed, and the profile is smooth.

mobility in polymeric materials has been recently discussed by Berger and Sauer [27]. Again, the subject of the study is the tiny fibrils of PS crazes. Macromolecular chains in a craze are in a very peculiar situation: it can be calculated that the specific surface area within a PS craze can be as high as $10^8\,\text{m}^2/\text{m}^3$, and a large fraction of the polymeric chains must reside at the polymer air/interface. The polymer mobility within the fibrils has been measured by thermally stimulated depolarization currents (TSC). Basically, the samples are polarized by means of an electric field applied at a given rate between two temperatures (T_p and T_0). The field is then turned off and the depolarization current recorded while the sample is heated from T_0 to T_p is measured. The striking results obtained in this experiment are shown in Figure 2.6 [27]. The TSC curve of the uncrazed sample is almost featureless, as expected for nearly immobile polymer chains. On the other hand, the TSC spectrum of the crazed sample exhibits a strong increase in depolarization current, which reveals the presence of significant polymer mobility. It is astounding to observe that the increase of the depolarization current of the crazed sample over that of the uncrazed one begins at a temperature of about 5 °C, that is nearly 100 °C below the T_g of PS (note that, in the original paper [27], the authors provide several proofs which rule out the occurrence of experimental artefacts). The inset of Figure 2.6 shows that the enhanced mobility of surface chains is destroyed after annealing above T_g, a treatment which brings about the complete healing of the crazes and changes the status of the macromolecules of the fibrils from surface chains to bulk chains.

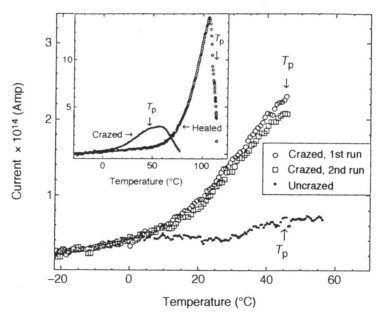

Figure 2.6 TSC spectra for crazed and uncrazed PS samples, polarized at T_p (see text for explanation). Shown in the inset are spectra taken from a crazed and a fully healed film [27]. (Reprinted with permission from Berger and Sauer, *Macromolecules*, **24**, 2096. Copyright (1991) American Chemical Society.)

The work of Berger and Sauer is one of the most direct and conclusive pieces of experimental evidence of the enhanced mobility of polymer surfaces. As compared to craze healing measurements, it must be noted that the latter require irreversible changes in the fibril microstructure, that is mobility must be high enough to allow polymer diffusion. On the other hand, TSC measurements can measure segmental mobility at temperatures much lower (actually, as shown by Figure 2.6, the time–temperature required by the measurement makes possible reproducible TSC spectra, which means that in the first run the fibril structure does not change). The enhanced depolarization current observed on the crazed PS sample at temperatures as low as 5 °C clearly indicates that, as far as the dynamic behaviour is concerned, surface and bulk chains display completely different characteristics.

The second point which was discussed in the previous section was the environmental dependence of the orientation of surface groups. Given the experimentally proven enhanced mobility of surface segments and side chains, it is expected that polymer surfaces can change their surface structure (in terms of orientation of surface functionalities, concentration of surface groups, etc.) in response to a change in the interfacing environments. The thermodynamic driving force for surface restructuring is the minimization of the interfacial free energy between the polymer and the environment: the enhanced kinetics of polymer surfaces makes it possible to observe restructuring with several experimental techniques [1]. Most reports are based on contact angle or interfacial energetics measurements, which will be introduced only in Chapter 4. Thus, in this chapter, the most important results will be only briefly discussed, assuming an elementary knowledge of contact angle and interfacial energetics. Some examples of the dynamic behaviour of polymer surfaces will be discussed later on in this book.

Pennings and Bosman [28] have measured the evolution of the surface free energy of poly (vinyl chloride–vinyl acetate) (PVC–PVAc) copolymers compression moulded against gold foil at 403 K. The rationale for this experiment is to prepare the surfaces under one set of conditions (in particular, against a high energy environment, i.e. gold) and to measure how the polymer surface relaxes in another environment (air). The measured surface tension of the copolymer showed the expected trends, i.e. it decreased as a function of ageing time. When measurements were performed at different temperatures, it was found that the higher the temperature the quicker the relaxation, according to the expected increased mobility at higher temperature. From the time–temperature data Pennings and Bosman calculated an activation energy for the relaxation process of about 43 KJ/mole, which is close to the activation energy of the β-relaxation for pure PVAc. This result suggests that the mechanism responsible for the observed relaxation is the motion of vinyl acetate side chains, and that, contrary to previous findings, the mobility of surface side chains is close to that of bulk chains. However, no direct proof that the observed surface free energy change is due to β-relaxation exists, and the uncertainties on the surface free energy calculation from contact angle data (see Chapter 4) prevents definite conclusions.

Ratner and coworkers observed interesting modifications of the surface composition of hydroxylethymethacrylate (HEMA) grafted silicones [5, 29]. Surface grafted samples were analysed by X-ray photoelectron spectroscopy (XPS, Chapter 3) in the hydrated state (at $-160\,°C$). The XPS spectrum revealed the typical features of the hydrophilic grafted layer, along with some signals of the hydrophobic substrate. The samples were dehydrated *in situ*, by warming to ambient temperature in the ultra-high vacuum of the XPS chamber. The analysis showed that the surface composition of the sample was completely changed, and only the silicone substrate was observed. The results indicates that, upon dehydration, either the poly(HEMA) chains migrate into the silicone matrix or low molecular weight silicone chains outdiffuse to minimize the surface free energy.

Similar results were observed in the papers of Lavielle and coworkers [17–20]. The hydrophilic acrylic graft is not observed when the polymer surface is analysed by XPS. However, after prolonged ageing in water, the typical features of the acrylic graft can be detected.

Ter-Minassian-Saraga and coworkers observed, by XPS, the effect of the interfacing environment on sulphur containing functionalities introduced by oxidizing $KMnO_4$–H_2SO_4 treatment of polyethylene surfaces [30–32] (Chapter 7). The sulphur peak disappears after ageing the treated samples at high temperature in air, but is only slightly reduced if the thermal treatment is performed in water, suggesting that in a high surface energy environment oxidized groups tend to remain at the interface.

As discussed in the introduction to this chapter, Holly and Refojo concluded [4, 33], by contact angle measurement, that PHEMA surface groups can overturn according to the characteristics of the interfacing medium (Chapter 4). Other surface chemistry reversal, based on contact angle measurements, has been reported by Ruckenstein and coworkers [34–37] and Andrade and coworkers [38, 39].

Polymer surface dynamics is the origin of the ageing of surface treated polymers. As discussed later on in this book, the surface properties of polymers can be improved and adapted to a given application by many different surface modification techniques. It is often observed that the properties imparted by the treatment worsen as time from treatment elapses, a phenomenon which is commonly called 'ageing'. While surface contamination, blooming of additives and adsorption of ubiquitous contaminants can explain some cases of ageing, a growing body of experimental evidence shows that the dynamics of polymer surfaces is often responsible for this behaviour. In many cases, hydrophobic polymer surfaces are made hydrophilic by some kind of treatment. The latter surface configuration, however, has a greater surface energy than the former, and, when the sample is stored in air, a driving force exists to restore the original structure or, at least, to lower the surface energy of the treated surface. As a result, the high energy, polar groups introduced by the treatment can be buried away and the properties imparted by the treatment are lost. As an example, Figure 2.7 shows the effect of ageing at room temperature in laboratory air on the advancing water contact angle of several polymers, whose surface was made hydrophilic by

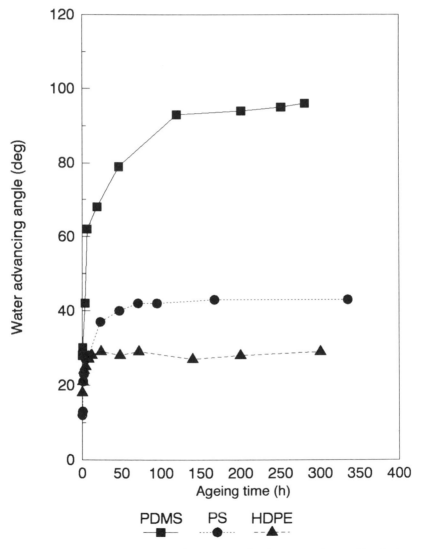

Figure 2.7 Ageing of several oxygen plasma treated polymers, as detected by the increase of the advancing water contact angle as the time for treatment elapses. PDMS = poly (dimethylsiloxane), PS = poly(styrene), HDPE = high density poly(ethylene).

an oxygen plasma treatment. It can be seen that the details of the recovery (i.e. the kinetic and the final value of the contact angle) depend on the nature of the polymer (in addition to the treatment conditions). A huge amount of literature on the subject exists [40–51], in particular concerning the ageing of plasma-treated polymers. A review paper was recently published [52]. Because of the theoretical and practical importance of ageing phenomena, this topic will be discussed more fully in Chapter 11. At this time, it is important to state that research shows that ageing of plasma treated polymers is a thermally activated phenomenon which involves, depending on the details of the parent polymer and the ageing conditions, reorientation of side groups, short-range motions and long-range outdiffusion of untreated chains.

Van Damme, Hogt and Feijen discussed the contact angle hysteresis of a series of poly(n-alkylmethacrylate) in terms of the reorganization of macromolecules side chains or segments at the polymer surface [53]. Their results nicely document the effect of the side-chain length and of the temperature on the measured contact angle hysteresis. If surface dynamics is taken into account, a convincing explanation of the observed behaviour, which is otherwise very difficult to rationalize, results.

Finally, Everhart and Reilley discussed the effect of functional groups mobility on the results of surface analysis [6]. From their results, it is clear that functional groups can migrate back and forth on the surface and subsurface region, depending on the environmental conditions (in particular, Everhart and Reilley studied the effect of different kinds of permeable solvents). Again, an extremely dynamic picture of the polymeric surface arises, whose effect on the surface properties cannot be anticipated by the conventional description of rigid and immobile surfaces.

In summary, recent findings, both at the theoretical and experimental level, suggest that macromolecular dynamics, which deeply affect the bulk properties of polymers, must also be taken into account in the description of surface properties. Actually, the surface regions of polymers appear as zones of enhanced mobility as compared to bulk ones, as shown by the dynamic simulation of Mansfield and Theodorou [12]. As a consequence, the use of bulk values, such as the T_g, to gauge the occurrence and the extent of surface dynamic phenomena can be highly misleading.

Despite the technical difficulties involved in the characterization of polymer surface dynamics, experimental evidence convincingly shows that surface chains and chain segments are quite mobile and that, due to their enhanced mobility, they can yield to the thermodynamic thrust towards the minimization of the interfacial free energy in times short enough to affect surface properties. Polymer surfaces are very sensitive to processing conditions and re-equilibrate in the use environment. The effects of this dynamic attitude on the performance of polymeric surfaces are manifold: the orientation of chemical groups on polymer surfaces and the effect of a surface modification treatment are time, temperature and environment-dependent. A surface modification must be conceived taking a look at the working environment of the material. A surface analysis must take into account the occurrence of artefacts arising from the reorientation of surface groups during analysis [1, 5, 6, 54]. In short, the further degree of freedom granted by the enhanced short and large scale macromolecular mobility must always be considered in order to properly interpret polymer surfaces properties.

REFERENCES

[1] *Surface and Interfacial Aspects of Biomedical Polymers*, J. D. Andrade (Ed.), Plenum Press, New York (1985), Vol. 1.
[2] *Polymer Surface Dynamics*, J. D. Andrade (Ed.), Plenum Press, New York (1988).
[3] I. Langmuir, *Science*, **87**, 493 (1938).
[4] F. J. Holly and M. F. Refojo, *J. Biomed. Mater. Res.*, **9**, 315 (1975).
[5] B. D. Ratner, P. K. Whatersby, A. S. Hoffman, M. A. Kelly, and L. H. Scharpen, *J. Appl. Polym. Sci.*, **22**, 643 (1978).
[6] D. S. Everhart, and C. N. Reilley, *Surf. Interface Anal.*, **3**, 126 (1981).
[7] P. G. de Gennes, *J. Chem. Phys.*, **55**, 572 (1971).
[8] P. G. de Gennes, *Phys. Today*, June, 33 (1983).
[9] P. G. de Gennes, *Scaling Concepts in Polymer Physics*, Cornell University Press, Ithaca, NY (1979).
[10] J. D. Ferry, *Viscoelastic Properties of Polymers*, 3rd edn, John Wiley & Sons, New York (1980).
[11] M. Doi and S. F. Edwards, *Polymer Dynamics*, Oxford University Press, London (1986).
[12] K. F. Mansfield and D. N. Theodorou, *Macromolecules*, **24**, 6283 (1991).
[13] K. F. Mansfield and D. N. Theodorou, *Macromolecules*, **23**, 4430 (1990).
[14] M. P. Allen and D. J. Tildesley, *Computer Simulation of Liquids*, Clarendon Press, Oxford (1987).
[15] J. P. Hansen and I. R. McDonald, *Theory of Simple Liquids*, 2nd edn, Academic Press, London (1986).
[16] P. J. Flory, *Principles of Polymer Chemistry*, Cornell University Press, Ithaca (1953).
[17] J. Schultz, A. Carre, and C. Mazeau, *Int. J. Adhes. Adhes.*, **4**, 163 (1984).
[18] L. Lavielle and J. Schultz, *J. Colloid Interface Sci.*, **106**, 438 (1985).
[19] L. Lavielle, J. Schultz, and A. Sanfield, *J. Colloid Interface Sci.*, **106**, 446 (1985).
[20] L. Lavielle, G. Lischetti, A. Sanfed, and J. Schultz, *J. Colloid Interface Sci.*, **138**, (1990).
[21] A. Sanfeld, *Introduction to the Thermodynamics of Charged and Polarized Layers*, Wiley, London/New York (1968).
[22] A. Sanfeld, *Nuovo Cimento Soc. Ital. Fis. D*, **12D**, 901 (1990).
[23] R. P. Kambour, *J. Polym. Sci.*, **2**, 4165 (1964).
[24] A. C. M. Yang and E. J. Kramer, *J. Polym. Sci., Polym. Phys.*, **23** 1353 (1985).
[25] G. Giannotta and M. Morra, unpublished results.
[26] H. R. Brown and E. J. Kramer, *J. Macromol. Sci. Phys.*, **B 19**, 487 (1981).
[27] L. L. Berger and B. B. Sauer, *Macromolecules*, **24**, 2096 (1991).
[28] J. F. M. Pennings and B. Bosman, *Colloid Polym. Sci.*, **257**, 720 (1979).
[29] B. D. Ratner, in *Surface and Interfacial Aspects of Biomedical Polymers*, J. D. Andrade (Ed.), Plenum Press, New York (1985), Chapter 10, Vol. 1.
[30] A. Baszkin and L. Ter-Minassian-Saraga, *Polymer*, **15**, 759 (1974).

[31] B. Catoire, P. Bouriot, A. Baszkin, L. Ter-Minassian-Saraga, and M. M. Boisson- nade, *J. Colloid Interface Sci.*, **79**, 143 (1981).
[32] J. C. Eriksson, C. G. Golander, A. Baszkin, and L. Ter-Minassian-Saraga, *J. Colloid Interface Sci.*, **100**, 381 (1984).
[33] F. J. Holly and M. F. Refojo, in *Hydrogels for Medical and Related Applications* J. D. Andrade (Ed.), ACS Symposium Series N. 31, pp. 252–66, Amer. Chem. Soc., Washington, D.C. (1976).
[34] E. Ruckenstein and S. V. Gourisankar, *J. Colloid Interface Sci.*, **107**, 488 (1985).
[35] E. Ruckenstein and S. V. Gourisankar, *J. Colloid Interface Sci.*, **109**, 557 (1986)ː
[36] E. Ruckenstein and S. H. Lee, *J. Colloid Interface Sci.*, **120**, 153 (1987).
[37] S. H. Lee and E. Ruckenstein, *J. Colloid Interface Sci.*, **120**, 529 (1987).
[38] J. D. Andrade and W. Y. Chen, *Surface Interface Anal.*, **8**, 253 (1986).
[39] K. G. Tingey, J. D. Andrade, R. J. Zdrahala, K. K. Chittur, and R. M. Gendrau, in *Surface Characterization of Biomaterials*, pp. 255–70, B. D. Ratner (Ed.), Elsevier, Amsterdam (1988).
[40] H. Yasuda, A. S. Sharma, and T. Yasuda, *J. Polym. Sci., Polym. Phys. Ed.*, **19**, 1285 (1981).
[41] A. K. Sharma, F. Millich, and E. W. Hellmuth, *J. Appl. Polym. Sci.*, **26**, 2205 (1981).
[42] T. Yasuda, K. Yoshida, T. Okuno, and H. Yasuda, *J. Polym. Sci., Polym. Phys. Ed.*, **26**, 2061 (1988).
[43] T. Yasuda, T. Okuno, K. Yoshida, and H. Yasuda, *J. Polym. Sci., Polym. Phys. Ed.*, **26**, 1781 (1988).
[44] Y. L. Hsieh and E. Y. Chen, *Ind. Eng. Chem. Prod. Res. Dev.*, **24**, 246 (1985).
[45] H. S. Munro and D. I. McBriar, *J. Coatings Technol.*, **60**, 41 (1988).
[46] W. J. Brennan, W. J. Feast, H. S. Munro, and S. A. Walker, *Polymer*, **32**, 527 (1991).
[47] M. Morra, E. Occhiello, and F. Garbassi, *J. Colloid Interface Sci.*, **132**, 504 (1989).
[48] E. Occhiello, M. Morra, P. Cinquina, and F. Garbassi, *ACS Polym. Preprints*, **31**, 308 (1991).
[49] M. Morra, E. Occhiello, R. Marola, F. Garbassi, P. Humphrey, and D. Johnson, *J. Colloid Interface Sci.*, **137**, 11 (1990).
[50] H. J. Griesser, J. H. Hodgkin, and R. Schmidt, in *Progress in Biomedical Polymers*, C. B. Gebelein and R. L Dunn (Eds), Plenum Press, New York (1990), pp. 205–15.
[51] H. J. Griesser, D. Youxian, A. E. Hughes, T. R. Gegenbach, and A. W. H. Mau, *Langmuir*, **7**, 2484 (1991).
[52] M. Morra, E. Occhiello, and F. Garbassi, *First International Conference on Polymer Solid Interfaces*, Namur, Sept. 1991, IOP Publishing, pp. 407–428 (1992).
[53] H. S. Van Damme, A. H. Hogt, and J. Feijen, *J. Colloid Interface Sci.*, **114**, 167 (1986).
[54] F. Garbassi, E. Occhiello, M. Morra, L. Barino, and R. Scordamaglia, *Surf. Interf. Anal.*, **14**, 595 (1989).

Part II

CHARACTERIZATION METHODS

Chapter 3

Spectroscopic Methods

In the last 25 years, as new spectroscopic methods able to analyse solid surfaces were conceived and developed, and old methods were adapted to new tasks, they were applied when possible to polymer surfaces. A classical example is that of X-ray photoelectron spectroscopy (XPS), developed in the 1960s by the group of Siegbahn [1] at Uppsala, Sweden, on the basis of the Einstein [2] theory of photoelectric effect and basic concepts published in 1914 [3]. Applications to polymers began to appear around 1974 [4, 5] and a great impetus to the field was given at that time by the group of D. T. Clark at Durham University, England.

On the other hand, other surface sensitive techniques demonstrated unsuitable for polymers, like Auger electron spectroscopy (AES), were under development in the same period [6]. In fact, AES uses an electron beam for probing the surface, and the poor ability of polymers to dissipate the local electrical charge precludes the collection of a useful spectrum.

Nevertheless, many techniques are now available to study surfaces and/or interfaces in polymeric materials.

In this chapter, spectroscopic techniques are considered. The presentation reflects, in some depth, the increasing penetration of the physical probe. First, ions are discussed (ISS, RBS, SIMS), then electrons (XPS) and finally photons (ultraviolet, visible, infrared, radiofrequency, etc.).

In each case physical principles and instrumentation are briefly described, followed by some applications, chosen from recent work, in order to point out the potential and limits of each one.

A list of the most commonly used acronyms for instrumental techniques is provided in Table 3.1.

3.1 ION SCATTERING SPECTROSCOPY

Spectroscopic techniques for polymer surface analysis using ion beams can be divided into two groups. The first one comprises ion scattering spectroscopies (low

Table 3.1   List of acronyms

ATR	Attenuated total reflectance
DRS	Diffuse reflectance spectroscopy
EM	Emission (spectroscopy)
ERS	External reflection spectroscopy
IETS	Inelastic electron tunnelling spectroscopy
ISS	Ion scattering spectroscopy
LEIS	Low energy ion scattering
NMR	Nuclear magnetic resonance (spectroscopy)
NRA	Nuclear reaction analysis
PAS	Photoacoustic spectroscopy
RA	Reflection adsorption (spectroscopy)
RBS	Rutherford backscattering
SERS	Surface enhanced Raman scattering
SIMS	Secondary ion mass spectroscopy
SSIMS	Static secondary ion mass spectroscopy
T	Transmission (spectroscopy)
UPS	Ultraviolet photoelectron spectroscopy
XPS	X-ray photoelectron spectroscopy
CMA	Cylindrical mirror analyser
IR	Infrared
UV	Ultraviolet
UHV	Ultra high vacuum
mw	Microwave
rf	Radiofrequency

energy ion scattering, Rutherford backscattering spectroscopy), the second one secondary ion mass spectroscopies. They are distinguished by the fact that in ISS ion energies are measured, while in SIMS the amounts of emitted ions are detected.

The first demonstration of ISS as a surface analytical technique has been published by Smith [7], who bombarded a metallic target using noble gas positive ions at energies of 0.5–3.0 keV. The technique has been reviewed by Buck [8], more recently by Benninghoven [9] and Gardella [10], who dedicated specific attention to polymers.

In ISS an ion beam, usually of He^+, Ne^+, Ar^+, etc., is focused on the surface. Collisions of incident ions with surface atoms cause variations in their state of motion and energy [8] as represented in Figure 3.1. To a good approximation, the collision of the incident ion with a surface atom can be considered elastic. As a consequence, the energy of the incident ion after the collision is given by equation (3.1):

$$E_1 = [\cos\theta + (M_2^2/M_1^2 - \sin^2\theta)^{1/2}/(1 + M_2/M_1)]^2 E_0 \qquad (3.1)$$

where E_0 is the original energy of the incident ion, M_1 its mass, M_2 the mass of the surface atom and θ the angle of observation.

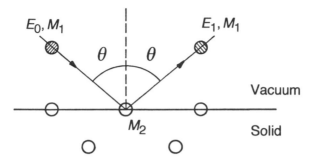

Figure 3.1 The physical basis of ISS: an ion with energy E_0 and mass M_1 is elastically scattered by a surface atom of mass M_2.

The version of ISS which is most commonly reported uses low energies of the incident ion (E_0), normally in the 500–5000 eV range. In this case, a widely used acronym is LEIS (low energy ion scattering). In these conditions only the first layer of the material is observed, so it is a very surface-sensitive technique.

Most collisions are single, binary and elastic, so that they obey equation (3.1) and relate to the first monolayer of the target. Multiple and inelastic collisions induce a decrease of resolution, but are less important. In fact, the probability of neutralization increases with the number of collisions and neutral atoms are not detected.

Equation (3.1) indicates that the resolution, $M/\delta M$, is a function both of angle of incidence and of the ratio of the incident ion mass to that of the target. Resolution worsens with increasing M_2/M_1 and decreasing the angle of incidence [8].

To obtain quantitative information about surface composition it is necessary to know how the observed intensity at a given energy of the scattered ion depends on physical and instrumental factors (equation (3.2)) [8]:

$$I_i = I_0 N_i A (d\sigma_i/d\Omega) \delta\Omega T P_i = k_i N_i \tag{3.2}$$

where N_i is the surface concentration of the given element i, σ_i is the cross-section, $\delta\Omega$ is the solid angle, T an instrumental factor, A the bombarded area, and P_i the probability that the incident ion will not be neutralized. Both the cross-section and the neutralization probability are functions of the angle of incidence and the ion-to-atom interaction potential.

Absolute quantification by means of equation (3.2) is difficult, since many of the relevant factors are not well known. However, it is possible to obtain a calibration using appropriate standards [8, 11]. ISS can provide semiquantitative information about surface composition, while information about the chemical state of the observed element is not usually available.

The experimental arrangement is outlined in Figure 3.2. The sample is placed

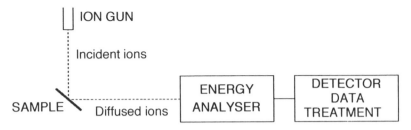

Figure 3.2   The experimental arrangement of ISS.

in an UHV chamber (*ca* 10^{-8} Pa). The ion gun can be similar to those used for ion etching in connection with other techniques, such as XPS, in order to clean the surface or perform depth profiling. Apart from noble gases, the use of alkali metal ions has been suggested for the lower tendency to neutralization. Their main defect is the loss of specificity, since ions which undergo multiple collisions are detected. On the other hand, charging phenomena in insulating materials are less likely. The energy of scattered ions is analysed by a 127° electrostatic analyser or by a cylindrical mirror analyser (CMA). The latter usually offers better sensitivity, since it is able to detect an atomic concentration at the surface of some 10^{-5} atoms/surface, an order of magnitude better than the 127° electrostatic analyser.

The application of ISS to polymeric materials has not been extensive, largely because it has been subjected to the competition of other techniques, like static SIMS, offering a larger amount of information. Another problem was the possibility of surface damage under the ion beam, low-damage conditions have been consequently developed [11, 12]. Adhesion problems [13, 14], glass/polymer interfaces [15], and polymer surfaces [11, 12, 16–18] have been studied.

A combined ISS/dynamic SIMS study of glass/polymer interfaces [15] showed that ISS provides lower resolution with respect to SIMS, because energies instead of masses are analysed.

A LEIS study of several polymer surfaces has been reported [16]. The technique is extremely sensitive to the presence of surface impurities. For instance, inorganic or organometallic compounds used as stabilizers in polymers tend to segregate at the surface. When examining poly(vinyl chloride) (PVC) containing Sn compounds, a large amount of tin was detected at the surface [16]. The corresponding LEIS spectrum is reported in Figure 3.3. The intense broad peak at low energy was found to be a common feature of polymer LEIS spectra, and was tentatively attributed to sputtered H^+ ions or He backscattered.

Concentrations of polar groups such as —OH, —COOH, —O—, in polymers have been found to be higher in the bulk of the polymer than at the surface. The conformation of polymer chains near the surface appears to depend on the minimization of surface energy. In poly(oxymethylene) the stoichiometric O/C

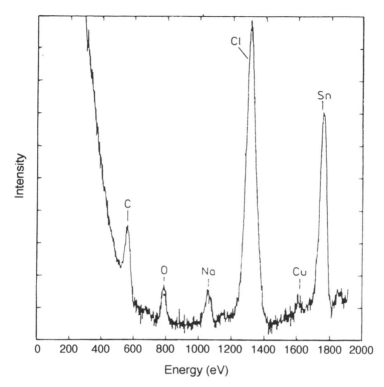

Figure 3.3 ISS spectrum of PVC. (Reproduced by permission of Elsevier Science Publishers from ref. [16].)

ratio is 1, while the surface ratio detected by ISS is 0.15 [16]. The same trend has been observed in other polymers, copolymers and blends [11, 12, 16–19]. Structural sensitivity of ISS was recently reported [10, 19]. Analysis of stereoregular poly(methyl methacrylates) (PMMA) and random copolymers of PMMA and poly(methacrylic acid) (PMAA) showed differences in measured C/O atomic ratios that were interpreted on the basis of shadowing and orientation effects [10]. Figure 3.4 shows that ISS is able to discriminate among isotactic, syndiotactic and atactic PMMA in terms of the scattered intensity C/O ratios. The analysis was extended to Ar/H_2O plasma-modified poly(methyl methacrylates) [20] and copolymers with methacrylic acid [21]. The sputtering occurring during the ISS experiments was used to obtain a qualitative depth profile of plasma modified sample surfaces. The observed increase of the C/O ratio towards the value of unmodified polymers supports the idea that oxygen-rich functions have been incorporated, following plasma treatments, in the near-surface region of polymers [22].

Another ISS technique is RBS (Rutherford backscattering spectroscopy),

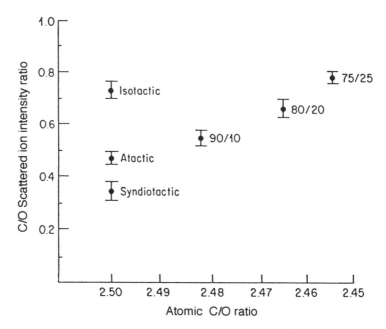

Figure 3.4   C/O scattered ion intensity ratio from ISS versus bulk C/O ratio for PMMA and copolymers with methacrylic acid. (Reproduced by permission of Elsevier Science Publishers from ref. [10].)

which uses a beam of energetic He$^+$ ions in the MeV range. Theory and instrumentation are similar to those of LEIS and has been reviewed by Gossett [22]. Because of surface elastic scattering problems, RBS is not able to detect efficiently elements with mass lower than 28 (silicon). Furthermore, RBS is more penetrating than LEIS; the thickness of the observed layer is a function of the energy of the ion beam and usually several hundred nanometres.

One of the rare studies with RBS concerns the diffusion of Cu(II) ions in a polypropylene film oxidized between Cu foils [23]. The diffusion rates of copper carboxylate complexes in polymeric matrices have been studied [24–25], as well as polymer surfaces which have undergone ion bombardment [26, 27]. The weathering of the surface of bisphenol A polycarbonate has been determined by RBS after the derivatization of phenol and carboxylic acid groups using thallous ethoxide [28].

Nuclear reaction analysis (NRA) is an ion-beam technique based on the ^{2}H(^{3}He, ^{4}He)^{1}H nuclear reaction. A 700 keV ^{3}He beam from a Van de Graaff accelerator is directed on a polymer deuterated sample at a 15° angle of incidence. ^{3}He ions undergo the above nuclear reaction with deuterons ^{2}H, producing ^{4}He (α) particles that are detected by an energy dispersive detector. A magnetic filter allows the separation of the α particles from elastically scattered ions. An energy spectrum of the α particles is collected, containing information on the composition profile of the deuterated moiety in the sample [358]. This is based on the fact that the energy of a specific α particle depends on the depth at which it was produced, as high the depth, as low the energy. An advantage of NRA is to have intrinsically high resolution [359]. A film of about 13 nm of deuterated PS results in an experimentally determined profile having a width of 14 nm [360]. The resolution decreases with depth, maintaining acceptable values (20–30 nm) for depths below 300 nm. Both forward recoil spectroscopy and Rutherford backscattering given, on equivalent samples, higher spatial resolution. NRA can be used to study the interfacial broadening of polymer/polymer systems and was applied, for instance, to PS/deuterated PS bilayer films [360].

3.2 SECONDARY ION MASS SPECTROSCOPY

SIMS (secondary ion mass spectroscopy) techniques have acquired an ever-increasing importance in the last few years, particularly in the analysis of semiconductors and polymers. Some books have been published on the general subject [29, 30] in addition to useful collections of standard spectra [31, 32].

In a SIMS experiment, the particles (secondary ions) emitted from the surface as a consequence of ion or atom bombardment are mass analysed. In the latter case the technique is also called FABMS (fast atom bombardment mass spectroscopy). The potential of SIMS in surface analysis emerged when it was demonstrated by Benninghoven [33] that it is possible to reduce the primary beam density to a very low level, so developing static SIMS near to dynamic SIMS already used in elemental analysis. The difference between static and dynamic SIMS has to be emphasized. In static SIMS (SSIMS) the ion current incident on the surface is very limited (some 10^{-10} ampere), secondary ions come from the first one or two layers of the material and the surface is not substantially damaged. In dynamic SIMS the ion current is somewhat higher (up to 10^{-7} ampere), the surface is rapidly eroded and there is not a true specificity to the first layers [34]. While SSIMS can be used for surface studies, dynamic SIMS has found applications in depth profiling studies, e.g. the ion implantation effects on semiconductors.

The processes following from the impact of an accelerated ion or atom with the surface are sketched in Figure 3.5 [13, 34]. The energy of the incident ion is dissipated in a series of collisions and secondary particles are emitted even at a distance from the primary impact site. Such a process is commonly called sputtering. A fraction of emitted particles is submitted to ionization and can be analysed. If the latter process is simultaneous or consecutive to the former it is still controversial [31]. The intensity i_s^M of the signal corresponding to the secondary ion M^+ as a function of the primary ion flux (i_p), is described by equation (3.3) [34]:

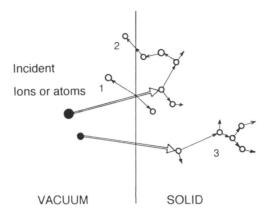

Figure 3.5 Schematic representation of the sputtering process: (1) ion emission near the impact point; (2) ion emission at some distance from the impact point; (3) no emission, only secondary phenomena.

$$i_s^M = i_p S R^+ \theta_M \beta \tag{3.3}$$

where θ_M is the surface concentration of M, expressed as coverage, β an instrumental factor, S the sputtering yield (i.e. the number of particles M emitted after each primary ion impact) and R^+ is the probability of particle M to be emitted as an ion and detected. S and R^+ determine the spectrum, since they depend on emission and ionization mechanisms, which are functions of the electronic structure of the material. Various models have been proposed for the above mechanisms and a short but exhaustive review is reported in ref. [31]. The complication of SIMS theory causes difficulties in the interpretation of spectra, particularly quantitatively. In fact, sensitivities have to be defined for each fragment and are characteristic of each material.

A typical SIMS instrument includes an ion source, a sample holder and a mass analyser. Several levels of sophistication are available and include computer facilities for handling and treatment of data. The sample is analysed in an UHV chamber, as in ISS. Ion guns using noble gases such as Ne, Ar, Xe are commonly used. Ions are accelerated by an electric potential which has to be chosen for each ion; e.g. 4 keV has been used for Ar^+ and 2 keV for Xe^+ [35, 36]. Currents used in SSIMS are very low (e.g. around 1 nA) to minimize the surface damage.

Liquid metal guns (Ga, In, Cs) have also been used to obtain a very brilliant and easily collimated primary ion beam [37].

The application of SSIMS on insulating materials, e.g. polymers, gives rise to important charging problems, causing a sensitivity decrease and alterations in relative intensities of peaks. To overcome such a problem, an electron gun (flood gun) hitting a broader area than the ion gun is used. The aim is to neutralize the positive charge induced on the material [35]. Unfortunately the flood gun can itself have some effect on the observed spectrum [38]. Using a flood gun to excite secondary ions, the mass spectrum of poly(butylmethacrylate) has been obtained [36]. This phenomenon is called electron stimulated ion emission and has to be minimized in static SIMS experiments. This was achieved by using low electron currents. Vickerman and coworkers demonstrated that charging problems are remarkably reduced by using atoms instead of ions as primary beam [39, 40]. In FABMS, the ion beam, already accelerated and collimated, is passed through a chamber containing a relatively high pressure of inert gas [41, 42]. A fraction of the ions (up to 20%) is neutralized, preserving kinetic energy and direction. An atom beam is thus obtained, while residual ions are removed electrostatically.

Polystyrene has been studied using both ions and atoms as primary beam [42]. Quantitative and qualitative differences were observed. The FABMS spectra showed higher intensities and the decrease with time was slower (the rate of decay was 25% of that observed when using ions as primary beam). Also, the peaks corresponding to high masses were more intense than in SIMS.

Of the analysers used in static SIMS, the most popular is the quadrupole mass analyser, which is easy to handle and relatively cheap. Magnetic and electrostatic sector analysers are more effective, particularly for very high masses, but their transmission performance is rather poor. Recently, time-of-flight (TOF) analysers have gained considerable popularity since they allow a consistent increase in sensitivity. It has been shown that a gain in sensitivity of a factor 100–1000 can be achieved over the quadrupole analyser [43]. The ion transmission is constant over the entire mass range and the simultaneous detection of all masses with the same polarity (positive or negative) facilitates and speeds the data averaging [17].

As in conventional mass spectrometry, in SSIMS it is possible to obtain both positive and negative ion spectra. Most of the data now available refer to positive ion spectra, since negative ion spectra were considered much less informative [37]. However, negative ion spectra are particularly useful in detecting groups containing electronegative elements, such as oxygen and halogens. In Figure 3.6, the positive and negative SSIMS spectra of poly(ethylene) are shown. As in other polyolefins, in the positive ion spectrum groups of peaks corresponding to C_nH_{2n-1} and C_nH_{2n+1} masses appear at m/z 27–29, 41–43, 55–57 etc. At a value of m/z around 150, only the background remains visible. In the negative ion spectrum, only peaks corresponding to C^-, CH^-, CH_2^-, C_2^- and C_2H^- are clearly visible. Two keV Xe^+ ions were used as primary beam to obtain the negative ion mass spectrum of poly(ethylene oxide) (PEO) [35], reported in Figure 3.7. Many mass peaks were observed in this case, up to a value of 350 D. Negative ion spectra have been used to determine the degree of polymerization of 3-methacryloxypropylsilane on various fillers [44], considering the relative intensity variations of specific mass peaks.

Even if performed in the static mode, SIMS experiments can cause some surface damage [45]. Variations of chemical composition can be observed during

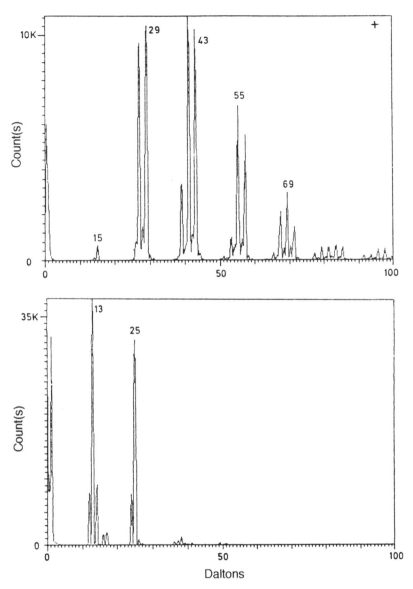

Figure 3.6 Positive and negative SIMS spectra of PE.

an analysis, since the energy introduced by primary particles favours a differential sputtering of the species present at the surface. Breakage or formation of chemical bonds, enrichment or depletion of particular functional groups and changes in oxidation state can occur under ion bombardment as a function of time. The case

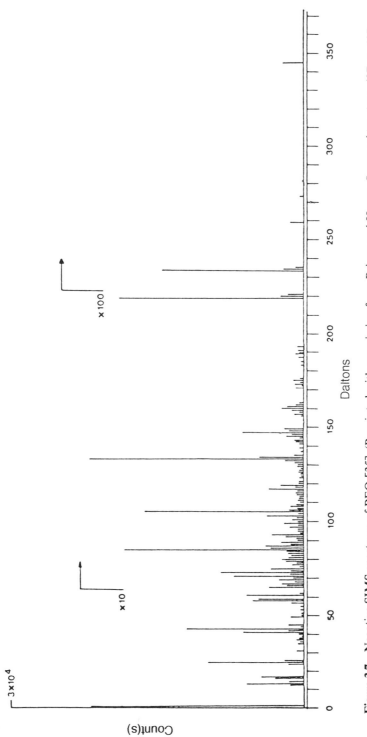

Figure 3.7 Negative SIMS spectrum of PEO [35]. (Reprinted with permission from Briggs and Hearn, *Spectrochim. Acta*, **40B**, p. 407. Copyright (1985), Pergamon Press Ltd.)

of polystyrene (PS) is well known, where the XPS shake-up peak associated with the presence of aromatic rings disappears after the absorption of some 10^{+14} ions/cm^2 [38]. A study on the dependence of the XPS C/O ratio as a function of ion dose in PMMA demonstrated that a value lower than 10^{+13} ions/cm^2 is necessary to contain the damage within tolerable limits [35]. The effect is illustrated in Figure 3.8. At higher ion dose values, SSIMS spectra cannot be interpreted on the basis of the structure of the parent polymer, since they tend to have a common appearance, due to $(C_n H_m)^+$ ions. This is accompanied by a strong decrease of peak intensities. Finally, morphological changes can occur, in particular amorphization, and charging effects. Studies on damage effects on

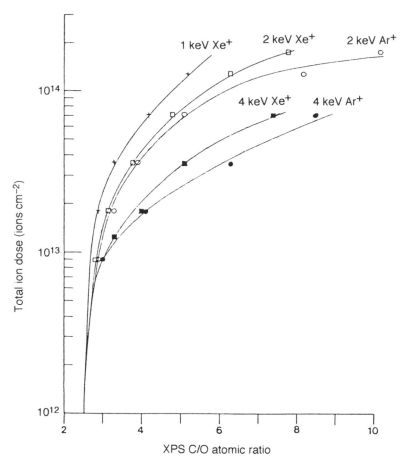

Figure 3.8 Ion beam damage of PMMA, assessed by XPS C/O atomic ratio variations [35]. (Reprinted with permission from Briggs and Hearn, *Spectrochim. Acta*, **40B**, p. 407. Copyright (1985), Pergamon Press Ltd.)

polymers have also been carried out in the case of polyethylene (PE), poly (ethyleneoxide) (PEO) [46, 47], poly(tetrafluoro ethylene) (PTFE), poly(ethylene terephthalate) (PET) [48] and poly(vinyl chloride) (PVC) [47, 49]. It was observed that polymers are damaged more rapidly under an ion beam than an atom beam. This implies particle–surface electronic interactions, involving the transfer of an electron to the approaching ion from the solid via a resonant or Augar process [50]. The appearance in the SIMS spectrum of PVC of peaks attributed to polycyclic aromatic structures was attributed to bond dissociation created by the occurrence of antibonding excited states at the polymer surface, due to the above processes [49].

Static SIMS is particularly effective in providing a variety of qualitative information about the surface, even if the interpretation is not always straightforward. Its sensitivity is the highest among surface techniques (about 10^{-6} coverage can be detected) even though sensitivity can vary widely depending on the chemical nature of the observed species [37].

Many common polymers have been studied by SSIMS. Standard positive ion and negative ion spectra have been published [31].

The fragmentation observed by SSIMS is typical of each polymer and can be considered as a sort of fingerprint. The observed ions are interpretable in terms of the repetitive unit of the polymer. Fragmentation mechanisms have much in common with those observed using electron impact sources [35]. The most relevant features and recent results for some polymers are summarized.

3.2.1 Polyamides

Spectra of polylactam (Nylon-6) and dicarboxylic acid-diamine polyamides (Nylon-6, 6) have been discussed in detail by Briggs [51]. In positive spectra, the $30\,D$, $41\,D$ and $55\,D$ peaks dominate, corresponding to $CH_2{=}NH_2$, $CH{=}C{=}O/C_3H_5$ and $CH_2{=}CH{-}C{=}O$, respectively. Groups containing N, like CN (26 D) and $O{=}C{=}N$ (42 D) are better seen in negative ion spectra. Using a time-of-flight analyser, the TOF–SIMS spectrum of Nylon-6 has also been published [52]. The spectrum is very similar to that obtained by a quadrupole analyser, but the higher mass resolution capability provided a unique identification of peaks (for instance K and C_3H_3 at 39 D are distinguishable) and a better understanding of fragmentation mechanism.

3.2.2 Poly (methacrylates)

Poly(methacrylates) have been studied using SSIMS by various groups [11, 12, 36, 53]. Both the polymer skeleton and pendant groups produce a characteristic fragmentation [36], giving crowded positive and negative ion spectra. The part of the positive ion spectrum with masses over 100 Daltons has been found to be dominated by peaks relative to the polymer skeleton.

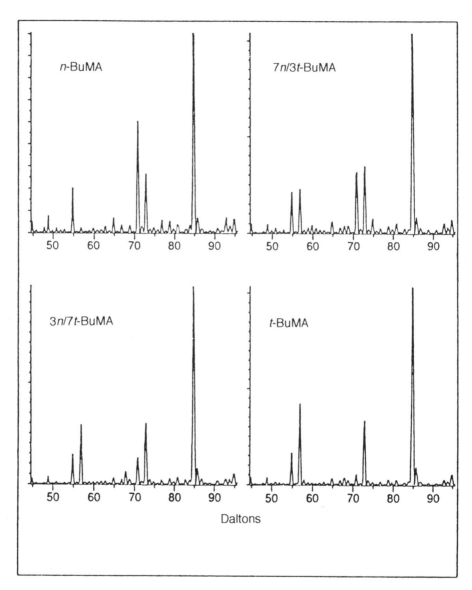

Figure 3.9   Negative SIMS spectra (50–90 amu region) of poly(*n*-butyl methacrylate) (*n*-BuMA), poly(*t*-butyl methacrylate) (*t*-BuMA) and their copolymers. (Reproduced with permission from ref [55].)

Table 3.2 Static SIMS peak intensities for relevant peaks in Daltons of poly(butyl methacrylates (BuMA) (from ref. [55])

	Positive			Negative				
	55	57	69	55	57	71	73	85
n-BuMA	52	128	100	20	2	48	28	100
7*n*/3*t*-BuMA	47	198	100	17	20	30	31	100
3*n*/7*t*-BuMA	41	288	100	14	30	14	29	100
t-BuMA	37	353	100	15	37	5	32	100
i-BuMA	40	186	100	16	6	42	33	100

In the case of PMMA (polymethylmethacrylate), a very strong peak at 15 Daltons (CH^{3+}) is obtained, while in PEMA (polyethylmethacrylate) the peak at 29 (C^2H^{5+}) is dominant.

A sort of fingerprint of polymethacrylates is the presence of a particularly intense peaks at 69 Daltons, prominent in polymethacrylates but not in polyacrylates. It was recently attributed to the methacryloyl ion $CH_2=C(CH_3)CO$ [54]. A complete study on the fragmentation of acrylic and methacrylic polymers has been carried out. The negative ion spectra were found a powerful structural probe of this class of polymers, since many ions retained the intact ester side chain [54]. Poly(hydroxyalkyl methacrylates) had a distinctively different fragmentation pattern.

A study of butyl methacrylate polymers demonstrated that different isomers can be distinguished by static SIMS and the percentage of each isomer in random copolymers can be determined to within 10% [55]. The isomers examined were normal-, iso- and *tert*-butyl methacrylate (*n*-, *i*- and *t*-BuMA, respectively). Two *n*/*t*-BuMA copolymers were also studied. Large differences were observed mainly in the m/z = 50–90 range of negative ion spectra (Figure 3.9). The above results were obtained by taking into consideration the positive m/z = 55, 57 and 69 peak intensities (normalized to m/z = 69) and the negative m/z = 55, 57 and 71, 73 and 85 peak intensities (normalized to m/z = 85). Measured intensities are reported in Table 3.2, from which several trends of peak intensity ratios were drawn [55].

3.2.3 Polypropylene and other Polyolefins

The main difference in the positive ion spectra of PE (Figure 3.6) and polypropylene (PP) is the enhanced intensity, in the latter case, of the peak at m/z = 69. It has been observed in other polymers with methyl pendant groups, and has been attributed to the formation of a stable dimethylcyclopropyl carbocation [36]:

$$
\begin{array}{cc}
H_3C & CH_3 \\
\diagdown \; \diagup \\
C\!-\!C \\
\diagdown \diagup \\
H\!-\!C^+\!-\!H \\
| \\
H
\end{array}
$$

Various other aliphatic hydrocarbon polymers have been studied by van Ooij and coworkers [56, 57] and returned to by Briggs [58] who noted that, especially for unsaturated polymers, ion beam damage can give differences in the spectra.

3.2.4   Other Polymers of Industrial Relevance

The positive and negative ion SSIMS spectra of poly(tetrafluoroethylene) (PTFE) are rather simple. The former is dominated by the peak at 31 Daltons (CF^+) and, with the exception of C^+, corresponds to $C_nF_m^+$ ions; similarly, in the latter, peak due to $C_nF_m^-$ occur, together with those at 19 Daltons (F^-) and at 38 Daltons (F_2^-), which dominate the spectrum [31].

A characteristic of PS is the presence of a very strong peak at 91 D in the positive ion spectrum. This was attributed to the cyclic tropylium ion $C_7H_7^+$ [31]. Peaks are visible up to m/z = 300 D and are generally attributed to polycyclic cations. With respect to PS, polycarbonate (PC) and PET contain aromatic groups in the backbone of the polymer. Furthermore, they also contain in the backbone a heteroatom (oxygen). In spite of this, their positive ion spectra are not very different from that of PS. Negative ion spectra are more specific, with several peaks derived from the monomeric repeating units [59, 60].

PVC has a positive ion spectrum very similar to that of PE. In the negative ion spectrum, only Cl^- and Cl_2^- appear [31].

3.2.5   Multicomponent Systems (Polymer Blends and Block Copolymers)

Static SIMS has been successfully used to determine the surface composition of polymer blends. Blends of PS with poly(vinylmethylether) [61] or two different PCs [62] have been studied. In the latter study, carried out with a TOF mass analyser, a full quantitative analysis was demonstrated to be possible, suggesting that matrix effects may be negligible.

Block copolymers can have important applications where surfaces and interfaces are important, such as adhesives, surfactants, compatibilizers, biomedical materials, etc. Surface segregation in segmented poly(etherurethanes) has been determined comparing three polymers that have in common the hard segment (formed by MDI, i.e. methyl-diphenyl-diisocyanate and ethylene diamine) while the soft segment (PPG, polypropylene glycol) has an increasing molecular weight, in the range 425–2000 D [63, 64]. An accurate comparison of poly (etherurethanes) spectra with those of pure PPG and a model of the hard segment allowed the observation that at low molecular weight MDI dominates at the surface; conversely, preferential surface segregation of PPG 2000 was found.

3.2.6 Use of Isotopes

In the analysis of polymer surfaces, the unique capability of SIMS to discriminate mass isotopes has been widely used. Different approaches have been adopted, such as polymerization of labelled monomers and functionalization of surfaces with isotope containing reagents. In the first case, deuterated PS has often been considered [65–68], provided that labelled styrene can be easily purchased and polymerized in a conventional way. In a different approach, labelled reagents were used to modify the polymer surface, for instance by plasma treating with ^{18}O [63, 69, 70].

The concentration profile at the surface of deuterated PS (d-PS)/hydrogenated PS (h-PS) blends has been studied by ISS and SIMS [65]. SIMS was operated in the dynamic mode, following the intensity of several signals (corresponding to D, C, CH and CD ions) during erosion. Profiles consistent with mean-field theories were observed, except in the near surface region where the gradient was found to be much smaller than predicted by theory. This suggested that surface interactions are longer ranged than assumed in the strictly local interaction mode.

Operating with symmetric bilayers of d-PS and h-PS in a large range of molecular weights and several annealing temperatures, interdiffusion coefficients were estimated [66]. A relatively easy measurement of self-diffusion coefficients was obtained by dynamic SIMS, consistent with other methods like neutron reflection. Good agreement with reptation scaling law predictions for the static and dynamic properties of polymer melts [71] was found in some conditions.

In order to investigate the surface-induced orientation of symmetric diblock PS/PMMA copolymers, perdeuteration of either PS or PMMA blocks was considered [67]. Also in this study SIMS depth profiling was used, following the intensities of C, H and D ion peaks. While as-cast films showed a constant intensity of the above peaks, characteristic oscillations of H and D intensities were observed on films annealed at 170 °C (Figure 3.10). This behaviour was due to the lamellae morphology with orientation parallel to the surface assumed by blocks after annealing. Thickness of lamellae of 37.0 + 1.5 nm was measured for d-PS/PMMA and slightly more for PS/d-PMMA. A preference of PS for the free surface and of PMMA for the copolymer/substrate interface was also observed.

In conventional SIMS it is impossible to distinguish hydrocarbon fragments from oxygen-containing fragments (for instance, $C_3H_5^+$ and $CHCO^+$ at 41 D or $C_3H_7^+$ and CH_3CO^+ at 43 D). The use of labelled precursors to create plasma-deposited polymer films provided an unambiguous identification, for the further

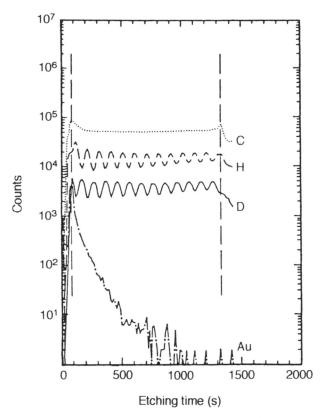

Figure 3.10 SIMS profile of a 500 nm thick deuterated poly(styrene) (d-PS)/PMMA sample annealed 72 h at 170 °C. C: carbon signal ; H: hydrogen signal; D: deuterium signal; left dashed line: air/copolymer interface; right dashed line: copolymer/substrate interface [67]. (Reprinted with permission from Coulan *et al.*, *Macromolecules*, **22**, 2581. Copyright (1989) American Chemical Society.)

understanding of the structural features of polymeric substrates. Acetone, ^{13}C-acetone and perdeuterated acetone were used for a very accurate analysis [72]. The study assigned peaks to fragments with certainty and compiled tables of relative intensities (Table 3.3). These latter can be used in structural studies of parent polymers.

Hydrophobic recovery of oxygen plasma treated PS was studied by static SIMS, operating with d-PS as a substrate and ^{18}O as reactant [68]. The dependence on time of peaks at 18 D and 19 D (corresponding to $^{18}O^-$ and $^{18}OH^-$, respectively) showed the complete recovery of the oxygen plasma treated surface after annealing at 140 °C, so confirming the contact angle results. The occurrence of different hydrophobic recovery mechanisms was confirmed by using labelled blends, i.e. d-PS/PS$_{2700}$ and d-PS/PS$_{50000}$, where d-PS has a

Table 3.3 Relative SIMS intensities of oxygen containing secondary positive ions for the $(1, 2, 3-{}^{13}C_3)$ acetone plasma deposited film [72]. (Reprinted with permission from Chilkoti, Ratner and Briggs, *Anal. Chem.*, **63**, 1612. Copyright (1991) American Chemical Society.)

m/z (Daltons)	Assignment	Intensity (relative units)
43	${}^{13}CH\,{}^{13}CO^+$	4.9
45	${}^{13}CH_3\,{}^{13}CO^+$	100.0*
47	${}^{13}C_2H_5O^+$	2.8
58	${}^{13}C_3H_3O^+$	2.8
60	${}^{13}C_3H_5O^+$	4.2
62	${}^{13}C_3H_7O^+$	1.4
71	${}^{13}C_4H_3O^+$	2.1
73	${}^{13}C_4H_5O^+$	8.5
75	${}^{13}C_4H_7O^+$	9.2
77	${}^{13}C_4H_9O^+$	1.4
86	${}^{13}C_5H_5O^+$	2.1
88	${}^{13}C_5H_7O^+$	4.9
90	${}^{13}C_5H_9O^+$	2.1

*Base peak in the positive ion SIMS spectrum of $(1, 2, 3-{}^{13}C_3)$ acetone plasma deposited film.

molecular weight of 42 700. The intensities of 91 D and 98 D peaks in the positive ion spectra were used in order to determine the relative percentage at the surface of d-PS (the 98 D peak is due to the perdeuterated tropylium ion $C_7D_7^+$). A good agreement between surface and bulk d-PS concentration was found in the d-PS/PS$_{50000}$ blend, while a prevalence at the surface of PS$_{2700}$ was observed in the d-PS/PS$_{2700}$ blend. A further large decrease of d-PS was observed after annealing at $100\,°C$ (Figure 3.11). Results were interpreted on a basis of two mechanisms of hydrophobic recovery, i.e. macromolecular reorientation and diffusion.

Analogous recovery studies were carried out on other ${}^{18}O$ plasma-treated polymers, like PP [69] and polydimethylsiloxane (PDMS) [70]. In both studies, negative ion spectra were demonstrated to be the most informative. For PDMS, peaks centred around 119 D ($Si(CH_3)_2OSiOH^-$), 135 D ($Si(CH_3)_2OSiO_2H^-$) and 211 D ($Si(CH_3)_2O_2Si(CH_3)_2O_2SiOH^-$) were used to understand the attack site of oxygen, showing that a preferential interaction between oxygen and silicon atoms occurs.

3.2.7 Applicative Studies

Static SIMS has been used to characterize quantitatively the bonding of octa-decyldimethylchlorosilane (ODS) to silicas used as chromatographic materials

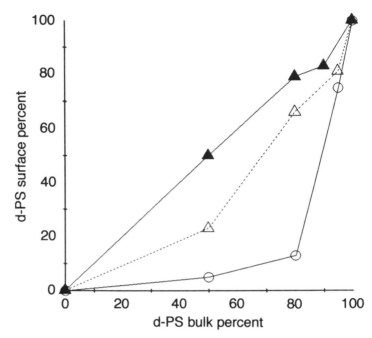

Figure 3.11   d-PS surface versus bulk concentration for d-PS/PS$_{50000}$ blend (▲); d-PS/PS$_{2700}$ blend (△); d-PS/PS$_{2700}$ blend after plasma treatment and annealing at 393 K for 72 h (○). (Reproduced by permission of Elsevier Science Publishers BV from ref. [68].)

[73]. The SiOH$^+$/SiO$^+$ ratio has been found to decrease with increasing ODS coverage, suggesting its reaction with surface silanol groups.

Several applications of SSIMS to thin (monomolecular) layers of Langmuir–Blodgett films have been reported [74–75]. The degree of saturation of the fatty acid has been readily determined by the analysis of SSIMS spectra, as well as the approximate depth from which molecular and quasi-molecular ions originate in alternating Langmuir–Blodgett multilayers.

Charged and sterically stabilized polymer colloids have also been studied [76]. The detection of sulphate-derived peaks and the presence of peaks attributable to PEG portions on the surface provided new evidence on the structure of colloids and the stabilization mechanism of particles.

A great deal of applicative SIMS studies have been given to modified polymer surfaces, mainly by plasma or corona discharges [68–70, 77], but also by other techniques like flame [78]. The surface specificity of SSIMS was demonstrably unique when changes were limited to the first few molecular layers and in order to understand the chemistry. Studies on plasma treated PS [68], PP [69] and PDMS [70] have been referred to previously. Studies on plasma-treated PE and

PP have also been made by van Ooij [77], in order to elucidate the chemistry of the treatment process. Incorporation of oxygen and/or nitrogen, depending on the plasma used, and formation of unsaturations was observed. Methyl loss from PP, forming a PE-like surface, and cross-linking phenomena were also confirmed. The interaction of nitrogen and ammonia plasma with PS and PC was studied [79]. The interpretation of results was helped by the use of derivatization reactions with salicylaldehyde and 5-Br-salicylaldehyde. An easy chain scission of PC at the carbonate bond was observed with the formation of bisphenol A terminated chains. Surface modification of polymers is reported more fully in Part III of this book.

3.2.8 Advances in SIMS Instrumentation

In recent years instrumentation research has followed two main directions, i.e. the obtainment of ion images and the improvement of sensitivity and resolution particularly at high masses. The use of time-of-flight analysers has constituted a great advance towards the latter objective.

The development of microanalytical SSIMS using ions [35, 37] or atoms [80] was suggested by the relatively easy collimation of ion beams to obtain good spatial resolution. With noble gas ions, a resolution of some micrometres is obtainable, while with liquid metal ion beams like Cs^+ the spatial resolution drops down to fractions of a micrometre [37].

In atom beam imaging (FABMS imaging) FAB guns have been modified to improve collimation. Using Ar atoms resolutions comparable to those obtained in SSIMS imaging have been obtained. Atom beams avoid charging problems. In the usual operating mode, the analyser is fixed to observe a fragment of given mass. The ion (atom) beam is then rastered on the surface by computer-guided electrostatic apparatus. The intensity recorded at each observation point is plotted in two dimensions. The use of a time-of-flight analyser has improved the effectiveness of SSIMS imaging [31].

By choosing peaks typical of different compounds, the surface distribution of various components is obtained in heterogeneous systems. The image given by secondary electrons emitted as a consequence of heavy particle bombardment [35, 37] can also be registered.

Most applications of SIMS imaging techniques concerns the analysis of microcircuits, but images of polymer surfaces have also been obtained [35, 37]. In an interesting application, the fracture surface of a carbon fibre composite material has been examined [35]. By SEM, where a metal coating is necessary to obtain the image, information about the chemistry of the fracture surface is not available. SSIMS imaging, using peaks typical of both graphite and polymer and of graphite alone, has shown that some fibres present residues of the polymeric matrix. Such information was important to evaluate the fracture mechanism at the surface.

Images of a drug delivery system, consisting of theophylline laden polymer beads, have been obtained, centering on negative ions, aspecific like O^- and specific of the drug, like CN^- and CNO^- [81]. An even distribution of the drug over the beads surface was established.

Polymer blends have been studied by SIMS imaging. By depth profiling a miscible blend of polyvinylidenefluoride and PMMA, a vertical composition distribution was observed [82]. Chemical imaging of the surface of PMMA/PS blends [83] and polyethersulphone containing 6% of dimethylsiloxane [84] have been published. In a study on a fractured surface of PP/nylon-6 immiscible blend

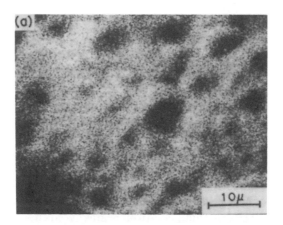

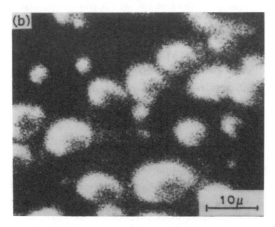

Figure 3.12 SIMS chemical images of (a) C_2^- and (b) CN^- ions for a PP/polyamide 6 blend. (Reproduced by permission of the Society of Polymer Science, Japan from ref. [85].)

[85], the effect of polyacrylic was elucidated, showing the different images obtained in the presence or absence of the compatibilizer (Figure 3.12).

Another space resolved technique suitable for studying polymer surfaces is the LAser Microprobe MicroAnalysis (LAMMA) [17]. Ion emission is induced by an intense laser beam, with a lateral resolution in the 2–100 μm range. It has been applied to the study of dye spots on PS [17] and of various other organic materials [86].

One of the problems is SIMS is the lack of data concerning secondary ion fragmentation. In order to study the secondary ion formation and fragmentation in more detail, a tandem SIMS system has been proposed and developed [87–90]. The system consists of a triple quadrupole instead of a single quadrupole analyser, as shown in Figure 3.13. It records a conventional SIMS spectrum, when quadrupoles two and three (Q2 and Q3) are in a radiofrequency-only mode and Q1 is scanned. In MS–MS or daughter mode, Q1 is tuned to transmit one particular ion of interest, Q2 is filled with a target gas and Q3 is scanned. The mass-selected ion (parent ion) passes from Q1 into Q2, collides with the target gas atoms or molecules. Owing to collisions, collisionally activated dissociation (CAD) occurs, producing fragments that are scanned in Q3, giving the daughter spectrum. Finally, in neutral loss mode, Q1 and Q3 are scanned together, but with a fixed mass difference between them, equal to the mass of the suspected neutral being lost [87–89]. The system was applied to various polymers, like PTFE [87], LDPE [89], PET and PS [90]. It was observed that LDPE fragment ions obey quite strictly the even electron rule (i.e., an ion with all of its electrons paired, may only fragment to give another even-electron ion, while a radical cation may fragment to give either an even- or odd-electron ion) [89]. In PET low energy collisional events were found responsible for fragmentation, while PS exhibits a highly complex behaviour, resulting in the formation of polyaromatic ions [90]. An example of fragmentation steps in PS from 191 D to 89 D is given in Figure 3.14.

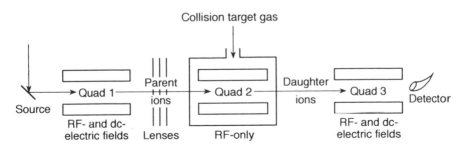

Figure 3.13 The experimental arrangement of a tandem SIMS analyser. (Reproduced with permission from ref. [90].)

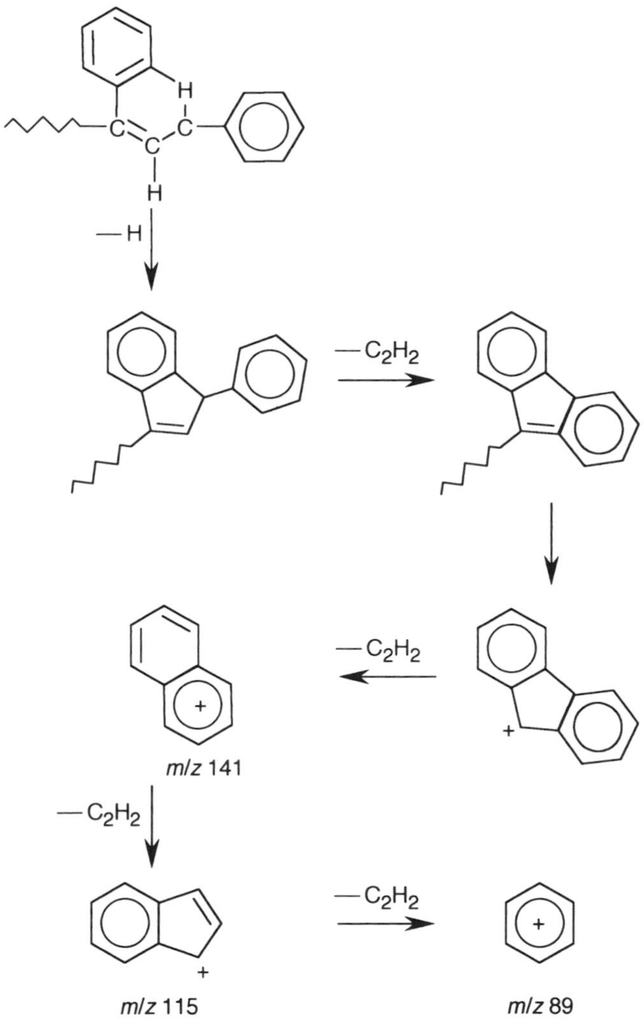

Figure 3.14 Fragmentation steps in PS, starting from 191 Daltons ion. (Reproduced by permission of the Royal Society of Chemistry from ref. [90].)

3.3 X-RAY PHOTOELECTRON SPECTROSCOPY

X-ray photoelectron spectroscopy (XPS) became one of the most popular spectroscopic techniques available for surface analysis of polymers. Several books [91–93] describe the technique and its applications. Reviews concerning its application to polymers [91, 92, 94–96] have been published.

The XPS spectrum is obtained by irradiating the sample with a monochromatic (or quasi-monochromatic) X-ray source. Photons can collide with electrons, giving them their energy. Emitted electrons acquire a kinetic energy which is equal to the difference between the energy of the incident photon and the binding energy of the electron to the nucleus. For each element and each core level, atomic sensitivity factors, ASF (i), which account for instrumental factors and for the cross-section of the photon–electron collision have been defined. They must be calculated for each instrument, a compilation valid for the cylindrical mirror analyser (CMA) has been published [93]. When the experimental XPS intensity, $I(i)$, corresponding to the various elements **i** present at the surface is known the relative concentrations $C(i)$, expressed as atomic per cent, can be calculated using the equation:

$$C(i) = \frac{I(i)/\text{ASF}(i)}{\Sigma_j I(j)/\text{ASF}(j)} \tag{3.4}$$

Emitted electrons are characterized by the inelastic mean free path (IMFP), the value of which depends on the electron kinetic energy. Only the electrons generated near the surface (within a range of few nanometres) have a finite probability of escaping from the material without changing their state of motion [91–93]. These electrons originate the XPS signal, while those which have changed their state of motion contribute to the background. In a first approximation, when electron kinetic energies exceed about 100 eV, the IMFP depend only on the square root of the kinetic energy [91, 92]. The problem of IMFP in inorganic and organic compounds has been discussed [97].

The actual constant which characterizes the probability of the electron leaving the material without collisions is the escape depth (l). It is given in equation (3.5) as a function of the IMFP (Φ) and the angle θ between the axis normal to the surface and the axis of the analyser:

$$l = \Phi \cos \theta \tag{3.5}$$

The XPS intensity can be expressed in a simplified form as a function of the number of atoms of the element **i** in the examined layer, $N(i)$, the atomic sensitivity factor and the electron escape depth $l(i)$ [91]:

$$I(i) = \text{ASF}(i) \int_0^\infty N(i) \exp[-z/l(i)]dz \tag{3.6}$$

z is the axis normal to the sample surface, its origin lies at the surface. A consequence of the exponential dependence expressed in equation (3.6) is that more than 95% of the observed electrons come from a layer which has a thickness three times the electron escape depth, which in its turn depends on the IMFP. As the latter has a value, at the commonly used kinetic energies, of 0.5–4 nm [97], the observed layer is 2–12 nm thick. This means in practice that XPS is rather less

surface sensitive than ion spectroscopies, like ISS and SIMS. Another consequence of equation (3.6) is that angular resolved XPS (i.e. taking XPS experiments
at different sampling angles) is a non-destructive way of making depth profiling.
For instance, a sampling angle of 15° corresponds to a sampling depth of 2.2 nm,
while a sampling angle of 90° corresponds to a sampling depth of 8.6 nm [10].

An alternative method of depth profiling is ion bombardment of the surface
with an ion gun, collecting XPS spectra at opportune intervals. Such a technique
is not suitable for polymeric materials, since it alters the surface layer, depleting or
enriching it in particular functional groups [91, 92].

The XPS instrumentation consists of a UHV chamber (usually at a background vacuum of 10^{-7}–10^{-8} Pa) containing the sample holder, an X-ray gun
and an electron analyser. The most popular X-ray guns use the Kα lines of Mg
(1253.6 eV) and Al (1486.6 eV), which ensure both good resolution and sensitivity.
Other radiation lines like Zr Kα (2042.4 eV) and Ti Kα (4510.0 eV) have also been
used, with the main purpose of avoiding overlap with Auger lines, to detect new
peaks and to extend non-destructive depth profiling up to 20 nm [97]. In modern
spectrometers, emitted electrons are detected by a cylindrical mirror (CMA) or
hemispherical analyser. A sketch of instrumental arrangement is shown in
Figure 3.15.

XPS resolution is not particularly high (about 0.5 eV) and its sensitivity is low
with respect to SIMS (10^{-3} coverages can be detected) [91]. The performance of
conventional X-ray sources, which emit a radiation spurious for satellite lines,
can be improved by monochromatization, using a monochromator crystal. In
this way the full width at half maximum (FWHM) carbon peak (C1s) decreases
from 1.0 eV or more to 0.6 eV.

Using XPS all elements, except H and He, are observed. From photoemission
spectra, quantitative (surface concentrations) and qualitative (functional groups)
information can be obtained. Core electrons, with high binding energies, originate the most important spectral features. These electrons are less sensitive to the
chemical state than valence electrons, nevertheless they offer information, even if

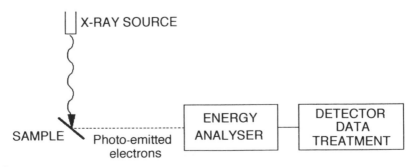

Figure 3.15 Experimental set-up of XPS.

approximate and sometimes ambiguous, on the chemical state of detected elements.

The XPS spectrum of a clean hydrocarbon polymer is rather simple since only the C1s peak is present (Figure 3.16). Contaminants and/or surface oxidation are often observed (Figure 3.17), but unlike in SIMS surface cleanliness is not a life or death problem. Thus, most of the information about polymer surfaces has been drawn from C1s peaks, since they are usually a sum of components relative to the different chemical states of the carbon atom at the surface. Line fitting of C1s spectra allows, semiquantitatively, an evaluation of the relative amounts of different functional groups [91–95] at the surface. Large chemical shift effects are given by fluorocarbon polymers, owing to the electron attractive effect of fluorine (Figure 3.18). A list of chemical shifts corresponding to common functional groups is given in Table 3.4. Carbon bonded to other carbon atoms or hydrogen provides a component at 284.6 eV, carbon singly bonded to oxygen at 286.2 eV, carboxyl carbon at 288.8 eV, etc.

Shake-up peaks, which originate from a two-electron process, provide other types of information [92, 94, 95]. In addition to the emission of a core electron, a simultaneous excitation of a valence electron to an unfilled level can occur, giving rise to the shake-up peak. They appear at the lower kinetic energy side of main peaks and are expected when a polymeric system contains unsaturation. They are particularly prominent in polymers containing aromatic pendant groups [94] like PS (Figure 3.19) and therefore their occurrence is an indication of the aromaticity of a material [98].

Several studies of X-ray photoelectron spectra of valence electrons, usually accompanied by theoretical calculations of the polymer valence bands have been reported [99–101]. Obviously, valence electrons are not characteristic of any element, but they reflect the structure of electron bands of a layer which is thicker than that invaded by core electrons, and thus can be considered representative of the bulk of the polymer. The study of valence bands is generally performed by solid state UPS (ultraviolet photoelectron spectroscopy), but are otherwise rarely used in polymer studies [102, 103].

Recently, valence band spectra have been proposed for identification purposes using a fingerprint approach [104]. In fact, even if the valence band spectrum consists of an envelope structure due to many closely spaced energy levels, they show specific features for each polymer. The valence band spectra of PE and PP are shown in Figure 3.20. The former also shows many contributions, at about 11 eV and 17 eV, attributed to the antibonding and bonding molecular orbitals C2s, respectively. The broad weak structure below 10 eV is due to C2p. In the case of PP, the methyl pendant group gives a strong contribution represented by the feature at 14.5 eV. Also isomeric structures, like normal-, iso- and *tert*-butyl methacrylates give valence band spectra different one to another [55]. Valence bands have been used to determine structural changes in plasma-treated polymers [104, 105].

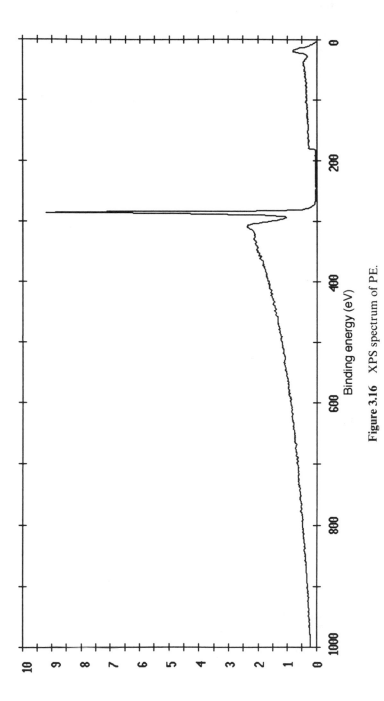

Figure 3.16 XPS spectrum of PE.

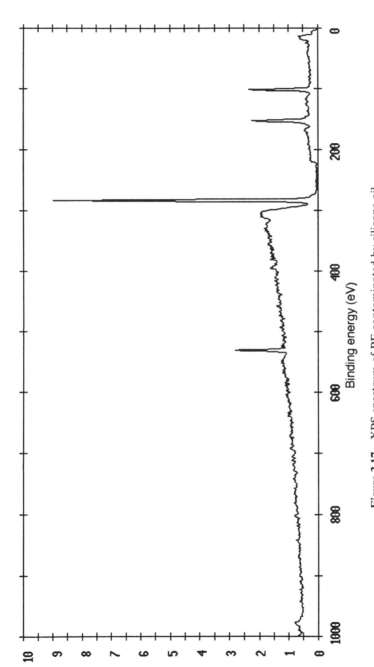

Figure 3.17 XPS spectrum of PE contaminated by silicone oil.

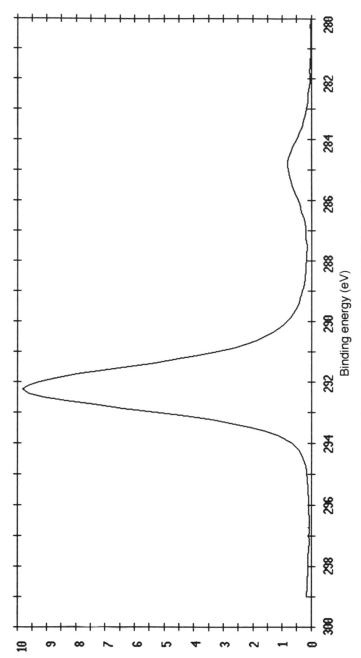

Figure 3.18 Carbon C1s peak of PTFE. The small peak at 285 eV is hydrocarbon contamination.

Table 3.4 Chemical shifts of C1s photoemission peak (adapted from ref. [93]), assuming hydrocarbon C at 285.0 eV

Compound	Binding energy (eV)
TiC	281.6
WC	283.1
C (graphite)	284.7
$(CH_2)_n$	285.0
$MeCH_2NH_2$	285.6
$MeCH_2Cl$	286.3
$MeCH_2OH$	286.5
$MeCH_2OEt$	286.8
$MeCH_2OOCMe$	287.1
CS_2	287.2
$Fe(CO)_5$	288.3
Me_2CO	288.2
$(NH_2)_2CO$	288.7
C_6F_6	288.7
MeCOONa	289.0
MeCOOEt	289.4
MeCOOH	289.5
Na_2CO_3	289.7
$NaHCO_3$	290.2
CO	290.6
CO_2	292.1
$(CHFCH_2)_n$	288.0
$(CHFCHF)_n$	288.4
$(CHFC_{F2})_n$	289.3
$(CF_2CH_2)_n$	290.7
$(CF_2CHF)_n$	291.5
$(CF_2)_n$	292.3
CF_3COONa	292.4
CCl_4	292.6
CF_3COMe	292.7
CF_3COOEt	293.1

Surfaces of polymers [92–95], copolymers [18, 106, 107] and blends [108, 109] have been widely studied by XPS. Modifications of polymer surfaces, by chemical or physical methods (mainly plasma, but also laser, flame, etc.) have been followed by XPS analysis, often in connection with other techniques [110–115]. Grafting studies on to polymer surfaces have also been carried out [116–119]. In such studies, observed chemical shifts are sometimes not sufficient to understand the chemistry at the surface, for instance it is impossible to distinguish ketones from aldehydes, acids from esters, etc. Since each group has a specific chemical reactivity, many reactions have been proposed able to introduce a new, easy

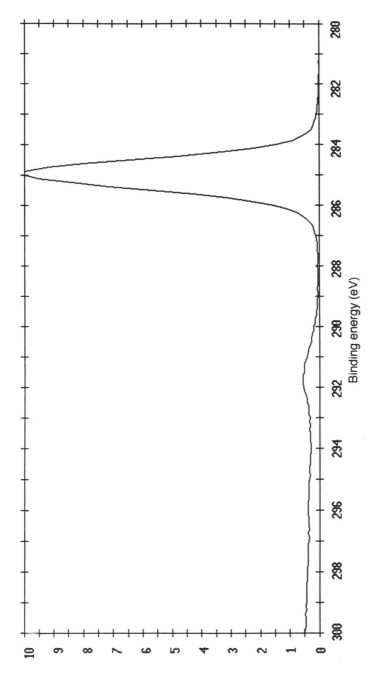

Figure 3.19 Carbon C1s peak of PS. The shake-up peak is visible.

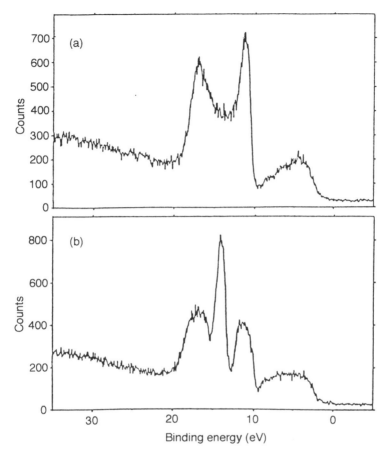

Figure 3.20 Valence band spectra of (a) PE; (b) PP. (Reproduced with permission from ref. [104].)

detectable element at the surface. This method has been called 'tagging' or 'derivatization' and has been excellently reviewed by Andrade [96] and Batich [120]. Derivatization reactions proposed in the literature are listed in Table 3.5. Instead of tagging a functional group, a fraction of a polymer having a different reactivity can be evidenced. For instance, amorphous and crystalline regions of PET have been distinguished, based on the fact that the former is better swelled in dimethylsiloxane, consequently reaction with OsO_4 is much more evident than for crystalline PET [120]. Owing to the absence of standards and the lack of knowledge of the crystallinity phenomenon, this approach must be considered only qualitative.

Table 3.5 Labelling species for derivatization studies of polymer surfaces [120]
(Reproduced by permission of Elsevier Science Publishers BV)

Functional group	Labelling species	Reference
C—OH	$(CF_3CO)_2O$	[121, 123–125]
	$(i\text{-}Pr\text{-}O)_2Ti\,(acac)_2$	[122, 126]
	CF_3COCl	[127]
	CBr_3CO_2H	[121]
	$(NO_2)_2C_6H_3COCl$	[121]
	$CF_3(CH_2)_2Si(CH_3)_2NCOCH_3$	[128]
	ClC_6H_4NCO	[129]
	$(NO_2)_2C_6H_3F$	[130]
—NH$_2$	C_6F_5CHO	[121, 125, 128]
	CS_2	[121]
	HCl	[128, 131]
	$CH_3CH_2SCOCF_3$	[121]
—SH	$AgNO_3$	[121]
	$[C_6H_3(NO_2)_2(CO_2Na)S]_2$	[121]
	$HgCl_2$	[130, 132]
—OOH	SO_2	[126]
Si—OH	$CF_3(CH_2)_2Si(CH_3)_2NCOCH_3$	[128]
—COOH	$NaOH$	[121, 126, 128, 130]
	$BaCl_2$	[121, 128]
	$AgNO_3$	[121, 123, 128, 129]
	$TlOC_2H_5$	[121, 123]
	CF_3CH_2OH	[121, 128]
	$C_6F_5CH_2Br$	[121, 128]
	$(CF_3CO)_2O$	[121, 128]
	$Ca(NO_3)_2$	[133]
	H_2NR	[124]
	$Ba(OH)_2$	[130]
	$^{45}CaCl_2$	[134]
C=C	Br_2	[121, 128, 135]
	$Hg(CH_3CO_2)_2O$	[121]
	$Hg(CF_3CO_2)_2O$	[121, 128]
	OsO_4	[121]
C—C	HCl	[121, 128]
O	$(CF_3CO)_2O$	[128]

XPS is generally considered a non-destructive technique. However, some spectral changes induced by X-ray irradiation have been published and reviewed [136]. On polymers, a detailed study on beam damage of PTFE has been reported [137], interpreting the observed fluorine depletion as caused by X-rays rather than electron bombardment. Results have been recently confirmed using a monochromatic X-ray beam, and extended to other polymers, like PMMA, PVC, PET and polyacrylonitrile (PAN) [138]. It was concluded that non-fluorine containing polymers undergo considerably less damage than PTFE at equal exposure times. Furthermore, the risk of damage appeared limited using the current low flux instruments [138].

Selected examples of polymer surface studies by XPS are given below.

3.3.1 Polymers and Copolymers

Most of studies of polymer surfaces by XPS have been performed in the 1970s, starting from the studies carried out at Durham University [4, 5]. These studies are accounted for in several books and reviews [91–96]. Acrylate and methacrylate polymers have been characterized by XPS and SIMS, finding that photoelectron spectra are too similar to clearly differentiate each element of either series [139]. In methacrylate polymers, four species contributed to the C1s peak, at 284.8, 285.5, 286.6 and 288.7 eV, assigned to $C-C$, $C-C$ β-shifted, $C-O$ and $O=C-O$, respectively (Figure 3.21 (a)). The O1s spectra of studied polymers are rather similar and show the presence of two peaks of equivalent intensity, assigned to carbonyl oxygen (532 eV) and ether oxygen (533 eV) (Figure 3.21 (b)). In polyacrylates, the contribution at 285.5 eV is absent in the C1s peak.

The behaviour of diblock copolymer surfaces are of interest to study since they depend strongly on the connectivity of constituents. Differences between the domain structure at the surface with respect to the bulk were recognized early. In PDMS-PS copolymer films having different morphologies, the surface was found to consist of an essentially pure PDMS [140]. The intensities of Si2p and O1s peaks were considered as representative of PDMS moiety, that of the shake-up satellite of PS. In PMMA–PS diblock copolymers, assuming a lamellar periodic morphology [67], XPS was used to determine the surface excess of PS [141]. Experimental results confirmed the mean-fields theory calculations. A segmented poly(etherurethane) block copolymer was also studied [142]. This material consists of a 4, 4′–diphenylmethane diisocyanate (MDI) reacted in a 1:1 ratio with a 1, 4 diol chain extender to form the hard urethane segment, which in turn reacted with a poly(tetramethylene)ether glycol (soft segment). In spite of surface contamination problems mainly due to unreacted low molecular weight compounds, the surface segregation of the polyether blocks was observed.

Variations of the surface composition in block copolymers with respect to the bulk have been interpreted as a result of the required minimization of surface energy.

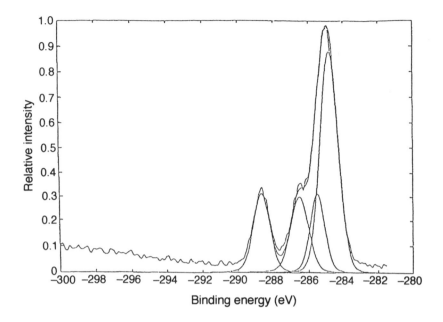

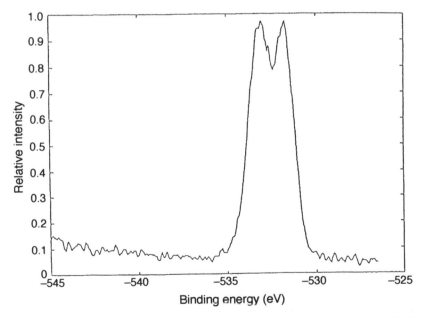

Figure 3.21 XPS peaks of methacrylate polymers: (a) C1s, from right to left C—C, C—C β-shifted, C—O, and O=C=O; (b) O1s, from right to left, carbonyl and ether, (Reproduced by permission of the American Institute of Physics from ref. [139].)

3.3.2 Modified Polymer Surfaces

Some unfavourable properties of polymers, connected for instance to wetting and adhesion, stimulated extensive studies on their modification using both wet chemical and physical methods. While the subject is specifically treated in Chapters 9 and 10, this section is devoted to the use of XPS as a diagnostic method of modification treatment.

Treating PE (polyethylene) with a chromic acid/sulphuric acid mixture mainly produces carboxyl groups [143]. Treating a poly(chlorotrifluoroethylene) surface with methyllithium removes $-CF_2-$ groups near the surface, with an increase of the hydrocarbon component, by the introduction of methyl groups and insaturations [144]. Etching of PTFE with a solution of sodium naphthalene in tetrahydrofuran (THF) results in deep changes of C1s peak, occurrence of O1s and Na1s peaks and a decrease of F1s intensity. Rinsing experiments with H_2O and THF determined minor changes and the disappearance of sodium from the surface. The latter observation was attributed to the removal of NaF, while the polymer surface was found oxidized, with the presence of carboxyl, carbonyl and hydroxyl bonds [145]. Oxidation of a PP film with chromium(VI) oxide in acetic acid/acetic anhydride was found to introduce to a thin layer (*ca* 10 nm) hydroxyl groups, olefins, ketones and esters [146]. Derivatization experiments were used to unambiguously identify the groups formed. A gradual dissolution of the modified layer was also observed.

Various physical methods have been used to improve the adhesion properties of polymers. Using XPS it has been shown that flame treatments induce the formation of hydroxyl, carbonyl and carboxyl groups both on PE [147], PP [114, 148] and ethylene-propylene copolymers [149]. On the basis of the relative amount of functional groups at the surface a step-by-step oxidation mechanism of the methyl pendant group has been proposed [114].

Electrical discharge treatment in the presence of oxygen [150, 151] and electron beam treatment in air also favour the formation of oxygen containing functions [152] on the surface of polyolefins.

Cold plasma, a technology developed in the electronic industry, became a very popular method for treatment of polymer surfaces. XPS has been applied in the study of water or oxygen plasma treatments of PE surfaces [153]; these processes provide high densities of carboxylic acid groups, which can be converted to alcohols or esters. Studies on oxygen or air plasma treated surfaces have been carried out on PP [154], PS [155–157], PET [158], PTFE [159, 160], PC [161], PDMS [70], polyimides [162], copolymers [111], and many other materials. Studies have also been made using other gases like argon [112], nitrogen and ammonia [110, 79], H_2O [163], with the general aim of understanding the modification mechanism and the suitability of the technique for applicative purposes.

Surface halogenation has also been widely studied [92]. PP has been treated in

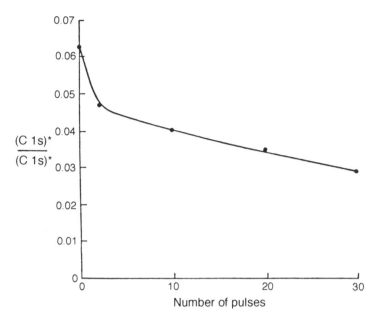

$\dfrac{(C\ 1s)^*}{(C\ 1s)^*}$

Figure 3.22 Loss of aromaticity in poly(α-methylstyrene), expressed as shake-up/C1s peaks ratio as a function of fluence. (Reproduced by permission of the American Institute of Physics from ref. [113].)

a cold plasma of halogenomethanes [164]; CF_3Cl and CF_3Br cause respectively the chlorination and bromination of PP surfaces, while CF_3H produces a film of fluoropolymer on the surface. Surface fluorination with fluorine gas has also been studied [165].

The effect of laser treatment on PET [166] and PMMA [113, 166] has been discussed. With 193 nm radiation, ablative photodecomposition of the surface occurs. For PET, the treatment causes oxygen depletion, suggesting a preferred removal of carboxyl groups. For PMMA, no preferential removal of oxygen was observed, indicating decomposition to monomer units [166]. Generally speaking, laser irradiation seems more effective on surface morphology than chemical composition [113, 167]. However, in poly(α-methylstyrene), loss of aromaticity was observed, in connection with the decrease of the intensity ratio between shake-up and C1s peaks (Figure 3.22) [113].

3.3.3 Polymer Blends

The surface-energy minimization can also produce surface enrichment in blends, as observed by XPS [108–109]. In PVC/PMMA blends, surface enrichment of PMMA was found at all compositions, if blends were cast from THF, while a

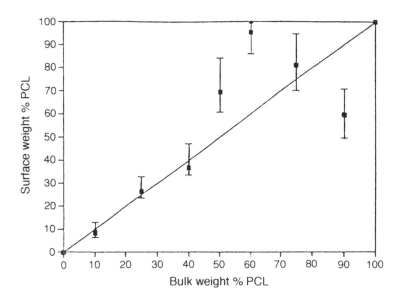

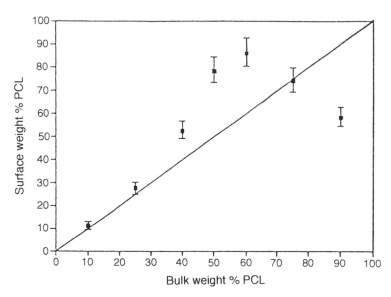

Figure 3.23 Surface composition versus bulk composition in PCL/PVC blends. Surface composition was calculated from C/O peak area ratios (above) or C1s components (below) [170]. (Reprinted with permission from Clark *et al.*, *Macromolecules*, **22**, 4495. Copyright (1989) American Chemical Society.)

composition equivalent to that of bulk was exhibited by blends cast from methylethylketone [168]. Angular resolved studies on PVC/poly(vinyl-methylether) (PVME) blends suggested a composition gradient profile, with surface enrichment of PVME, the component with minor surface tension [109]. Also in PS/PVME blends PVME was preferentially found at the surface, for the same reasons [169]. It was possible to model the concentration gradient and to demonstrate that the surface concentration depends on molecular weight of the constituents.

The surface composition of polymer blends can be determined by XPS provided that each component in the blend has a unique element or functional group present. On this basis, Thompson proposed the advantages of SSIMS with respect to XPS, studying polycarbonate/PS blends [62]. When possible, it can be

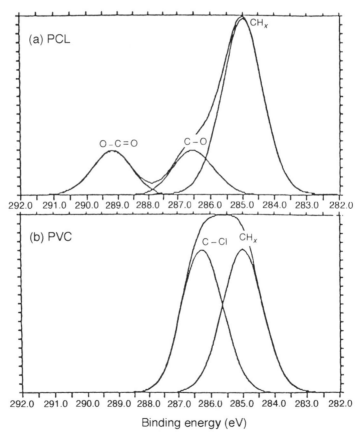

Figure 3.24   C1s components of: (a) PCL; (b) PVC [170] (Reprinted with permission from Clark *et al.*, *Macromolecules*, **22**, 4495. Copyright (1989) American Chemical Society).

useful to confirm results by different methods. For instance, in a study on poly (ε-caprolactone) (PCL)/PVC, the surface composition versus bulk composition was plotted both from C/O peak area ratios and the suitable curve-fitted C1s peaks (Figure 3.23) [170]. The latter method was possible providing that the C1s peaks of the two moieties are very different (Figure 3.24). In both cases no surface enrichment was found for blends with less than 50% by weight of PCL, while for compositions in the range 50–90% of PCL, a surface excess of PCL was observed. Finally, the 90% PCL blend showed an enrichment of PVC. While the excess of PCL is easily explained by the surface tension of homopolymers, the behaviour below 50% of PCL can be attributed to the fact that those blends, where PCL is amorphous, are miscible, while crystalline PCL, present at higher concentrations, is immiscible. This aspect has been deepened in a further study on the same system, where the number-average molecular weight of PVC was varied [171]. It was found that the surface composition of a PCL/PVC blend is governed by a combination of the homopolymer molecular weight effects and degree of crystallinity. When miscible and immiscible blends can be prepared with the same components, different surface compositions can be observed, as in the case of the PVC/poly(α-methylstyrene-co-acrylonitrile) system [172], where the surface enrichment of PVC is more pronounced in immiscible blends (Figure 3.25).

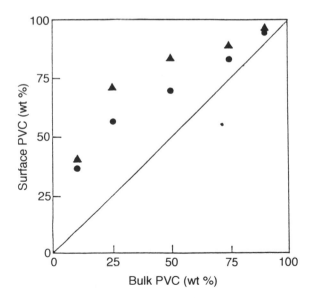

Figure 3.25 Surface composition versus bulk composition in PVC/poly(α-methyl-styrene-co-acrylonitrile) miscible (●) and immiscible (▲) blends. (Reproduced by permission of Elsevier Science Publishers BV from ref. [172].)

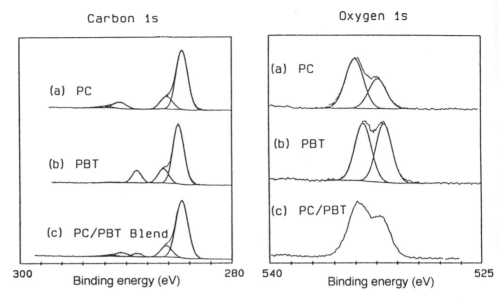

Figure 3.26 C1s and O1s spectra of bisphenol A-polycarbonate (PC), poly(butylene terephthalate) (PBT) and a PC/PBT (50/50 wt %) blend. (Reproduced by permission of Elsevier Science Publishers from ref. [173].)

Spectral simulation can be used to better quantify the surface composition in blends. This scheme has been applied to PC/PBT blends, observing an enrichment of PC [173]. The simulation has been made starting from XPS spectra of pure homopolymers, as shown in Figure 3.26 for carbon and oxygen, respectively.

The variation in blend composition in the outer first layers can be very important, for example in biomedical materials,' where surface composition determines the biocompatibility of the material. In polyurethanes, enrichment of ether groups at the surface has been observed [174, 175].

Complex systems like blends of copolymers have also been successfully studied [176], i.e. PDMS with bisphenol A–PC or bisphenol A–polysulphone, also containing diurethane segments.

3.3.4   Grafting Studies

Techniques able to modify polymer surfaces by photoinduced graft polymerization have been recently developed by Ranby [177]. By grafting water-soluble monomers onto the surface of hydrophobic polymer substrates like polyolefins and PET, permanently hydrophilic surfaces are produced. Preferred monomers were acrylic acid, methacrylic acid, acrylamide [119, 177, 178], glycidyl acrylate and methacrylate [179], etc. Since the latter produce a surface with reactive

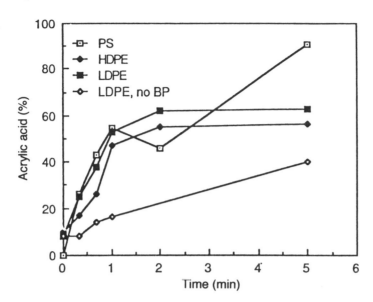

Figure 3.27 Percentage of grafted acrylic acid from XPS spectra as a function of UV irradiation time. (Reproduced by permission of John Wiley & Sons, Inc. from ref. [178].)

epoxy groups, further reactions are allowed, like those with poly (ethyleneglycol) (PEG), heparin and antibodies [118]. In such studies, XPS was used to determine the amount of grafted polymer on to the substrate surface. For instance, grafting PS, PE or PP with acrylic acid, since the latter contains oxygen and the former not, the O1s/C1s intensity ratio allowed the calculation of the percentage of grafted molecules versus irradiation time by UV (Figure 3.27). After 5 min, values from 56% to 90% were measured, depending on the nature of substrate [178].

3.3.5 Conductive Polymers

Polyacetylenes doped with bromine [180] or sodium [181] have been studied. From variations of the C1s profiles, two components were found in both systems. In Br-doped polyacetylene, the low binding energy component was attributed to carbon atoms in the low conductive (undoped) domain, while the high energy component was attributed to the carbon atoms in the metallic domain, where they are positively charged since an electron was transferred to form a Br_3^- ion [180]. In Na-doped polyacctylene, the low and high binding energy components were attributed to negatively charged and neutral carbon atoms respectively [181]. In Ag-doped poly(pyridyl derivative) conductive films, XPS was found able to detect low concentrations of Ag (< 1 at. %), and also concentration gradients under sputtering [182]. An accurate XPS study has been performed on

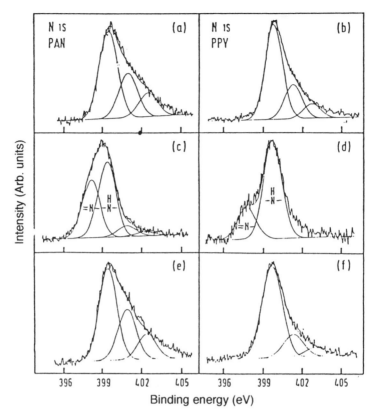

Figure 3.28   N1s spectra of: (a) polyaniline/perchlorate complex (PAN); (b) polypyrrole/perchlorate complex (PPY); (c) emeraldine base (EM); (d) deprotonated PPY (DP-PPY); (e) reprotonated EM; (f) reprotonated DP-PPY. (Reproduced with permission from ref. [183].)

polypyrrole and polyaniline and their N-substituted derivatives [183]. Several nitrogen components were observed and their chemical nature identified (Figure 3.28).

3.3.6   Miscellaneous

The adsorption of molecules on surfaces has been studied by XPS. Physical adsorption of glycine on graphitic surfaces has been observed [184], with the aim to model the interaction of proteins on biomaterials. A partially oriented double layer, having a structure similar to crystalline β-glycine, is stable on the surface.

The surface of fillers for composite materials has also been characterized. Carbon fibres [185–187] and coated glass fibres [188–189] have been studied.

Oxide surfaces treated with alkylamine silanes, normally used as promoters to improve the filler-to-matrix adhesion in composites, have been examined [189]. The bonding of 3-methacryloxypropyltrimethoxysilane (3-MPS) to various fillers has been studied by XPS and SSIMS [44]. The fraction of surface covered by the silane and the thickness of the silane layer were estimated by using simplified models based on XPS intensities. Both coverage and thickness were found to influence the mechanical and transport properties of the composite material.

The reaction of epoxides [185] and amines [186] with carbon fibres after activation of the fibres by oxidation with sulphuric and nitric acids has been studied. The reaction of epoxide or amine groups at the surface was considered as a model of the surface reaction of resins which constitute the matrix in carbon fibre composites. The formation of both epoxide and amine layers on carbon fibres has been demonstrated. The bonding is assumed to be covalent in both cases. Average layer thicknesses can be calculated from XPS data.

A great impetus to XPS studies in heterogeneous systems like composite materials is expected from progress in building spectrometers with a largely better lateral resolution.

Using techniques based on the limitation of the photon source size or, alternatively, on the limitation of the field of view of the electron collection or detection system, sample regions as small as $150 \mu m$ can be analysed [190]. On this basis, imaging systems have been proposed, complementary to imaging SIMS [191].

Metal/polymer interfaces, particularly in their fracture behaviour, have been studied quite extensively by XPS. A review of these applications has been published [192], and results are discussed in Chapter 10 on adhesion.

Finally, studies by XPS have been carried out on coatings [193] and polymer colloids [76].

3.4 INTERNAL REFLECTION SPECTROSCOPY

The previously discussed techniques require ultra-high vacuum for operation. Conversely, all the spectroscopies based on electromagnetic radiation, from radiofrequency to near ultraviolet (UV), do not need such a facility. The problem has been to set up experimental methods able to observe with sufficient sensitivity a layer of limited thickness. Internal reflection spectroscopy (IRS), also known as attenuated total reflectance (ATR) has been examined for such a purpose. Several books and reviews have been published about IRS [194–199].

The physical phenomenon on which IRS is based has long been known. An electromagnetic wave incident at the interface between two different media, with refraction indexes n_1 and n_2 respectively ($n_1 > n_2$), is totally reflected. The angle of incidence must be higher than the critical angle θ_c, defined as:

$$\theta_c = \arcsin(n_2/n_1) = \arcsin(n_{21}) \tag{3.7}$$

The theoretical treatment of IRS is not easy; for the sake of simplification, the medium 2 is assumed to have a very low absorption [194]. In this approximation, the time-averaged energy flux is zero in medium 2. Instantaneously, the so-called evanescent wave occurs; this is an electromagnetic wave with components in all space orientations, having an intensity decreasing as a function of the distance from the interface between media 1 and 2 (Figure 3.29). The amplitude of the evanescent wave E depends on such a distance (z), the amplitude of the electric field at the interface (E_0), the angle of incidence (θ), the radiation wavelength in free space (τ) and the refraction index of medium 1:

$$E = E_0 \exp \left[-(2\pi n_1/\tau)(\sin^2\theta - n_{21}^2)^{1/2} z \right] \tag{3.8}$$

In the literature, the penetration is often given as $d(p)$, that is the reciprocal of Γ, which represents the electric field amplitude decay coefficient:

$$\Gamma = [2\pi n_1 (\sin^2\theta - n_{21}^2)^{1/2}/\tau] \tag{3.9}$$

At this thickness, $E = E_0 \times (1/e)$. The meaning of $d(s) = 3/\Gamma$, in that case $E = E_0 \times 0.05$.

If the hypothesis of zero absorption of medium 2 is correct, the intensity of the radiation after reflection should be equal to that striking the sample surface. Internal reflection spectroscopy is based on the fact that medium 2 has a net, even if low, absorption [194]. The reflectivity R is less than 1:

$$R = I/I_0 = \exp \left[-\alpha d(e) \right] = 1 - \alpha d(e) \tag{3.10}$$

In equation (3.10), α is the absorption coefficient and $d(e)$ the effective penetration of the evanescent wave. The absorption coefficient is a function both of the refractive index of medium 2 and of the wavelength. With the usual absorption coefficients the reflectivity loss is less than 1%.

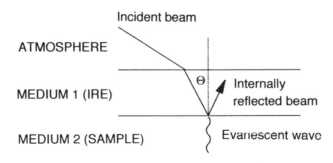

Figure 3.29 Physical basis of IRS (IRE: internal reflection element). (Reproduced by permission of the Society for Applied Spectroscopy from ref. [194].)

In the approximation for low absorption, the absorbance A is given by:

$$A = \alpha d(e) = (n_{21}\alpha/\cos\theta)\int_0^t E^2 dz \qquad (3.11)$$

where $d(e)$ is the effective thickness [194]. The final expression is:

$$A = \alpha d(e) = (n_{21}\alpha E_0^2)/(2\Gamma\cos\theta) \qquad (3.12)$$

and the effective thickness of the observed layer is given by:

$$d(e) = (n_{21}E_0^2)/(2\Gamma\cos\theta) \qquad (3.13)$$

More sophisticated theoretical treatment [198] does not substantially modify the final result. The factors to be considered in calculating the thickness of the observed layer are the angle of incidence, the refractive index of medium 1 and the radiation wavelength. The angle of incidence is particularly important because it influences the amplitude of the electric field at the interface between media 1 and 2 (Figure 3.30), altering both penetration and sensitivity, and is present in equations (3.11)–(3.13). Given all the effects, $d(e)$ decreases with increasing the angle of incidence.

Medium 1, in contact with medium 2 (the sample), is called the internal reflection element (IRE). Its choice determines the critical angle (equation (3.7)), and n_{21} is present in equations (3.11)–(3.13). An increase of n_1 induces a decrease

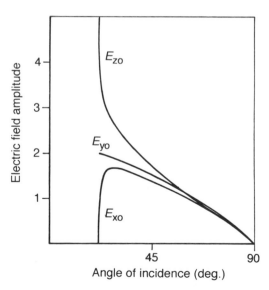

Figure 3.30 Dependence of components of electrical field amplitude on angle of incidence in IRS. (Reproduced by permission of the Society for Applied Spectroscopy from ref. [194].)

of d(e). Within the same technique (for example infrared) the IRS spectrum is affected by variations of n_1 as a function of wavelength [194].

The wavelength is present in equation (3.9) for Γ, which in turn is present in equation (3.13) for d(e). A decrease in the wavelength means a decrease in the thickness of the observed layer.

IRS has been used in IR, Raman and fluorescence modes. IR and Raman modes give vibrational information, allowing identification of functional groups in the examined layer. Fluorescence spectroscopy, through the analysis of the spectrum and of the lifetime of excited states, enables the study of the electronic states of the species present in the examined layer.

The wavelengths used in fluorescence and Raman studies are in the 300–700 nm range; a layer with a thickness of hundredths or tenths of a micrometre can be observed [194]. In IR mode, a 2.5–16 μm range is scanned and the thickness of the observed layer is several micrometres.

Absorption is measured in IRS–IR, while emission is measured in Raman and fluorescence modes, as a consequence the respective instrumental arrangements are different, as shown in Figure 3.31. The sample must be in intimate contact with the IRE, which is usually a suitably cut single crystal. In an IRS–IR study on powders, alterations of the relative intensities of bands have been observed with respect to those obtained analysing a plaque [200].

The radiation strikes a face of the IRE, then undergoes multiple reflections at the IRE/sample interface, before reaching the detector.

The penetration of the IR radiation can be decreased below the limits allowed by variations in angle of incidence and refractive index of the IRE, by depositing on the examined surface a non-absorbing film of controlled and uniform thickness [197–199]. Of course, this decreases not only the thickness of the observed layer but also the sensitivity, because the intensity of the evanescent field is an exponential function of the dimension normal to the surface (equation (3.8)). Besides penetration, sensitivity and spectral contrast affect the choice of IRE. Increasing n_1, decreases the penetration, sensitivity and spectral contrast. Something similar could be said about the choice of the angle of incidence [194].

The lower limit of the surface layer measurable by FTIR–IRS has been determined [201]. Operating with thin layers of polystyrene deposited on polyurethane, it was observed that a thickness of 5 nm or even less can be detected using spectral subtraction. Good detection was also observed in a FTIR–IRS study of Langmuir–Blodgett films of stearic acid 1–9 monolayers thick [202].

Methods have been proposed to enhance IRS–IR spectra based on the deposition of thin metal films [203] or islands [204]. Ag or Ni have been used, by evaporation on the polymer surface to be analysed, or directly to the IRE. Metal film thickness was in the range of 3–6 nm. In Figure 3.32, the IRS–IR spectrum of a PET film is compared with that of a PET film coated with a 2.5 nm layer of PDMS; spectra are rather similar and in the difference spectrum only the bands characteristic of PDMS (at 1260, 1095, 1020 and 800 cm^{-1}) are barely observed

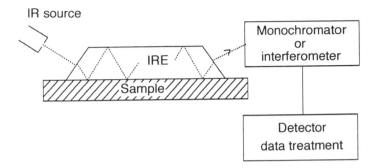

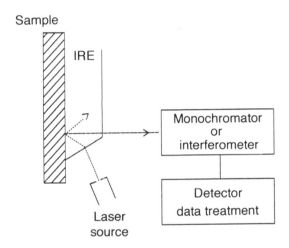

Figure 3.31 Experimental set-up for IRS: infrared mode (above); Raman mode (below). In both cases the emitted radiation passes through a monochromator (or interferometer) to a detector.

[204]. Conversely, the same are well visible (Figure 3.33) in the difference spectrum of the same PDMS film when an Ag film was deposited (6 nm thick) on the Ge crystal used as IRE. A modified method, known as CIRCLE ATR, has been proposed to increase surface sensitivity in investigations on flexible films or fibres [205]. In this arrangement, depicted in Figure 3.34, IRE is fully surrounded by the sample, so increasing the surface in contact with sample and consequently the signal-to-noise ratio.

An ASTM procedure for obtaining IRS–IR spectra has been established [206]. The practical aspects of the technique and the related instrumental accessories have also been described [207].

The experimental arrangement outlined in Figure 3.31 does not allow a

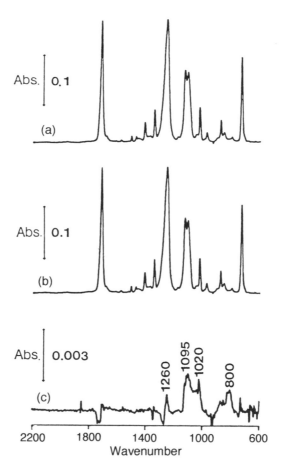

Figure 3.32 FTIR/ATR spectra of: (a) PET film coated with a PDMS layer (2.5 nm thick); (b) uncoated PET film; (c) difference spectrum between (a) and (b). (Reproduced by permission of the Society for Applied Spectroscopy from ref. [204].)

double-beam study. With modern instruments, allowing computer-controlled spectral subtraction, it is possible to record a spectrum without the sample and remove possible spurious contributions [194].

From a quantitative point of view, the Lambert–Beer law is not generally valid in IRS–IR. There is some absorbance/concentration proportionality, particularly when absorption coefficients are low, but the use of intensity ratios of bands present in the same spectrum seems more suitable [208]. Calibration curves based on samples of known concentration have also been suggested [208].

IRS in Raman mode is less popular than IRS–IR. The reasons are its lower sensitivity and increased instrumental problems [209]. Emission spectra are collected, in the layout shown in Figure 3.31. The exciting radiation is monochromatic, generally provided by a laser. The observed layer is shallower than in

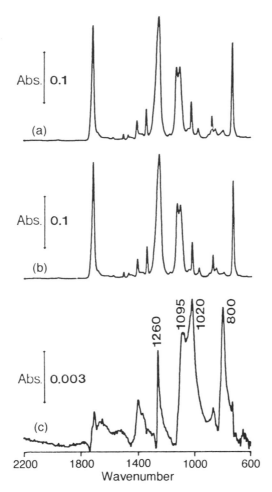

Figure 3.33 FTIR/ATR spectra of same samples of Figure 3.32 in the presence of a 6 nm Ag film. (Reproduced by permission of the Society for Applied Spectroscopy from ref. [204].)

IRS–IR (10–500 nm), depending on wavelength, incidence angle and IRE [209–211]. The choice of the internal reflection element is critical: it must be transparent, with high refractive index and without Raman transitions in the range of interest. In this frame, sapphire seems the best choice [209 211]. To achieve the best sensitivity, the incidence angle is usually kept near the critical angle. Again, the IRE–sample contact should be as intimate as possible. There have been few applications of this technique to polymer surfaces, because of its complexity and lack of sensitivity.

The instrumental problems met in IRS–Raman are repeated in IRS–fluorescence. This technique has found many applications in biology [212], and

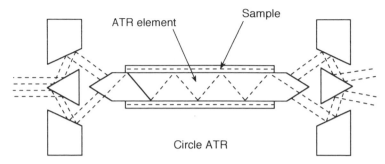

Figure 3.34 Experimental set-up for CIRCLE ATR. (Reproduced by permission of the Society for Applied Spectroscopy from ref. [205].)

has also been used to characterize polymer surfaces [213]. The experimental apparatus, observed layer and IRE are as in IRS–Raman.

IRS–IR has found a large number of applications in very different fields of surface analysis of polymers, concerning for instance, blends [168, 170, 171, 214, 215], block copolymers [142, 163], and grafting [177]. Part of its popularity can be ascribed to the availability of accessories to perform IRS–IR measurements on commercial IR spectrometers. Since depth profiles between 1 and 10 μm can be obtained, results are not related to the surface in the strictest sense. In fact, IRS–IR is less surface specific than XPS, but also less sensitive to surface contamination. In most cases, a combination of techniques has been used, such as XPS and IRS–IR [170]. Depth profiles of incompatible blends for biomedical applications have been obtained by changing the angle of incidence [216, 217]. For Avcothane, a commercial incompatible blend of polyurethane and poly-dimethylsiloxane, it has been observed that increasing the incidence angle (i.e. decreasing the radiation penetration) the intensity of bands attributed to PDMS increases, suggesting its segregation at the surface [217]. In quantitative studies of poly(ethylene oxide)/poly(propylene oxide) blend surfaces, computer techniques have been used to fit experimental spectra with composite spectra obtained adding in different proportions of spectra of pure polymers. An enrichment of the surface with PEO was observed [214]. In blends of an acrylate copolymer with a fluoro-copolymer, the surface segregation of the acrylic moiety was observed [215], notwithstanding its surface tension is higher than that of the fluoro-copolymer. This result has been attributed to the immiscibility between the two polymers and the large difference between their surface tensions.

A depth profiling study of three component copolymers (styrene, butadiene, 4-vinylpyridine), again used as biomedical materials, showed the surface segregation of 4-vinylpyridine and that macroscopic properties such as hydrophilicity and thrombogenicity depend on the method of preparation [218]. A depth profiling study of a PVC membrane with crown ethers anchored on it has been performed [219].

In biomaterials, adsorption phenomena are also important. The interaction of polymeric biomaterials with various proteins and blood have been studied [220].

Surface treatments of polymers such as photo-oxidation [221], chemical oxidation [146, 222] and fluorination [223] have been studied. Through peroxide intermediates, —OH, C—O—C, —C=O and —COOR groups are formed during the photo-oxidation of PP [221]. Chemical oxidation [222] produces the same oxygen-containing functions. Ordinary compression or extrusion processes can introduce a limited quantity of oxidized function on polymer surfaces [221, 224]. The effect of treating PE with sulphuric acid fumes, in order to obtain a biomedical material with special anticoagulant properties, has been reported [225]. Plasma and corona treatments [226] have also been studied extensively. In a study on plasma modified PMMA [20], consequences of treatments were enhanced by using positive and negative subtraction (Figure 3.35). The positive spectral subtraction refers to the unmodified spectrum subtracted from the modified spectrum, with the resulting intensity difference between them positively maximized. The negative spectral subtraction refers to the unmodified spectrum subtracted from the modified spectrum, with the resulting intensity difference between them negatively maximized [20]. Polysiloxane plasma polymers have been studied, particularly in their ageing processes [227, 228].

A study of the fracture surface of PE has been reported [229]. The surface terminal group concentration was found to be an order of magnitude higher than in bulk. This result was attributed to breakages of macromolecular chains near the stressed region.

The conformational difference between surface and bulk in PE and polyvinylalcohol has been studied [224]. More *trans* isomers were observed at the surface than in the bulk. Polarization techniques have been used to study surface orientation in PP [230]. The application of infrared dichroism allowed the surface orientation to be quantified [230]. The influence of uniaxial draw on surface crystallinity and orientation in PET films was examined by polarized IRS–IR [231]. Two different IRE were used (Ge and KRS-5), a common practice to change the analysis depth, and measurements in directions parallel (MD) and perpendicular (TD) to stretching direction were made. Intensity ratios of bands at 1340 or 1370 cm^{-1} (assigned to CH$_2$ wagging vibrations from the *trans* or *gauche* glycol segments, respectively) with respect to the ring vibration band at 1410 cm^{-1} (assumed as reference), as well as MD/TD dichroic ratios, were plotted versus draw ratio and angle of incidence. The *trans* conformer was found to exhibit a higher orientation and concentration near the surface of the oriented films. Such quantities also increased with increasing uniaxial draw ratio. Temperature gradient within the film that occurred during stretching has been assumed to be responsible for the above behaviour [231].

An interesting application of IRS–IR to composite materials has been presented [232]. A PP sheet was printed using mica plaques, to simulate a PP/mica

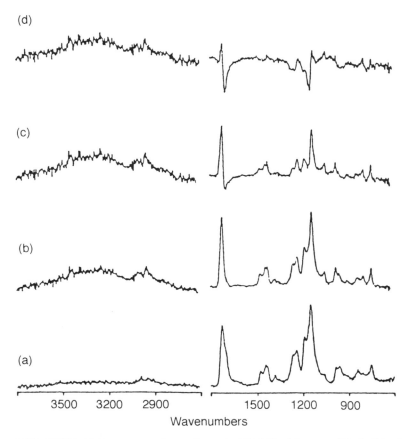

Figure 3.35 FTIR-ATR spectra of syndiotactic PMMA: (a) unmodified; (b) plasma treated PMMA; (c) positive spectral subtraction; (d) negative spectral subtraction. (From ref. [20].)

interface. The PP film was detached and examined by IRS–IR dichroism, observing that at the surface the polypropylene chains assume a preferential orientation, as if the crystallization had been nucleated by mica.

3.5   DIFFUSE REFLECTANCE SPECTROSCOPY

Diffuse reflectance spectroscopy (DRS) has long been used, mainly in the ultraviolet-visible (UV–Vis) region [233]. In recent years, diffuse reflectance in the IR region has become available. The introduction of Fourier transform instruments and improved optics have increased its attraction [234, 235]. Diffuse reflectance occurs when the spectrum is a function of both absorption and scattering events, when strongly scattering samples, such as powders, are examined. The intensity of the radiation arriving on the detector is a function of both the imaginary and real part of refractive index. Since an exact theory of the phenomenon is very onerous, various simplified treatments have been developed, that proposed by Kubelka and Munk [236, 237] has been the most successful. The Kubelka–Munk expression in its simplest form is:

$$K/S = (1 - R)^2/2R \tag{3.14}$$

where R is the diffuse reflectance as measured, and K and S are the empirically estimated absorption and scattering coefficients, respectively.

The Kubelka–Munk model has several limiting conditions:

(1) The incoming radiation is monochromatic and completely diffused (actually a low absorption and a low specular reflectivity are sufficient);
(2) The horizontal extent of the sample is infinite, as well as its thickness (again this condition is not essential);
(3) The sample is macroscopically uniform;
(4) The sample is not fluorescent;
(5) The optical medium in contact with the specimen has the same refractive index as the layer under examination.

In these conditions, since S is considered constant in the usual wavelength range, K is proportional to the absorption coefficient and consequently to the concentration of the examined species.

Other models have been proposed to interpret DRS spectra, particularly those of Pitts–Giovannelli and Rozenberg [238], but the Kubelka–Munk equation remains the most used, particularly because it is easy to handle.

Diffuse reflectance spectroscopy is not a surface technique, because it also reflects the bulk composition. It becomes a surface technique if there is a phase deposited on a substrate which does not give a DRS spectrum in the same region. The use of a shallow layer of reflecting powder (usually KBr) to make DRS partly surface-specific has been reported [239]. This is possible because of an initial scattering of the incident radiation on KBr. The radiation then reaches the material from many random incident angles, thus increasing the scattering at the KBr/sample interface. In coated materials there is a further scattering at the second interface. Because the radiation passes many times through the top layer, the sensitivity is rather higher than that of transmission techniques [240].

DRS can be used in the UV-Vis and IR wavelength ranges. In the latter case it is often called diffuse reflectance infrared Fourier transform spectroscopy (DRIFT). DRIFT, transmission infrared and FTIR-PAS (photoacoustic spectroscopy, see below) are suitable for similar problems. Transmission FTIR is particularly appropriate for studying adsorption phenomena (such as the silane promoter/filler interface in composite materials). DRIFT has the disadvantage of being instrumentally more complex and less direct in data handling, but it is more sensitive and does not require particular sample treatments. FTIR-PAS and DRIFT have in common the feature that samples can be observed without pretreatment. FTIR-PAS has some surface specificity (the observed layer is some μm thick) and is applicable to any type of sample. DRIFT provides more reproducible quantitative data and is by far more sensitive. In the UV-Vis region electronic states are studied, while in DRIFT vibrational states are observed.

The experimental arrangements relative to UV-Vis and DRIFT modes are outlined in Figure 3.36. The radiation source is similar to those used in the corresponding transmission spectroscopic techniques. The difference between the UV-Vis and IR arrangements lies in the collection of the scattered radiation. In the former (Figure 3.36 (a)) the reflecting sphere is coated internally with a highly reflecting material (e.g. MgO, NaF). The scattered radiation is then collected by a suitable photomultiplier.

In DRIFT, since the reflecting sphere is not efficient enough, more sophis-

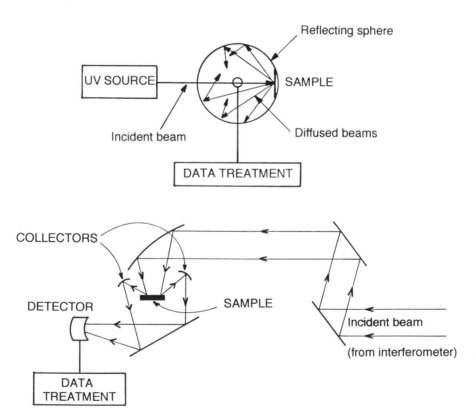

Figure 3.36 Experimental set-up for DRS: UV-visible mode (above); Fourier transform IR mode (below).

ticated devices are necessary. The use of hemispherical or ellipsoidal mirrors, with high focusing power, is now common (Figure 3.36 (b)).

For the presentation of results, it is necessary to account for the fact that Kubelka–Munk assumptions are not completely respected experimentally. From a quantitative point of view, it is important that the scattering coefficient is not constant, but depends on the wavelength, particularly when it becomes comparable to the dimension of the particles studied. Standards are necessary to correct deviations from the ideal Kubelka–Munk behaviour. The ratio of the reflectances observed for sample and for standard is used at each wavelength: $R' = R$ (sample)$/R$ (standard).

In UV-Vis DRS, different standards have been used (MgO, NaF, NaCl, etc.). In IR-DRS, KCl and KBr have been most used; small particles are preferred to optimize scattering.

In normal applications, the sample is examined in powder form. Applications to fibres and polymers (as films or plaques) have also been proposed. In Figure 3.37, the transmission IR spectrum of polycarbonate is compared with the first-derivative-like DRS spectrum of the same material, showing that intensity values are not maintained. In the case of powders, spectral characteristics (e.g. bandwidths and relative intensities) have been shown to depend on particle dimensions. Plots of K/S versus concentration are not linear when the sample is dispersed in a reflecting material (KCl) [234]. This effect has been verified at high concentrations. However, the technique can be used for quantitative studies [241].

The most common application to polymeric materials is the study of silane adhesion promoters on fillers for composite materials. Even if such an application is not strictly on polymers, it is useful to describe potential and limits of DRS. Because silanes increase the filler/matrix compatibility, it is important to study the silane-to-filler bonding, in order to understand the silane action and its effect on the properties of the material.

Such interactions have also been studied by transmission FTIR [242]. With respect to such a technique, DRIFT is much more sensitive and samples should not be compacted in KBr pellets, so preserving the physicochemical properties of the adsorbed layer. Furthermore, samples with surface specific area (SSA) lower than $0.5\,\mathrm{m^2 g^{-1}}$ can be easily studied. The observation of 3.68 mg of silane adsorbed on 1 g of glass with $0.01\,\mathrm{m^2 g^{-1}}$ SSA has been reported [235]. The possibility of studying small quantities of silane allows operation with silane concentrations similar to those normally used in application.

A study of silanes on glass fibres [243] has shown the importance of the orientation of the sample with respect to the incident radiation. Fibres scatter radiation anisotropically, so that the intensity of scattered radiation shows an angular dependence. The latter can be partly eliminated by averaging spectra obtained at different orientations. However, the most efficient way to overcome the problem is to cover the fibres with a shallow layer of reflecting powder [239].

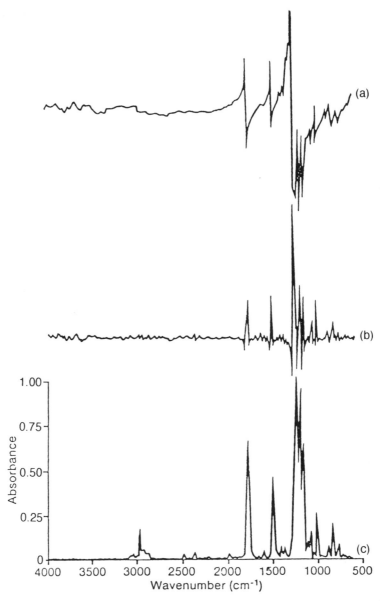

Figure 3.37 DRS spectrum of polycarbonate (PC): (a) first-derivative-like reflection spectrum of PC; (b) differentiated spectrum of (a); (c) transmission IR spectrum of PC. (Reproduced from ref. [248] by permission of Butterworth Heineman Ltd (c).)

In a study of the adsorption of 3-aminopropyltriethoxysilane (APS) on oxidized silica, quantitative data about the bonding of the silane have been obtained [244]. Of particular interest are the following spectral ranges: $3800-3000\,cm^{-1}$ (SiO–H and N–H stretch), $1800-1600\,cm^{-1}$ (NH_2 bending) and $1200-800\,cm^{-1}$ (Si–O–Si stretch). An adsorbate coverage lower than a monolayer has been detected. The amino group of APS is important, because it forms hydrogen bonds with surface hydroxyl groups. An important feature of DRIFT spectra is the observation of Si–O–Si bridges between silane and surface. At higher coverages silane–silane Si–O–Si bridges are also formed. The two situations have been distinguished analysing the spectrum of APS physisorbed on KBr [244].

A comprehensive series of organic functional groups supported on silica gel have been observed [245]. The matrix background was eliminated using spectral subtraction techniques, with acceptable semiquantitative results.

DRIFT spectroscopy is not really a surface specific technique, but for polymeric films some variation of sensitivity with depth can be obtained adding at the surface a reflecting powder, e.g. KBr [239–240]. In the case of superimposed films, e.g. polyvinylfluoride (PVF) on PET, the spectrum of the top layer was found to predominate when this technique was used [240]. By varying the

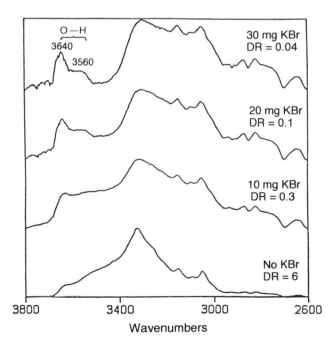

Figure 3.38 DRS spectra of Kevlar^R fibres recorded with increasing amounts of KBr. (Reproduced by permission of the Society for Applied Spectroscopy from ref. [246].)

amount and particle size of KBr for a PMMA film the movement of surface carbonyl groups to lower frequencies was observed to be due to hydrogen bonds with adsorbed water [240]. The effect of increasing the KBr concentration allowed the observation of the water absorption on Kevlar fibres exposed to water vapour (Figure 3.38) [246].

DRIFT has some capability for depth profiling, but the dependence of intensity with depth is not clear. Furthermore, strong alterations of the relative intensity of bands have been observed. The progressive oxidation of a poly(dimethylfulvalene) powder showed an increase with time of oxygen containing functions (OH, C=O, etc.) [234]. Studies by DRIFT of organic compounds in industrial materials, without altering the sample, have been reported [247]. The method was particularly effective for powders and foams.

In contrast, the analysis of polymer films can be difficult because specular reflection is stronger than diffuse reflectance, so that the Kubelka–Munk model cannot be used for interpretation. PMMA films on PE as thin as 10 nm have been detected [248]. Because the log–log plot of the absorbance of the PMMA carbonyl band at 1730 cm^{-1} as a function of the PMMA layer thickness was found to be linear, some chances to use DRIFT for quantitative analysis of thin layers remain open [248].

Diffuse reflectance has been used for orientation studies of drawn PET. Even when the specular component predominates, information has been obtained on the molecular orientation of film or fibre surfaces [249].

3.6   PHOTOACOUSTIC SPECTROSCOPY

Photoacoustic spectroscopy (PAS) is based on evaluation of the acoustic signals produced by the conversion to thermal energy of the energy absorbed by a solid irradiated with a modulated electromagnetic wave [250–253]. The technique is mostly used for solid, thick and opaque samples. The sample is introduced in a gas-filled cell, one wall of which is transparent to incident radiation. The photoacoustic signal produced is detected by a microphone.

The modulated electromagnetic wave hits the sample and is partly absorbed. The attenuation can be expressed according to the Lambert–Beer law. The optical penetration is usually expressed by the reciprocal of the absorption coefficient at the particular wavelength. The absorbed energy is totally converted into heat in the solid. The theory of PAS deals with the quantity of thermal energy arriving at the surface of the solid [250–252]; this energy flux heats the gas layer in contact with the surface. Because the incident radiation is modulated with a frequency Ω, an acoustic wave with the same frequency is generated.

The thickness of the solid layer contributing to the photoacoustic signal is characterized by the thermal diffusion length μ:

$$\mu = (2\tau)/(\sigma C\Omega) \tag{3.16}$$

where τ is the thermal conductivity, σ the density, C the specific heat and Ω the modulation frequency. The original theory provided an unidimensional model of the periodic heat flux from the solid to the gas [254]. The measured signal is given by:

$$P = Q \exp[-i(\delta + \theta)] \qquad (3.17)$$

where

$$Q = q \exp(-i\Phi) \qquad (3.18)$$

$$q = K\alpha\mu[2/(\alpha\mu + 1)^2 + 1]^{1/2} \qquad (3.19)$$

$$\Phi = \arctan(1 + 2/\alpha\mu) \qquad (3.20)$$

Only Q determines the spectrum; δ and θ are phase factors related to the production of the signal and to the instrument, and their values can be obtained examining a standard material; q is the modulus (equation (3.19)) and Φ the phase (equation (3.20)). K is a complex constant accounting for various factors. Both modulus and phase are functions of the absorption coefficient, α, and of the thermal diffusion length, μ.

It is possible to obtain information from a photoacoustic spectrum if spectral parameters, namely intensity q and phase Φ are functions of α, i.e. of the incident wavelength. Saturation occurs when q is not a function of the wavelength: if $\alpha\mu \gg 1$, equations (3.19) and (3.20) transform to equations (3.21) and (3.22):

$$q = K\sqrt{2} \qquad (3.21)$$

$$\Phi = \arctan(1) = \pi/4 \qquad (3.22)$$

When a saturating material is used, the signal intensity can be normalized, thus eliminating possible variations of incident radiation intensity as a function of wavelength. The instrumental phase factor $(\delta + \theta)$ can also be obtained.

Because the absorption coefficient is greater in the presence of strong chromophores and for electronic transitions, the photoacoustic effect is more frequently used in the IR range. Furthermore, FTIR instruments offer advantages both in calibration and speed of data collection. The measured quantity is usually q. If $\alpha\mu \ll 1$, the valid equation is:

$$q = K\alpha\mu[1/(1 + \alpha\mu)]^{1/2} = K\alpha\mu \qquad (3.23)$$

Plots of q versus α show a linear trend, implying that q is linearly related to the concentration of the examined species. This is valid over limited ranges.

In materials presenting strong scattering, the assumption that the absorption of radiative energy of the material depends only on the complex absorption coefficient is not valid. Instead of q, a complex function taking into account scattering events, which alter the radiative intensity inside the sample has been proposed [255]:

$$K\alpha\mu = 2^{1/2}q/[(1 + R)\sin(\Phi - \pi/4)] \tag{3.24}$$

where R is the diffuse reflectance expressed according to Kubelka and Munk [236, 237].

The thermal diffusion length depends only on the physicochemical characteristics of the sample and the modulation frequency (equation (3.16)). In a polymeric material, at normal modulation frequencies, thermal diffusion lengths are micrometres or tens of micrometres. The surface specificity is equal to or less than in infrared IRS. As the thermal diffusion length is inversely proportional to the modulation frequency, increasing the latter decreases the depth of the observed layer. However, to observe a layer of material 100 nm thick, the modulation frequency would have to be in the kHz range.

The above information refers to solid, thick and opaque samples. Thin samples and thick transparent samples have also been considered. In the first case, if the thermal diffusion length is greater than the film thickness, the spectrum depends on film thickness, absorption coefficient and modulation frequency. In the second case, observed intensities depend on thermal diffusion length, absorption coefficient and the value of $\Omega^{-3/2}$ [254].

Obviously, depending on the wavelength used, different information is obtained. The more common case concerns infrared radiation, in this case vibrational information is obtained.

The radiation source for photoacoustic spectroscopy is not particularly different from that used for transmission or reflection IR spectroscopy. In fact, the photoacoustic cell is commonly sold as an accessory of IR spectrometers, both dispersive and Fourier transform. FTIR can collect and average spectra very quickly, overcoming sensitivity problems. The heart of the equipment is the photoacoustic cell itself, the manufacture of which is a very delicate task. A typical apparatus is outlined in Figure 3.39. The cell is built with the minimum volume allowing the introduction of the sample, since the amplitude of the photoacoustic signal is inversely proportional to its dimensions. The cell should be as free as possible from mechanical vibrations, that cause spikes in the spectrum. The electric signal detected by a sensitive microphone is amplified and processed to yield the spectrum.

The purity of the transport gas is important. Problems are posed by the presence of water or carbon dioxide that are not easy to remove, and which have strong infrared absorptions and give large photoacoustic responses. Several gases have been used to transport the signal, good results have been obtained with helium, which offers a good thermal contact with the sample [256].

Numerous applications of PAS to polymer surfaces have been reported [196, 250, 252, 256, 257]. An important feature is that the sample surface can be analysed in any form. The dependence of the spectrum on the sample shape [258] and, in the case of powders, on the particle size [259] has been studied.

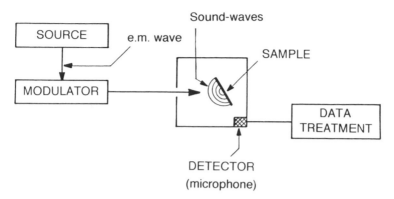

Figure 3.39 Experimental set-up for FTIR–PAS.

The spectra of a commercial polymer sample observed in the form of smooth surface, pellets and powder were found to be similar [258], some increase of intensity has been observed in powder samples. A clear dependence of the photoacoustic spectrum on particle size was demonstrated, particularly when the latter is comparable with the thermal diffusion length [259]. In general, spectral contrast is better with smaller particle size, but the reproducibility worsens. The reason lies both in the occurrence of scattering phenomena and in the dependence of the signal intensity on the radiation pathways. In fact, the radiation induces a photoacoustic signal while both entering and leaving.

From a quantitative point of view, problems arise from the limited range of linearity of q versus concentration. The use of carbon black as a standard to normalize spectra can itself induce errors, because such material tends to adsorb water [260]. Recently, linear computation of FTIR–PAS spectra has been proposed in order to allow better quantitative results [261]. Spectra similar to those of transmission infrared spectra can be obtained, and the presence of surface contaminants is easily detected.

When compared with infrared IRS and DRIFT, FTIR–PAS is less direct from an instrumental point of view. However, any type of sample, even thick ones, with weak scattering or with a rough surface can be examined. A comparative study of infrared IRS and PAS dichroism techniques has been reported, for determining surface orientation in films of drawn PET. FTIR–PAS allowed the measurement of dichroic ratios, although values were smaller than in infrared IRS–IR, perhaps because PAS gives more penetration [262]. In a combined study of incompatible blends by the above two spectroscopic techniques, it was shown that photoacoustic spectra do not depend on sample morphology, but pose problems if used for quantitative purposes [217].

As in infrared IRS, it is possible to perform depth profiling by changing the modulation frequency. A study of two-layer samples of epoxy resin on PP and

PET has been reported [263]. Some results are reported in Figure 3.40, at different thicknesses of the epoxy top layer. At a thickness of 5 μm the substrate bands are clearly visible. Spectral subtraction enabled the spectra of both the top layer and the substrate to be obtained. For depth profiling purposes, the use of step-scan acquisition of FTIR–PAS spectra was reported [264]. Because of in step-scan FTIR a single modulation frequency is used, this implies that the thermal diffusion length remains constant during the spectrum collection. Increasing the modulation frequency, signals coming from the top layer are maximized. Furthermore, lock-in detection allows an easy extraction of the

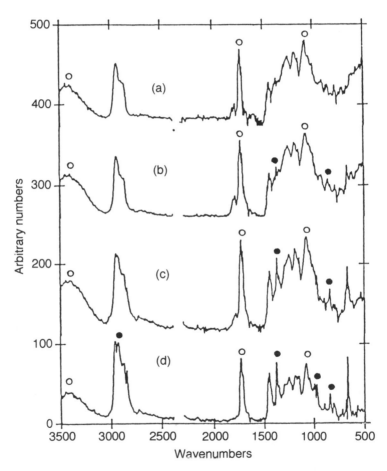

Figure 3.40 FTIR/PAS spectra of an epoxy coating on PP film; the thickness of the coating is: (a) 23 μm; (b) 9 μm; (c) 5 μm; (d) 3 μm. Open and full circles represent the characteristic peaks of epoxy and PP, respectively. (Reproduced by permission of the Society for Applied Spectroscopy from ref. [263].)

signal phase, thereby allowing more precise information on the depth of a specific layer [264]. Surface layers of 5 μm or less can be conveniently separated.

The effects of thermal oxidation on phenolic resins have been studied [265]. Bands assigned to carbonyl groups appear, because of the oxidation of methylene groups. Transmission spectra suggest a constant increase of oxidation with time, while infrared PAS spectra indicate saturation of the surface after a specific length of treatment. Analogous effects have been observed in the plasma oxidation of a styrene-divinylbenzene copolymer [266]. The occurrence of carbonyl and/or carboxyl (1750–$1700\,cm^{-1}$) and ether-ester (1350–$1000\,cm^{-1}$) signals was observed.

Ultraviolet PAS studies of silicas on which functional groups have been anchored have been reported [255, 267]. In silica functionalized with trimethyl-silyl groups, the ratios of the CH_3 rocking bands to the Si–O stretch were shown to be proportional to the coverage [268]. Ultraviolet PAS has also been used in studying the chemical effect of mechanical degradation of PVC. The formation of conjugated double bonds as a consequence of HCl elimination by a radical mechanism induced by mechanical stresses was shown [269].

Photothermal spectroscopy is strictly related to photoacoustic spectroscopy [251]. In this case, the energy of the incident electromagnetic radiation is converted to thermal energy. The latter induces radiative emission at the surface, which is detected.

3.7 OTHER VIBRATIONAL TECHNIQUES

Several other vibrational techniques have been devised to solve problems, in surface analysis, particularly for the study of thin films. Polymer films have been studied by using external reflection infrared techniques and emission infrared techniques. Inelastic electron tunnelling spectroscopy and surface-enhanced Raman spectroscopy are also relevant.

Finally, the possibility to apply electron-induced vibrational spectroscopy to polymer surfaces has been recently demonstrated.

3.7.1 External Reflection Spectroscopy

Like IRS, external reflection spectroscopy (ERS) can be applied in different regions of the spectrum. In the case of polymer films on metals, mainly infrared ERS has been applied. Several books and reviews have been published on this argument [270, 271]. The sample arrangement and the optical phenomena involved in obtaining a spectrum are outlined in Figure 3.41. The electromagnetic wave passes from the first medium (air atmosphere) to the second medium (polymer) by refraction. This is followed by a reflection at the metal/polymer interface and finally by a second refraction at the polymer/air interface.

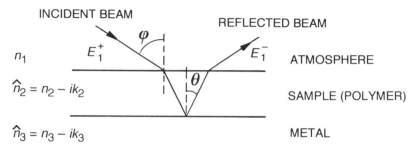

Figure 3.41   The physical basis of ERS.

The theory of ERS (also called reflection-absorption spectroscopy) involves the expression of reflectance (E_1^- / E_1^+, Figure 3.41) as a function of several parameters, including the refractive index of the polymer, the frequency of the incident radiation and the thickness of the polymer film [272–274]. Given the correct values for such entities, a theoretical calculation of reflectance is possible. Conversely, when the experimental spectrum is obtained, the thickness and optical constants of the observed layer can be evaluated, within some precision limits. Ellipsometry has been used to follow the thickness and refractive index of growing thin films [96, 275, 276]. In ERS the whole layer is observed, so the technique is not specific for surfaces. An important phenomenon, particularly when very thin films (less than 10 nm) are examined, is the phase variation of the normal component of the radiation caused by the reflection on the metal. Such a variation, of 180°, induces suppression of the radiation component normal to the metal surface. The dependence of reflectance on incidence angle is shown in Figure 3.42. No particular sensitivity problems are encountered in the analysis of transparent films. Non-polarized radiation is directed at an angle of about 45° [270, 271]. In the case of very thin films, it is necessary to work at maximum reflectance, with an incidence angle near 90°. At such angles, illumination of the sample tends to decrease, so that the best compromise between illumination and maximal reflectance must be sought [270]. In Figure 3.43, the PMMA band at 1730 cm^{-1} is shown, compared with the analogous ones obtained by transmission infrared and ATR spectroscopies. The thickness of the film was about 60 nm, the substrate a copper foil [271].

In an infrared ERS study of polyacrylonitrile on silver, the optical constants were determined and the importance of using polarized radiation for studying thin films ($< 0.5\,\mu$m) was emphasized [277].

ERS requires only conventional infrared instrumentation, apart from the optics necessary to ensure that the radiation follows the appropriate pathway (Figure 3.41). Suitable devices have been developed to study *in-situ* reactions [278–280]. A double-beam arrangement has been suggested [281], in order to subtract the clean substrate contribution (Al in the specific case) from the sample spectrum.

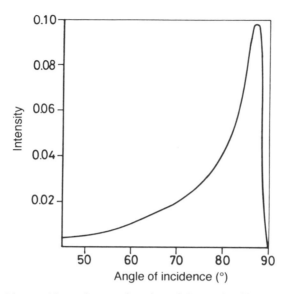

Figure 3.42 Observed intensity as a function of the angle of incidence in IR–ERS.

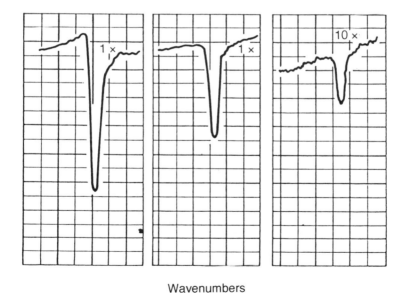

Wavenumbers

Figure 3.43 Band at $1730\,\text{cm}^{-1}$ of a PMMA film (thickness 57 nm) obtained with transmission FTIR (right), ATR–IR (centre); IRS–IR (left). (Reproduced by permission of Elsevier Science Publishers BV from ref. [271].)

The problem of sensitivity has been worked out by different approaches, using more sensitive detectors or multiple reflections. Operation at a correct angle is essential. Experimental problems encountered at angles near 90° in grazing-angle infrared reflection spectroscopy have been discussed [282, 283].

Among the applications of infrared ERS, a study of an epoxide adhesive on aluminium produced no evidence for strong aluminium/resin interaction [281]. In the case of a methacrylate adhesive on Al, the shape and position of the $C=O$ band were shown to depend on the film thickness. This result was attributed to interactions of the carbonyl group with the aluminium oxide/hydroxide layer present at the interface [284].

Raman spectra have been obtained with similar optics. When thin films are anisotropically oriented, this technique can be used to obtain information about the degree of conformational order/disorder in polymer films on metals [285].

3.7.2 Emission Spectroscopy

The emission of radiation by polymer films deposited on metals has been studied mainly in the infrared region, although it also occurs in the UV-visible range [286]. The experimental set-up is sketched in Figure 3.44. A polymer film is heated through the metal on which it is deposited. The thermal energy favours transitions from the fundamental energy level to excited levels. The consequent decay can cause the emission of a photon. The spectrum is very similar to that obtained by transmission infrared spectroscopy [287, 288]. Emitted photons are collected and analysed conventionally. The theory of this type of emission spectroscopy has been discussed [289, 290]. Waves emitted and reflected by the metal surface can interfere destructively, so that the emission intensity depends on the angle of observation. This is analogous to the effect observed in ERS (Figure 3.42), but the curve is smoother. Reasonable emission intensities are obtained even at angles near 50° [290–292]. It has been calculated [290] that the maximum emission occurs at an angle of 85°, while at 55° the emission is decreased by one-third.

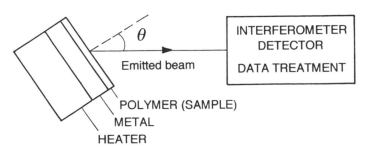

Figure 3.44 Experimental set-up for IR emission spectroscopy.

Another analogy with ERS is the dependence of the observed intensity and polarization of the emitted wave on the conformation of functional groups with respect to the surface. As in ERS, functional groups parallel to the surface do not provide detectable emission [290]. As a consequence, information about the orientation of macromolecules on the surface can be obtained [291].

Emission experiments can be done with both conventional infrared [293] and FTIR instruments. In the latter case, sensitivity problems are reduced by fast collection and averaging of a large number of spectra. Greater sensitivity can be achieved by using more efficient detectors, such as mercury cadmium telluride at liquid nitrogen temperature [294]. For thick films there are no problems of sensitivity, but self-absorption of emitted radiation has been reported [294], altering the relative intensities of the peaks.

Stray radiation can be eliminated by subtracting the spectrum of the beam splitter (when a FTIR instrument is used) and by using the spectrum of the clean metal surface as the standard. The emission as a function of the intensities pertaining to the sample (I_s), the beam splitter (I_b), and the standard (I_M) is given [292] by:

$$E = (I_s - I_b)/(I_M - I_b) \tag{3.25}$$

Emission spectroscopy is competitive with infrared ERS, with respect to obtainable information and the objects studied. In fact, it provides information even in the presence of a rough metal surface. Disadvantages are its lower sensitivity and increased instrumental complexity. Examples of applications are a semiquantitative determination of the thickness of a styrene-acrylonitrile copolymer film on aluminium [262] and the study of the spectrum of PET on aluminium [288].

3.7.3   Inelastic Electron Tunnelling Spectroscopy

Inelastic electron tunnelling spectroscopy (IETS) is an electroanalytical technique which can provide vibrational, rotational and electronic information, even though it has been used mainly in the vibrational mode.

Figure 3.45 shows how a sample for IETS is set up. A layer of polymeric material (i.e. adsorbate, adhesive, coating or else) is deposited on an insulator, at both sides of which suitable metal electrodes are bound. An increasing potential is applied to the electrodes, usually in the 0–500 MeV range. The system is kept at low temperature (< 4.2 K) to maintain all functional groups at the vibrational ground state and to allow superconductivity of the electrodes [295]. When the potential is applied, electrons flow between the electrodes. Elastic tunnelling occurs when electrons pass through the insulating layer without interactions, that is without energy loss. On the other hand, inelastic tunnelling is achieved by losing energy for the interaction with a vibrational mode of energy $h\Omega$; equation

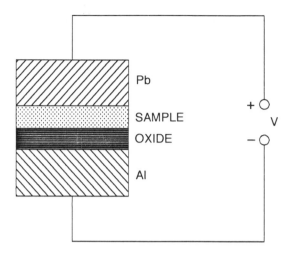

Figure 3.45 Schematic representation of sample for IETS.

(3.26) must be satisfied:

$$E_1 + eV = E_2 + h\Omega \tag{3.26}$$

The physical basis of the phenomenon is sketched in Figure 3.46.

Inelastic tunnelling induces limited (0.01–1%) variations of the cell conductance. It has been shown [296] that for electron/electric dipole interactions, the second derivative of the current I with respect to the potential V is equivalent to the vibrational spectrum:

$$d^2I/dV^2 \in \| <i\|\mu_z\|0> \|^2\delta(eV - h\Omega) \tag{3.27}$$

In the case of electron-induced dipoles, there is an analogy with the Raman spectrum. It is possible to measure d^2V/dI^2, which is proportional to d^2I/dV^2, by adding a low oscillating potential (some 0.3–1.5 MeV) to the ramp potential.

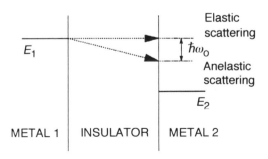

Figure 3.46 Physical basis of IETS.

The measured quantity is the second harmonic of the potential. Plotting d^2I/dV^2 versus V gives a spectrum containing both infrared and Raman bands [295, 296]. Because the bandwidth is proportional to kT, experiments are generally carried out at liquid He temperature. Sample preparation is critical. The thickness of the insulator-coating layer is very important because it governs the junction resistance [295, 297]. The first electrode, usually Al, is vapour-deposited on a support. One of the surfaces is oxidized to produce an oxide layer of controlled thickness (a few nanometres). The sample coating is then applied, either from the vapour phase or from solution (the thickness is even less than that of the oxide). Lastly, a lead electrode is deposited from the vapour phase, electric contacts are attached and the sample cell is dipped in liquid He.

The sensitivity of the technique is quite good for studies on very thin films. However, the effects of the deposition of lead electrodes are not well known. Equation (3.26) provides the relationship between the meV and cm^{-1} scales. Both infrared and Raman information can be obtained and both resolution and sensitivity are good, particularly in comparison with ERS or emission techniques.

The silane/aluminium oxide interaction has been studied, with regard to the promoter/filler bonding topic [298–300]. Studies of the adsorption of methylphosphonic acid on alumina are also reported [301]. In the case of 3-aminopropyltriethoxysilane (APS), the NH stretching band was found to be broadened and shifted to lower frequencies, suggesting a hydrogen bond of the amino group with an oxygen atom either of the substrate or of a Si–O–Si bridge [298]. The IET spectrum of APS is compared in Figure 3.47 with the corresponding infrared spectrum [302]. The Si–C stretching bands were particularly intense, suggesting that Si–C bands are normal to the surface. In fact, in IETS the exciting electric field is longitudinal [295], not transverse as in infrared. For 3-glycidyloxypropyltrimethoxysilane [299, 303], bands apparently related to C$=$C and C$=$O double bonds were observed; they were attributed to the reaction of the epoxide function, giving double bonds with water elimination. It has been shown that monoalkoxysilanes are not adsorbed on alumina whereas trialkoxysilanes are readily adsorbed [300]. The reaction in anhydrous conditions involves the surface hydroxyl groups present on alumina. The formation of Al–O–Si bonds is probably stabilized by the subsequent condensation of chemisorbed silane molecules to form siloxane oligomers [300]. Similar studies were extended to several sylanes containing C$=$C bonds, showing a different behaviour in dependence on the method of preparation of the insulating film [304].

IETS studies on adhesion phenomena, by analysing thin adhesive films on aluminium have become quite popular. The use of epoxide films [305], methacrylates [306, 307], and polyvinylalcohol [307] has been reported. In the reaction between a multifunctional amine and bifunctional epoxide mixture, it has been shown that the amine nitrogen tends to pick up hydrogen atoms from the oxide surface. Al–O–C bands are formed, with the elimination of Al–OH groups [305]. Polymer films of poly(vinylacetate), poly(vinylalcohol), PMMA and several

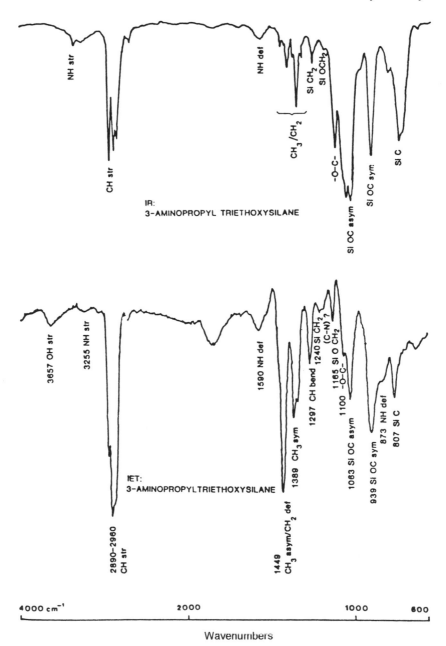

Figure 3.47   IR (above) and IETS (below) spectrum of a silane. (Reproduced by permission of Elsevier Science Publishers BV from ref. [302].)

polymers obtained by plasma polymerization [308] have been studied on aluminium oxide. Poly(vinylacetate) undergoes hydrolysis of ester groups, and the polymer chemisorbs as poly(vinylalcohol). Presumably carboxyl absorption is due to acetates formed by hydrolysis [306]. For the interaction with aluminium oxide, the formation of Al–O–C bridges has been suggested, even though bonding by hydrogen bonds between hydroxyl groups of the polymer and at the oxide surface has not been completely ruled out [307]. Polymers with ester side groups were found to be subjected to a high level of ester cleavage, producing a carboxylate anion and an alcohol [309].

3.7.4 Electron-induced Vibrational Spectroscopy

Electron-induced vibrational spectroscopy is basically an ultra-high vacuum external reflection spectroscopy. It is generally known as high resolution electron energy loss spectroscopy (HREELS) and studies the backscattered fraction of a monochromatized low energy (1–10 eV) electron beam [310]. A very high performance electron energy analyser, analogous to those used in XPS, is necessary for collecting the back-scattered beam, which contains vibrational peaks. Widely used in studies on the adsorption of small molecules on metal or semiconductor surfaces, HREELS is very promising in polymer surface studies. Its surface sensitivity is high (it probes the first 2.5 nm), it is especially sensitive to the light elements involved in chemical bonds, hydrogen included [311]. Resolution and charging problems of the specimen under irradiation might be solved in order to increase its field of application. Thin polymer films deposited on metal substrates were demonstrated as suitable for analysis.

HREELS is able to give unique information on the molecular structure of the surface of polymers. Chain-end segregation at the surface has been demonstrated in the case of poly(ethylene) by the strong intensity of the bending vibration of CH_3 at $1370 \, cm^{-1}$ [312]. Useful information was also obtained for chain segregation at the surface in the case of poly(styrene) (PS) and deuterated-PS [311], for molecular orientation during metallization of a poly(imide) [311] and for polymer surface tacticity. The latter has been studied in the case of PS, showing a different amount of the C–H stretch band of the aromatic ring, which suggests a relative enrichment of phenyl groups at the surface of the isotactic material [313], as well as in the case of isotactic and syndiotactic PMMA. Electron-induced vibrational spectra of both polymers are shown in Figure 3.48. For a carboxyl stretch band of similar intensity at 210 MeV (1 MeV = $8.066 \, cm^{-1}$), the syndio PMMA presents a lower C–H stretch band at 370 MeV and also a lower intensity of the CH_3 and CH_2 deformation band regions around 175 MeV. This suggests a larger $C{=}O/CH_n$ ratio on the syndiotactic polymer surface [313].

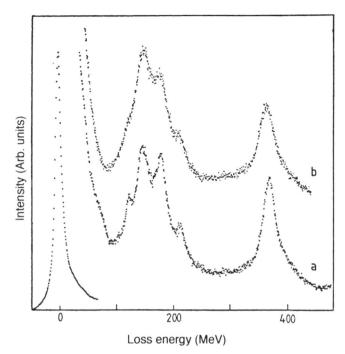

Figure 3.48 Electron-induced vibrational spectra of PMMA: (a) isotactic; (b) syndiotactic [311]. (Reprinted with permission from Pireaux *et al.*, *Langmuir*, **7**, 2433. Copyright (1991) American Chemical Society.)

3.7.5 Some Rarely Applied Vibrational Techniques

Several other techniques have been developed for the study of thin films. Infrared surface electromagnetic wave spectroscopy (SEWS) has been applied to the study of organic adsorbates on metal surfaces [314]. Surface picosecond Raman gain spectroscopy and waveguide Raman spectroscopy have found similar applications [315].

Surface-enhanced Raman scattering (SERS) uses the strong increment in intensity of Raman bands associated with functional groups of molecules near (5–20 nm) a rough metal surface, usually constituted by silver [316–318]. The applicability of SERS to polymer surfaces has been reported in a study on poly-*p*-nitrostyrene [316], and in a study on the effects of chromic/sulphuric acid etching on poly(ethylene) films [319]. The desired surface sensitivity is obtained, but the sample preparation is complex. Thin films of pthalocyanine complexes on silver have been also studied [318].

Another technique, rather difficult from an instrumental point of view, is waveguide Raman spectroscopy [320, 321]. In order to achieve satisfactory sensitivity, it uses the Raman resonance effect and the conduction of radiation obtained with a suitable sample preparation [320]. A study of a cyanine dye on poly(vinylalcohol) showed a strong interaction of the hydroxyl groups of the polymer with nitrogen atoms of the quinoline ring of the cyanines. Such an interaction has been considered responsible for the polymer/dye adhesion [321]. More details of Raman studies on structure and orientation in thin films can be found in a review recently published [322].

3.8 TRANSMISSION SPECTROSCOPIES

The theory and instrumentation related to transmission vibrational spectro-scopies will not be discussed here. Both the infrared and Raman techniques are well known and widely diffused, and are now mainly in the Fourier transform version (FTIR) [323, 324].

Transmission techniques observe the whole material, consequently their appli-cation to surface problems is limited to the cases where surfaces and/or interfaces are very broad. In fact, they have been applied in studies of the bonding of silanes on fillers with high surface specific area or in compatible blends. Silanes have been studied by infrared [242, 325–330], Raman [331] and ultraviolet modes [332]. In most cases the substrate material had a surface specific area greater than 100 m²/g. Samples investigated were mostly in the form of KBr pellets or thin layers of silica compacted under high pressure. In the latter case scattering problems could arise.

The structure of 3-aminopropyltriethoxysilane (APS) on silica has been studied. The silane, according to the preparation conditions, is present as a chelate ring or as an extended chain. In the first case amino groups interact with silica [325]. The structure of APS on silica is particularly sensitive to the pH of the treating solution. At low pH values, a form with protonated amine predomi-nates, while at high pH the previously described structures are present [242].

The use of FTIR instead of DRIFT means less experimental complexity and a more easily interpretable spectrum. Disadvantages are lower sensitivity and the necessity to use compact samples; such a treatment can alter the original sample.

Characterization of the reaction at the silane/polymeric matrix interface by FTIR has been reported [333]. The reaction was studied with and without filler. The reaction of APS is strongly affected by the degree of condensation of the silane at the interphase. Interface studies are important also for compatible blends. FTIR was used to examine why a blend is compatible and which intermolecular interactions are involved in compatibility. The study of conforma-tional and morphological effects is also important.

FTIR is much more precise than dispersive infrared. The signals obtained are particularly suitable for digital treatment, so several methods of data treatment have been developed. Spectral addition and subtraction, curve fitting, factorial analysis and derivative spectroscopy [334] can be mentioned. Least-squares fitting of spectra has been used to calculate the sum of single component spectra which best reproduces the spectrum of a polymer blend [335]. Factorial analysis, in a substantially equivalent approach, has been used to determine the different components producing a spectrum [336]. Derivative spectroscopy [337] has been used to facilitate spectral interpretation, particularly when a partial overlapping of bands occurs.

Finding intermolecular interactions in compatible blends is easier and unequivocal whenever the observed functional groups have uncoupled vibrational modes. Much work has been focused on blends where one of the moieties contains a carbonyl or carboxyl group [334]. In poly (caprolactone)/poly(vinyl chloride) (PCL–PVC) blends, the carbonyl group of PCL has been found to interact with the methine hydrogen of PVC, to form a hydrogen bond. Much more evident is the interaction in systems with hydroxyl groups, as in poly-(vinylphenol) (PVPh) [334]. Actually, poly(propiolactone) forms blends with PVPh but not with PVC.

It is always necessary to be cautious in the interpretation, because intermolecular interactions can cause band shifts of some cm^{-1} and/or broadenings. Furthermore, miscibility also depends on other factors, such as molar concentration of interacting species, distribution of molecular weights, steric considerations, etc. In blends of polystyrene with poly(ethylene oxide) or poly(vinyl methyl ether), which do not contain functional groups with vibrational modes uncoupled with the skeleton, interpretation is difficult. It is not possible to neglect morphological effects or to draw unambiguous conclusions about the presence of intermolecular interactions. Several studies on blend morphology have been reported [334]. A broad range of materials with different fractions of amorphous and crystalline domains has been examined when particular bands can be associated with crystallinity.

3.9   NUCLEAR MAGNETIC RESONANCE SPECTROSCOPY

Nuclear magnetic resonance (NMR) spectroscopy is not a surface technique, however, applications to surface problems have been made under conditions similar to those for transmission infrared spectroscopy. NMR can be used when there is a clear-cut distinction between surface and bulk signals. As NMR has poor sensitivity, samples with a large surface specific area are necessary.

Application of NMR to surface problems has been reviewed [338, 339]. The silane/filler interaction has been treated in detail [339]. As with infrared spectroscopy, NMR can be applied where a diffuse interface is present, as in compatible blends.

NMR spectroscopy can provide very rich information with regards to vibrational and electronic structure and to stereochemistry, as well as to the state of motion in the system. Of course, different nuclei can be observed and different experimental techniques are available [340–344]. Both NMR theory [340, 341] and instrumentation [342] are well known, equation (3.28) is just a reminder of the expression of the spin hamiltonian, showing the possibility of obtaining various NMR parameters:

$$\hat{H}_{NMR} = \hat{H}_{CS} + \hat{H}_J + \hat{H}_{DD} + \hat{H}_Q \qquad (3.28)$$

Terms reported in equation (3.29) refer to chemical shielding (CS), indirect coupling (J), direct coupling (DD) and quadrupolar coupling (Q). All physical phenomena determining NMR parameters need a tensorial description, particularly in the solid state. Also relaxation times, i.e. time constants characterizing the return of the system to equilibrium after excitation, are rather informative. Because the order of magnitude of NMR parameters (as frequencies) is similar to that of molecular motions, NMR can provide information about the state of motion.

Solid-state NMR is more complex than solution NMR, and relaxation times are less favourable in the solid state than in solution. Generally, a solid state NMR spectrum looks like a broad (even tens of kHz) and not easily interpretable band. The use of high power decoupling, magic angle sample spinning (MASS), and cross-polarization (CP) allows solid-state spectra to be obtained which approximate the spin hamiltonian valid for solution experiments [343]. Figure 3.49 shows a solid-state NMR spectrum of an ethylene/carbon monoxide copolymer.

High power decoupling, in the presence of protons coupled to the observed nucleus, helps to eliminate the effect of direct and indirect coupling constants. High power is necessary for solid samples to cover the bandwidth found in proton spectra [339].

MASS devices [343, 344] spin the sample at high rotation frequencies (3–8 kHz) around an axis forming an angle of 54.7° (magic angle) with the magnetic field direction. Several interactions (shielding tensor anisotropy, quadrupolar coupling, direct coupling) are thus averaged to zero and do not influence the spectrum. High power decoupling and MASS techniques give a 'liquid-like' spectrum, but some information is lost. Cross-polarization provides increased sensitivity by means of the Hartmann–Hahn effect [343]. Conventional NMR instruments can be adapted to solid-state studies by using CP-MASS without important modifications, while special instruments like wide-line NMR are required without CP-MASS [341].

Several NMR studies have been devoted to the bonding of silanes to silica gel surfaces, used as a model for the behaviour of glass fibres. Its specific surface area is high (> 100 m^2/g), suitable for NMR studies [339]. CP-MASS techniques with ^{29}Si have been used to study untreated silica gel surfaces [345, 346]. Hydration/dehydration phenomena were used to probe surface reactivity. Bulk/surface discrimination is provided by cross-polarization, which affects only

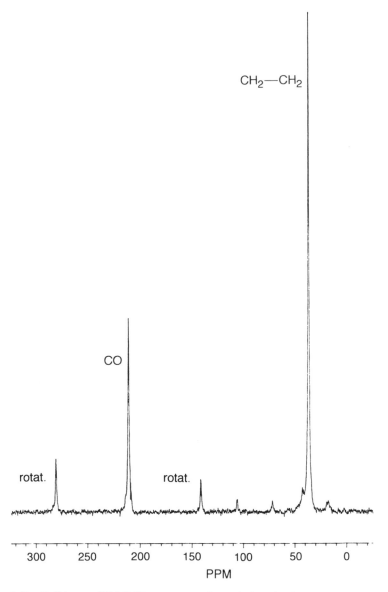

Figure 3.49   Solid state ¹H NMR spectrum of an ethylene/carbon monoxide polymer.

surface Si atoms vicinal to hydroxide groups. Such Si atoms can be observed with good sensitivity owing to their preferential relaxation, and their bonding to 0, 1 or 2 OH groups can be distinguished by the different chemical shift. The organic side of chemisorbed silanes has been studied by ^{13}C NMR [347, 348]. Measuring the chemical shifts in the organic side-chain of the silane made it possible to predict their values on the basis of simple addition rules, when the functional groups present in the chain are known, as in solution NMR [339]. Because NMR is particularly sensitive to molecular motions, side-chain motion has been studied. Various techniques have been used, such as taking into account the entity of cross-polarization, chemical shift anisotropy or T_1 and T_2 relaxation times [339, 347, 349, 350]. The amount of cross-polarization decreases for very long side-chains. The motion of end groups is rapid, making cross-polarization less efficient [347]. The motion of phenyl groups anchored on silanes bound to silica has been studied rigorously by using chemical shift anisotropy [349]. Proton T_1 and T_1 (spin locking) relaxation times for silanes with different side-chain lengths (8 and 18 carbon atoms) have been measured [350]. Shorter T_1 's and longer T_1 (spin locking) were found in the case of the shorter side-chain, suggesting a freer motion in short chains. Longer chains probably interact either with the silica surface or with other alkyl chains. Measurements of T_1 were also made in order to check the polymer chain mobility at the surface of labelled substrates [351].

An interesting application of imaging NMR to the absorption of water in composite materials has been reported [352]. Two epoxide matrices, reinforced with glass fibres, were kept under water for long periods and then analysed by ^{1}H-imaging NMR. The absorption processes observed depended on the properties of the composite material, particularly on the presence of defects providing penetration pathways for water. Imaging NMR, with improvement of lateral resolution, will offer many opportunities for non-destructive studies of interfaces.

Another application of NMR to surfaces/interfaces in polymeric materials is to compatible blends (diffuse interface). This topic has been reviewed [353, 354]. Miscibility in polymeric blends has not been studied using traditional parameters (chemical shift, quadrupolar coupling), probably because interactions between polymer chains are feeble, inducing only small variations. Relaxation phenomena are used [353], because the spin diffusion mechanism allows dispersion to be studied in limited ranges of the magnetization introduced in the system.

The use of proton T_1 (spin locking), obtained by ^{13}C measurements, has proved quite successful [355]. For polystyrene-poly(phenylene oxide) blends it has been observed that, when the blend is compatible, spin diffusion is quick and homogeneous in both polymers, giving rise to a single T_1 (spin locking is found). The latter is a one-dimensional NMR technique. 2-D techniques have been introduced, achieving spectra correlating two dimensions, on the basis of cross-relaxation caused by spin diffusion [356]. In polystyrene–poly(vinyl methyl ether) (PS–PVME) blends, when the blend was compatible, cross-peaks between aromatic protons of PS and methyl and methine protons of PVME were found, suggesting a reciprocal interaction [356].

^{13}C spin diffusion has been used to study PET/bisphenol A–polycarbonate blends [357]. Compared with ^{1}H spin diffusion, which has a range of a few nm, ^{13}C spin diffusion is more selective (tenths of nm), depending on the lower ^{13}C gyromagnetic ratio. Both ^{13}C T_1 's and 2-D spectra have been used to characterize the blend, suggesting the occurrence of some intermolecular reactions of these polymers [357].

3.10 COMPARISON OF SPECTROSCOPIC METHODS

Due to the large number of techniques now available for the analysis of polymer surfaces, a short discussion seems necessary in order to compare their advantages

Table 3.6 Comparison of spectroscopic techniques for the analysis of polymer surfaces

Technique	Qualitative information	Quantitative information	Depth profiling	Observed layer	Sensitivity*
ISS	Surface elements	Semiquantitative surface composition	No	1–2 layers	10^{-4}–10^{-6}
SSIMS	Surface functional groups	Difficult	No	< 1 nm	10^{-7}
XPS	Surface elements and functional groups	Surface composition, functional groups (semi-quantitative)	Yes, angle resolved	4–10 nm	10^{-2}–10^{-3}
IRS-IR	Vibrational spectrum	Semiquantitative using band ratios	Yes, change incident angle or IRE†	1–10 μm	about 10^{-2}
IRS-Raman	Vibrational spectrum	Semiquantitative using band ratios	Yes, change incident angle or IRE	0.01–0.5 μm	10^{-1}–10^{-2}
DRS-IR	Vibrational spectrum	Semiquantitative	Not controlled	Whole	10^{-3}–10^{-4}
IR-PAS	Vibrational spectrum	Not very dependable	Yes, change modulation frequency	1–50 μm	10^{-1}–10^{-2}
ERS-IR	Vibrational spectrum	Semiquantitative	No	Whole	about 10^{-2}
IR-Emiss	Vibrational spectrum	Semiquantitative	No	Whole	10^{-1}–10^{-2}
IETS	Vibrational spectrum (IR + Raman)	Semiquantitative	No	Whole	about 10^{-2}
SERS	Raman spectrum	Semiquantitative	No	5–20 nm	about 10^{-2}
IR	Vibrational spectrum	Quantitative	No	Whole	about 10^{-2}
NMR	Chemical shift, coupling constants, relaxation times	Difficult	No	Whole	depends on nature of nucleus

*For the definition of sensitivity, see text
†Internal Reflection Element

and disadvantages. This task is not simple for two main reasons. The first one is the continuous progress in instrumentation and computer treatment of data, which is changing constantly the figures used in comparisons, i.e. sensitivity, resolution, etc. Furthermore, the suitability of a particular technique depends on the specific problem under examination, for instance the characterization of a rough, real surface of a massive specimen must be treated differently with respect to the analysis of a thin polymer film accurately prepared in the laboratory on a proper substrate, or of an interface in a composite material or polymer blend.

Generally speaking, true surface problems must be tackled by a technique of sufficient surface sensitivity, i.e. an ion or electron spectroscopy; vibrational spectroscopies are suitable for thin films and sometimes interface problems; the latter can also be studied with NMR.

In laboratory practice, it is highly desirable that the same specimen be examined by different techniques. This can offer precious cross data and help interpretation, for this purpose XPS and SSIMS are frequently used together. Or, choosing techniques having a different surface sensitivity, for instance XPS and IRS, a sort of depth profiling can be accomplished.

In Table 3.6, a technical comparison is reported for the most frequently used techniques, considering the information provided (qualitative, quantitative, depth profiling) and some physical constraints (thickness of the observed layer, sensitivity).

Table 3.7 Practical considerations for using spectroscopic techniques in characterization of polymer surfaces

Technique	Instrumentation	Specimen preparation problems	Surface damage
ISS	Efficient	UHV analysis	Yes, limited
SSIMS	Efficient	UHV analysis Charging problems	Yes, limited
XPS	Efficient	UHV analysis	Very limited
IRS-IR	Medium difficulty	Good contact with IRE is essential	No
DRS-IR	Medium difficulty	Well distributed sample required	No
IR-PAS	Cell efficiency critical	None	No
ERS-IR	Incidence angle important	None	No
IR-Emission	Efficient, diffic. elimination of stray radiation	Specimen heated	No
IETS	Efficient	Critical	No
SERS	Efficient	Critical	No
IR-Transmission	Limited	Powders must be compacted	No
NMR	Efficient	Sample rotation, frequency tuning, cross polarization critical	No

Because it is difficult to find a uniform sensitivity indication, the content (in atomic terms) of the layer of material observed by a particular technique was taken as unity, expressing sensitivity as the fraction, always atomic, that can be estimated. Thus in XPS a sensitivity of $10^{-2}-10^{-3}$ means that a $1-0.1\%$ atomic concentration of an element in the few nanometres observed can be detected. Conversely, the indication of an analogous sensitivity in NMR would refer to a concentration in the whole sample, thus absolute quantities would be quite different.

Table 3.7 summarizes some practical criteria on instrumentation, problems in working conditions and in sample preparation and degree of surface damage for all the techniques considered.

REFERENCES

[1] K. Siegbahn, C. Nordling, A. Fahlman, R. Nordberg, K. Hamerin, J. Hedman, G. Johansson, T. Bergmark, S.-E. Karlsson, I. Lindgren, and B. Lindberg, *Electron Spectroscopy for Chemical Analysis—Atomic, Molecular and Solid State Structure Studies by Means of Electron Spectroscopy*, Almqvist and Wiksell, Boktryekeri AB, Stockholm (1967).

[2] A. Einstein, *Ann. Physik*, **17**, 132 (1905).

[3] H. Robinson and F. W. Rawlinson, *Phil. Mag.*, **28**, 277 (1914).

[4] D. W. Dwight and W. M. Riggs, *J. Colloid Interface Sci.*, **47**, 650 (1974).

[5] D. T. Clark, W. T. Feast, I. Ritchie, W. K. R. Musgrave, M. Modena, and M. Ragazzini, *J. Polym. Sci. Polym. Chem. Ed.*, **12**, 1049 (1974).

[6] R. E. Weber and W. T. Peria, *J. Appl. Phys.*, **38**, 4355 (1967).

[7] D. P. Smith, *J. Appl. Phys.*, **38**, 340 (1967).

[8] T. M. Buck, in *Methods of Surface Analysis*, A. W. Czanderna (Ed.), Elsevier, Amsterdam, Ch. 3 (1975).

[9] A. Benninghoven, F. G. Rudenauer, and H. W. Werner, *Secondary Ion Mass Spectroscopy*, Wiley–Interscience, New York (1986).

[10] J. A. Gardella, Jr. *Appl. Surf. Sci.*, **31**, 72 (1988).

[11] J. A. Gardella, Jr. and D. M. Hercules, *Anal. Chem.*, **53**, 1879 (1981).

[12] J. A. Gardella, Jr. and D. M. Hercules, *Anal. Chem.*, **52**, 226 (1980).

[13] W. L. Baun, *Pure Appl. Chem.*, **54**, 323 (1982).

[14] W. L. Baun, *J. Adhes.*, **7**, 261 (1976).

[15] A. T. DiBenedetto and D. A. Scola, *J. Colloid Interface Sci.*, **64**, 480 (1978).

[16] G. E. Thomas, G. C. J. van der Ligt, G. J. M. Lippits, and G. M. M. van der Hei, *Appl. Surf. Sci.*, **6**, 204 (1980).

[17] D. M. Hercules, *Mikrochim. Acta*, **Suppl. 11**, 1 (1985).

[18] R. L. Schmitt, J. A. Gardella, Jr., J. H. Magill, L. Salvati, Jr., and R. L. Chin, *Macromolecules*, **18**, 2675 (1985).

[19] T. J. Hook, R. L. Schmitt, J. A. Gardella, Jr., L. Salvati, Jr., and R. L. Chin, *Anal. Chem.*, **58**, 1285 (1986).

[20] T. J. Hook, J. A. Gardella, Jr., and L. Salvati, Jr., *J. Mater. Res.*, **2**, 117 (1987).

[21] T. J. Hook, J. A. Gardella, Jr., and L. Salvati, Jr., *J. Mater. Res.*, **2**, 132 (1987).

[22] C. R. Gossett, in *Industrial Applications of Surface Analysis*, L. A. Casper and C. J. Powell (Eds), ACS Symp. Series n. 199, Ch. 4 (1982).

[23] H. H. G. Jellinek, in *Physicochemical Aspects of Polymer Surfaces*, K. L. Mittal (Ed.), Plenum Press, New York (1983), vol. I.

[24] D. L. Allara, C. W. White, R. L. Meek, and T. H. Briggs, *J. Polym. Sci. Chem. Ed.*, **14**, 93, (1976).

[25] D. L. Allara and C. W. White, *Adv. Chem. Ser.*, **169**, 273 (1978).

[26] T. Venkatesan, S. R. Forrest, M. L. Kaplan, C. A. Murray, P. H. Schmidt, and B. J. Wilkens, *J. Appl. Phys.*, **54**, 3150 (1983).

[27] T. Venkatesan, T. Wolf, D. Allara, B. J. Wilkens, and G. N. Taylor, *Appl. Phys. Lett.*, **43**, 934 (1983).

[28] S. J. Valenty, J. J. Chera, G. A. Smith, W. Katz, R. Argani, and H. Backhru, *J. Polym. Sci. Chem. Ed.*, **22**, 3367 (1985).

[29] J. C. Vickerman, A. Brown, and N. M. Reed, *Secondary Ion Mass Spectrometry. Principles and Applications.* Internat. Ser. Monographs on Chemistry, No. 17, Oxford Science Publ., Oxford (1989).

[30] A. W. Czanderna and D. M. Hercules (Eds), *Ion Spectroscopy for Surface Analysis*, Plenum Press, New York (1991).

[31] D. Briggs, A. Brown, and J. C. Vickerman, *Handbook of Static Secondary Ion Mass Spectrometry (SIMS)*, Wiley, Chichester (1989).

[32] J. G. Newman, B. A. Carlson, R. S. Michael, J. F. Moulder, W. J. van Ooij, and T. A. Hohlt, *Static SIMS Handbook of Polymer Analysis*, Perkin-Elmer Corp., Physical Electronics Div., Eden Prairie, MN (1991).

[33] A. Benninghoven, *Z. Physik*, **230**, 403 (1970).

[34] J. A. McHugh, in *Methods of Surface Analysis*, A. W. Czanderna, (Ed), Elsevier, Amsterdam, Ch. 6 (1975).

[35] D. Briggs and M. J. Hearn, *Spectrochim. Acta*, **40B**, 707 (1985).

[36] D. Briggs, M. J. Hearn, and B. D. Ratner, *Surf. Interf. Anal.*, **6**, 184 (1984).

[37] A. Brown and J. C. Vickerman, *Surf. Interf. Anal.*, **6**, 1 (1984).

[38] D. Briggs and A. B. Wootton, *Surf. Interf. Anal.*, **4**, 109 (1982).

[39] D. J. Surman and J. C. Vickerman, *Appl. Surf. Sci.*, **9**, 108 (1981).

[40] D. J. Surman, J. A. van der Berg, and J. C. Vickerman, *Surf. Interf. Anal.*, **4**, 160 (1982).

[41] F. M. Devienne and J. C. Roustan, *Compt. Rend. Acad. Sci.*, **276C**, 923 (1973); **283B**, 397 (1976).

[42] A. Brown, J. A. van der Berg, and J. C. Vickerman, *Spectrochim. Acta*, **40B**, 871 (1985).

[43] P. Steffan, E. Niehuis, T. Friese, D. Greifendorf, and A. Benninghoven, in *Secondary Ion Mass Spectroscopy SIMS IV*, Springer Series in *Chem. Phys.* Vol. 36 (A. Benninghoven, J. Okano, R. Shimizu, H. W. Werner, Eds), Springer-Verlag, Berlin (1984).

[44] F. Garbassi, E. Occhiello, C. Bastioli, and G. Romano, *J. Coll. Interf. Sci.*, **117**, 258 (1987).

[45] S. Storp, *Spectrochim. Acta*, **40B**, 745 (1985).

[46] D. Briggs, and M. J. Hearn, *Int. J. Mass Spectrom. Ion Proc.*, **67**, 47 (1985).

[47] D. Briggs and M. J. Hearn, *Vacuum*, 36, 1005 (1986).

[48] G. J. Leggett and J. C. Vickerman, *Anal. Chem.*, **63**, 561 (1991).

[49] G. J. Leggett and J. C. Vickerman, *Appl. Surf. Sci.*, **55**, 105 (1992).

[50] P. T. Murray and J. W. Rabalais, *J. Am. Chem. Soc.*, **103**, 1007 (1981).

[51] D. Briggs, *Org. Mass. Spectr.*, **22**, 91 (1987).

[52] W. J. van Ooij and A. Sabata, *Surf. Interf. Anal.*, **19**, 101 (1992).

[53] J. E. Campana, J. J. DeCorpo, and R. J. Colton, *Appl. Surf. Sci.*, **8**, 337 (1981).

[54] M. J. Hearn and D. Briggs, *Surf. Interf. Anal.*, **11**, 198 (1988).

[55] D. G. Castner and B. D. Ratner, *Surf. Interf. Anal.*, **15**, 479 (1990).

[56] W. J. van Ooij and R. H. G. Brinkhuis, *Surf. Interf. Anal.*, **11**, 430 (1988).

[57] W. J. van Ooij and M. Nahmias, *Rubber Chem. Technol.*, **62**, 656 (1989).

[58] D. Briggs, *Surf. Interf. Anal.*, **15**, 734 (1990).

[59] J. Lub, F. C. B. M. Van Vroonhoven, D. Van Leyen, and A. Benninghoven, *Polymer*, **29**, 998 (1988).

[60] J. Lub, F. C. B. M. Van Vroonhoven, E. Bruninx, and A. Benninghoven, *Polymer*, **30**, 40 (1989).

[61] Q. S. Bhatia and M. C. Burrell, *Surf. Interf. Anal.*, **15**, 388 (1990).

[62] P. M. Thompson, *Anal. Chem.*, **63**, 2447 (1991).

[63] M. J. Hearn, D. Briggs, B. D. Ratner, and S. C. Yoon, *Surf. Interf. Anal.*, **10**, 384 (1987).

[64] M. J. Hearn, B. D. Ratner, and D. Briggs, *Macromol.*, **21**, 2950 (1989).

[65] X. Zhao, W. Zhao, J. Sokolov, M. H. Rafailovich, S. A. Schwarz, B. J. Wilkens, R. A. L. Jones, and E. J. Kramer, *Macromol.*, **24**, 5991 (1991).

[66] S. J. Whitlow and R. P. Wool, *Macromol.*, **24**, 5926 (1991).

[67] G. Coulon, T. P. Russell, V. R. Deline and P. F. Green, *Macromol.*, **22**, 2581 (1989).

[68] E. Occhiello, M. Morra, F. Garbassi, D. Johnson, and P. Humphrey, *Appl. Surf. Sci.*, **47**, 235 (1991).

[69] E. Occhiello, M. Morra, F. Garbassi, P. Humphrey, and J. C. Vickerman, *SIMS VII* (Ed P. Benninghoven), p. 789 (1989).

[70] M. Morra, E. Occhiello, R. Marola, F. Garbassi, P. Humphrey, and D. Johnson, *J. Coll. Interf. Sci.*, **137**, 11 (1990).

[71] P. -G. de Gennes, *J. Chem. Phys.*, **55**, Suppl. 2, 572 (1971).

[72] A. Chilkoti, B. D. Ratner, and D. Briggs, *Anal. Chem.*, **63**, 1612 (1991).

[73] S. J. Simko, M. L. Miller, and R. W. Linton, *Anal. Chem.*, **57**, 2448 (1985).

[74] J. H. Wandass and J. A. Gardella Jr., *J. Am. Chem. Soc.*, **107**, 6192 (1985).

[75] R. W. Johnson, Jr. and J. A. Gardella Jr., *ACS Polym. Prepr.*, **32**, 200 (1991).

[76] A. Brindley, M. C. Davies, R. A. P. Lynn, S. S. Davis, J. Hearn, and J. F. Watts, *Polymer*, **33**, 1112 (1992).

[77] F. Garbassi, E. Occhiello, F. Polato, and A. Brown, *J. Mater. Sci.*, **22**, 1450 (1987).

[78] W. J. van Ooij, R. H. G. Brinkhuis, and J. Newman, *Proc. SIMS VI*, 1987.

[79] J. Lub, F. C. B. M. van Vroonoven, E. Bruninx, and A. Benninghoven, *Polymer*, **30**, 40 (1989).

[80] F. Degreve and J. M. De Long, *Surf. Interf. Anal.*, **7**, 177 (1985).

[81] M. C. Davies, A. Brown, J. M. Newton, and S. R. Chapman, *Surf. Interf. Anal.*, **11**, 591 (1988).

[82] R. Chujo, T. Nishi, Y. Sumi, T. Adachi, H. Naitoh, and H. Frenzel, *J. Polym. Sci., Polym. Lett. Ed.*, **21**, 487 (1983).

[83] S. J. Simko, S. R. Bryan, D. P. Griffis, R. W. Murray, and R. W. Linton, *Anal. Chem.*, **57**, 1198 (1985).

[84] A. Wirth, S. P. Thompson, and G. Gregory, *SIMS VI*, John Wiley & Sons, New York, p. 639 (1988).

[85] S. Hosoda, *Polymer J.*, **4**, 353 (1989).

[86] D. M. Hercules, F. P. Novak, and Z. A. Wilk, *Anal. Chim. Acta*, **195**, 61 (1987).

[87] G. J. Leggett, D. Briggs, and J. C. Vickerman, *J. Chem. Soc. Faraday Trans.*, **86**, 1863 (1990).

[88] G. J. Leggett, D. Briggs, and J. C. Vickerman, *Surf. Interf. Anal.*, **16**, 3 (1990).

[89] G. J. Leggett, J. C. Vickerman, and D. Briggs, *Surf. Interf. Anal.*, **17**, 737 (1990).

[90] G. J. Leggett, J. C. Vickerman, D. Briggs, and M. J. Hearn, *J. Chem. Soc. Faraday Trans.*, **88**, 297 (1992).

[91] D. Briggs and M. P. Seah (Eds), *Practical Surface Analysis*, John Wiley, New York–London (1983).

[92] D. Briggs (Ed.), *Handbook of X-ray and Ultraviolet Photoelectron Spectroscopy*, Heyden, London (1977).

[93] G. E. Muilenburg (Ed), *Handbook of X-ray Photoelectron Spectroscopy*, Physical Electronics Div., Perkin Elmer Corp., Norwalk (1979).

[94] A. Dilks, in *Electron Spectroscopy—Theory, Techniques and Applications*, C. R. Brundle, A. D. Baker (Eds), vol. 4, Academic Press, London (1981).

[95] D. T. Clark, in *Characterization of Metal and Polymer Surfaces*, L. H. Lee (Ed.), Academic Press, London (1977).

[96] J. D. Andrade (Ed), *Surface and Interfacial Aspects of Biomedical Polymers*, Plenum Press, New York (1985).

[97] T. G. Vargo and J. A. Gardella, Jr., *J. Vac. Sci. Technol.*, **A7**, 1733 (1989).

[98] D. T. Clark and A. Dilks, *J. Polym. Sci., Chem. Ed.*, **14**, 533 (1976).

[99] J. Riga, J. P. Boutique, J. J. Pireaux, J. J. Verbist, in *Physico-chemical Aspects of Polymer Surfaces*, K. L. Mittal (Ed.), vol. 1, Plenum Press, New York (1983).

[100] J.-M. André, J. Delhalle, and J. J. Pireaux, in *Photon, Electron and Ion Probes of Polymer Structure and Properties*, D. W. Dwight, T. J. Fabish, H. R. Thomas (Eds), ACS Symp. Series, **162**, 151 (1981).

[101] J. J. Pireaux, J. Riga, R. Caudano, and J. Verbist, ibid., p. 169.

[102] W. R. Salaneck, ibid., p. 121.

[103] C. Shang Xian, K. Sekim, H. Inokuchi, S. Hashimoto, N. Ueno, and K. Sugita, *Bull. Chem. Soc. Japan*, **58**, 890 (1985).

[104] R. Foerch, G. Beamson, and D. Briggs, *Surf. Interf. Anal.*, **17**, 842 [1991].

[105] M. Morra, E. Occhiello, and F. Garbassi, unpublished results.

[106] H. R. Thomas and J. J. O'Malley, *Macromolecules*, **12**, 323 (1979).

[107] D. T. Clark, J. Peeling, and J. M. J. O'Malley, *J. Polym. Sci., Chem. Ed.*, **14**, 543 (1976).

[108] D. H. -K. Pan and W. M. Prest, Jr., *J. Appl. Phys.*, **58**, 2861 (1985).

[109] H. J. Busscher, W. Hoogsten, L. Dijkema, G. A. Sawatsky, A. W. J. van Pelt, H. P. de Jong, G. Challa and J. Arends, *Polymer Comm.*, **26**, 252 (1985).

[110] R. Foerch and D. Johnson, *Surf. Interf. Anal.*, **17**, 847 (1991).

[111] M. A. Golub and R. D. Cormia, *Polymer*, **30**, 1576 (1989).

[112] Y. Nakayama, F. Soeda, A. Ishitani, and T. Ikegami, *Polym. Engineer. Sci.*, **31**, 812 (1991).

[113] M. C. Burrell, Y. S. Liu, and H. S. Cole, *J. Vac. Sci. Technol.*, **A4**, 2459 (1986).

[114] F. Garbassi, E. Occhiello, and F. Polato, *J. Mater. Sci.*, **22**, 207 (1987).

[115] I. Sutherland, D. M. Brewis, R. J. Heath, and E. Sheng, *Surf. Interf. Anal.*, **17**, 507 (1991).

[116] R. D. Goldblatt, J. M. Park, R. C. White, L. J. Matienzo, S. J. Huang, and J. F. Johnson, *J. Appl. Polym. Sci.*, **37**, 335 (1989).

[117] Zhang Pei Yao, and B. Ranby, *J. Appl. Polym. Sci.*, **41**, 1459 (1990).

[118] K. Allmer, J. Hilborn, P. H. Larsson, A. Hult, and B. Ranby, *J. Polym. Sci., Polym. Chem.* **28**, 173 (1990).

[119] E. Uchida, Y. Uyama, H. Iwata, and Y. Ikada, *J. Polym. Sci., Polym. Chem.*, **28**, 2837 (1990).

[120] C. D. Batich, *Appl. Surf. Sci.*, **32**, 57 (1988).

[121] C. N. Reilley, D.S. Everhart, in *Applied Electron Spectroscopy for Chemical Analysis*, H. Windawi and F. Ho (Eds), Wiley–Interscience, New York (1982).

[122] W. G. Collier and T. P. Tougas, *Anal. Chem.*, **59**, 396 (1987).

[123] R. A. Dicke, J. S. Hammond, J. E. de Vries, and J. W. Holubka, *Anal. Chem.*, **54**, 2045 (1982).

[124] C. D. Batich and A. Yahiaoui, *J. Polym. Sci., Polym. Chem.*, **25**, 3479 (1987).
[125] W. Gombotz, A. Hoffmann, and Y. Sun, *Trans. Soc. Biomaterials, 13th Annual Meet.* (1987), p. 27.
[126] D. Briggs and C. R. Kendall, *Intern. J. Adhesion Adhesives*, Jan. 1982, 13.
[127] J. F. M. Pennings and B. Bosman, *Colloid Polym. Sci.*, **258**, 1099 (1980).
[128] D. E. Williams, 182nd ACS Natl. Meet. (1981), unpublished result.
[129] T. Ohmichi, H. Tamaki, H. Kawasaki, and S. Tatsuta, in *Physicochemical Aspects of Polymer Surfaces*, K. L. Mittal (Ed), Plenum Press, New York (1983).
[130] H. Boehm, *Advan. Catalysis*, **16**, 179 (1966).
[131] P. R. Moses, L. M. Wier, J. C. Lennox, H. O. Finklea, J. R. Lenhard, and R. W. Murray, *Anal. Chem.*, **50**, 576 (1978).
[132] D. T. Clark, *Pure Appl. Chem.*, **54**, 415 (1982).
[133] G. W. Simmons and B. C. Beard, *Abstr. Pittsburgh Conf. Analytical Chemistry and Applied Spectroscopy*, p. 212 (1981).
[134] B. Leclercq, M. Sotton, A. Baszkin, and L. Ter-Minassian-Saraga, *Polymer*, **18**, 675 (1977).
[135] H. Spell, C. Christenson, *Proc. Paper Synthetic Conf. (TAPPI)*, Atlanta, p. 283 (1978).
[136] S. Storp, *Spectrochim. Acta*, **40B**, 745 (1985).
[137] D. R. Wheeler and S. V. Pepper, *J. Vac. Sci. Technol.*, **20**, 226 (1982).
[138] R. Chaney and G. Barth, *Fresenius Z. Anal. Chem.*, **329**, 143 (1987).
[139] W. F. Stickle and J. F. Moulder, *J. Vac. Sci. Technol.*, **A9**, 1441 (1991).
[140] D. T. Clark, J. Peeling, and J. M. O'Malley, *J. Polym. Sci., Polym. Chem.*, **14**, 543 (1976).
[141] P. F. Green, T. M. Christensen, and T. P. Russell, *Macromol.*, **24**, 252 (1991).
[142] T. G. Vargo, D. J. Hook, J. A. Gardella, Jr., M. A. Eberhardt, A. E. Meyer, and R. E. Baier, *Appl. Spectrosc.*, **45**, 448 (1991).
[143] S. Randall Holmes-Farley, R. H. Reamey, T. J. McCarthy, J. Deutch, and G. M. Whitesides, *Langmuir*, **1**, 725 (1985).
[144] A. J. Dias and T. J. McCarthy, *Macromolecules*, **18**, 1826 (1985).
[145] J. Lukas, L. Lochmann, and J. Kalal, *Angew. Makromol. Chem.*, **181**, 183 (1990).
[146] K.-W. Lee and J. McCarthy, *Macromol.*, **21**, 309 (1988).
[147] D. Briggs, D. M. Brewis, and M. B. Konieczko, *J. Mater. Sci.*, **14**, 1344 (1979).
[148] I. Sutherland, D. M. Brewis, R. J. Heath, and E. Sheng, *Surf. Interf. Anal.*, **17**, 507 (1991).
[149] E. Sheng, I. Sutherland, D. M. Brewis, and R. J. Heath, *Surf. Interf. Anal.*, **19**, 151 (1992).
[150] D. Briggs, C. R. Kendall, A. R. Blithe, and A. B. Wootton, *Polymer*, **24**, 27 (1983).
[151] D. Briggs and C. R. Kendall, *Polymer*, **20**, 1053 (1979).
[152] D. Klee, D. Gribbin, and D. Kirch, *Angew. Makromol. Chem.*, **131**, 145 (1985).
[153] R. G. Nuzzo, and G. Smolinsky, *Macromolecules*, **17**, 1013 (1984).
[154] E. Occhiello, M. Morra, G. Morini, F. Garbassi, and P. Humphrey, *J. Appl. Polym. Sci.*, **42**, 551 (1991).
[155] A. J. Dilks, *J. Polym. Sci., Polym. Chem.*, **19**, 1319 (1981).
[156] E. Occhiello, M. Morra, P. Cinquina, and F. Garbassi, *Polym. Prepr.*, **31**, 308 (1990).
[157] E. C. Onyiriuka, L. S. Hersh, and W. Hertl, *J. Coll. Interf. Sci.*, **144**, 98 (1991).
[158] E. Occhiello, F. Garbassi, and G. Morini, *MRS Symp. Proc.*, **119**, 155 (1988).
[159] M. A. Golub, T. Wydeven, and R. D. Cormia, *Polymer*, **30**, 1571 (1989).
[160] M. Morra, E. Occhiello, and F. Garbassi, *Langmuir*, **5**, 872 (1989).
[161] E. Occhiello, M. Morra, F. Garbassi, and J. Bargon, *Appl. Surf. Sci.* **36**, 285 (1989).
[162] M. A. Golub, T. Wydeven, and R. D. Cormia, *Polymer Comm.*, **29**, 285 (1988).

[163] T. G. Vargo, D. J. Hook, J. A. Gardella, Jr., M. A. Eberhardt, A. E. Meyer, and R. E. Baier, *J. Polym. Sci., Polym. Chem.*, **29**, 535 (1991).
[164] M. Strobel, S. Corn, C. S. Lyons, and G. A. Korba, *J. Polym. Sci., Polym. Chem.*, **23**, 1125 (1985).
[165] G. A. Corbin, R. E. Cohen, and R. F. Baddour, *Macromolecules*, **18**, 98 (1985).
[166] R. Srinivasan and S. Lazare, *Polymer*, **26**, 1297 (1985).
[167] E. Occhiello, F. Garbassi, and V. Malatesta, *J. Mater. Sci.*, **24**, 569 (1989).
[168] J. J. Schmidt, J. A. Gardella, Jr., and L. Salvati, Jr., *Macromol.*, **22**, 4489 (1989).
[169] Q. S. Bhatia, D. H. Pan, and J. T. Koberstein, *Macromol.*, **21**, 2166 (1988).
[170] M. B. Clark, C. A. Burkhardt, and J. A. Gardella, Jr., *Macromol.*, **22**, 4495 (1989).
[171] M. B. Clark, C. A. Burkhardt, and J. A. Gardella, Jr., *Macromol.*, **24**, 799 (1991).
[172] S. H. Goh, H. S. O. Chan, and K. L. Tan, *Appl. Surf. Sci.*, **52**, 1 (1991).
[173] M. C. Burrell and J. J. Chera, *Appl. Surf. Sci.*, **35**, 110 (1988–89).
[174] V. Sa Da Costa, D. Brier-Russel, E. W. Salzmann, and E. W. Merrill, *J. Coll. Interf. Sci.*, **80**, 445 (1981).
[175] M. D. Lelah, L. K. Lombrecht, B. R. Young, and S. L. Cooper, *J. Biomed. Mater. Res.*, **17**, 1 (1983).
[176] A. J. Pertsin, M. M. Gorelova, V. Yu. Levin, and L. I. Makarova, *J. Appl. Polym. Sci.*, **45**, 1195 (1992).
[177] B. Ranby, Z. M. Gao, A. Hult, and P. Y. Zhang, in *Chemical Reactions on Polymers*, J. L. Benham (Ed.), ACS Symp. Ser. no. 364 (1988).
[178] K. Allmer, A. Hult, and B. Ranby, *J. Polym. Sci., Polym. Chem.*, **26**, 2099 (1988).
[179] K. Allmer, A. Hult, and B. Ranby, *J. Polym. Sci., Polym. Chem.*, **27**, 3405 (1989).
[180] I. Ikemoto, Y. Cao, M. Yamada, H. Kuroda, I. Harada, H. Shirakawa, and S. Ikeda, *Bull. Chem. Soc. Japan*, **55**, 721 (1982).
[181] I. Ikemoto, T. Ichihara, C. Egawa, K. Kikuchi, H. Kuroda, Y. Furukawa, I. Harada, H. Shirakawa, and S. Ikeda, *Bull. Chem. Soc. Japan*, **58**, 747 (1985).
[182] S. G. MacKay, M. Bakir, I. H. Musselman, T. J. Meyer, and R. W. Linton, *Anal. Chem.*, **63**, 60 (1991).
[183] K. T. Kang, K. G. Neoh, and K. L. Tan, *Surf. Interf. Anal.*, **19**, 33 (1992).
[184] W. R. Salaneck, I. Lundstroem, and B. Liedberg, *Progr. Coll. Polym. Sci.*, **70**, 83 (1985).
[185] K. Waltersson, *Composites Sci. Techn.*, **22**, 223 (1985).
[186] K. Waltersson, *Composites Sci. Techn.*, **23**, 303 (1985).
[187] C. Kozlowski and P. M. A. Sherwood, *J. Chem. Soc. Faraday Trans. I*, **81**, 2745 (1985).
[188] G. D. Nichols, D. M. Hercules, R. C. Peek, and D. J. Vaughan, *Appl. Spectrosc.*, **28**, 219 (1974).
[189] P. R. Moses, L. M. Wier, J. C. Lennox, H. O. Finklea, J. R. Lenhard, and R. W. Murray, *Anal. Chem.*, **50**, 576 (1978).
[190] C. D. Wagner and A. Joshi, *Surf. Interf. Anal.*, **6**, 215 (1984).
[191] R. L. Chaney, *Surf. Interf. Anal.*, **10**, 36 (1987).
[192] W. J. van Ooj, in *Physicochemical Aspects of Polymer Surfaces*, K. L. Mittal (Ed.), Plenum Press, New York, vol. 2, p. 1035 (1983).
[193] D. Schwamm, J. Kulig, and M. H. Litt, *Chem. Mater.*, **3**, 616 (1991).
[194] F. M. Mirabella Jr., *Appl. Spectrosc. Rev.*, **21**, 45 (1985).
[195] N. S. Harrick, *Internal Reflection Spectroscopy*, Wiley, New York (1967).
[196] T. Nguyen, *Progr. Org. Coat.*, **13**, 1 (1985).
[197] F. M. Mirabella Jr., *J. Polym. Sci., Phys. Ed.*, **21**, 2403 (1983).
[198] C. B. Hu and C. S. P. Sung, *Polym. Prepr.*, **21**, 156 (1980).
[199] K. Ohta and R. Iwamoto, *Appl. Spectrosc.*, **39**, 418 (1985).

[200] A. J. Barbetta, *Appl. Spectrosc.*, **38**, 29 (1984).
[201] K. Otha and R. Iwamoto, *Anal. Chem.*, **57**, 2491 (1985).
[202] F. Kimura, J. Umemura, and T. Takenaka, *Langmuir*, **2**, 96 (1986).
[203] M. Hirano, R. Nemori, Y. Nakao, and Y. Yamada, *Mikrochim. Acta*, **II**, 43 (1988).
[204] Y. Nishikawa, Y. Ito, K. Fujiwara, and T. Shima, *Appl. Spectrosc.*, **45**, 752 (1991).
[205] A. M. Tiefenthalar and M. W. Urban, *Appl. Spectrosc.*, **42**, 163 (1988).
[206] *Standard Practices for Internal Reflection Spectroscopy*, ASTM Annual Book 42, E573.
[207] J. K. Barr and P. A. Fluornay, in *Physical Methods in Macromolecular Chemistry*, B. Carroll (Ed), vol. I, Dekker, New York (1969).
[208] F. M. Mirabella Jr., *J. Polym. Sci., Phys. Ed.*, **23**, 861 (1985).
[209] R. Iwamoto, M. Miya, K. Ohta, and S. Mima, *J. Chem. Phys.*, **74**, 4780 (1981).
[210] R. Iwamoto, M. Miya, K. Ohta, and S. Mima, *J. Am. Chem. Soc.*, **102**, 1212 (1980).
[211] R. Iwamoto, M. Miya, K. Ohta, and S. Mima, *Appl. Spectrosc.*, **35**, 584 (1981).
[212] B. K. Lok, Y. Cheng, and C.R. Robertson, *J. Coll. Interf. Sci.*, **91**, 87 (1983).
[213] H. Masuhara, N. Mataga, S. Tazuke, T. Murao, and I. Yamazaki *Chem. Phys. Lett.*, **100**, 415 (1983).
[214] J. G. Bots, L. van der Does, and A. Bantjees, *Brit. Polym. J.*, **19**, 527 (1987).
[215] Y. Kano, K. Ishikura, S. Kawahara, and S. Akiyama, *Polymer J.*, **24**, 135 (1992).
[216] R. Iwamoto and K. Ohta, *Appl. Spectrosc.*, **38**, 359 (1984).
[217] J. A. Gardella, Jr., G. L. Grobe III, W. L. Hopson, and E. M. Eyring, *Anal. Chem.*, **56**, 1169 (1984).
[218] M. D. Lelah, S. L. Cooper, H. Ohnuma, and T. Kotaka, *Polymer J.* **17**, 841 (1985).
[219] R. Kellner, G. Fischboeck, G. Goetzinger, E. Pungor, K. Toth, L. Polos, and E. Lindner, *Z. Anal. Chem.*, **322**, 151 (1985).
[220] S. Winters, R. M. Gendreau, R. L. Leininger, and R. J. Jacobsen, *Appl. Spectrosc.*, **36**, 404 (1982).
[221] D. J. Carlsson and D. M. Wiles, *Macromol.*, **2**, 587, 597 (1970); **4**, 174 (1972).
[222] A. Narebska and Z. Bukowski, *Makromol. Chem.*, **186**, 1411 (1985).
[223] C. S. Blackwell, P. J. Degen, and F. D. Osterholtz, *Appl. Spectrosc.*, **32**, 480 (1978).
[224] A. E. Tshmel, V. I. Vettegren, and V. M. Zolotarev, *J. Macromol. Sci., Phys.*, **B21**, 243 (1982).
[225] G. Fonseca, J. M. Perena, J. G. Fatou, and A. Bello, *J. Mater. Sci.*, **20**, 3283 (1985).
[226] J. H. Lee, H. G. Kim, G. S. Khang, H. B. Lee, and M. S. Jhon, *J. Coll. Interf. Sci.*, **151**, 563 (1992).
[227] A. M. Wrobel, J. Kryszewski, and M. Gazicki, *J. Macromol. Sci., Chem.*, **A20**, 583 (1983).
[228] A. M. Wrobel, *J. Macromol. Sci., Chem.*, **A22**, 1089 (1985).
[229] A. E. Tshmel and V. I. Vettegren, *Eur. Polym. J.*, **12**, 853 (1976).
[230] F. M. Mirabella Jr., *J. Polym. Sci., Phys. Ed.*, **22**, 1283, 1293 (1984).
[231] D. J. Walls and J. C. Coburn, *J. Polym. Sci., Polym. Phys.*, **30**, 887 (1992).
[232] A. Garton, S. W. Kim, and D. M. Wiles, *J. Polym. Sci., Lett. Ed.*, **20**, 273 (1982).
[233] G. Koertum, *Reflectance Spectroscopy*, Springer-Verlag, New York (1969).
[234] M. P. Fuller and P. R. Griffiths, *Anal. Chem.*, **50**, 1906 (1978).
[235] H. Maulharkt and D. Kunath, *Appl. Spectrosc.*, **34**, 383 (1980).
[236] P. Kubelka and F. Munk, *Z. Tech. Physk.*, **12**, 593 (1931).
[237] P. Kubelka, *J. Opt. Soc. Am.*, **38**, 448 (1948).
[238] H. G. Hecht, *Appl. Spectrosc.*, **37**, 348 (1983).
[239] S. R. Culler, H. Ishida, and J. L. Koenig, *Appl. Spectrosc.*, **38**, 1 (1984).
[240] S. R. Culler, M. T. McKenzie, L. J. Fina, H. Ishida, and J. L. Koenig, *Appl. Spectrosc.*, **38**, 791 (1984).

[241] M. T. McKenzie and J. L. Koenig, *Appl. Spectrosc.*, **39**, 408 (1985).
[242] S. Naviroj, S. R. Culler, H. Ishida, and J. L. Koenig, *J. Coll. Interf. Sci.*, **97**, 308 (1984).
[243] M. T. McKenzie, S. R. Culler, and J. L. Koenig, *Appl. Spectrosc.*, **38**, 786 (1984).
[244] S. R. Culler, H. Ishida, and J. L. Koenig, *J. Coll. Interf. Sci.*, **106**, 334 (1985).
[245] J. A. Davies and A. Sood, *Makromol. Chem.*, **186**, 1631 (1985).
[246] E. G. Chatzi, H. Ishida, and J. L. Koenig, *Appl. Spectrosc.*, **40**, 847 (1986).
[247] J. M. Chalmers and M. W. MacKenzie, *Appl. Spectrosc.*, **39**, 634 (1985).
[248] J. A. J. Jansen and W. E. Hass, *Polymer Comm.*, **29**, 77 (1988).
[249] G. Xue, *Makromol. Chem., Rapid Commun.*, **6**, 811 (1985).
[250] A. Rosencwaig, *Photoacoustics and Photoacoustic Spectroscopy*, New York (1980).
[251] S. O. Karstad and P. E. Nordall, *Appl. Surf. Sci.*, **6**, 372 (1980).
[252] A. Rosencwaig, *Anal. Chem.*, **47**, 592A (1975).
[253] E. P. C. Lai, B. L. Chan, and M. Hadjmohammadi, *Appl. Spectrosc. Rev.*, **21**, 179 (1985).
[254] A. Rosencwaig and A. Gersho, *J. Appl. Phys.*, **47**, 64 (1976).
[255] L. W. Burgraaf and D. E. Leyden, *Anal. Chem.*, **53**, 759 (1981).
[256] J. F. McClelland, *Anal. Chem.*, **55**, 89A (1983).
[257] K. Krishnan, *Appl. Spectrosc.* **35**, 549 (1981).
[258] D. W. Vidrine, *Appl. Spectrosc.*, **34**, 314 (1980).
[259] N. L. Rockley, M. K. Woodward, and M. G. Rockley, *Appl. Spectrosc.*, **38**, 329 (1984).
[260] S. M. Riseman and E. M. Eyring, *Spectrosc. Lett.* **14**, 163 (1981).
[261] R. O. Carter III, *Appl. Spectrosc.*, **46**, 219 (1992).
[262] K. Krishnan, S. Hill, J. P. Hobbs, and C. S. P. Sung, *Appl. Spectrosc.*, **36**, 257 (1982).
[263] N. Teramae and S. Tanaka, *Appl. Spectrosc.*, **39**, 797 (1985).
[264] R. M. Dittmar, J. L. Chao, and R. A. Palmer, *Appl. Spectrosc.*, **45**, 1104 (1991).
[265] N. Teramae, M. Hiroguchi, and S. Tanaka, *Bull. Chem. Soc. Japan.* **55**, 2097 (1982).
[266] D. J. Gerson, *Appl. Spectrosc.*, **38**, 436 (1984).
[267] D. E. Leyden, D. S. Kendall, L. W. Burggraf, F. J. Pern, and M. DeBello, *Anal. Chem.*, **54**, 101 (1982).
[268] R. W. Linton, M. L. Miller, G. E. Maciel, and B. L. Hawkins, *Surf. Interf. Anal.* **7**, 196 (1986).
[269] M. E. Abu-Zeid, E. E. Nofal, L. A. Tahseen, and F. A. Abdul-Rasoul, *J. Appl. Polym. Sci.*, **30**, 3791 (1985).
[270] D. L. Allara, in *Characterization of Metal and Polymer Surfaces*, L. H. Lee (Ed.), Academic Press, London (1977).
[271] H. G. Tompkins, in *Methods of Surface Analysis*, A. W. Czanderna (Ed.), Elsevier, Amsterdam (1975).
[272] R. G. Greenler, *J. Chem. Phys.*, **44**, 310 (1966).
[273] W. G. Golden, D. S. Dunn, and J. Overend, *J. Catal.*, **71**, 395 (1981).
[274] D. L. Allara, A. Baca, and C. A. Pryde, *Macromolecules*, **11**, 1215, (1978).
[275] R. P. Netterfield, P. J. Martin, W. G. Sainty, R. M. Duffy, and C. G. Pacey, *Rev. Sci. Instrum.*, **56**, 1995 (1985).
[276] R. T. Graf, J. L. Koenig, and H. Ishida, *Anal. Chem.*, **58**, 64 (1986).
[277] C. A. Sergides, A. R. Chughtai, and D. M. Smith, *J. Polym. Sci. Phys. Ed.*, **23**, 1573 (1985).
[278] H. H. G. Jellinek, H. Kachi, and I. Chodak, *J. Polym. Sci. Chem.*, **23**, 2291 (1985).
[279] S. C. Lin, J. Bulkin, and E. M. Pearce, *J. Polym. Sci. Chem. Ed.*, **17**, 3121 (1979).
[280] C. A. Sergides, A. R. Chughtai, and D. M. Smith, *Appl. Spectrosc.*, **39**, 735 (1985).
[281] G. J. Kemeny and P. R. Griffiths, *Appl. Spectrosc.*, **35**, 128 (1981).

[282] J. F. Rabolt, M. Jurich, and H. D. Swalen, *Appl. Spectrosc.*, **39**, 269 (1985).
[283] W. Knoll, M. R. Philpott, and W. G. Golden, *J. Chem. Phys.*, **77**, 219 (1982).
[284] I. Kusaka and W. Suetaka, *Spectrochim. Acta*, **36A**, 647 (1980).
[285] J. F. Rabolt, F. C. Burne, N. E. Schlotter, and J. D. Swalen, *J. Chem. Phys.*, **77**, 219 (1982).
[286] P. V. Huong, *Adv. Infrared Raman Spectrosc.*, **4**, 85 (1978).
[287] D. Kember, D. H. Chenery, N. Sheppard, and J. Fell, *Spectrochim. Acta*, **35A**, 455 (1979).
[288] J. L. Lauer, and L. E. Keller, *Appl. Surf. Sci.*, **9**, 175 (1981).
[289] R. G. Greenler, *Surf. Sci.*, **69**, 647 (1977).
[290] D. L. Allara, D. Teicher, and J. F. Durana, *Chem. Phys. Lett.*, **84**, 20 (1981).
[291] K. Wagatsuma, K. Honma, and W. Suetaka, *Appl. Surf. Sci.*, **7**, 281 (1981).
[292] Y. Nagasawa and A. Ishitani, *Appl. Spectrosc.*, **38**, 168 (1984).
[293] J. Derkosch and W. Mikenda, *Mikrochim. Acta*, **8**, 101 (1985).
[294] P. R. Griffiths, *Appl. Spectrosc.*, **26**, 73 (1973).
[295] B. F. Lewis, M. Mosesman, and W. H. Weinberg, *Surf. Sci.*, **41**, 142 (1974).
[296] J. Lambe and R. C. Jaklevic, *Phys. Rev.*, **165**, 821 (1968).
[297] M. G. Simonsen, R. V. Coleman, and P. K. Hansma, *J. Chem. Phys.*, **61**, 3789 (1974).
[298] A. F. Diaz, U. Hetzler, and E. Kay, *J. Am. Chem. Soc.*, **99**, 780 (1977).
[299] D. M. Brewis, J. Comyn, D. P. Oxley, and R. G. Pritchard, *Surf. Interf. Anal.*, **6**, 40 (1984).
[300] J. D. Alexander, A. N. Gent, and P. N. Henriksen, *J. Chem. Phys.*, **83**, 5981 (1985).
[301] M. Higo, Y. Owaki, and S. Kamata, *Chem. Lett.*, 1309 (1985).
[302] J. Comyn, in *Adhesion 9*, K. W. Allen (Ed.), Elsevier, London (1984), p. 147.
[303] J. Comyn, D. P. Oxley, R. G. Pritchard, C. R. Werrett, and A. J. Kinloch, *Int. J. Adhesion and Adhesives,* **9**, 201 (1989).
[304] J. Comyn, D. P. Oxley, R. G. Pritchard, C. R. Werrett, and A. J. Kinloch, *Int. J. Adhesion and Adhesives*, **10**, 13 (1990).
[305] N. M. D. Brown, R. J. Turner, S. Affrossman, I. R. Dunkin, R. A. Pethrick, and C. J. Shields, *Spectrochim. Acta*, **40B**, 847 (1985).
[306] S. Reynolds, D. P. Oxley, and R. G. Pritchard, *Spectrochim. Acta*, **38A**, 103 (1981).
[307] R. R. Mallik, R. G. Pritchard, C. C. Harley, and J. Comyn, *Polymer*, **26**, 551 (1985).
[308] J. Comyn, C. C. Horley, R. R. Mallik, and R. G. Pritchard, *Int. J. Adhesion and Adhesives*, **6**, 73 (1986).
[309] J. Comyn, C. C. Horley, R. R. Mallik, and R. G. Pritchard, in *Adhesion 11*, K. W. Allen (Ed), Elsevier, London (1987), p. 38.
[310] H. Ibach and D. L. Mills, *Electron Loss Spectroscopy and Surface Vibrations*, Academic Press, New York (1982).
[311] J. J. Pireaux, C. Gregoire, R. Caudano, M. Rei Vilar, R. Brinkhuis, and A. J. Schouten, *Langmuir*, **7**, 2433 (1991).
[312] J. J. Pireaux, P. A. Thiry, R. Caudano, and P. Pfluger, *J. Chem. Phys.*, **84**, 6462 (1986).
[313] M. Rei Vilar, M. Schott, J. J. Pireaux, C. Gregoire, P. A. Thiry, R. Caudano, A. Lapp, A. M. Botelho di Rego, and J. Lopez da Silva, *Surf. Sci.*, **189/190**, 927 (1987).
[314] Y. J. Chabal and A. J. Sicvers, *Phys. Rev. Lett.*, **44**, 944 (1980).
[315] J. P. Heritage and D. L. Allara, *Chem. Phys. Lett.*, **74**, 507 (1980).
[316] D. L. Allara, C. A. Murray, and S. Bodoff, in *Physicochemical Aspects of Polymer Surfaces*, K. L. Mittal (Ed.), Plenum Press, New York (1983), vol. I.
[317] H. Ishida, and A. Ishitani, *Appl. Spectrosc.*, **37**, 450 (1983).

[318] C. Jennings, R. Aroca, A.-M. Hor, and R. O. Loutfy, *Spectrochim. Acta*, **41A**, 1095 (1985).

[319] L. M. Siperko, *Appl. Spectrosc.* **43**, 226 (1989).

[320] J. F. Rabolt, R. Santo, and J. D. Swalen, *Appl. Spectrosc.*, **34**, 517 (1980).

[321] J. F. Rabolt, N. E. Schlotter, J. D. Swalen, and R. Santo, *J. Polym. Sci. Phys. Ed.*, **21**, 1 (1983).

[322] J. F. Rabolt and J. D. Swalen, in *Spectroscopy of Surfaces*, R. J. H. Clark and R. E. Hester (Eds), John Wiley, Chichester (1988), p. 1.

[323] J. R. Ferraro and L. J. Basile, *Fourier Transform Infrared Spectroscopy*, Academic Press, New York (1982).

[324] D. A. Lang, *Raman Spectroscopy*, McGraw-Hill, New York (1977).

[325] C. Chiang, H. Ishida, and J. L. Koenig, *J. Coll. Interf. Sci.*, **74**, 396 (1980).

[326] H. Ishida and J. L. Koenig, *J. Coll. Interf. Sci.*, **64**, 555 (1978).

[327] H. Ishida and J. L. Koenig, *J. Coll. Interf. Sci.*, **64**, 565 (1978).

[328] H. Ishida, S. Naviroj, K. Tripathy, J. J. Fitzgerald, and J. L. Koenig, *J. Polym. Sci. Phys. Ed.*, **20**, 701 (1982).

[329] S. R. Culler, H. Ishida, and J. L. Koenig, *J. Coll. Interf. Sci.*, **106**, 334 (1985).

[330] K. Tsutsumi and H. Takahashi, *Coll. Polym. Sci.*, **1**, 506 (1985).

[331] J. L. Koenig and P. T. K. Shih, *Mater. Sci. Eng.*, **20**, 127 (1975).

[332] M. Kawaguchi, M. Komiya, and A. Takahashi, *Polym. J.*, **14**, 563 (1982).

[333] S. R. Culler, H. Ishida, and J. L. Koenig, *J. Coll. Interf. Sci.*, **109**, 1 (1986).

[334] M. M. Coleman and P. C. Paynter, *Appl. Spectrosc. Rev.*, **20**, 255 (1984).

[335] G. Ramana Rao, C. Castiglioni, M. Gussoni, G. Zerbi, and E. Martuscelli, *Polymer*, **26**, 811 (1985).

[336] M. K. Antoon, K. H. Koenig, and J. L. Koenig, *Appl. Spectrosc.*, **31**, 518 (1977).

[337] G. Leonard, J. L. Halary, and L. Monnerie, *Polymer*, **26**, 1507 (1985).

[338] T. M. Duncan and C. Dybowski, *Surf. Sci. Rep.*, **1**, 157 (1981).

[339] A. M. Zaper and J. L. Koenig, *Adv. Coll. Interf. Sci.*, **22**, 113 (1985).

[340] A. Abragam, *The Principles of Nuclear Magnetism*, Oxford University Press, Oxford (1961).

[341] M. Mehring, *High Resolution NMR Spectroscopy in Solids—NMR Basic Principles and Progress*, vol. 11, Springer-Verlag, Berlin (1976).

[342] R. K. Harris, B. E. Mann (Eds), *NMR and the Periodic Table*, Academic Press, New York (1978).

[343] A. Pines, M. G. Gibby, and J. S. Waugh, *J. Chem. Phys.*, **59**, 569 (1973).

[344] E. R. Andrew, *Progr. NMR Spectrosc.*, **8**, 1 (1971).

[345] D. W. Sindorf and G. E. Maciel, *J. Am. Chem. Soc.*, **105**, 1487 (1983).

[346] D. W. Sindorf and G. E. Maciel, *J. Phys. Chem.*, **87**, 1487 (1983).

[347] G. R. Hays, A. D. H. Clague, R. Huis, and G. Van der Velden, *Appl. Surf. Sci.*, **10**, 247 (1982).

[348] C. Chiang, N. Liu, and J. L. Koenig, *J. Coll. Interf. Sci.*, **86**, 26 (1982).

[349] D. W. Sindorf and G. E. Maciel, *J. Am. Chem. Soc.*, **105**, 1848 (1983).

[350] D. Slotfeldt-Ellingsen and H. A. Resing, *J. Phys. Chem.*, **84**, 2204 (1980).

[351] J. Van Alsten and D. W. Ovenall, *Polymer Comm.*, **32**, 18 (1991).

[352] W. P. Rothwell, D. R. Holecek, and J. A. Vershaw, *J. Polym. Sci. Lett. Ed.*, **22**, 241 (1984).

[353] J. R. Havens and J. L. Koenig, *Appl. Spectrosc.*, **37**, 226 (1982).

[354] V. J. McBrierty and D. C. Douglass, *J. Polym. Sci. Macromol. Rev.*, **16**, 295 (1981).

[355] E. O. Steidkal, J. Schaefer, M. D. Sefcik, and R. A. McKay, *Macromol.*, **14**, 275 (1981).

[356] P. Caravatti, P. Neuenschwander, and R. R. Ernst, *Macromol.*, **18**, 119 (1985).

[357] M. Linder, P. M. Henrichs, J. M. Hewitt, and D. J. Massa, *J. Chem. Phys.*, **82**, 1585 (1985).

[358] W. K. Chu, J. W. Mayer, and M. A. Nicolet, *Backscattering Spectroscopy*, Academic Press, Orlando (1978).

[359] U. K. Chaturvedi, U. Steiner, O. Zak, G. Krausch, G. Schatz, and J. Klein, *Appl. Phys. Lett.*, **56**, 1228 (1990).

[360] U. Steiner, U. K. Chaturvedi, O. Zak, G. Krausch, G. Schatz, and J. Klein, *Makromol. Chem., Macromol. Symp.*, **45**, 283 (1991).

Chapter 4

Surface Energetics and Contact Angle

The physical origin of surface tension is the unfavourable state of matter at interfaces. When atoms or molecules are exposed at an interface, they are no longer surrounded from every side by like molecules. They must either lose some of the interaction energy in the case of an ideal surface against vacuum or share some of the interaction energy with the molecules of the surrounding medium. Hence the thrust of liquids to minimize their surface area, or, in general, the occurrence of capillary phenomena (capillarity is the collective name of phenomena occurring on surfaces that are mobile enough to modify their shape according to surface tension requirements [1]).

Returning to Chapter 1, the origin of the surface energy of materials can be understood by thinking of the pairwise summation between the atoms of one medium with all the atoms of the other medium (equation (1.7), Figure 1.5). If the summation is extended to all atoms, including atoms in the same medium, two additional energy terms appear [2], namely:

$$W = -\text{constant} + A/12\pi D_0^2 \tag{4.1}$$

where the constant is the bulk or cohesive energy of the atoms of a given phase in their equilibrium position ($D = D_0$), A is the Hamaker constant (Chapter 1) and the second term arises from the existence of the two surfaces whose atoms have fewer neighbours than those in the bulk. Equation (4.1) shows that the total energy per unit area of two interacting surfaces separated by a distance D can be written, apart from cohesive energy, as follows:

$$W = (A/12\pi)(D_0^{-2} - D^{-2}) \tag{4.2}$$

For two isolated surfaces ($D = \infty$), equation (4.2) becomes:

$$W = A/12\pi D_0^2 = 2\gamma \tag{4.3}$$

whence:

$$\gamma = A/24\pi D_0^2 \qquad (4.4)$$

where γ is the parameter generally called surface tension.

Equation (4.4) could be used, in principle, to calculate the surface tension of condensed phases held together by van der Waals' forces (or, as discussed below, of the van der Waals' component of the total surface tension). The main problem with equation (4.4), however, is that it is not clear how the interfacial contact separation D_0 should be determined. This problem has been discussed by several authors [2–5]. Israelachvili found that a universal cut-off distance of 0.165 nm yields values for γ in good agreement with measured values for liquids and solids.

Generally, surface chemistry textbooks introduce surface tension with the observation that work is required to extend a thin soap film. While this approach is of great help in the case of liquids, where all the work exerted is used to create new surface, a more thermodynamic approach is required for solids, which, by definition, are portions of matter able to withstand stress without deformation [1].

Thermodynamics handles the state of surface species by a quantity to be added to the energy of the system, therefore the energy balance must contain a further work term, γdA, related to the surface contribution. For a reversible process and for a planar single interface:

$$dE = TdS - PdV + \gamma dA + \sum \mu \, dn \qquad (4.5)$$

where A is the area of the interface. In a more general case the interfacial contribution is constituted by a summation over all interfaces.

With reference to the definition of the Gibbs (G) and Helmholtz (F) free energy it is possible to write:

$$dG = -SdT + VdP + \gamma dA + \sum \mu \, dn \qquad (4.6)$$

$$dF = -SdT - PdV + \gamma dA + \sum \mu \, dn \qquad (4.7)$$

so that the surface tension γ can be defined by the following equation:

$$\gamma = \left(\frac{dE}{dA}\right)_{S,V,n} \qquad (4.8)$$

$$\gamma = \left(\frac{dG}{dA}\right)_{T,P,n} \qquad (4.9)$$

$$\gamma = \left(\frac{dF}{dA}\right)_{T,V,n} \qquad (4.10)$$

In the thermodynamics of interface-containing systems the surface energy, Gibbs surface free energy and Helmholtz surface free energy are defined as excess quantities, drawing an imaginary and arbitrary dividing mathematical surface (Gibbs surface) between the two phases separated by the interface. The two phases are considered homogeneous up to the dividing surface:

$$E_s = e_s A = (E - E_\alpha - E_\beta) \tag{4.11}$$

$$G_s = g_s A = (G - G_\alpha - G_\beta) \tag{4.12}$$

$$F_s = f_s A = (F - F_\alpha - F_\beta) \tag{4.13}$$

where:

E_s = surface energy
e_s = specific surface energy
A = area of the Gibbs dividing surface
E = energy of the whole system
$E_{\alpha, \beta}$ = energy of the phases separated by the interface

and the same nomenclature holds for the Gibbs and Helmholtz free energy.

From the previous equations it is clear that surface quantities are those parts of the total that cannot be accounted for by treating the system as homogeneous up to the dividing mathematical surface of zero volume. A similar definition also holds for the adsorption Γ: the number of moles of a given component that must be attributed to the unit surface area, frequently quoted as surface excess.

As shown, for instance in reference [6], the specific quantities of equations (4.11) to (4.13) are connected to the surface tension in this way:

at constant V and S $\qquad e_s = \gamma + \sum \Gamma \mu \tag{4.14}$

at constant P and T $\qquad g_s = \gamma + \sum \Gamma \mu \tag{4.15}$

at constant T and V $\qquad f_s = \gamma + \sum \Gamma \mu \tag{4.16}$

where the summation is extended to all components and Γ is the adsorption or surface excess of a given component.

From the above equations, the numerical value of surface tension and, for instance, Gibbs specific surface free energy coincide only in special cases. The above discussion was developed following the Gibbs suggestion of a mathematical dividing surface of zero volume. This formulation is discussed in many books, for instance [6,7], but some authors find this concept unnatural and prefer alternative approaches. Interfacial thermodynamics developed in the frame of the Guggenheim formulation of finite volume interface [8] are discussed in reference [9].

From the previous discussion it follows that surface tension is the thermodynamic parameter which controls the surface composition of materials. Clearly, as a thermodynamic parameter, it will not tell us if kinetics are quick enough to observe the surface composition predicted by surface tension, and, as a parameter describing a specific point of the interaction between bodies (the point of contact), it will not tell us if two separated bodies will touch (see Chapter 1). However, in the case of freely contacting bodies, surface tension will control all physical interfacial events, from the surface composition of an isolated material (atmosphere–solid contact), to wetting phenomena (liquid–solid contact), to whether two separate surfaces should be fused in a single interface (adhesion).

The previous discussion clearly highlights the importance of the measurement of the solid surface tension. Unfortunately, unlike liquids, whose surface tension can be measured by direct methods [1], simply measuring the work required in order to create a new surface, solids must be probed by indirect methods (see, however, the section on surface force measurements in Chapter 7).

Here we will discuss the two main indirect methods of solid surface tension measurement, that is contact angle and inverse gas chromatography methods. The former will be discussed in greater length, since it is by far the most frequently used and most accessible technique. Its deceptive simplicity, moreover, makes it a very interesting subject for discussion and reflection. The latter is a comparatively new technique, with every indication of a bright future.

4.1 CONTACT ANGLES AND THE YOUNG EQUATION

Contact angle measurement is probably the most common method of solid surface tension measurement. Contact angle data, especially in the case of polymeric materials, can be obtained with low price instruments and with simple techniques. As discussed below, however, the interpretation of data is not always straightforward, and the correct use of data requires knowledge of the thermodynamic status of the observed angles [1, 9–11].

At the basis of the measurement of solid surface tension by contact angle there is the equilibrium at the three-phase boundary, shown in Figure 4.1. The drop of liquid that is put on a solid surface will modify its shape under the pressure of the different surface/interfacial tensions, until reaching equilibrium. In 1805, Thomas Young described the three-phase equilibrium in terms of the vectorial sum shown in Figure 4.1, resulting in the following equation of interfacial equilibrium

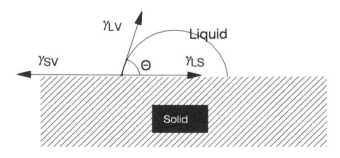

Figure 4.1   The Young equation.

[12, 13]:

$$\gamma_{sv} - \gamma_{sl} = \gamma_{lv} \cos \Theta \qquad (4.17)$$

where γ_{ij} is the interfacial tension between phases i and j, subscripts s, e and v refer to solid, liquid and vapour respectively and Θ is the equilibrium (Young) contact angle. The solid–vapour interfacial tension is linked to the intrinsic solid surface tension, or the surface tension of the solid in equilibrium with its own vapour or in vacuum, by the following relationship:

$$\pi_e = \gamma_s - \gamma_{sv} \qquad (4.18)$$

and π_e is called the spreading pressure, which represents the decrease of solid surface tension due to vapour adsorption. Equation (4.18) shows that, if π_e is negligible, the Young equation (4.17) can be written in terms of the true or intrinsic solid surface tension. Intuitively, it can be guessed that the role of the spreading pressure will increase, for a given liquid, with increasing solid surface tension. The problem is much debated [14, 15], although it is generally accepted that, if the contact angle is larger than about 10°, the spreading pressure can be safely neglected [16–20].

The experimental demonstration of the Young equation is very difficult, since there is practically no way of obtaining the solid surface and interfacial free energy contained in equation (4.17) by other independent, assumption-free techniques and this experimental uncertainty also had consequences on the theoretical side of the problem. In the past, the correctness and the generality of the Young equation was seriously questioned. However, as stated by Neumann [10], the Young equation results from the application of mathematical analysis and capillarity principles, much as the common Laplace equation of pressure drop across curved interfaces, therefore the two equations should have the same status and degree of acceptance. However, contrary to the Laplace equation, which can be experimentally checked and was never questioned, the experimental uncertainty concerning the former also leads to doubts on the theoretical side. Mathematical demonstrations of equation (4.7) were given for instance by Gibbs [21] (neglecting the effect of the gravitational field) and by Johnson [22] (where the effect of gravity is included. The latter paper also gives some interesting information on the debate on the correctness of the Young equation).

From the practical side, the calculation of solid surface tension data from contact angle measurement and the use of the Young equation must face two major problems, as follows.

Perhaps the greatest challenge which is posed to people trying to exploit the Young equation is the phenomenon called contact angle hysteresis: while, according to equation (4.17), the vectorial sum of interfacial tension should yield only one contact angle, it is commonly observed that a drop of liquid on a solid surface exhibits a range of allowed angles. The observed range is usually characterized by the measurement of the maximum (advancing angle) and the minimum (receding angle) allowed value. The problem is, of course, which of the observed angles (a range of some tens of degree is not unusual) should be used in equation (4.17). Contact angle hysteresis, or the difference between the contact angle of a drop of liquid advancing or receding on a solid surface, first discussed in the past century [23, 24], cannot be explained solely by the Young equation, and has greatly contributed to cloud by uncertainty the underlying theory and the experimental data. However, the dramatic development of contact angle hysteresis theories which occurred in the last part of this century have completely reversed the problem, and the correct interpretation of contact angle hysteresis data is now considered one of the most intriguing and fruitful features of contact angle measurement as a surface analytical technique.

The other main problem stems from the fact that the Young equation (4.17) actually contains two unknowns: in fact, while the contact angle is experimentally measured, as described below, and the liquid surface tension can be measured by several direct means [1], neither of the two other terms can be measured independently. Therefore a further relationship must be found, if it exists, but this search is plagued by the same problem which affected the debate on the Young equation. On the other side, which makes contact angle measurement so appealing, there is almost no other way of obtaining data on the interfacial energetics of solids. As a consequence, many different theories have been proposed, and no general agreement on the best one exists.

In the following sections, contact angles, hysteresis and solid surface tension will be discussed. First, the more common experimental ways of contact angle measurement will be presented. Then contact angle hysteresis theories will be briefly reviewed, and finally the contact angle–interfacial energetics relationship will be discussed.

4.2 EXPERIMENTAL MEASUREMENT OF CONTACT ANGLES

The three most commonly used methods of contact angle measurement are the sessile drop, the captive bubble and the Wilhelmy plate technique [1, 9, 10, 25].

In the sessile drop experiment, a droplet of a properly purified liquid is put on the solid surface by means of a syringe or a micropipette. The droplet is generally observed by a low magnification microscope, and the resulting contact angle, according to Figure 4.1, is measured by a goniometer fitted in the eyepiece. Contact angle hysteresis is measured by increasing or decreasing the drop volume until the three-phase boundary moves over the solid surface, as shown in Figure 4.2(a). It is important to avoid vibration and distortion of the droplet during the volume change, and this is best obtained by keeping the capillary pipette of the microsyringe immersed in the drop during the entire measurement. The contact angle is not affected by the presence of the pipette in the drop of liquid [9, 25].

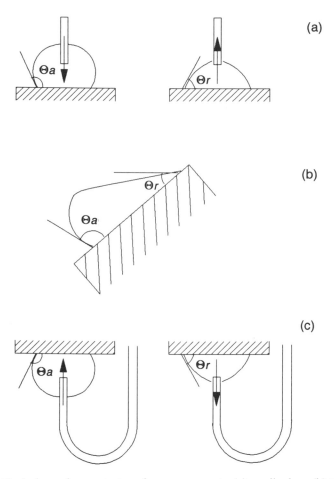

Figure 4.2 Techniques for contact angle measurements: (a) sessile drop (b) tilting plate (c) captive bubble.

Very often, in published papers and reports, it is possible to find data on a contact angle measured by the sessile drop technique, simply putting a drop of liquid on the surface. This type of measurement is usually called static, as opposed to the dynamic contact angles which are measured, for instance, by the Wilhelmy plate technique (discussed below). If the surface shows contact angle hysteresis this static angle should not be reported since, as discussed below, it bears no more information that any of the other allowed values included between the advancing and the receding angle. The correct measurement in this case requires that both the advancing and receding angle should be measured, as previously described, and reported. The latter are, in a sense, dynamic, since they are obtained by measuring the contact angles of a slowly moving drop. However, as discussed by Andrade [26], there is no difference between static and dynamic values of the contact angle for low velocities of the liquid front with normal low viscosity liquids.

Another way of measuring the advancing and receding angles is the so-called tilting plate, where the probed surface is tilted until the drop of liquid moves (Figure 4.2(b)). In a further technique the receding angle is measured leaving the liquid to evaporate from the deposited drop. A definite drawback of this approach is the large amount of time taken by the measurement, which means that all surface active impurities, either from the atmosphere or from the sample, can accumulate at the interface and affect the result [26].

In the captive bubble technique, an air or liquid drop is put on the sample surface immersed in a liquid medium by means of a U-shaped needle. If two liquids are used they must be, of course, immiscible. Typical examples are hydrocarbon bubbles in water, an approach which is used when the solid surface has to be probed in an aqueous environment. Advancing and receding angles are measured as in the sessile drop technique, increasing or decreasing the bubble volume until moving the three-phase boundary (Figure 4.2(c)).

In the Wilhelmy plate technique the advancing and receding angles are calculated from the force exerted as the sample is immersed or withdrawn from a liquid. A typical instrumental setup is shown in Figure 4.3: it is composed of a microbalance and a movable stage, on which the liquid container is fitted. Usually, a computer controls the stage velocity and movements and provides the software required for calculations.

The results are generally described by the following equation:

$$F = F_w + F_{ie} + F_b = mg + \gamma_L \, p \cos \Theta + (x - x_0) A \, dg \qquad (4.19)$$

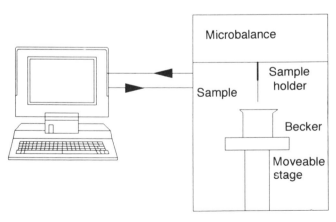

Figure 4.3 Typical experimental set-up for the Wilhelmy plate technique.

where the three contributions to the overall force have been emphasized. In equation (4.19), F_w represents the weight contribution (m is the sample mass and g is the local gravity constant), usually zeroed during the calibration routine in the experimental practice, and F_{ie} is the contribution of interfacial energetics, that is the product of the liquid surface tension (γ_L) times the sample perimeter (p) times the cosine of the contact angle. Finally, F_b is the buoyancy contribution, given by the product of the immersed volume (the stage displacement, x, minus the displacement at zero depth of immersion, x_0, times the sample area) times the liquid density, d, times g.

It is clear that, in the typical case of a sample of constant perimeter, equation (4.19) yields a straight line, where $F = F_w + F_{ie}$ at $x = x_0$ (hence the contact angle can be calculated), and whose slope increases with increasing liquid density. In real cases, contact angle hysteresis gives rise to parallel advancing and receding tracks, and the finite time interval required for an advancing or receding meniscus to fully develop gives a smooth appearance to discontinuities (the point of immersion and where relative solid–liquid motion is reversed), as shown in Figure 4.4.

Equation (4.19) has recently been written in a more general form in order to take into account the shear stress effect which can be exerted by the fluid on the sample surface [27, 28], which, from basic fluid mechanics, can be written as,

$$F_v = p(x - x_0)\tau_0 \tag{4.20}$$

where the subscript v indicates that this contribution arises from the shear stress (τ_0) exerted by the viscous flow of the liquid on the solid. The shear stress

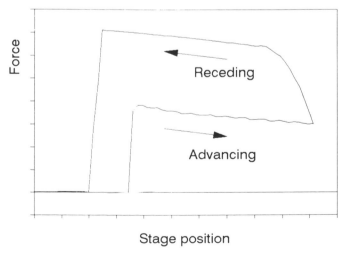

Figure 4.4 Typical wetting cycle obtained by the Wilhelmy plate technique.

contribution has been modelled and quantitated in the case of the wetting of carbon fibre by Newtonian fluids [27, 28]. The qualitative effect is the same also for non-Newtonian fluids. Its effect on the general shape of the recorded tensiogram is shown in Figure 4.5: in the case of wetting of the carbon fibre by a dilute water/poly(ethyleneglycol) (PEG) solution of low viscosity (0.01% PEG,

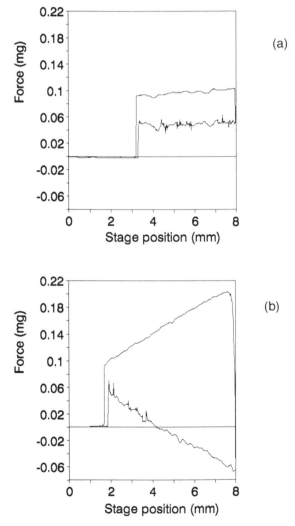

Figure 4.5 Effect of liquid viscosity on the wetting force: (a) carbon fibre in 0.01% PEG_{3400}/water solution. (b) Carbon fibre in 1% PEG_{3400}/water solution. Stage velocity is 150 μm/s in both cases.

$M_w = 3400$) (Figure 4.5(a)), the recorded force is explained by equation (4.19), in which the buoyancy contribution is negligible due to the small cross-section of the fibre. On the other hand, when the same fibre is wetted by a viscous water/PEG solution (1%), a marked slope is observed. Note that the sign of the slope changes when the direction of motion is reversed (on the other hand, the slope due to buoyancy is always of the same sign, yielding parallel advancing and receding tracks, as shown in Figure 4.4).

This shear stress effect plays a measurable role only when highly viscous fluids are used: it has been calculated [28] that in the case of wetting of the same carbon fibre by water, a stage velocity of 1 cm/s would be required in order to obtain a slope of 1 μg/mm.

The main advantages of the Wilhelmy plate technique over more conventional measurements are the control of the interfacial velocity, the possibility of performing sequential loops [26], as discussed in the next section, and the fact that measurements can also be performed on small diameter fibres. On the other hand, its main disadvantages are that the two surfaces of the sample must be identical and that plots resulting from samples of complex shape are difficult to interpret.

4.3 CONTACT ANGLE HYSTERESIS

4.3.1 Thermodynamic Hysteresis

As discussed previously, a drop of liquid on a solid surface usually does not follow the behaviour predicted by the Young equation, since, instead of only the one angle resulting from the balance of interfacial energetics, it usually displays a range of allowed contact angles. The reason for this disagreement between theory and practice is the non-ideal nature of real surfaces. According to the hypothesis underlying the Young equation, solid surfaces should be homogeneous and smooth. Moreover the liquid–solid interaction should not result in swelling of the solid, liquid penetration or liquid-induced surface restructuring or deformation. In other words, the drop of liquid should behave as a neutral probe, only characterized by its interfacial tension parameters, and should interact with the smooth, homogeneous and rigid solid surface only by interfacial energetics. Clearly, real surfaces very often do not fulfil all these requirements. Contact angle hysteresis results from the relaxation of these constraints.

Before discussing the physical meaning of measurable contact angles on non-ideal surfaces, it is important to consider how the Young equation itself and the equilibrium contact angle are modified by surface defects. In 1936, Wenzel reasoned that within a measured unit area of a rough surface there is more surface than in the same measured unit area on a smooth surface [29]. While specific surface quantities are, of course, the same on the two surfaces, the relative magnitude of the vectors composing the Young equation are modified, as shown in Figure 4.6. Wenzel proposed the following equation:

$$r(\gamma_s - \gamma_{sl}) = \gamma_{lv} \cos \Theta_w \tag{4.21}$$

where r is the so-called roughness factor:

$$r = \text{actual surface/geometric surface} \tag{4.22}$$

as shown in Figure 4.6. The subscript 'w' indicates that the angle appearing in equation (4.21) (usually called the Wenzel angle) is different from that of the Young equation (4.17) (Young angle, or Θ_y). Their relationship is the following:

$$\cos \Theta_w = r \cos \Theta_y \tag{4.23}$$

Since the roughness factor is always greater than unity, equation (4.23) shows that the effect of roughness is to magnify the wetting properties of the solid, that is the Wenzel angle will increase with roughness if the Young angle is greater than 90° and will decrease if it is less than 90°.

The left side of equation (4.23) is generally called effective adhesion tension, and it couples intrinsic material properties (the Young angle for a given liquid) with physical conditions (r).

A rigorous derivation of the Wenzel equation was later given by Good [30]. It is important to remember that the physical feature affecting the Wenzel angle is the increased surface area of rough surfaces, so that no simple correlation is expected with surface roughness alone, as measured, for instance, by a profilometer, as discussed by Wenzel in a famous debate with Bikermann [31, 32].

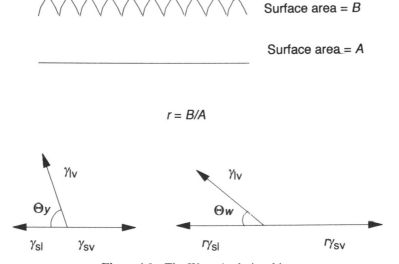

Figure 4.6 The Wenzel relationship.

As previously discussed, the Young equation assumes that surfaces are homogeneous. On the other hand, real surfaces can be heterogeneous, that is they can be composed by domains of different composition and hence different wetting properties. This domain structure can be either a consequence of surface contamination or the result of surface phase separation (Chapter 8). The effect of a patchy structure on the Young equation was described by Cassie, who proposed the following equation (Cassie equation) for a two-component, heterogeneous surface [33]:

$$\cos \Theta_c = Q_1 \cos \Theta_1 + Q_2 \cos \Theta_2 \tag{4.24}$$

where subscripts 1 and 2 refer to the two components of the surface, Q_1 is the fractional coverage of component i and Θ_1 its Young contact angle. The angle Θ_c is the Cassie angle, or the equilibrium contact angle of a given liquid on the heterogeneous surface.

The Cassie treatment of heterogeneous surfaces has recently been reviewed by Israelachvili and Gee [34], who modified equation (4.24) in order to account for heterogeneities close to atomic or molecular dimensions:

$$(1 + \cos \Theta_c)^2 = Q_1 (1 + \cos \Theta_1)^2 + Q_2 (1 + \cos \Theta_2)^2 \tag{4.25}$$

which replaces equation (4.24) whenever the size of the chemically heterogeneous patches approach molecular or atomic dimensions. In the latter treatment, instead of cohesion energies, polarizabilities, dipole moments and surface charges are averaged, owing to the very low dimension of heterogeneity. This treatment yields equilibrium angles lower than those predicted by equation (4.24).

The previous discussion focused on the effect of roughness and heterogeneity on the equilibrium contact angle, that is on the minimum of the free energy–contact angle relationship. Contact angle hysteresis, at this stage, is still unexplained. In this respect, it has been shown that roughness and heterogeneity have another major effect on the free energy–contact angle relationship, which is schematized in Figure 4.7: in the case of the non-ideal surface a large number of accessible metastable states (local minima) are introduced. Their physical origin stems from the fact that roughness and heterogeneity perturb the behaviour of a drop of liquid in the same way, by hindering the droplet motion [35]. As a consequence, free energy barriers are created and the system can reside in one of the many free energy local minima. Curves such as the one shown in Figure 4.7(b) have been calculated by many researchers, from the pioneering and fundamental works of Johnson and Dettre on highly idealized model surfaces to more general and comprehensive treatments [36–47]. Results of the first models have often been confirmed by more sophisticated treatments.

The quoted models predict that roughness and heterogeneity leads to a large number of metastable configurations separated by energy barriers, as shown in Figure 4.7. The lowest free energy configuration corresponds to the angle calculated from the Wenzel equation in the case of roughness and by the Cassie

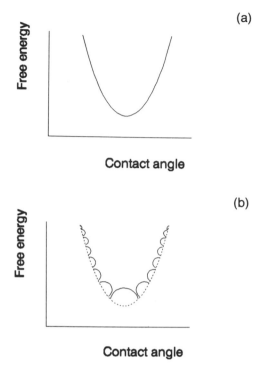

Figure 4.7 The effect of a non-ideal surface (heterogeneous or rough) on the free energy–contact angle relationship: (a), ideal surface; (b), non-ideal surface.

equation in the case of heterogeneity. The system will reside in a metastable state, depending on the height of the energy barriers and the macroscopic vibrational energy of the droplet, so that it is impossible to distinguish the true minimum of the free energy from the other metastable states. The two main consequences of this feature are that it is impossible to measure the Wenzel or the Cassie angles (they are indistinguishable from the other allowed angles) and that, as already discussed, it makes no sense to characterize a surface showing contact angle hysteresis by a static angle, since the measured value bears no more information than any of the other allowed angles. On the other hand, the measurement and the analysis of contact angle hysteresis, that is of the maximum (advancing) and the minimum (receding) angle can give useful information on the surface structure of the sample. Figure 4.8 shows the theoretical behaviour of the advancing and receding angle as a function of roughness for surfaces whose Young contact angle (that is the equilibrium contact angle on the corresponding smooth surface) is lower or greater than 90° respectively. In both cases the advancing angle increases and the receding angle decreases as roughness increases. Above a certain degree of roughness, depending on roughness dimension and geometry, wettability shifts

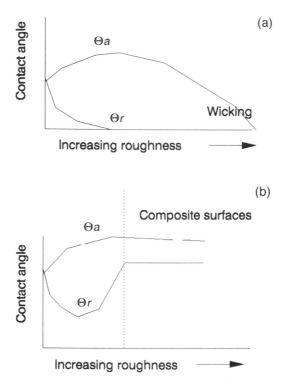

Figure 4.8 Theoretical behaviour of advancing and receding angles as a function of roughness (a) $\Theta_Y < 90°$ (b) $\Theta_Y > 90°$.

from an interfacial energetics to a capillary phenomenon [48]. In surfaces whose Young angle is lower than 90° wicking, liquid penetration promoted by capillarity occurs, and the drop of liquid is adsorbed in the textured solid (the precise geometry of the surface texture, however, plays an important role. As discussed in Chapter 9, even very small contact angles can prevent water from entering pores of suitable fabrics). On the other hand, on hydrophobic surfaces characterized by $\Theta_Y > 90°$, the liquid can no longer penetrate the cracks and crevices of the surface, and the drop of liquid is in contact with a surface composed in part by the solid and in part by air. These composite surfaces are an extreme case of a heterogeneous surface, and their equilibrium contact angle can be expressed by equation (4.6), taking the contact angle of the liquid on air as 180°, that is:

$$\cos \Theta_{cb} = Q_1 \cos \Theta_1 - Q_2 \tag{4.26}$$

this is called the equation of Cassie and Baxter [49, 50]. Theory shows that the free energy barrier between metastable states is greatly reduced in composite surfaces, hence the large reduction of hysteresis shown in the right-hand part of

Figure 4.8 (b). Composite surfaces, whose physical nature greatly enhances their non-wetting characteristics, exist in nature in furs, feather and leaves. An example of transition to a composite surface, in full agreement with the theoretical behaviour of Figure 4.8(b), is shown in Figure 4.9 [15]. Here, a poly(tetra-fluoroethylene) (PTFE) sample has been subjected to an oxygen plasma treat-

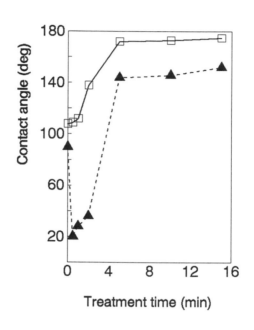

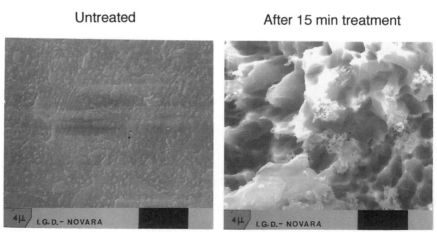

Figure 4.9   The effect of time of O_2 plasma etching on water contact angles and surface morphology of PTFE.

ment, that, as shown in the figure, has a market etching effect on PTFE. The hydrophobic nature of PTFE (the contact angle of water on smooth PTFE is about 108°) does not allow water to penetrate the rough surface structure created by etching, and a composite surface, characterized by very high contrast angles, low hysteresis and freely rolling drops of water, is ultimately produced.

The effect of surface heterogeneity on advancing and receding angles is shown in Figure 4.10(a), for the model case of a surface composed of two different

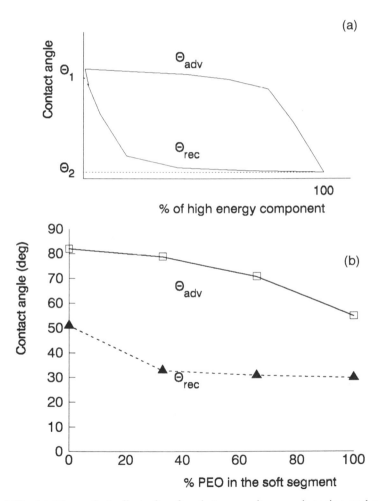

Figure 4.10 (a) Theoretical effect of surface heterogeneity on advancing and receding contact angles. Θ_1 is the Young angle of the low energy component, Θ_2 the Young angle of the high energy component. (b) Water contact angle of polyurethanes with different amounts of poly(ethyleneoxide)–poly(tetramethyleneoxide) in the soft segment. Data plotted from Table 2 of ref. [52].

domains. The advancing angle reaches the saturation value of the less wet component at low fractional coverages, while the receding angle reflects more closely the wet portion of the surface. In Figure 4.10(b), experimental results by Andrade and coworkers [52] on phase separated polyurethanes with different amounts of poorly and highly wettable segments are compared with the theoretical prediction.

4.3.2   Kinetic Hysteresis

The quoted theories and examples refer to thermodynamic hysteresis, that is a time independent difference between the advancing and receding angle. It is theoretically similar to other examples of hysteresis existing in nature, all of them promoted by the existence of a large number of metastable states accessible to the system [53]. Conversely, it is often possible to observe a kinetic hysteresis, caused by time dependent liquid–solid interactions. As a consequence of these interactions the solid or the liquid or both are modified and, if the modification time is comparable with the measurement time, the observed angle will appear time dependent.

A very simple example of kinetic hysteresis is given by water contact angle measurement on poly(dimethylsiloxane) (PDMS). The extremely high advancing water contact angle (113°) reflects the very hydrophobic nature of methyl groups. The receding angle, on the other hand, is rather low (about 80° against greater than 90° observed on the methyl-groups covered polypropylene) and decreases as the contact time between the water drop and the PDMS surface increases (Figure 4.11) [54]. Probably, the behaviour of the receding PDMS angle reflects two different time dependent mechanisms: the first one is the reorientation of the very flexible siloxane backbone, which allows the interaction to be maximized between water and the high energy Si–O–Si units [55]. The second mechanism involves water diffusion in the open silicone structure, both entropically (by the infinite gradient of concentration at the water–PDMS interface) and enthalpically (by the Si–O–Si/water interaction) driven.

The most well known example of kinetic hysteresis is given by the water wettability of poly(2-hydroxyethylmethacrylate) (PHEMA) hydrogels, described by Holly and Refojo in a classical paper [56]. They observed a very large hysteresis and a high value of the advancing angle, despite the large water content of these materials. The reason for the observed behaviour was attributed to structural changes at the gel–air phase boundary, the hydrophilic groups being buried in the aqueous phase within the gel when the surface was exposed to air, but able to quickly reorientate in a water environment. From a contact angle measurement point of view, this means that the surface probed by the advancing angle is much more hydrophilic than the one probed by the receding angle (Figure 4.12), hence the large contact angle hysteresis measured. More generally, this finding opened the way towards the idea of the dynamic behaviour of

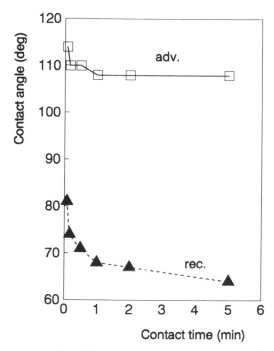

Figure 4.11   Advancing and receding water contact angles on poly(dimethylsiloxane) as a function of the contact time of the water drop.

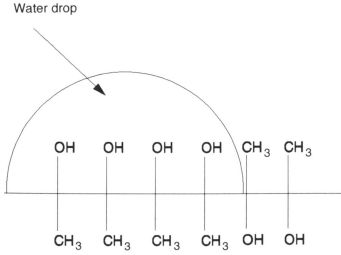

Figure 4.12   Schematic of the Holly and Refojo model of the poly(hydroxyethylmeth-acrylate) surface.

polymer surfaces (Chapter 2) and the recognition that surface properties of polymers can be environment dependent [26].

Swelling and water penetration are a general cause of time dependent hysteresis. These effects can be readily observed changing the measurement time. In this respect, the Wilhelmy plate technique, where the stage speed and the measurement time can be carefully and reproducibly controlled, is the technique of choice. Figure 4.13 shows sequential loops at different stage velocities on a PP surface grafted (see Chapter 7) with the rather hydrophobic methylmethacrylate (MMA) and the hydrophilic hydroxyethylacrylate (HEA). The grafting routine yields a bumpy surface texture [57], which, in the present case, is very similar on the two samples. The Young angle of both PMMA and PHEA is below 90° so that,

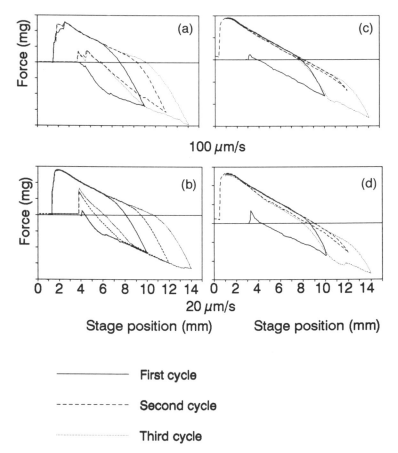

Figure 4.13 Sequential wetting loops for MMA grafted (a, b) and HEA grafted (c, d) polypropylene surfaces for two different surface velocities.

according to Figure 4.8(a), water penetrates into the pores. Thus, in the second and third loops, the water front advances over a wetted surface, and, accordingly the advancing angle is reduced. However, if the stage velocity is decreased, the water contained into the pores has enough time to partially evaporate from the more hydrophobic PMMA, and sequential loops become more and more reproducible as the stage velocity is decreased, while only slight effects are observed on the very hydrophilic PHEA grafted surface.

Another intriguing time dependent effect was observed by Bayramli and coworkers when measuring the contact angle, by the Wilhelmy plate technique, of high energy fibres as a function of stage velocity [58]. They observed a sawtooth shape of the advancing portion of the tensiogram above a critical stage velocity, below which the conventional shape was observed. This behaviour was attributed to the adsorption of surface active impurities (SAI) at the three-phase boundary. At sufficient low velocity the advancing water front always sees a contaminated surface, hence the smooth tensiogram recorded. On the other hand, at higher velocities, the combined effect of the three-phase line motion and SAI diffusion gives rise to the observed behaviour.

Recently, similar behaviour was observed in the study of the water wettability of thin PHEMA films [28, 59], Figure 4.14. In this case, a possible explanation involves the previously quoted Holly and Refojo model of PHEMA surfaces: at low stage velocity, the overturning of hydrophobic groups has enough time to keep pace with the advancing water front. However, as the stage velocity is increased, the water front must advance over a surface which is much less wettable than the already wet surface. Thus, the unwetted portion of the surface behaves like a poorly wettable surface with classic contact angle hysteresis as previously described, except that, in accordance with the dynamic approach, it reverts to a more wettable surface when covered by water. Another possible explanation of the observed behaviour is the deformation induced by the vertical component of the liquid surface tension, which causes a type of highly localized (and velocity dependent) surface roughness. As discussed in the book edited by Andrade [26] this effect should play a role in hydrated, low modulus surfaces.

More generally, the sawtooth behaviour seems the experimental output of the time dependent obstacle to liquid motion on the solid surface. In classic, time-independent, contact angle hysteresis theory, these obstacles, even if of a different nature (either morphological or chemical), create the same free energy barrier to the liquid motion and give rise to the same macroscopic behaviour (widely different advancing and receding angle). In the same way, in the present dynamic studies, velocity-dependent defects of a different nature (surface heterogeneity due to reorientation or accumulation of SAI, or localized surface roughness) create the same velocity dependent free energy barriers, resulting in a temporarily hindered motion above a certain critical liquid–solid speed.

From the previous discussion it is clear that the correct analysis of contact angle hysteresis data, the different wetting behaviour in response to a different

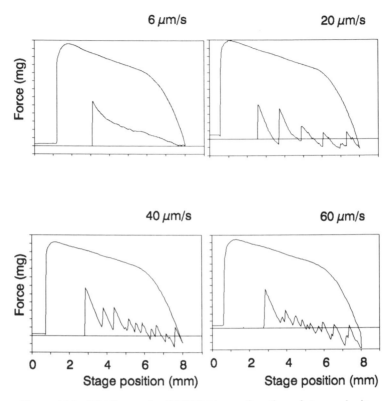

Figure 4.14   Wetting cycle of PHEMA as a function of stage velocity.

environment, and the time dependence of contact angles can offer valuable help in understanding the nature of polymer surfaces. On the other hand, the use of contact angle data within the Young equation is sometimes a delicate matter, since among all the observable angles, only a few have the thermodynamic status of Young angles. A careful analysis of contact angle behaviour is always required before deciding if the measured macroscopic angle is the result of surface energetics vectors in the Young sense, or rather the combined effect of surface energetics and morphological features, liquid or environment induced effects, and so on. If the measured angle is recognized as a Young angle, the problem is now, as discussed before, how to the handle the basic one equation–two unknown relationship (equation (4.17)). This topic will be discussed in the next section.

4.4   FROM CONTACT ANGLE TO SURFACE TENSION

As discussed previously, only two of the four variables of the Young equation can be readily measured, that is the liquid interfacial tension and the contact angle. In order to obtain the solid surface tension a further relationship must be found, if it exists such as:

$$\gamma_{sl} = f(\gamma_{sv}, \gamma_{lv}) \tag{4.27}$$

The search for such a relation has been and remains the battleground of many researchers. The history of its development is at least as long and controversial as that of contact angle hysteresis.

In the following, we will discuss the main theories, starting from the well-known critical surface tension.

4.4.1 Critical Surface Tension

The concept of critical surface tension was first introduced by Fox and Zisman in 1950 [60–63]. They observed a quasi-linear relationship between the cosine of the advancing angle and the liquid surface tension for a series of homologous liquids. If non-homologous liquids were used, experimental points formed a narrow rectilinear band. A typical Zisman plot on a hydrophobic polymer is shown in Figure 4.15.

Strictly speaking, this relationship is not linear and we will reconsider its functional dependence when dealing with other methods. Water and other hydrogen-bonding liquids usually appreciably deviate from linearity. However, the operational definition of the critical surface tension is the extrapolation to $\cos \Theta = 1$ of the straight-line plot of the said variables, and in this way the minimum value of liquid surface tension required for spreading is obtained. The plot of advancing contact angles as a function of the liquid surface tension is usually called the Zisman plot. The critical surface tension concept met great success, allowing the difficulties in searching equation (4.27) to be overcome. Yet, the existence of the relationship depicted in the Zisman plot suggests that such an equation does exist.

The meaning of the critical surface tension can be understood by substituting its definition in the Young equation:

$$\gamma_c = \gamma_{sv} - \gamma_{sl} \tag{4.28}$$

or

$$\gamma_c = \gamma_s - \gamma_{sl} - \pi_e \tag{4.29}$$

thus the critical surface tension is smaller or equal to the true surface tension. The concept of critical surface tension is widely used and Zisman and coworkers must be credited for having offered the first workable theory of measurement of interfacial energetics in systems containing solid phases. The introduction of the critical surface tension greatly benefited many surface-related technological fields. Actually, its main drawback arises from too broad an application of the method. While Zisman and coworkers highlighted that the critical surface

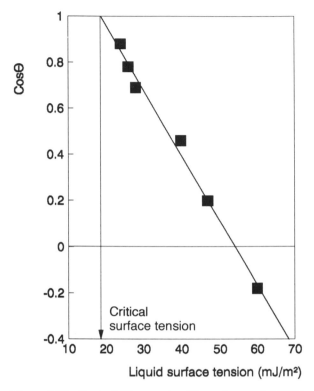

Figure 4.15   Typical Zisman plot. The substrate is PTFE.

tension is not the solid surface tension and took care to work on surfaces as ideal as possible (in the sense of the Young equation previously discussed), too often the two parameters are considered identical and researchers consider it sufficient to measure only the advancing angle, in this way losing all the information from hysteresis data. Moreover, as already stated, water and other liquids usually deviate appreciably from linearity when the surface that is measured contains functional groups which can establish hydrogen-bonding with the liquid. This observation suggests that one parameter is not enough to describe the interactions going on at the interface and clearly, thinking of the several different forces operating at interfaces, as discussed in Chapter 1, it would sound strange if, for instance, the interaction between a surface bearing hydrophilic groups and a van der Waals' liquid or a hydrogen-bonding liquid, could be described by the same parameter (this idea is, however, the basis of the Neumann equation of state theory, see below). The inherent complexity of the interfacial interaction is the idea beyond the development of subsequent theories.

4.4.2 Other Methods

In the second half of the 1950s, Good *et al.* formulated a theory of interfacial tension using microscopic (statistical mechanics) considerations [64–69], as reviewed by Good in an interesting paper [70].

The starting point of the theory is the Berthelot relation for the attractive constant between like and unlike molecules [71], Good *et al.* tried to relate the interfacial tension between two phases (1 and 2) to the geometric mean of the surface tension of each phase:

$$\gamma_{12} = \gamma_1 + \gamma_2 - 2(\gamma_1\gamma_2)^{1/2} \tag{4.30}$$

Equation (4.30) was tested and rejected by Rayleigh some fifty years earlier, but the wealth of data collected by Good showed the reason for its apparent failure, because it is a particular case of a more general equation, now known as the Good–Girifalco equation, namely:

$$\gamma_{12} = \gamma_1 + \gamma_2 - 2\Phi(\gamma_1\gamma_2)^{1/2} \tag{4.31}$$

where Φ is the so-called interaction parameter. Φ is not a mathematical trick, but a characteristic of a given system that can be evaluated from molecular properties of the two phases. The interaction parameter was found very close to unity when dominant cohesive and adhesive forces were of the same kind. Interfaces satisfying these requirements were called regular interfaces. To a good approximation, the interaction parameter is given in this case by:

$$\Phi = 4(V_1\,V_2)^{1/3}/(V_1^{1/3} + V_1^{1/3})^2 \tag{4.32}$$

where V's are molar volumes.

The reason why Rayleigh failed to recognize the correctness of equation (4.30) is that the interfaces he investigated were not regular, so the interaction parameter deviated appreciably from unity.

If equation (4.31) is substituted in the Young equation (4.17), the following relationship results:

$$\gamma_s = [\gamma_1(1 + \cos\Theta) + \pi_e]^2/4\Phi^2\gamma_1 \tag{4.33}$$

or, neglecting spreading pressure:

$$\gamma_s = \gamma_1(1 + \cos\Theta)^2/4\Phi^2 \tag{4.34}$$

In this way it is possible to calculate the value of the solid surface tension from a single contact angle measurement and the knowledge of the interaction parameter of the system.

The Good–Girifalco theory stimulated much work on the real meaning of Zisman's critical surface tension, its relationship with the true solid surface tension [72, 73] and how to express the interaction parameter in terms of molecular properties [74]. Wu exploited the critical surface tension and the

Good–Girifalco theory and proposed an equation of state [75] that gives accurate values of surface tension [9]. By substituting the definition of critical surface tension in (4.33) and expanding in a power series, a rapidly converging expression was obtained, so that:

$$\gamma_{c,\phi} = \Phi^2 \gamma_s - \pi_e \qquad (4.35)$$

where the second subscript of the critical surface tension indicates its dependence on the interaction parameter.

Combining (4.35) and (4.34):

$$\gamma_{c,\phi} = (\tfrac{1}{4})(1 + \cos \Theta)^2 \gamma_1 \qquad (4.36)$$

In this way a spectrum of critical surface tensions can be straightforwardly calculated and, if data are plotted as a function of liquid surface tensions, a smooth curve is obtained. Since the maximum value of Φ is unity (and in this case, neglecting spreading pressure, the critical surface tension equals the true surface tension) the maximum of this curve is the solid surface tension [74, 75].

From the previous relationships it is possible to obtain an analytical representation of the slope of the Zisman plot. The straight line used in the Zisman plot can be written as:

$$\cos \Theta = 1 + a(\gamma_c - \gamma_1) \qquad (4.37)$$

and the expression for the slope can be readily obtained resolving equation (4.36) with respect to $\cos \Theta$ and taking its derivative with respect to the independent variable liquid surface tension [9]:

$$a = (\gamma_c / \gamma_1^3)^{1/2} \qquad (4.38)$$

confirming that the relationship represented in the Zisman plot is nonlinear.

The Good–Girifalco theory is one of the last where interfacial parameters are expressed as a single component. The last one is the much debated Neumann theory, whose starting point is the demonstration of the existence of equation (4.27) [76]. This finding gives the Young equation in the following form:

$$\gamma_{sv} - f(\gamma_{sv}, \gamma_{lv}) = \gamma_{iv} \cos \Theta \qquad (4.39)$$

or

$$\cos \Theta = g(\gamma_{sv}, \gamma_{lv}) \qquad (4.40)$$

According to equation (4.40), the surface tension of a given solid can be obtained simply by measuring the equilibrium contact angle of a liquid of known surface tension. Now the problem is how to explicitly formulate equation (4.27), and it can be solved either by a statistical thermodynamics or empirical approach. Neumann *et al.* followed the second way [77], obtaining the following relationship:

$$\gamma_{sl} = [(\gamma_{sv})^{1/2} - (\gamma_{lv})^{1/2}]^2 / \{1 - [0.015(\gamma_{sv}\gamma_{lv})^{1/2}]\} \qquad (4.41)$$

Equation (4.41) is not defined for all values of surface tension and the use of contact angles and proper algorithms bypasses these mathematical difficulties. Combining equation (4.41) with the Young equation, a third order equation in the solid surface tension is obtained. Computer programs were written [77–78] which allow the computation of the solid surface tension and solid–liquid interfacial tension from the liquid surface tension and the equilibrium contact angle.

The main theoretical consequence of the Neumann approach is that the established thermodynamic fact of the existence of equation (4.27) leaves no room for the possibility of obtaining surface tension components, that is the relative contribution of interaction of different physical origin, from contact angle measurement. According to Neumann, the Young equation belongs to the world of macroscopic thermodynamics, therefore it cannot give information on microscopic phenomena. This is in complete disagreement with all the theories developed after Good–Girifalco, as we will see below. According to Neumann, the surface tension components are adjustable parameters which are a function of the total solid and liquid surface tension and play the same role as the interaction parameter Φ of equation (4.31) in correcting the insufficient description of interfacial tension given by equation (4.30) [79].

Many comments and replies have been published on this topic [80–83]. In general, opposition to the equation of state approach has been manifested.

4.4.3 Surface Tension Components

The starting point of theories which split the interfacial parameter in several different terms is the classical paper of Fowkes published in 1962 [84]. In this paper it is suggested that the surface tension should be considered as a sum of independent terms, each representing a particular intermolecular force. For instance, the surface tension of water (w), whose main intermolecular forces are hydrogen bonding (h) and dispersion forces (d), can be written as:

$$\gamma_w = \gamma_w^h + \gamma_w^d \tag{4.42}$$

On the other hand, a saturated hydrocarbon (c), will be characterized by a surface tension wholly constituted by a dispersion term, so that at a water–hydrocarbon interface only dispersion interaction will be operative. In this way, applying the Good–Girifalco equation, the interfacial tension between water and a saturated hydrocarbon can be expressed as:

$$\gamma_{cw} = \gamma_c + \gamma_w - 2(\gamma_c\gamma_w^d)^{1/2} \tag{4.43}$$

This idea proved immediately useful in different fields of surface chemistry [85, 86], including among others contact angle measurement. When a drop of water is put on the surface of a solid which can interact only by dispersion forces, such as polymeric hydrocarbons, only dispersion forces will be operative at the water–solid interface. Neglecting spreading pressure, the Young equation then becomes:

$$\gamma_w(1 + \cos \Theta) = 2(\gamma_w^d \gamma_s)^{1/2} \tag{4.44}$$

so that the solid surface tension, in this particular case, can be readily calculated. Equation (4.44) is sometimes called the Young–Good–Girifalco–Fowkes equation. Fowkes's suggestion brought interfacial tension closer to the microscopic world. This theory stimulated a great deal of work and has been frequently reviewed ([87]), for instance). From the point of view of contact angle measurement, it gave rise to a number of different equations, in order to calculate, beside the dispersion component, the other components of the interfacial tension. Hydrogen bonding and interactions between permanent dipoles were, in general, lumped together in a so-called polar component. Several different treatments extended the geometrical mean approach of equation (4.44) also to the polar term, as follows:

$$\gamma_{sl} = \gamma_s + \gamma_l - 2(\gamma_s^d \gamma_l^d)^{1/2} - 2(\gamma_s^p \gamma_l^p)^{1/2} \tag{4.45}$$

where the superscript p indicates, of course, the polar components. It must be noted that Fowkes always opposed this generalization of equation (4.43). The geometrical mean approach was suggested for instance by Owens and Wendt [88] and by Kaelble [89, 90]. In the former case the two unknown solid surface tension components can be calculated by measuring the equilibrium contact angle of two different liquids on the surface. This allows, combining equation (4.45) with the Young equation, to write a system of two equations and two unknowns (the liquid surface tension component can, of course, be measured with conventional methods and are tabulated [9]. The latter reference also gives a complete quantitative comparison of the different methods).

Another approach was suggested by Wu, the so-called harmonic (reciprocal in the original paper) mean method [91]. The dispersion and the polar interaction terms of equation (4.45) are substituted by:

$$I^d = 2(\gamma_l^d \gamma_s^d)/(\gamma_l^d + \gamma_s^d) \tag{4.46}$$

and

$$I^p = 2(\gamma_l^p \gamma_s^p)/(\gamma_l^p + \gamma_s^p) \tag{4.47}$$

Again a system of two equations and two unknowns can be written and the measurement of the equilibrium contact angle of two liquids is required. Wu's suggestion is based on the empirical observation that the harmonic mean method gives better results for polymers. Both methods are approximated descriptions of the dispersion term, and it is difficult to decide a priori which approximation works better. As to the polar part, the treatment given in [91] leads to the geometric mean, but the approach was found inaccurate, while good results were obtained with the harmonic mean. Water and methylene iodide are generally used as measuring liquids, whichever the equation used afterwards.

A detailed account of the importance of the correct choice of the liquids is given by Dalal [92]. Basically, he shows the capital importance of using liquids with different polarities since, if the polarity of the liquids is similar, small errors in the input data lead to large errors in the results. In this sense, since water and methylene iodide have very different polarities, their use is correct. Better results can be obtained by using many liquids in order to obtain an overdetermined set of equations, so that a simultaneous best-fit solution can be obtained. Dalal shows also that the geometric and the harmonic mean methods give very close results, contrary to the suggestion of Wu [9]. This disagreement apparently results from the different input value of the liquid surface tension component used by the two authors.

Another intriguing contribution is given by Erbil and Meric [93], which used nonlinear programming methods to obtain surface tension components and the critical surface tension.

Many treatments, based on the captive bubble technique, have been proposed for the calculation of interfacial energetics in liquid environments [94–98]. This approach is of capital importance in those cases where surface mobility of polymers allows restructuring processes when the interfacing phase is changed (Chapter 2) [26]. In the quoted methods, interfacial parameters are calculated either according to the geometric mean or to the harmonic mean approach. A comparison of different theories is presented in [99]. In this paper the critical surface tension, the geometric mean and the harmonic mean were used to characterize the polymer–air interface while a geometric mean method was used to characterize the polymer–water interface. Surface tension components appeared as functions both of the coupling of liquids and the approximation used, while the calculation of the overall surface tension yielded consistent results. The underwater measurements suggested the presence of significant conformational changes, confirming that the environment is a further variable in surface characterization of polymers.

Recently, two important developments appeared in the literature. The first comes from the work of Fowkes [100–102], who reviewed his previous theory and suggested a different approach. According to Fowkes, even if the language of intermolecular forces in condensed phases is dominated by the term polar, this kind of force plays an important role only in gas phases [100]. In fact, in the gas phase nearly all interactions involve no more than two molecules at any time, while in condensed phases the number of nearest neighbours is much higher, providing conflicting local fields which minimize dipole interactions. On the other hand, while in the gas phase the mean intermolecular distance is too great to allow acid–base bonding, Lewis acid–base interactions occur among the polar groups in liquids and in solids. As discussed in Chapter 1, hydrogen-bonding is explicitly considered a sub-set of acid–base interactions. In this way equation

(4.42) can be rewritten as:

$$\gamma = \gamma^{ab} + \gamma^{d} \tag{4.48}$$

where the superscripts ab denote the contribution of acid–base interactions and the superscript d the dispersion forces. The latter can be determined by measuring the contact angle with a liquid having only dispersion force interaction. Methylene iodide is usually used for this purpose, since it has a high surface tension, which prevents it from spreading, and no intermolecular acid–base interaction. The acid–base interactions cannot be predicted by the geometrical mean equation; however, careful predictions of interaction enthalpies can be obtained by the equation of Drago [103, 104], as discussed in Chapter 1. The acid–base work of adhesion can be measured as follows. The work of adhesion between a solid and a liquid is generally defined as:

$$W_{sl} = \gamma_{s} + \gamma_{l} - \gamma_{sl} \tag{4.49}$$

According to the Young equation and the Fowkes theory, the acid–base work of adhesion then is written as:

$$W_{sl}^{ab} = W_{sl} - W_{sl}^{d} = \gamma_{1}(1 + \cos\Theta) - 2(\gamma_{s}^{d}\gamma_{1}^{d})^{1/2} \tag{4.50}$$

and can be evaluated by the measurement of contact angles of methylene iodide (which gives the last term on the right side of the equation) and acidic or basic liquids (in the Lewis sense). Typical results, which refer to measurements on copolymers of ethylene and acrylic acid, are shown in Figure 4.16 (from ref. [101]). With increasing basicity of the test liquid, the magnitude of the acid–base work of adhesion obviously increases. Interestingly, the acidic test liquid) 35% phenol in tricresylphosphate) gives no measurable work of adhesion. This is in agreement with the Fowkes theory (no acid–base interaction occurs between acidic media). On the other hand, earlier theories would have predicted polar interactions between these dipoles.

The second recent contribution is due to Good, van Oss and Chaudhury [105, 106]. As discussed in Chapter 1, the three electrodynamic forces (namely London dispersion, Keesom dipole–dipole and Debye dipole-induced dipole forces) decay with distance at the same rate, and obey the same combining rules. Chaudhury pointed out that the surface tension components arising from the three electrodynamic interactions should be put together. In agreement with Fowkes's results, it was found that dipole–dipole interactions make a small contribution to surface and interfacial tension in condensed phases so that electrodynamic interactions are overwhelmingly dominated by the dispersion forces term. Taken together, they give rise to a surface tension component that is called the Lifshitz–van der Waals (LW) component. If suitable groups are present, significant contributions to interfacial energetics can arise from hydrogen bounding, that is Lewis acid–base or electron–acceptor/electron–donor interactions. These kind of forces were at first called short range (SR) [105], due

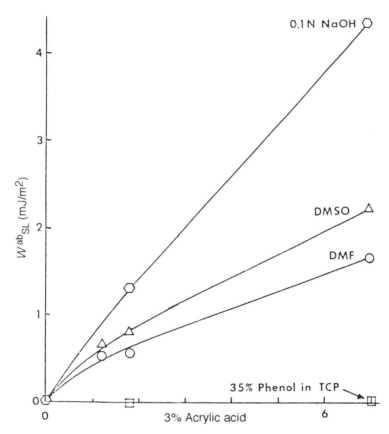

Figure 4.16 The acid–base contribution to the work of adhesion of acidic or basic liquids on copolymers of ethylene and acrylic acid. DMSO = Dimethylsulphoxide; DMF = Dimethylformamide; TCP = Tricresylphosphate. (Reproduced by permission of Plenum Publishing Corp., from ref. [101].)

to the rapid decay with distance, but later [106] it was stated that in water (and probably also in other polar liquids) molecular ordering may extend over several molecular diameters, so that the term acid–base (AB) was preferred. As a result the surface tension can be expressed as:

$$\gamma = \gamma^{LW} + \gamma^{AB} \tag{4.51}$$

In [105] the two components of surface tension were expressed by a geometric mean rule, yielding an Owens–Wendt like equation (4.45), with the understanding that this way of expressing the acid–base interaction is purely empirical. Surface tension components can be readily calculated by measuring the equilibrium contact angle of two different liquids of known surface tension components.

A more complex approach was developed [106, 107], incorporating the electron–donor and the electron–acceptor parameters in the acid–base component. A substance can show both electron–donor and electron–acceptor properties, or can be overwhelmingly basic showing virtually only electron–donor behaviour, the opposite case being possible but far less common [106]. Materials exhibiting only electron–donor or electron–acceptor capacity are called monopolar materials. It must be noted that an acid-base parameter will be manifested only if the opposite parameter is present in another molecule or in another part of the same molecule; thus a single polar parameter of a monopolar material cannot contribute to its energy of cohesion, whose polar or acid–base component can be expressed, for the generic bipolar material i, as:

$$-2\gamma_i^{AB} = -4(\gamma_i^+ \gamma_i^-)^{1/2} \tag{4.52}$$

where the superscripts $+$ and $-$ indicate the electron–acceptor and the electron–donor parameters respectively [106]. The acid–base component of the interfacial tension between materials 1 and 2 is defined by:

$$\gamma_{12}^{AB} = \gamma_1^{AB} + \gamma_2^{AB} - 2(\gamma_1^+ \gamma_2^-)^{1/2} - 2(\gamma_1^- \gamma_2^+)^{1/2} \tag{4.53}$$

When this expression is combined with the Young equation, a Young–Good–Girifalco–Fowkes–van Oss–Chaudhury equation results:

$$\gamma_1(1 + \cos\Theta) = 2[(\gamma_s^{LW}\gamma_1^{LW})^{1/2} + (\gamma_s^+ \gamma_1^-)^{1/2} + (\gamma_s^- \gamma_1^+)^{1/2}] \tag{4.54}$$

Now the three parameters characterizing the solid surface tension and its behaviour towards the surroundings can be determined by measuring the equilibrium contact angles of at least three liquids, provided the relevant values of the surface tension components of these liquids are known.

It is interesting to note the increasing number of parameters required to describe interfacial energetics. The reason for this trend is the increasing weight of intermolecular, microscopic theories in explaining a relationship which started as a mechanical interaction. The above-mentioned results indicate that the choice of liquids strongly influences the process of obtaining surface tension components. Moreover, they suggest that the complexity of solid surfaces is such that measurements performed using single liquids or even two liquids are unable to provide complete information [108].

The previous discussion shows that, currently, no general agreement exists on which is the best way to progress from contact angle to surface tension. Some of these options show only minor differences, while in other cases, as in the Neumann, or macroscopic approach, versus the surface tension components, or microscopic, approach, the incompatibility is rooted in the theory.

The complexity of the physics involved in surface energetics currently precludes a unified approach to all problems. Therefore any experimental situation and any level of approximation can profit from the application of one or more of the approaches which have been described.

4.5 SURFACE THERMODYNAMIC PARAMETERS BY INVERSE GAS CHROMATOGRAPHY

The adsorption of vapours has been a classical method for investigating solid surfaces [109]. However, until recently, this technique was limited to surfaces characterized by high area and high affinity towards the vapour phase, such as metals or inorganic oxides. This situation changed with the development of gas-chromatographic (GC) methods for physical measurements [110, 111], which paved the way for the measurement of the adsorption behaviour of probe vapours at very low concentration, where lateral interactions (that is interactions among adsorbed molecules) are negligible and the thermodynamic functions depend only on adsorbate–adsorbent interactions. The technique which allows the evaluation of the thermodynamic parameters of solid surfaces by the GC method was called inverse gas chromatography (IGC) to stress that, contrary to conventional GC, the component of interest is the stationary phase. Thus, in GC, the solid packing materials, which is known and often specifically designed, serves to separate and identify the different components of an unknown gas or vapour. On the other hand, in IGC, specific vapour probes are utilized to elucidate the interaction behaviour, and hence the surface characteristics, of the packing material. Except for the role reversal of the stationary and mobile phase, the principle and instrumentation behind IGC are identical to that of GC, and will not be discussed here. The objective of this section is to describe how to obtain solid surface thermodynamics data from IGC measurements.

The basic parameter which can be measured in GC is the retention volume (V_N), which, in the limit of infinite dilution, where the Henry's law region is reached, can be written as :

$$V_N = KA \qquad (4.55)$$

where V_N is the net retention volume, K is the partition coefficient of the given vapour probe between the adsorbed and gaseous state, and A is the surface area of the solid [112]. Experimentally, the limit of infinite dilution is obtained by injecting very small amounts of gaseous solute: for instance, in the quoted paper [112], the partial pressures of injected n-alkanes were in all cases less than 0.1 Nm^{-2}, corresponding to a mole fraction of less than 10^{-5} in the N_2 carrier gas.

The variation of the standard free energy of adsorption is given by:

$$\Delta G^0 = RT \ln K + C_1 \qquad (4.56)$$

where C_1 is a constant depending on the reference state for the adsorbed molecule. Combining equations (4.55) and (4.56):

$$\Delta G^0 = RT \ln V_N + C_2 \qquad (4.57)$$

so that a direct relationship between the experimental parameter V_N and the thermodynamic quantity is obtained. The problem is now how to correlate interfacial energetics to the free energy of adsorption. In this respect, it is usually observed that ln V_N of n-alkanes varies linearly with their number of carbon atoms. Thus, by injecting an homologous series on n-alkanes probes and by measuring their net retention volume, it is possible to calculate the free-energy of adsorption corresponding to the repeating unit, that is one methylene group. Gray and coworkers reasoned that the free energy of desorption per unit area of methylene will be approximately equal to the work of adhesion of a liquid surface ideally composed only by methylene groups on the same adsorbent [112–116]. Since, in this case, interaction occurs only through dispersion forces, according to the previously described Fowkes theory, it is possible to write:

$$- \nabla G^{CH_2}/NA_{CH_2} = 2(\gamma_{(CH_2)}\gamma_s^d) \tag{4.58}$$

where the experimentally measured free energy of desorption per methylene group is divided by the molar area of methylene groups obtained by the product of the Avogadro number N times the area occupied by one methylene group A_{CH_2} (0.06 nm², according to Gray and coworkers. A discussion on this value is contained in ref. [112]). On the left side of equation (4.58), $\gamma_{(CH_2)}$ is the surface tension of the hypothetical liquid whose surface is composed only by methylene groups. This value can be estimated as 35.6 mJ/m² at 20 °C, either by extrapolation of surface tension data for polyethylene melts to room temperature, or by extrapolation of surface tension data for linear alkanes to infinite chain length [112, 116]. Anhang and Gray calculated the dispersion component of poly (ethylene terephthalate) by IGC [118], obtaining a value (40 mJ/m²) in good agreement with contact angle data [9]. Besides the basic assumptions discussed in the quoted references, from the experimental point of view it is important that measurements are performed in the limit of infinite dilution and that vapour solid interaction occurs only by surface adsorption. If bulk sorption occurs, this method cannot be employed, therefore it is best restricted to highly crystalline or polar polymers which do not sorb hydrocarbons into the bulk polymer in the time-scale of the GC experiment.

There is currently a great deal of interest in the measurement of acid–base, electron donor–acceptor (formerly called polar, as previously discussed) interactions by IGC. A striking example of the effectiveness of IGC in this field is shown by Figure 4.17, taken from ref. 117, where the logarithm of the retention volume of alkanes, alkenes and dienes is plotted as a function of the chain length. The solid phase is fumed silica. Clearly, the alkenes give a constant retention volume (and hence interaction energy) increment over the alkanes. Interestingly, this increment is doubled when dienes are injected. The increase of the interaction energy is due to the specific (that is non-dispersive) interaction between π electrons of double bonds with the silica surface. This difference was not found when adsorption was carried out on a non-specific adsorbent, like, for instance, graphitized thermal carbon black [117]. The quantitation of dispersion and specific interactions is described in several interesting papers [117–128]. In the method proposed by Lavielle and Schultz, the relationship between the retention volume of alkanes

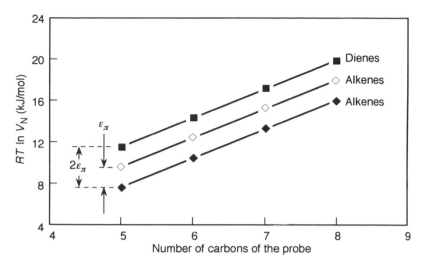

Figure 4.17 Variation of $RT \ln V_N$ versus the number of carbon atoms of the injected probes used to test non-treated silica. (Reproduced by permission of Friedr. Vieweg & Sohn, from ref. [117], with permission.)

and the dispersion component of the solid surface free energy is given by [121]:

$$RT \ln V_N = 2N(\gamma_s^d)^{1/2} A(\gamma_l^d)^{1/2} + K \qquad (4.59)$$

where N is the Avogadro number, A the surface area of the adsorbed probe molecule and γ_l^d the dispersive component of the surface tension of the probe in the liquid state. The constant K depends on the chosen reference state. With probes having an acidic, basic or amphoteric character, supplementary specific interactions lead to an increase of the retention time, and, knowing the acid–base surface properties of the probe, the solid surface may be characterized. In particular, the dispersive component γ_s^d is calculated from the slope of the straight line described by equation (4.59) ($RT \ln V_N$ versus $A(\gamma_l^d)^{1/2}$. With acid or base probes, the difference of ordinates between the experimental point and the reference line corresponding to the alkanes gives an evaluation of the specific free enthalpy of adsorption–desorption.

The quoted papers also discuss another intriguing feature of IGC, that is its sensitivity to surface sites of different nature. Thus, IGC measurements can give information on surface heterogeneity of polymers and fillers.

REFERENCES

[1] A. W. Adamson, *Physical Chemistry of Surfaces*, 5th edn, Wiley, New York (1990), Ch. 1.
[2] J. N. Israelachvili, *Intermolecular and Surface Forces*, Academic Press, London (1985), Ch. 11.

[3] J. N. Israelachvili, *J. Chem. Soc. Faraday Trans. 2*, **69**, 1729 (1973).
[4] R. Aveyard and S. M. Saleem, *J. Chem. Soc. Faraday Trans. 1*, **72**, 1609 (1976).
[5] D. B. Hough and L. R. White, *Adv. Coll. Interface Sci.*, **14**, 3 (1980).
[6] G. N. Lewis and M. Randall, *Thermodynamics*, rev. by K. S. Pitzer and L. Brewer, 2nd edn, McGraw-Hill, New York (1961).
[7] H. C. O. Lupis, *Chemical Thermodynamics of Materials*, North-Holland, New York (1983), Ch. 23.
[8] E. A. Guggenheim, *Trans. Faraday Soc.*, **36**, 397 (1940).
[9] S. Wu, *Polymer Interface and Adhesion*, Marcel Dekker, New York (1982).
[10] A. W. Neumann, *Adv. Coll. Interface Sci.*, **4**, 105 (1974).
[11] M. Morra, E. Occhiello, and F. Garbassi, *Adv. Coll. Interface Sci.*, **32**, 79 (1990).
[12] T. Young, *Phil. Trans.*, **95**, 65 (1805).
[13] T. Young, *Phil. Trans.*, **95**, 82 (1805).
[14] H. J. Busscher, G. A. M. Kip, A. Van Silfhout, and J. Arends, *J. Colloid Interface Sci.*, **114**, 307 (1986).
[15] A. W. Adamson, *Physical Chemistry of Surfaces*, 5th edn, Wiley, New York (1990), Ch. 10.
[16] H. W. Fox and W. A. Zisman, *J. Colloid Interface Sci.*, **5**, 514 (1950).
[17] D. P. Graham, *J. Phys. Chem.*, **68**, 2788 (1964).
[18] D. P. Graham, *J. Phys. Chem.*, **69**, 4387 (1965).
[19] W. H. Wade and J. W. Whalen, *J. Phys. Chem*, **72**, 2898 (1968).
[20] J. W. Whalen, *J. Colloid Interface Sci.*, **28**, 443 (1968).
[21] *The Collected Works of J. W. Gibbs*, Yale Univ. Press, New Haven (1928), vol. 1, p. 314.
[22] R. E. Johnson, Jr., *J. Phys. Chem.*, **63**, 1655 (1959).
[23] Lord Rayleigh, *Phil. Mag.*, **30**, 397 (1890)
[24] H. L. Sulman, *Trans. Inst. Mining and Metallurgy*, **29**, 44 (1920).
[25] A. W. Neumann and R. J. Good, in *Surface and Colloid Science*, R. J. Good and R. R. Stromberg (Eds), Plenum Press, New York (1979), vol. 2.
[26] J. D. Andrade, L. M. Smith, and D. E. Gregonis, in *Surface and Interfacial Aspects of Biomedical Polymers*, J. D. Andrade (Ed.), Plenum Press, New York (1985), vol. 1, Ch. 7
[27] G. Giannotta, M. Morra, E. Occhiello, F. Garbassi, and L. Nicolais, *J. Colloid Interface Sci.*, **148**, 571 (1992).
[28] M. Morra, E. Occhiello, and F. Garbassi, *J. Adh. Sci. Tech.*, **6**, 653 (1992).
[29] R. N. Wenzel, *Ind. Eng. Chem.*, **28**, 988 (1936).
[30] R. J. Good, *J. Amer. Chem Soc.*, **74**, 5041 (1952).
[31] J. J. Bikermann, *Abstract of Papers*, 116th Meeting, American Chemical Society, Atlantic City, New Jersey, September 1949, p. 8G.
[32] R. N. Wenzel, *J. Phys. Chem.*, **53**, 1466 (1949).
[33] A. B. D. Cassie, *Discuss. Faraday Soc.*, **3**, 16 (1948).
[34] J. N. Israelachvili and M. L. Gee, *Langmuir*, **5**, 288 (1989).
[35] R. J. Good, *J. Amer. chem. Soc.*, **74**, 5041 (1952).
[36] R. E. Johnson, Jr. and R. H. Dettre, *Adv. Chem. Ser.*, F. M. Fowkes (Ed.), American Chemical Society, Washington D. C. (1964), *43*, pp. 112–144.
[37] R. E. Johnson, Jr. and R. H. Dettre, *J. Phys. Chem.*, **68**, 1744 (1964).
[38] R. H. Dettre and R. E. Johnson, Jr., *J. Phys. Chem.*, **69**, 1507 (1965).
[39] R. E. Johnson, Jr. and R. H. Dettre, *Surface and Colloid Science*, E. Matijevic (Ed.), Wiley–Interscience, New York (1969), vol. 2, pp. 85–153.
[40] A. W. Neumann and R.J. Good, *J. Colloid Interface Sci.*, **38**, 341 (1972).
[41] J. D. Eick, R. J. Good, and A. W. Neumann, *J. Colloid Interface Sci.*, **53**, 235 (1975).

[42] C. Huh and S. G. Mason, *J. Colloid Interface Sci.*, **60**, 11 (1977).
[43] M. A. Fortes, in *Physicochemical Aspects of Polymer Surfaces*, K. L. Mittal (Ed.) Plenum Press, New York, 1983, vol. 1, p. 119.
[44] J. F. Joanny and P. G. de Gennes, *J. Chem. Phys.*, **81**, 552 (1984).
[45] L. W. Schwartz and S. Garoff, *Langmuir*, **1** 219 (1985).
[46] Y. Pomeau and J. Vannimemenus, *J. Colloid Interface Sci.*, **104**, 477 (1985).
[47] L. W. Schwartz and S. Garoff, *J. Colloid Interface Sci.*, **106**, 422 (1985).
[48] A. W. Adamson, *Physical Chemistry of Surfaces*, 5th edn, Wiley, New York (1990), Ch. 13, p. 495.
[49] A. B. D. Cassie and S. Baxter, *Trans. Faraday Soc.*, **40**, 549 (1944).
[50] A. B. D. Cassie and S. Baxter, *J. Text. Inst.*, **36**, T67 (1945).
[51] M. Morra, E. Occhiello, and F. Garbassi, *Langmuir*, **5**, 872 (1989).
[52] K. G. Tingey, J. D. Andrade, C. W. McGary Jr., and R. J. Zdrahala, in *Polymer Surface Dynamics*, J. D. Andrade (Ed.), Plenum Press, New York (1988), p. 153.
[53] R. E. Johnson, Jr., R. H. Dettre, and D. A. Brandreth, *J. Colloid Interface Sci.*, **62**, 205 (1977).
[54] M. Morra, E. Occhiello, F. Garbassi, R. Marola, and P. Humprey, *J. Colloid Interface Sci.*, **137**, 11 (1990).
[55] J. Owen, *Ind. Eng. Chem. Prod. Res. Dev.*, **19**, 97 (1980).
[56] F. J. Holly and M. F. Refojo, *J. Biomed. Mater. Res.*, **9**, 315 (1975).
[57] B. D. Ratner, in *Surface and Interfacial Aspects of Biomedical Polymers*, J. D. Andrade (Ed.), Plenum Press, New York (1985), vol. 1, Ch. 10.
[58] E. Bayramli, T. G. M. Van de Ven, and S. G. Mason, *Colloids and Surfaces*, **3**, 131 (1981).
[59] M. Morra, E. Occhiello, and F. Garbassi, *J. Colloid Interface Sci.*, **149**, 84 (1992).
[60] H. W. Fox and W. A. Zisman, *J. Colloid Sci.*, **5**, 514 (1950).
[61] H. W. Fox and W. A. Zisman, *J. Colloid Sci.*, **7**, 109 (1952).
[62] H. W. Fox and W. A. Zisman. *J. Colloid Sci.*, **7**, 428 (1952).
[63] W. A. Zisman, *Adv. Chem. Ser.*, **43**, F. M. Fowkes (Ed.), American Chemical Society, Washington, D.C. (1964), pp. 1–51.
[64] L. A. Girifalco and R. J. Good, *J. Phys. Chem.*, **61**, 904 (1957).
[65] R. J. Good, L. A. Girifalco, and G. Kraus, *J. Phys. Chem.*, **62**, 1418 (1958).
[66] R. J. Good and L. A.Girifalco, *J. Phys. Chem.*, **64**, 561 (1960).
[67] R. J. Good and C. J. Hope, *J. Chem. Phys.*, **53**, 540 (1970).
[68] R. J. Good and C. J. Hope, *J. Chem. Phys.*, **55**, 111 (1971).
[69] R. J. Good and C. J. Hope, *J. Colloid Interface Sci.*, **35**, 171 (1971).
[70] R. J. Good, *J. Colloid Interface Sci.*, **59**, 398 (1977).
[71] D. Berthelot, *Compt. Rend.*, **126**, 1703, 1857 (1898).
[72] J. R. Dann, *J. Colloid Interface Sci.*, **32**, 302 (1970).
[73] J. R. Dann, *J. Colloid Interface Sci.*, **32**, 321 (1970).
[74] S. Wu, *J. Adhes.*, **5**, 39 (1973).
[75] S. Wu, *J. Colloid Interface Sci.*, **71**, 605 (1979). Erratum in *J. Colloid Interface Sci.*, **73**, 590 (1980).
[76] C. A. Ward and A. W. Neumann, *J. Colloid Interface Sci.*, **49**, 286 (1974).
[77] A. W. Neumann, R. J. Good, C. J. Hope, and M. Sejpal, *J. Colloid Interface Sci.*, **49**, 291 (1974).
[78] C. P. S. Taylor, *J. Colloid Interface Sci.*, **100**, 589 (1984).
[79] J. K. Spelt and A. W. Newmann, *Langmuir*, **3** 588 (1987).
[80] J. K. Spelt and A. W. Neumann, *J. Colloid Interface Sci.*, **122**, 294 (1988).
[81] C. J. van Oss, R. J. Good, and M. K. Chaudhury, *Langmuir*, **4** 884 (1988).
[82] R.l E. Johnson and R. H. Dettre *Langmuir*, **5** 293 (1989).

[83] J. D. Morrison, *Langmuir*, **5**, 543 (1989).
[84] F. M. Fowkes, *J. Phys. Chem.*, **66**, 382 (1962).
[85] F. M. Fowkes, *Ind. Eng. Chem.*, **56**, 40 (1964).
[86] F. M. Fowkes, *Adv. Chem. Ser.*, **43**, F. M. Fowkes (Ed.), American Chemical Society, Washington, D.C. (1964), pp. 99–111.
[87] I. Israelachvili, *Quart. Rev. Biophys.*, **6**, 341 (1974).
[88] D. K. Owens and R. C. Wendt, *J. Appl. Polym. Sci.*, **13**, 1741 (1969).
[89] D. H. Kaelble, *J. Adhesion*, **2**, 66 (1970).
[90] D. H. Kaelble and E. H. Cirlin, *J. Polym. Sci.*, $A - 2$, **9**, 363 (1971)
[91] S. Wu, *J. Polym. Sci.*, *C*, **34**, 19 (1971).
[92] E. N. Dalal, *Langmuir*, **3**, 1009 (1987).
[93] H. Y. Erbil and R. A. Meric, *Colloids and Surfaces*, **33**, 85 (1988).
[94] W. C. Hamilton, *J. Colloid Interface. Sci.*, **40**, 219 (1972).
[95] J. D. Andrade, R. N. King, D. E. Gregonis, and D. L. Coleman, *J. Polym. Sci. Symp.*, **66**, 313 (1979).
[96] J. D. Andrade, S. M. Ma, R. N. King, and D. E. Gregonis, *J. Colloid Interface Sci.*, **72**, 488 (1979).
[97] R. N. King, J. D. Andrade, S. M. Ma, D. E. Gregonis, and L. R. Brostrom, *J. Colloid Interface Sci.*, **103**, 62 (1985).
[98] E. Ruckenstein and S. V. Gourisankar, *J. Colloid Interface Sci.*, **107**, 488 (1985).
[99] Y. C. Ko, B. D. Ratner, and A. S. Hoffman, *J. Colloid Interface Sci.*, **82**, 25 (1981).
[100] F. M. Fowkes, *Physicochemical Aspects of Polymer Surfaces*, vol. 2, K. L. Mittal (Ed.), Plenum Press, New York (1983), p. 583.
[101] F. M. Fowkes, *Surface and Interfacial Aspects of Biomedical Polymers*, vol. 1, J. D. Andrade (Ed.), Plenum Press, New York (1985) Ch. 9.
[102] F. M. Fowkes, *J. Adhesion Sci. Tech.*, **1**, 7 (1987).
[103] R. S. Drago, G. C. Vogel, and T. E. Needham, *J. Amer. Chem. Soc.*, **93**, 6014 (1970).
[104] R. S. Drago, L. B. Parr, and C. S. Chamberlain, *J. Amer. Chem. Soc.*, **99**, 3203 (1977).
[105] C. J. van Oss, R. J. Good, and M. K. Chaudhury, *J. Colloid Interface Sci.* **111**, 378 (1986).
[106] C. J. van Oss, M. K. Chaudhury, and R. J. Good, *Adv. Colloid Interface Sci.*, **28**, 35 (1987).
[107] C. J. van Oss, R. J. Good, and M. K. Chaudhury, *Separ. Sci. Technol.*, **22**, 1 (1987).
[108] H. J. Busscher and A. W. J. Van Pelt, *J. Mater. Sci. Lett.*, **6**, 815 (1987).
[109] A. W. Adamson, *Physical Chemistry of Surfaces*, 5th edn, Wiley, New York (1990), Ch. 16.
[110] R. J. Lamb and R. L. Pecsok, *Physicochemical Applications of Gas Chromatography*, Wiley, New York (1987).
[111] J. R. Conder and C. L. Young, *Physicochemical Measurements by Gas Chromatography*, Wiley, New York (1979).
[112] G. M. Dorris and D. G. Gray, *J. Colloid Interface Sci.*, **77**, 353 (1980).
[113] G. M. Dorris and D. G. Gray, *J. Colloid Interface Sci.*, **71**, 93 (1979).
[114] S. Katz and D. G. Gray, *J. Colloid Interface Sci.*, **82**, 318 (1981).
[115] S. Katz and D. G. Gray, *J. Colloid Interface Sci.*, **82**, 326 (1981).
[116] J. Anhang and D. G. Gray, *J. Appl. Polym. Sci.*, **27**, 71 (1982).
[117] M. Sidqi, G. Ligner, J. Jagiello, H. Balard, and E. Papirer, *Chromatographia*, **28**, 588 (1989).
[118] C. Saint Flour and E. Papirer, *Ind. Eng. Chem. Prod. Res. Dev.*, **21**, 337 (1982).
[119] C. Saint Flour and E. Papirer, *Ind. Eng. Chem. Prod. Res. Dev.*, **21**, 666 (1982).
[120] A. Vidal, E. Papirer, W. M. Jiao, and J. B. Donnet, *Chromatographia*, **23**, 121 (1987).
[121] J. Schultz, L. Lavielle, and C. Martin, *J. Adhesion*, **23**, 45 (1987).

[122] H. Balard, M. Sidqi, E. Papirer, J. B. Donnet, A. Tuel, H. Hommel, and A. P. Legrand, *Chromatographia*, **25**, 712 (1988).
[123] A. J. Vukow and D. G. Gray, *Langmuir*, **4**, 743 (1988).
[124] S. P. Wesson and R. E. Allred, *J. Adhesion Sci. Technol.*, **4**, 277 (1990).
[125] I. Tijburg, J. Jagiello, A. Vidal, and E. Papirer, *Langmuir*, **7**, 2243 (1991).
[126] J. B. Donnet and S. J. Park, *Carbon*, **7**, 29 (1991).
[127] L. Lavielle and J. Schultz, *Langmuir*, **7**, 978 (1991).
[128] A. C. Tiburcio and J. A. Manson, *J. Appl. Polym. Sci.*, **42**, 427 (1991).

Chapter 5

New and Emerging Methods

In this last chapter on surface characterization techniques, two classes of measurement will mainly be discussed, namely surface forces and electrokinetic measurements. In other words, the focus will be on how to measure the force (or energy) versus distance laws discussed in Chapter 1, and how to measure the basic parameter involved in underwater electrostatic interactions. Neutron scattering methods will finally be briefly discussed.

The importance of these techniques in the surface science of polymers is best shown by the continuous reference to surface forces or coulombic data which was made in the chapter on the basis of surface interactions. It is important to underline that, while electrokinetic measurement has a long history and can be considered to have a consolidated approach in colloid chemistry, its application to surfaces of non-finely dispersed polymers is a very different matter. The development of a commercial apparatus which allows the characterization of surfaces of bodies of almost any shape is very recent. As to direct force measurement, this powerful technique is still confined to a narrow group of specialists. The basic reason is that, until recently, the contraints imposed by the very nature of this measurement limited its application to highly specialized and idealized surfaces (typically, to mica surfaces immersed in selected solutions, the liquid, rather than the surfaces, being the main variable) or to polymeric chains adsorbed on mica. Technical developments, however, are changing this state of the art, and several papers where forces between polymer surfaces are the prime object of investigation have already been published [1–3].

The measurement of forces between surfaces will be the first topic discussed in the rest of this chapter.

5.1 SURFACE FORCE MEASUREMENTS

The basis of surface force measurement can be understood with reference to Figure 5.1, that shows two macroscopic objects, one fixed and the other

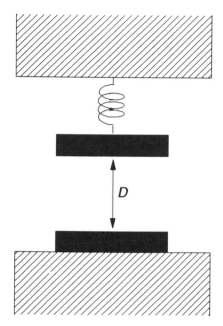

Figure 5.1 Schematic of the principle of surface force measurement.

mounted on a spring. It is clear that, if there is an attractive force between the two solids, on lowering the upper support, the spring will expand such that at any equilibrium separation the attractive force balances the elastic restorative force. The principle is very simple, much less so its practical exploitation, for several obvious reasons: as discussed in Chapter 1, if an experimental set up such as the one shown in Figure 5.1 has to be put in practice, the distance between surfaces should be of the order of nanometres. This means that the movable surface should be displaced by nanometre (or, better, sub-nanometre) steps and that the distance between surfaces should be measured with an adequate resolution. Moreover, the surfaces should be absolutely parallel, or else a signal averaged on continuously changing distances would be measured, and molecularly flat. Finally, a sensitive enough way of measuring the forces should be designed.

Despite this large number of difficulties, the measurement of surface forces has a more than fifty-year-old history [4]. Fundamental contributions have come from several different schools [5–7]. The present level of sophistication, however, is much more recent, since it dates back only to 1976, with the coming of age of the surface force apparatus (SFA) developed by Israelachvili [8, 9]. The SFA (which is commonly called the 'Jacob's box'), is by far the most versatile, productive and popular instrument for the measurement of forces between surfaces and will be discussed in some length in the next section. An accurate review of technical aspect of surface force measurement has been published by Lodge [10].

5.1.1 The Surface Force Apparatus

The classic SFA designed by Israelachvili is shown in Figure 5.2 [8, 9]. Some of the answers to the previously quoted problems are readily observed: first, forces are measured between surfaces of cleaved mica [11–13]. The reason is that mica can be cleaved to produce surfaces which are molecularly smooth over areas of the order of a square centimetre [14]. The two mica sheets are glued on to cylindrical glass formers and mounted opposite each other in a crossed cylinder configuration. This geometry is a very practical one experimentally, since the quoted problems of parallelism, as well as edge effects, are avoided. Moreover, a particularly useful aspect of making force measurements between two curved

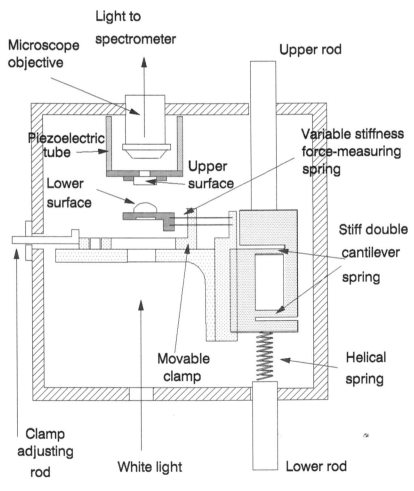

Figure 5.2 Simplified drawing of the Israelachvili first surface force apparatus.

surfaces is that the measured force can be directly related to the energy per unit area between two flat surfaces, according to the Derjaguin approximation discussed in Chapter 1 (equation (1.9), [15]).

The control of the separation between the mica surfaces is performed by a three-stage mechanism. An accuracy of about one μm is reached by the upper micrometre driven rod, which can be moved up and down by a two-way stepping motor. The lower micrometre-driven rod operates through a differential spring mechanism. The double cantilever spring is about a thousand times stiffer than the helical spring: this means that a one micrometre movement of the rod (which is driven by a two-way synchronous motor) is reduced to a one nanometre displacement between the two mica surfaces. Finally, a non-mechanical fine control is provided by a rigid piezoelectric crystal tube, which expands or contracts longitudinally by about 0.7 nm/V. In this way, the surfaces can be positioned to better than 0.1 nm.

Thus, this three-stage mechanism solves the problem of the control of surface separation. The measurement of the actual separation must be fine enough to accurately measure 0.1 nm distances, and it is done by use of multiple beam interference fringes [16–18]. Briefly, before glueing down the sheets, the two outer surfaces are coated with a 95% reflecting silver layer. The two silver layers form an optical interferometer: when white light is passed normally through it, the emerging light consists of discrete wavelengths, which can be separated and measured as sharp fringes of equal chromatic order (FECO) in an ordinary prism or grating spectrometer. Thus, when two mica sheets of the same thickness are separated from contact to a distance D, the fringes shift to longer wavelengths given by:

$$\tan(2\pi\mu D/l_n^D) = (2\mu_m/\mu \sin A)/[1 + \mu_m^2/\mu^2)\cos A \pm (\mu_m^2/\mu^2 - 1)] \qquad (5.1)$$

where l_n^D is the wavelength of fringe of order n at distance D, μ_m is the refractive index of mica at l_n^D, μ the refractive index of the medium between the two mica surfaces at l_n^D, + refers to odd order fringes and − to even order fringes, and A is given by:

$$A = (1 - l_n^o/l_n^D)/(1 - l_n^o/l_{n-1}^o)\pi \qquad (5.2)$$

where l_n^o is, of course, the wavelength of fringe of order n when the two surfaces are in contact.

Some observations on equations (5.1) and (5.3) can be made. First, the method actually measures the distance between the two silvered layers, on the outer side of the mica sheets. Then, equation (5.1) allows the calculation of both the distance and the medium refractive index independently, by measuring the shifts in wavelengths of an odd and an even fringe. It must be noted [9], that the accuracy is about ± (0.1–0.2) nm for measurements of D in the range 0–200 nm, while for μ it is better than 1% at large D but is less accurate as D falls below 10 nm. Finally, the correct use of the equations need an accurate prior determination of the refractive index of mica.

As to the measurement of the force–distance law, Figure 5.2 shows that the upper mica surface is fixed to the piezoelectric crystal while the lower mica sheet is suspended at the end of a steel cantilever leaf-spring. The forces are measured by suddenly reversing the voltage of the piezoelectric crystal, which expands or contracts by a previously calibrated amount. The principle of the method is shown in the scheme of Figure 5.3: if $x = 0$ is defined as the position of the lower

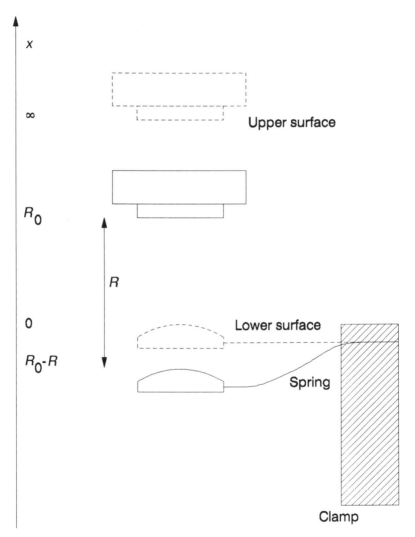

Figure 5.3   Basic coordinates of the method of measuring the law of force $F(D)$.

surface when the two surfaces are a large distance apart (no interaction), the lowering of the upper surface causes the lower surface to move to a new equilibrium position at $x = -(R - R_0)$. Since at any equilibrium separation the interaction force between the surfaces $[F(R)]$ is balanced by the restoring force of the spring, at equilibrium:

$$F(R) = K(R - R_0) \qquad (5.3)$$

If the piezoelectric crystal expands by a finite amount ΔR_0, the lower surface responds by a shift to the new equilibrium position $R - \Delta R$, so that:

$$F(R - \Delta R) = K(R - \Delta R - R_0 + \Delta R_0) = F(R) + K(\Delta R_0 - \Delta R) \qquad (5.4)$$

In this way, the previously described optical measurement of the separation between the surfaces also allows the evaluation of the forces between them. In fact, if $\Delta R = \Delta R_0$ there is no force between the surfaces, while, if a mismatch between the known expansion of the piezoelectric crystal and the shift of the position of the lower surface is detected, the force can be calculated by the known value of the spring constant. The latter is usually about 10^2 N/m, which means that a 1×10^{-7} N force causes a deflection of the spring of 1 nm. The correct calibration of the leaf-spring stiffness is, of course, very important, and this is usually done after each experiment, by placing small weights on the spring at the place where the mica surfaces were contacting and measuring the deflection by a travelling microscope [8, 9]. The other key point is the correct calibration of the voltage/displacement relationship of the piezoelectric crystal. This can be safely measured, before each run, at large separation, where no interactions occur.

A point to note is that, in this system, the force–distance law can give rise to some instability regions, where no equilibrium is possible and where no forces can be measured [9, 10]. If two surfaces are brought up to these instability points, either on approach or separation, they will jump to a region where equilibrium is allowed. It is possible to demonstrate that instability occurs when the first derivative of the force with respect to the distance is greater than the spring stiffness. Typically, for a force–distance law such as the classical DLVO curve, with primary and secondary minima (Chapter 1, Figure 1.12), jumps are observed, on approach, from the point where the inversion of the quoted derivative occurs into the secondary minimum and from the force barrier to the primary minimum contact. On separation, a large jump is observed from the primary minimum contact and from the secondary minimum well. Since the boundary of the instability region is ultimately controlled by the spring stiffness, it is clear that a close control of the experimental variable is needed to allow accurate estimates of the magnitudes and positions of the force barriers.

Recently a new SFA has been designed by Israelachvili and coworkers [19]. In this new version, four distance controls instead of three are employed, while the measurement of the separation and the geometry of the mica surfaces are the same. The four distance controls are respectively a micrometer, a differential micrometer, a differential spring and a piezoelectric tube.

5.1.2 Surface Forces and Polymer Surfaces

The SFA is undoubtedly an extraordinary tool for surface scientists. Unfortunately, the need for using molecularly smooth surfaces limits its usefulness in polymeric systems. The wealth of data on surfaces bearing adsorbed polymeric layers clearly highlights the most frequent application of the SFA to polymers.

Recently, several attempts to extend the application of the SFA in polymeric systems were published. Among them, particularly noteworthy is the paper by Cho and coworkers [1], who measured the force–distance law between plasma polymerized acrylic acid layers (PPAA). To this end, PPAA was deposited from plasma (see Chapter 6) on mica substrates, which were glued, as previously described, in a SFA. Measurements were performed in dry air, in humid air and in water. As an example of the results, Figure 5.4, taken from ref. [1], shows the force–radius ratio as a function of the distance between PPAA on mica in water. A strong repulsive force due to steric repulsion (Chapter 1) is observed below 60 nm separation, while at longer distances a double-layer repulsion, with a Debye length of about 120 nm, dominates the interaction. This contrasts with the strong adhesion measured in dry air, promoted, beside van der Waals' forces,

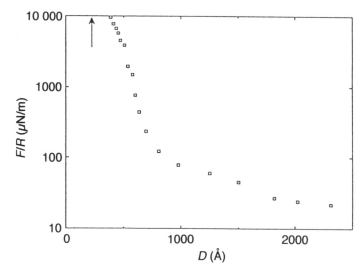

Figure 5.4 The force normalized by the radius as a function of distance between plasma polymerized acrylic acid deposited mica surfaces in water; the strong repulsive force at distance below 60 nm is due to steric repulsion. The arrow indicates the thickness of the layer under a high load. (Reproduced by permission of John Wiley & Sons, Inc. from ref. [1].)

also by non-dispersion interactions (possibly hydrogen bonding) and/or chain entanglement.

The paper quoted clearly shows the usefulness of SFA studies in the surface science of polymers. Unfortunately, only a few polymeric systems can be deposited on mica substrates. On the other hand, mica itself is poorly reactive, so that chemical reactions on its surface, for instance coupling or grafting of polymers, are not feasible [20]. An interesting alternative approach is the chemical activation of mica, for instance by exposure to a water vapour plasma [21]. The hydroxylate surface should be amenable to silane treatments.

A further way is to avoid the use of mica: the problem, of course, is to find an alternative transparent, molecularly smooth material. In this respect, Horn and coworkers have recently found methods of preparing two such materials, namely sapphire [22] and silica [23]. The former is prepared for growth by vapour phase transport above a solution of alumina in lead fluoride at 1300 °C [24]. Using this technique, large and thin platelets, with as much as 1 cm^2 of the basal surface exposed are formed. On the other hand, very thin sheets of silica can be prepared by rapidly blowing a large bubble from the melted end of a silica tube [23]. The smoothness of the surface is assured by the rapid setting of a liquid under high tension. The root-mean-square surface roughness, measured with a stylus, is less than 0.5 nm over a lateral distance of 30 μm, but measurement between silica sheets immersed in octamethylcyclotetrasiloxane (OMCTS) failed to reveal the periodic behaviour shown in Figure 1.16 [20]. This suggests that the fine scale surface roughness may be greater than the molecular diameter of OMCTS (about 0.8 nm).

Returning to polymers, molecularly smooth silica is an interesting substrate, since it allows the covalent linking of organic matrices through silane coupling. The best way, however, is to prepare molecularly smooth surfaces of transparent polymer, as done by Chaudhury and Whitesides in the direct measurement of interfacial interactions in poly(dimethylsiloxane) [3]. This kind of measurement allows the direct evaluation of the solid–solid interface interaction, and it is the lack of this type of measurement that was underlined in the chapter on contact angle and interfacial energetics. In the near future, the SFA is expected to play an important role in the theoretical and practical surface science of polymers.

5.2 THE EVALUATION OF ELECTROSTATIC INTERACTIONS

As discussed in Chapter 1, forces of an electrostatic nature play an important role in the control of interactions between bodies in aqueous fluids. Classic colloid chemistry literature is full of measurements on electrostatic interactions, and the underlying theoretical and experimental practice is well documented [25]. Electrostatic forces, of course, also deeply affect the interactions between objects of greater dimension: an example was shown in Chapter 1, where the adsorption of bacteria and proteins on solid surfaces was described in the frame of the DLVO theory. In Chapter 12, several results concerning the effect of the surface charge of artificial blood vessels on blood compatibility (that is on the interaction with formed elements of blood) will be reviewed.

In this section, we will take a look at some of the experimental techniques which are used to quantitate the parameters described in Chapter 1. Since they have been well documented in several books (e.g. [25–27]), only a brief reminder of the basis of these measurements and of the meaning of the measured quantities will be made. The emphasis will be rather on how it is possible to apply these techniques to macroscopic surfaces (thin plates), rather than to colloidal-size particles even if, as discussed below, in some instances the latter are the only measurable shapes.

In general, a complete description of the electrostatic characteristics of a surface requires a complete knowledge of the fine structure of the double layer, which, as discussed previously, is a difficult matter. Usually, the double-layer structure is discussed in terms of highly idealized models, such as those described in Chapter 1. In this case, a complete characterization requires the knowledge of the surface charge σ_0, the charge at the plane where the diffuse layer begins (i.e. the Stern layer or the outer Helmholtz plane, depending on the assumed model) σ_d, and the respective potentials. Specific applications may require knowledge of only one of these parameters, for instance, in the previously quoted example in Chapter 1, the evaluation of the electrostatic part of the interaction required only the knowledge of the potential. This is generally true when the problem is to evaluate the effect of the double layer on long-range interactions between bodies (equations (1.43) and (1.44)).

In any case, it is necessary to appreciate that the quoted parameters are pH and ionic strength dependent: ionizable surface groups, such as carboxylic groups, are, in general, weak acids, so that the amount of ionized moieties is strongly affected by pH. As discussed in Chapter 1, then, ionic strength affects the surface charge by screening. Also the potential at the plane where the diffuse layer begins is, of course, affected by the surface charge (and hence by the pH) and by the ionic strength. The obvious consequence is that the measurement of these parameters makes sense only if done at specified pH and ionic strength or, better, a full characterization requires the making of measurements over a range of pH and ionic strengths [28].

The measurement of the surface charge will be briefly discussed in the next section, afterwards more emphasis will be devoted to the measurement and meaning of the experimentally accessible potential.

5.2.1   Surface Charge Measurement

The measurement of the underwater surface charge of polymers can be done, in principle, rather straightforwardly with common titration techniques:

potentiometric or conductometric titrations are the most suited techniques [28, 29].

The basis of potentiometric titrations are well documented in several analytical chemistry textbooks. In the present case, it is important to realize that titrations involve a surface (whose nature will be discussed briefly below) within a solution, so that both of them are titrated at the same time. To measure the amount of charged species really belonging to the surface one must, then, subtract the contribution from the solution, which is done in a separate run.

From their side, conductometric titrations are based on the different mobilities of the titrated and titrant ions. A typical example is the titration of acidic groups with a base, such as NaOH. As the addition of the titrant increases the pH, protons bound to OH^-, and high mobility H^+ are replaced by slower Na^+ ions. It is clear that, if an ion with lower mobility replaces in the solution a more mobile ion, conductivity will be lowered. When no more ions can be replaced, conductivity will increase again, albeit at a different rate, due to the increase of the concentration of conductive species. The reversal of the slope of the conductivity–base concentration point is thus the equivalence point of the titration.

Even if the principles of these analytical techniques are fully consolidated for solution, homogeneous chemistry, their application to solid surface chemistry still presents some unresolved problems. Experimental difficulties found in conductometric titrations of solid samples are well documented in a series of interesting papers [30–36]. Anyway, the most basic problem is that the quoted measurements can be performed only on finely dispersed materials (lattices), or else the surface area is too low and the amount of titratable charges is below an acceptable degree of detection accuracy. Thus, these measurements are limited to colloidal particles, or, if the scope is to evaluate the surface charge of samples of lower surface-to-volume ratio, the results are bound to the assumption that the surface properties of finely dispersed polymers are identical to those of similarly prepared or treated macroscopic samples. The correctness of this assumption is a matter of debate, and no definite answer to this question exists [28]. Thus, it is always important to bear in mind that surfaces of finely dispersed powders may not be representative of surfaces of plates, and to question the correctness of the quoted assumption. A typical example is given by the comparison of the surface composition of a fine polypropylene (PP) powder, as polymerized, and of moulded PP. While XPS analysis of the former shows that it contains only carbon, it is often possible on the surface of the latter to detect a certain amount of oxygen (Figure 5.5). The likely origin of this is surface oxidation catalysed by the mould surface during melt processing [37]. The detailed analysis of the C1s peak shows that carbon-to-oxygen bonds are introduced, so that the surface of the moulded PP contains hydroxyl, carbonyl and carboxyl groups. Clearly, in this case, information based on the fine powder is of little help in understanding the underwater electrostatic behaviour of a moulded PP object.

Therefore, due to the lack of a deeper knowledge, the best answer to this

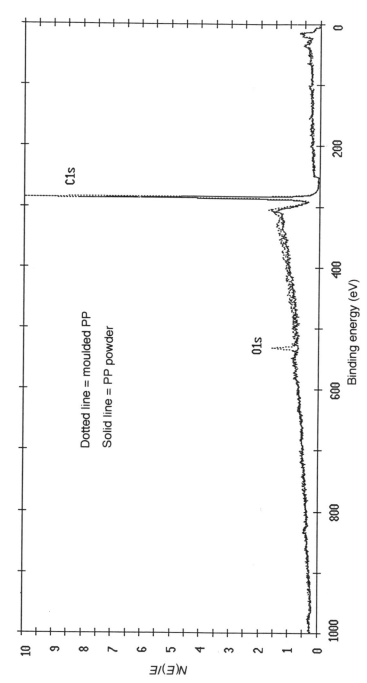

Figure 5.5 XPS survey spectra of a moulded polypropylene (PP) sheet and of PP powder as polymerized.

problem is to try to prepare the macroscopic sample and the microscopic powder in the same way (as much as possible) and subject them to the same pretreatment routine [28].

5.2.2 The Zeta Potential

The evaluation of the double-layer potential, an electrostatic quantity since it is related to double layers at rest, is based on electrokinetic phenomena. The quantity which is measured is the so-called zeta potential (ξ), a basic pillar of colloid chemistry [25]. Besides a quick description of experimental techniques, especially those techniques that probe thin plates instead of colloids, the scope of this section is to discuss how the kinetically determined quantity fits into the frame of the double layers discussed in Chapter 1.

There are basically three electrokinetic phenomena that are currently exploited to measure the zeta potential of surfaces of polymers. They are electrophoresis, electro-osmosis and streaming potential. In the latter case, depending on the design of the experimental set-up, as discussed below, a current instead of a potential can be measured, in which case the technique is called streaming current measurement. All these phenomena have in common that a moving and a stationary phase can be distinguished. In electrophoresis and electro-osmosis an electric field is applied and the movement of colloidal particles (electrophoresis) or of a solution (electro-osmosis) is observed. In streaming potential or streaming current measurements the liquid phase is moved by an applied pressure and the ensuing potential or current are measured.

In all the quoted techniques it is possible to write equations (such as equations (5.5) and (5.6) below) where a potential appears. This is the potential at the boundary between the moving and the stationary phase, or the the potential of the slipping plane, or the potential at the hydrodynamic shear plane, or the zeta potential. The key problem is, of course: what is the relationship between the electrokinetic zeta potential and the double-layer potentials described in Chapter 1? Despite the long history of the zeta potential in colloid chemistry, this problem is not yet solved [28]. Based on the theory developed in Chapter 1, the most adequate interpretation is that the zeta potential is the potential of the plane where the diffuse layer begins. Thus, it is satisfactory that it has been reported that ξ resembles the Stern potential (Chapter 1) more than ϕ_0 [28]. However, the basic point is that all double-layer theories are abstractions of very complex structures. The effort to design more realistic double layers, for instance by the introduction of the Stern plane or the inner or outer Helmholtz planes, important as they are, are only small steps forward as compared with the complexity of real systems. The picture is still more complicated when it comes to polymer surfaces: what looks like the double layer of a surface composed of grafted chains, freely moving, apart from the grafting site, in the liquid medium? To what plane belongs the measured potential? These questions are still waiting for an answer. The problem of the meaning of the zeta potential surfaces is discussed in many interesting papers [38–45].

Thus, the correct usage of the experimentally measured value requires a good deal of caution. Electrokinetic phenomena are a complex mixture of surface chemistry, electrochemistry and hydrodynamics. Even if general consensus exists that, in simple systems, the zeta potential can be equated to the potential of the Stern or the outer Helmholtz plane, this approach should not be taken for granted, especially when dealing with complex polymeric systems. Quoting Lyklema [28], ζ is an acceptable characteristic for those experiments in which only the charge or the potential in the outer double-layer part counts, like long-distance interactions.

Returning to the experimental techniques, we will discuss neither electrophoresis (which requires fine particles and which is fully described in many books) nor electro-osmosis. On the other hand, some attention will be paid to some recent advancements of streaming current and streaming potential measurements, especially in connection with the possibility of measuring surfaces of solids of different shapes.

The basic streaming potential experiment is shown in Figure 5.6. A solid

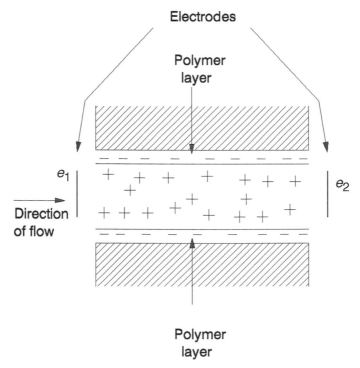

Figure 5.6 The principle of the creation of a streaming potential.

support, for instance a capillary, is the stationary phase. It can be made by the polymer under study, or it can be an inert support coated with a thin polymeric film. Let us assume that the surface is negatively charged and that an equal positive countercharge is in the solution. If a liquid is pumped through the capillary in the direction of the arrow, part of the positive charge is entrained and accumulates at electrode e_2. Thus a potential difference between the two electrodes exists $(D\phi_{sp})$, which is compensated by back flow of electricity through the electrolyte of the streaming cell. The streaming potential difference is measured by means of an high-impedance input voltmeter. Under steady state conditions:

$$\Delta\phi_{sp}/\Delta P = \varepsilon\varepsilon_0 \zeta k_0^{-1} \mu^{-1} \tag{5.5}$$

where ΔP is the applied pressure, ζ the zeta potential, μ the liquid viscosity, k_0 the specific conductivity of the solution and $\varepsilon\varepsilon_0$ the dielectric permittivity, as discussed in Chapter 1.

Streaming current measurements adopt the same set-up, but this time the two electrodes are short-circuited via a low-impedance input ammeter, resulting in a streaming current value. The current density in steady-state conditions (i_{sc}) is given by:

$$i_{sc}/\Delta P = \varepsilon\varepsilon_0 \pi a^2 \zeta l^{-1} \mu^{-1} \tag{5.6}$$

where a is the radius and l the length of the capillary. It is important to remember that the above equations refer to the most simple cases. In practical systems, very often complications arise, requiring modifications and extensions [25]. Anyway, streaming potential and streaming current experiments measure the zeta potential, whose knowledge allows the calculation, as shown in Chapter 1, of the charge at the same plane. Several hydrodynamic conditions must be satisfied in order to get a potential independent of flow geometry [25]: the fluid flow must be steady, incompressible, laminar or established. The last two conditions limits the applicability of streaming measurements to long capillary tubes or porous plugs of compacted particles or fibres. The former geometry is rather satisfactory, since a long capillary can be coated by polymer films from solution, so that no highly disperse phase is needed and no problem of transferability of the results exists. However, this geometry has also some problems: first, it is not always convenient or easy to prepare solutions of polymers. Therefore some instruments were developed which allow the characterization of flat sheets in the form of discs [25, 46, 47]. A more serious concern is that zeta potential measurement is only one of the surface analytical techniques which have to be applied in order to gain an understanding of the surface structure: thus, it would be convenient to have a streaming cell which allows the preparation of samples which can also be subjected to other analysis, such as XPS, contact angle or microscopy. A streaming cell system designed for the electrokinetic analysis of flat plate samples is probably the best answer to this problem. Such an approach was utilized by

DePalma [48] and by Van Wagenen and Andrade and coworkers [49, 50]. Recently, an instrument has been put on the market, which allows the measurement of the streaming potential/streaming current of solids of almost any shape [51, 52]. In the case of plates, the electrolyte solution is forced through a small channel built of two flat surfaces, as shown in Figure 5.7 [53].

From the previous discussion it is clear that, even if the zeta potential measurement is a rather old technique, many problems which are still open exist, especially in the case of the polymer–electrolyte interface. Apart from colloidal particles and lattices, the electrochemical characterization of polymer surfaces is still in its infancy even if, at least in principle, theories and (now) techniques are available. Experimental results show that zeta potential measurement, beside the description of long-range electrostatic interactions, can greatly contribute to our understanding of interfaces involving polymers. For instance, the quoted paper of Andrade coworkers [50], shows that synthetic neutral polymers (polystyrene (PS), polyvinylchloride (PVC) and polydimethylsiloxane (PDMS)) have a net negative zeta potential in 0.01 M KCl solutions. This behaviour was also observed on other polymers or polymeric dispersions [54–57]. The current explanation of this behaviour is the specific adsorption of ions such as OH^- or Cl^- from solution. The basic reason is that the smaller, more polarizable and less hydrated anions in an aqueous solution have a greater tendency to partition at an interface,

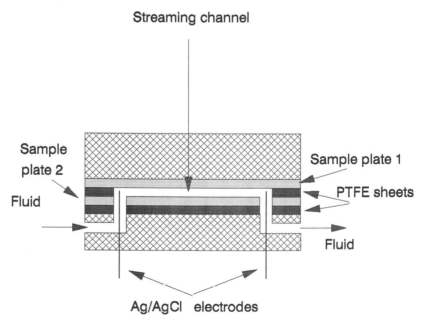

Figure 5.7 A commercial cell which allows measurement of streaming current–streaming potential on flat polymeric sheets.

especially when the polymer is hydrophobic. A strong relationship between the water contact angle and the zeta potential of polymers in KCl solutions (the higher the contact angle the more negative the potential) has been reported [58, 59]. An alternative explanation was invoked by Fowkes and Hielscher [60, 61], who suggested that the mechanism which provides negative electro-kinetic potential is electron injection from cationic water states into anionic states of PS. They were able to measure, by a field effect transistor (FET), a surface charge density of about 1×10^{12} electrons/cm^2. The surface charge calculated by Andrade and coworkers from the measured zeta potentials is about 3×10^{12} charges/cm^2, thus the agreement between the two independent measures is quite good. This kind of study clearly underlines the contribution that zeta potential measurement can bring to polymer science (beside its fundamental role in the description of electrostatic interactions).

Another example is given by the work of Voigt and coworkers [62]. Discussing some data of Andrade and coworkers on streaming potential measurement of copolymers of methylmethacrylate–hydroxyethylmethacrylate (MMA–HEMA) [63], they underlined a shortcoming of equation (5.5). A more general form of the latter is the following:

$$\Delta\phi_{sp}/\Delta P = \varepsilon\varepsilon_0\zeta(k_0 + k_s/b)^{-1}\mu^{-1} \tag{5.7}$$

where k_s is the surface conductivity and b a geometric parameter of the streaming cell configuration. It is common practice to neglect surface conduction, provided the bulk electrolyte volume to surface area ratio is high enough. This approxi-mation usually works with glassy polymers, but there is experimental evidence for a remarkable surface conduction of swollen porous or gel-like surfaces (such as HEMA surfaces) [64]. The result is that, in streaming potential measurement of this kind of material, surface conduction greatly contributes to the compensat-ing back electric current. As an example, Figure 5.8, taken from the data of ref. [63], shows a comparison between streaming current and streaming potential measurement of a glass and a swollen polyurethane as a function of the ionic strength. Differences in the results obtained from the two techniques are due to surface conduction: they are much more pronounced in the case of the swollen polymeric and in the lower ionic strength region. The conclusions are that polymers with swollen surfaces take up mobile ions that do not take part in the hydrostatic pressure induced flow of the test solution. Even if these 'captured' mobile electric charges are not moved together with the test solution, they contribute to the surface conduction by migration and electro-osmosis induced by the streaming potential difference generated. From an analytical point of view, these results show that it is important to check the effect of surface conduction when surface potentials are calculated from the streaming potential measuring technique (especially at low ionic strength). On the other hand, they show the extraordinary amount of data on the fine structure of charge distribution at the interface which is contained in electrokinetic measurements. To unravel the

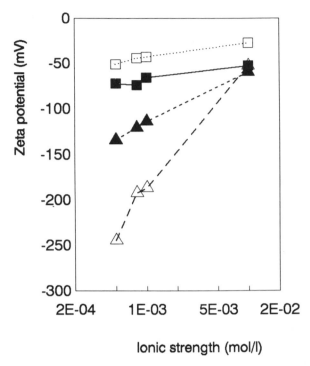

Figure 5.8 Comparison between surface potentials calculated from measured streaming potential (square) and streaming current (triangle) for glass (solid symbols) and swollen polyurethane (open symbols). (Redrawn from data of Table I of ref. [63].)

details of the fine structure is almost impossible by this technique alone. However, the recent advancements which were just described, i.e. the possibility of using the same surface for streaming potential and other surface sensitive techniques, are expected to magnify the role played by this 'old' technique in the surface science of polymers.

5.3 REFLECTIVITY OF NEUTRONS

In this last section, a short discussion on neutron reflectivity will be made.

Neutron reflectivity [65] provides the possibility of investigating several surface and interfacial phenomena with a spatial resolution in the range of nanometres. The technique is based on the fact that neutron reflectivity is a function of the component of the neutron momentum perpendicular to a film:

$$q_{z0} = (2\pi/\Gamma)\sin\theta \qquad (5.8)$$

where Γ is the neutron wavelength and θ the angle of incidence of the neutron beam with the surface.

In a medium composed of nuclei having neutron scättering amplitudes b,

$$q_{zi} = (q_{z0}^2 - 4\pi b/V)^{1/2} \qquad (5.9)$$

where V is the volume. The component of the neutron momentum parallel to the surface is identical to that in vacuum, thus the reflectivity can be described as a one-dimensional motion of a particle in a potential gradient. Although the reflectivity cannot be generally described in an explicit form, some cases accept an analytical expression of the potential as that of histograms of different thicknesses and different relative indexes, where the reflection at the boundary between two layers depends from the respective momenta:

$$r_{1,2} = (q_{z1} - q_{z2})/q_{z1} + q_{z2} \qquad (5.10)$$

and the total reflection of the system is a suitable combination of the products of the reflections at the single boundaries [66].

Based on the above assumptions, the concentration profiles of polymer chains in solution near the air–solution interface have been studied [67], as well as block copolymer solutions [68], immiscible polymer bilayers [69, 70], interdiffusion of two layers composed respectively by a hydrogenated and a deuterated moiety [71]. With respect to sharp interfaces, diffuse interfaces have the effect of lowering the reflectivity from its ideal value, thus interdiffusion of polymers can be studied [71].

Experimentally, a collimated beam of cold neutrons is directed towards the surface of the specimen (for instance constituted by a bilayer system) at a glancing angle of a fraction of a degree. The transmitted and reflected beams are detected by means of a two-dimensional position-sensitive detector. Because the neutron source has a pulsed nature, the measurement is taken in time-of-flight mode, utilizing a large range of neutron wavelengths (0.4–1.2 nm) in order to cover a wide spectrum of q_{z0}. Reflectivity profiles are collected as a function of q and experimental values are fitted starting from thickness, neutron scattering amplitudes [72] and volumes [73].

Variations of thickness of some nanometres can be observed, better than other techniques like NMR. Neutron reflectivity is suitable for establishing the equilibrium profile of polymers in solutions and melts, and measuring the kinetics of polymer diffusion in blends. In symmetric diblock copolymers, like P(S-b-MMA), lamellar morphology in the bulk can be investigated [74].

REFERENCES

[1] D. L. Cho, P. M. Claesson, C. G. Gölander, and K. Johansson, *J. Appl. Polym. Sci.*, **41**, 1373 (1990).

[2] M. Malmstem, P. M. Claesson, E. Pezron, and I. Pezron, *Langmuir*, **6**, 1572 (1990).
[3] M. K. Chaudhury and G. M. Whitesides, *Langmuir*, **7**, 1013 (1991).
[4] B. V. Derjaguin and M. M. Kusakov, *Izv. Akad. Nauk. SSSR, Ser. Khim.*, **5**, 741 (1936).
[5] J. Th. Overbeek and M. J. Spaarnay, *J. Coll. Sci.*, **7**, 343 (1952).
[6] G. Peschel and K. A. Adlfinger, *Naturwissenschaften*, **54**, 614 (1967).
[7] J. N. Israelachvili and D. Tabor, *Proc. R. Soc. London*, **A331**, 19 (1972).
[8] J. N. Israelachvili and G. E. Adams, *Nature*, **262**, 77 (1976).
[9] J. N. Israelachvili and G. E. Adams, *J. Chem. Soc. Faraday Trans.*, **I-74**, 975 (1978).
[10] K. B. Lodge, *Adv. Coll. Interface Sci.*, **19**, 27 (1983).
[11] A. I. Bailey and S. M. Kay, *Proc. Roy. Soc. Lond.*, **A301**, 47 (1967).
[12] A. I. Bailey and A. G. Price, *J. Chem. Phys.*, **53**, 3421 (1970).
[13] U. Breitmeier and A. I. Bailey, *Surface Sci.*, **89**, 191 (1979).
[14] S. Tolansky, *Multiple Beam Interferometry of Surface and Films*, Oxford University Press (1948).
[15] B. V. Derjaguin, *Kolloid Z.*, **69**, 155 (1934).
[16] D. Tabor and R. H. S. Winterton, *Proc. Roy. Soc. Lond.*, **A312**, 435 (1969).
[17] J. N. Israelachvili, *Nature*, **229**, 85 (1971).
[18] J. N. Israelachvili, *J. Colloid Interface Sci.*, **44**, 259 (1973).
[19] J. N. Israelachvili and P. M. McGuiggan, *J. Mater. Res.*, **5**, 2223 (1990).
[20] R. G. Horn and D. T. Smith, *J. Non-Cryst. Solids*, **120**, 72 (1990).
[21] J. L. Parker, D. L. Cho, and P. M. Claesson, *J. Phys. Chem.*, **93**, 6121 (1989).
[22] R. G. Horn, D. R. Clarke, and M. T. Clarkson, *J. Mater. Res.*, **8**, 413 (1988).
[23] R. G. Horn, D. T. Smith, and W. Haller, *Chem. Phys. Lett.*, **162**, 404 (1989).
[24] E. A. D. White and J. D. C. Wood, *J. Mater. Sci.*, **9**, 1999 (1974).
[25] R. J. Hunter, *Zeta Potential in Colloid Science*, Academic Press, London (1981).
[26] A. W. Adamson, *Physical Chemistry of Surfaces*, 5th edn, Wiley, New York (1990).
[27] P. C. Hiemenz, *Principles of Colloid and Surface Chemistry*, Dekker, New York (1977).
[28] J. Lyklema, in *Surface and Interfacial Aspects of Biomedical Polymers*, J. D. Andrade (Ed.), Plenum Press, New York (1985), vol. 1, Chapter 8.
[29] K. Furusawa, W. Norde and J. Lyklema, *Kolloid Z.Z. Polym.*, **250**, 908 (1972).
[30] J. W. Vanderhoff and H. J. van den Hull, *J. Macromol. Sci., Chem.*, **A7**, 677 (1973).
[31] J. Hen, *J. Colloid Interface Sci.*, **49**, 425 (1974).
[32] J. Stone-Masui and A. Watillon, *J. Colloid Interface Sci.*, **52**, 479 (1975).
[33] D. E. Yates, R. H. Ottewill, and J. W. Goodwin, *J. Colloid Interface Sci.*, **62**, 356 (1977).
[34] P. Bagchi, B. V. Gray, and S. Birnbaum, *J. Colloid Interface Sci.*, **69**, 502 (1979).
[35] D. H. Everett, M. E. Gultepe, and M. C. Wilkinsin, *J. Colloid Interface Sci.*, **71**, 336 (1979).
[36] M. E. Labib and A. Robertson, *J. Colloid Interface Sci.*, **77**, 151 (1980).
[37] S. Wu, *Polymer Surfaces and Adhesion*, Dekker, New York (1982), p. 327.
[38] J. Lyklema and Th. G. Overbeek, *J. Colloid Sci.*, **16**, 501 (1961).
[39] J. Lyklema, *J. Colloid Interface Sci.*, **58**, 242 (1977).
[40] L. K. Koopal and J. Lyklema, *Faraday Disc. Chem. Soc.*, **59**, 230 (1975).
[41] J. Lyklema, *Pure Appl. Chem.*, **46**, 149 (1976).
[42] D. G. Hall, *J. Chem. Soc. Faraday II*, **76**, 1254 (1980).
[43] D. G. Hall and H. M. Rendall, *J. Chem. Soc. Faraday I*, **76**, 2575 (1980).
[44] R. W. O'Brien and R. J. Hunter, *Can. J. Chem.*, **59**, 1878 (1981).
[45] II. Ohshima, T. W. Healy, and L. R. White, *J. Chem. Soc. Faraday Trans. 2*, **79**, 1613 (1983).

[46] N. Street, *Aust. J. Chem.*, **17**, 828 (1964).
[47] J. Lyons, M.Sc. Thesis, University of Melbourne (1979).
[48] V. A. DePalma, *Rev. Sci. Instrum.*, **51**, 1390 (1980).
[49] R. A. Van Wagenen and J. D. Andrade, *J. Colloid Interface Sci.*, **76**, 305 (1980).
[50] R. A. Van Wagenen, D. L. Coleman, R. N. King, P. Triolo, L. Brostrom, L. M. Smith, D. E. Gregonis, and J. D. Andrade, *J. Colloid Interface Sci.*, **84**, 155 (1981).
[51] D. Fairhurst and V. Ribitsch, *Polym. Mater. Sci. Eng.*, **62**, 57 (1990).
[52] D. Fairhurst and V. Ribitsch, *ACS Symp. Ser.*, **472**, 337 (1991).
[53] Brookhaven Instruments Corporation/PAAR, advertisement sheet.
[54] P. Benes and M. Paulenova, *Kolloid Z: Z. Polym.*, **251**, 766 (1973).
[55] H. P. M. Fromageot, J. N. Groves, A. R. Sears, and J. F. Brown, Jr., *J. Biomed. Mater. Res.*, **10**, 455 (1976).
[56] D. E. Brooks, *J. Colloid Interface Sci.*, **43**, 670 (1973).
[57] C. M. Ma, F. J. Micale, S. El-Aasser, and J. W. Vanderhoff, *Amer. Chem. Soc. Prepr.* **43**, 358 (1980).
[58] N. Kuehn, H. J. Jacobasch, and K. Lunkenheimer, *Acta Polymerica*, **37**, 394 (1986).
[59] M. Rätzsch, H. J. Jacobasch, and K. H. Freitag, *Adv. Colloid Interface Sci.*, **31**, 225 (1990).
[60] F. M. Fowkes and F. H. Hielscher, *Amer. Chem. Soc. Prepr.*, **43**, 358 (1980).
[61] F. M. Fowkes, in *Surface and Interfacial Aspects of Biomedical Polymers*, J. D. Andrade (Ed.), Plenum Press, New York (1985), vol. 1, Chapter 9.
[62] A. Voigt, R. Becker, and E. Donath, *J. Biomed. Mater. Res.*, **18**, 317 (1984).
[63] D. L. Coleman, D. E. Gregonis, and J. D. Andrade, *J. Biomed. Mater. Res.*, **16**, 381 (1982).
[64] J. Lyklema, *J. Electronal. Chem*, **18**, 341 (1968).
[65] S. A. Werner and A. G. Klein, in *Neutron Scattering*, R. Celotta, J. Levine, D. L. Price, and K. Skold (Eds), Ch. 4, Academic Press, New York (1987).
[66] O. S. Heavens, *Optical Properties of Thin Solid Films*, Dover, New York (1965).
[67] E. Bouchard, B. Farnoux, X. Sun, M. Daoud, and G. Jannink, *Europhys. Lett.*, **2**, 315 (1986).
[68] R. M. Richards, J. Penfold, S. J. Roser, and R. C. Ward, *ISIS Annual Report*, p. A39 (1987).
[69] J. S. Higgins, D. Walsh, J. Penfold, and R. C. Ward, *ISIS Annual Report*, p. A38 (1987).
[70] M. Stamm and C. F. Majkrzak, *ACS Polym. Prepr.*, **28**, 18 (1987).
[71] T. P. Russell, A. Karim, A. Mansour, and G. P. Felcher, *Macromolecules*, **21**, 1890 (1988).
[72] J. A. Ibers and W. C. Hamilton (Eds), *International Tables for X-Ray Crystallography*, Kynoch, Birmingham (1974).
[73] *Handbook of Chemistry and Physics*, CRC Press, Cleveland.
[74] H. Anastasiadis, A. Menelle, T. P. Russell, S. K. Satija, and C. F. Majkrzak, *Progr. Coll. Polym. Sci.*, **91**, 88 (1993).

Part III

MODIFICATION TECHNIQUES

Chapter 6

Physical Modifications

The activities aimed at physically modifying polymer surfaces can be divided into two main categories, the first involved with chemically altering the surface layer, the second with depositing an extraneous layer on top of the existing material, thereby generating a sharp interface.

Given the non-reactive character of polymer surfaces, the former requires generating high energy species, e.g. radicals, ions, molecules in excited electronic states, etc. In nature typical high energy media are fire and thunder. Flame treatments are themselves frequently applied in surface modification, while a number of methods, namely corona and plasma treatments, have been devised to mimic the physical characteristics of thunder, i.e. the ionization of air. An important side-effect of surface treatment might be etching, which is persistently pursued for electronics (resist stripping).

Coating by physical treatment involves itself in the generation, usually by high energy methods, of matter fundamentals, such as atoms or atomic clusters, to be deposited on polymer surfaces. Suitable techniques involve plasma (sputtering, plasma polymerization) or energy-induced sublimation (e.g. thermally or electron beam-induced evaporation).

A lot of applications have been devised, in both low cost (e.g. packaging) and high cost (e.g. electronics) markets. In the former case published information is mainly in patent and technical literature. In the latter, more academic studies have been performed, therefore extending the amount of scientific literature.

6.1 FLAME TREATMENTS

Flame treatments are widely applied in surface modification to introduce oxygen-containing functions at polyolefin surfaces, mainly to improve printability or paintability. Active species formed by the high temperature include radicals, ions and molecules in excited states.

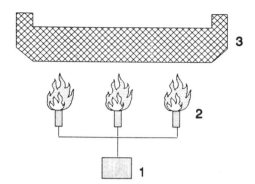

Figure 6.1 Schematic of a flame treatment process. An intelligent driving system (1) scans the flames produced by the burners (2) on the substrate (3), in this case an idealized bumper, holding them at fixed distance from the surface and following the irregularities of the shape.

The apparatus is conceptually straightforward (Figure 6.1), it consists in one or more flames which are held at a fixed distance from the sample and scanned over it at controlled speed. Treatment variables include flame composition and temperature, distance from the sample, and scan speed. The latter are particularly critical, since they control the extent of surface damage, which might result in loss of adhesion, due to weak boundary layer effects. To treat large areas or complicated shapes, special multiple or robotized systems have been devised.

Flame treatment studies have not attracted strong attention in both patent and academic literature [1–8]. In the first case a probable reason lies in the fact that whilst the physical phenomenon is well known, know-how on treatment optimization is essential and the latter is better defended by secrecy than by patenting. As to academic literature, university groups rarely have access to flame treatment units, therefore their interest has been limited.

Polyolefins have been most frequently dealt with in flame treatment studies. Most of them deal with surface characterization of treated polymers, trying to investigate which functional groups have been created by the treatment and how they affect either printability or adhesion.

A typical case history is the study of the effect of flame treatments on polypropylene by XPS, SSIMS, FTIR-PAS, contact angle measurements and peel tests [3–4]. The cited combination of techniques suggested surface enrichment in oxygen containing species, hydroxyls, carbonyls, carboxyls, land consequent improvements in wettability and adhesion. Pioneering SSIMS (FABMS) studies showed that, contrary to chemical intuition, the preferred site of interaction for active species was the methyl group of polypropylene [4].

6.2 CORONA TREATMENTS

Corona treatments exploit the corona effect, i.e. the formation of high energy electromagnetic fields close to charged thin wires or points, with consequent ionization in their proximity, even at atmospheric pressure and relatively low temperature [9–10]. In the ionized region excited species (ions, radicals, electrons, molecules in excited states, etc.) are present and the latter are active in surface modification, typically introduction of oxygen-containing functions.

Again the apparatus (Figure 6.2) is straightforward, it involves an electromagnetic field generator (d.c., a.c. or r.f.), thin wires or points and the possibility of scanning over the surface at a fixed distance [11]. Treatments parameters include electromagnetic properties (e.g. voltage, frequency), electrode shape, treatment atmosphere (most frequently air) and sample/electrode geometry.

For the easiest geometrical shape (i.e. flat films or sheets) bars or wires can be conveniently used, actually most polyolefin films for packaging purposes are routinely corona treated. For complex shapes, e.g. bumpers, robotized systems have been devised to optimize treatment uniformity.

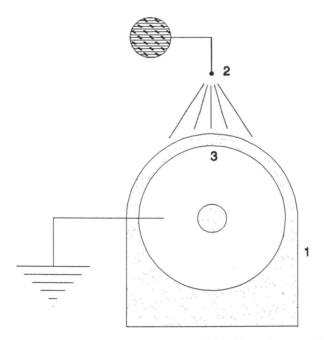

Figure 6.2 Schematic of a corona treatment manifold. The polymeric film (1) moves between a trip or filament (3) held at high electric potential (thousands of volts) and a grounded roll, the discharge induces the formation of excited species, which modify the surface.

Again the patent literature is not abundant [12], since trade secrets are predominant. Academic studies have been more frequent than for flame treatments [13–29], although with some prevalence of industrial groups.

Most of the work was performed in the late 1970s to early 1980s and it concentrated on the surface (XPS, wettability, etc.) and performance (adhesion) characterization of treated films. Lots of emphasis has been devoted to the effect of treatment parameters, namely electrical parameters (voltage, frequency, dissipated energy), mechanical characteristics (air gap, shape of the electrodes) and treatment atmosphere, including relative humidity [13, 15–17]. General features include the very low lifetime of radicals formed during the treatment, due to the reaction with atmospheric oxygen, as shown by ESR [15], the formation of oxygen containing groups and the dependence of the effect of the treatment upon ageing time.

More recently some more peculiar aspects have been dealt with, such as the formation of low molecular weight materials upon treatment, which can strongly affect the adhesion behaviour [22], and the use of corona treatment as promoters for further surface chemistry, namely graft polymerization [21].

6.3 'COLD' PLASMA TREATMENTS

Actual low temperature plasmas require low pressure to be sustained. An ionized region is formed, with its composition depending on the gas feed, always including high energy photons, electrons, ions, radicals and excited species. Low pressure plasmas can be used for surface activation by the introduction of oxygen-containing functions, etching by formation of gaseous species (e.g. carbon oxides or fluorides in CF_4/O_2 plasmas), or coating deposition by plasma polymerization (e.g. of fluorine or silicon-containing monomers) [30–36].

In terms of apparatus (Figure 6.3) the main difference as compared to corona treatments is the need of a vacuum chamber and gas feed, to maintain the appropriate pressure and composition of the gaseous mixture. Electromagnetic energy can be input by different coupling methods (inductive or capacitive), frequencies (d.c., a.c., r.f. or m.w.) and electrode configurations, depending on the application. The sample may be either present as a film or as a discrete object, altering geometrical requirements. Treatment parameters involve sample/chamber geometry, pressure, gas flow rate and electromagnetic parameters (e.g. frequency, power, etc.).

Plasma treatments are an important step in microelectronic fabrication technologics and a wide body of literature is involved with plasma etching of inorganics (e.g. silicon, silicon dioxide) and, much less frequently, organics (typically resists). Another important area is biomaterials, where the possibility of unusual chemistries when plasma polymerizing allowed lots of combinations in terms of coating chemistry.

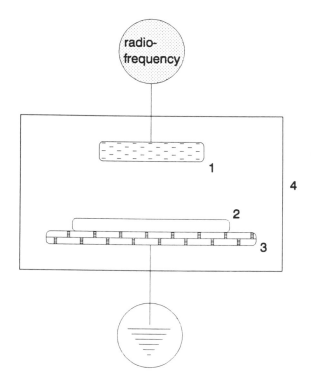

Figure 6.3 Schematic of a parallel plate system for cold plasma treatment. In a vacuum chamber (4), with pumping and gas-introduction systems, there are two electrodes: the anode (1), which is polarized with the radiofrequency, and the grounded cathode (3). The plasma is formed between the two electrodes and results in the formation of excited species which then act on the surface of the substrate (2).

In more trivial markets, such as packaging or fibres, the use of plasma is much less frequent, owing to the poor price-to-performace ratio. Yet very recently a few applications have been reported for promoting the adhesion of paints to bumpers, as a substitute for primers, which typically require the use of toxic solvents for the application.

Both the patent and the academic literature is very abundant, due to the importance of plasma treatments in 'frontier' applications and to the possibility of constructing a plasma equipment with a low amount of money, characteristics particularly important for academic institutions.

By far the most successful application of plasma treatment is in the electronic industries, mainly for etching purposes. Concerning polymers, discharges have been used mainly for stripping photoresists and for resist pattern delineation. The concepts which are most frequently found in papers dealing with this subject are etch, anisotropy, selectivity and texture of the etched surface, excellent reviews are available [37–38].

Plasma etchants for organic polymers include mostly oxygen and CF_4/O_2 mixtures, which act by the formation of oxygen and fluorine atoms, which then attack organic materials forming CO, CO_2, H_2O and fluorinated volatile compounds. Most commercial systems assist etching by accelerating ions on to the treated surface, this, further to enhancing etch rates [39], also improves anisotropy, which is very important for electronics processing [37–38]. Much of the recent work is involved with mechanistic aspects of etching phenomena [40–41], mainly with microwave plasmas, which are becoming increasingly important [42–44].

Many papers have been devoted to the effect of non-polymerizing plasmas on surface chemistry and properties, mostly printability and adhesion. As compared to corona, plasma treatments allow more degrees of freedom, including the choice of the reacting gas and of a number of parameters (pressure, gas flow rate, type of plasma apparatus, geometry, electrical parameters, etc.). In papers from polymer scientists, by far the most common occurrence is the comparison of different reacting gases based on their effect on surface chemistry. This sort of study originated in the early 1970s [45] and there is still plenty of activity to date.

A first series of papers is involved with oxygen and/or inert gas treatments, aimed at improving wettability and adhesion [45–69]. As observed in other instances, as time goes by the attention is being focused on the refinement of knowledge on various particular aspects of the treatment. Amongst published studies it is interesting to quote the work of Gerenser [47], who performed XPS characterization of *in-situ* plasma treated polyethylene (PE), showing that whereas oxygen and nitrogen plasmas created new chemical species on surfaces, inert gas plasmas did not have any detectable chemical effect. Yet upon exposure to the atmosphere, inert gas treated samples oxidized and on nitrogen treated samples nitrogen itself (mainly the one present as imine) was substituted by oxygen. This shows that in the application one should consider the treatment as a whole, including the passage from vacuum to atmosphere as one of the steps of the treatment.

Another very important issue is ageing [61–68]. It is known that the treatment-induced surface changes are partially reversible with time (hydrophobic recovery) and this is particularly important from an application point of view. It was shown that hydrophobic recovery depends on the polymer and on the treatment. It is accelerated by chain scission and slowed by cross-linking within the surface layer. In the absence of low molecular weight species and at low temperatures it occurs typically by reorganization within the surface modified layer. Thus ageing itself can be considered a step of the treatment, in the sense that it influences the outcome as determined by the properties imparted to the polymer surface (see also Chapter 9).

Finally, the oxidizing systems typically suggested for surface activation are actually etching media, therefore the balance between etching and surface modification should be considered. The two mechanisms have different time scales; therefore, while at long plasma treatments some sort of equilibrium between the surface and the active atmosphere is established, the first instances are essentially non-equilibrium and the resulting surface properties are very sensitive to plasma parameters such as gas flow rate [69].

Another series of papers have been devoted to fluorination, which is important to alter the barrier and friction behaviour of polymer surfaces [70–76]. Whilst these treatments are normally performed by contacting the surface with fluorine gas (see the appropriate section), the possibility of generating *in-situ* fluorine or fluorine containing species has been explored. Gases typically used in the etching of silicon have been suggested (e.g. CF_4, SF_6, CF_3Cl, etc.) [74–75], observing a balance between etching, fluorination and polymerization. As observed in etching studies performed on silicon [37–38], this depends critically on the composition of the plasma atmosphere, namely on the ratio between fluorine atoms and fluorocarbon radicals. This in turn depends on the establishment of an equilibrium between chemical activity of the plasma on the reactor walls and substrate and on parameters such as gas flow rate. It is one of the systems which better exemplify the strong dependence of all reported results on the type of apparatus and plasma parameters utilized, factors which are frequently under-estimated by the literature originated by chemists.

Plasma polymerization has attracted most attention from polymer scientists and has been extensively reviewed in a recent book [35]. Within a plasma gases or vapours such as fluorocarbons [77–82], silanes or siloxanes [83–91], vinyl monomers [92–96] and metal containing systems [97–100] are fragmented into smaller units which then coalesce on the substrate and reactor walls forming highly dense and cross-linked films, which have been claimed for improving characteristics such as antistaticity, wear resistance, permeation, etc. [101–105].

Having a polymer surface as a substrate is mostly an accident for plasma polymerization. The only characteristic which is strongly affected is interfacial adhesion, once the polymer surface is wholly covered the plasma chemistry is substrate-independent. Typically using polymer surfaces as a substrate has an advantage and a disadvantage. The former lies in the fact that chemical bonds at the interface can be formed. The latter hinges on the possible high cross-linking of the plasma polymerized layer, which can make them mechanically incoherent with the substrate, even more so when thermal shocks can be foreseen, owing to the dramatic difference in dilation coefficients of conventional and plasma polymers [35].

6.4 'HOT' PLASMA TREATMENTS

Plasmas generated at atmospheric pressure reach very high temperatures (5000–10 000 K). They have been extensively used in metallurgic operations, for depositing hard or passivating coatings by melting inorganic powder in the

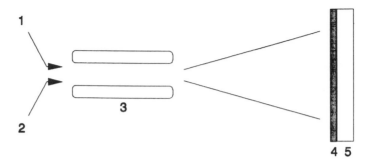

Figure 6.4   Schematic of a plasma torch for depositing ceramic layers. The treatment gas (1) is inserted in the torch (3) where the plasma is induced for instance by an electric arc. The temperature of the plasma at atmospheric pressure is very high (several thousands degrees). The ceramic powder (2) is injected in the region, it melts and is conveyed on the substrate (5), where it solidifies and forms the coating (4).

plasma itself and spraying molten droplets on metal surfaces [30–31, 36]. So far little application to polymer surfaces has been reported, due to the low heat resistance of most polymeric substrates.

A plasma torch (Figure 6.4) typically involves a powder inlet, a plasma gas inlet and electrodes to sustain the discharge. Possible applications include the deposition of wear resistant coatings, although care has to be taken to avoid excessive heat transfer to the polymer surface and delamination problems.

6.5   UV TREATMENTS

Photons, usually those with low wavelength, are energetic species which are used to activate many chemical reactions. A typical example of UV action on polymer surfaces is their degradation by sun exposure.

UV lamps are widely used for the treatment of polymer surfaces and the apparatus involves essentially a lamp and sample illumination devices, such as the possibility of selectively irradiating tiny areas (masks: microelectronics, see Figure 6.5) or moving the sample below the photon source (rollers: printing industries).

Most applications involve the photon-activated cross-linking (negative resists, paper coatings) or fragmentation (positive resists) of polymer coatings. Sensitizers are typically added to enhance the photon yield of the processes. Lamps operating between 250 and 400 nm wavelengths are most frequent.

There is a wide body of both academic and patent literature on UV-cured coatings, a rather extensive review can be found in ref. [106]. Similarly to other coating applications, the substrate does not really influence the curing of the coating by UV radiation.

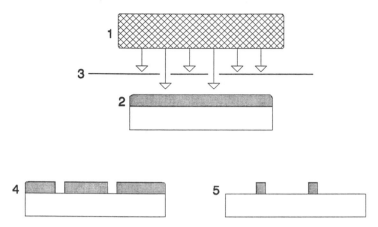

Figure 6.5 Schematic of a UV-treatment system, relative to photolithography. A UV source (1) irradiates a substrate with a photosensitive coating (2) through a mask (3). The coating is either depolymerized (positive resist) or cross-linked (negative resist) by the radiation. Further wet treatments are used for stripping the degraded positive resist (4) or the unaltered negative resist (5).

Much less prominent, but extensive, is the academic literature concerning the effect of photons (in the presence or absence of oxygen) on polymer surfaces [107–108]. Most commodity polymers have been extensively studied in terms of their photo-oxidative behaviour. Examples are PP [109–111], PET [112–113] and PC [114–115]. Irradiation in the presence of oxygen has been exploited to enhance wettability and related adhesion and antistaticity characteristics [108, 116]. Irradiation in the presence of polymerizable vapours affords a method to simulate plasma polymerization, as shown by Munro *et al.* [117].

6.6 LASER TREATMENTS

Lasers are photon sources characterized by energy and space coherence, with intensities which can be dramatically high (in the MW range). Similarly to conventional photon sources, they can be used to promote cross-linking or scission effects. Furthermore, the high energy density allows sintering effects.

The apparatus is conceptually similar to that used for UV treatments, with the directionality being exasperated. When used for cross-linking or sintering, lasers have been suggested for novel prototype-making activities, such as stereolithography and selective laser sintering. Scission effects (laser ablation) have been exploited for low-volume microlithography applications. Recently some applications to adhesion improvements have been reported, involving the use of IR lasers for surface cleaning.

The patent literature is relatively limited, while a number of papers have been devoted mainly to laser ablation [118–125]. The aim of these papers is in general to discriminate between photochemical and thermal effects in CO_2 and excimer laser-induced ablation, whilst measuring the dependence of the etch rate upon laser parameters (e.g. frequency, fluence). Surface composition studies showed that for oxygen-containing polymers oxygen depletion effects can occur, due to the efficient formation of stable carbon oxide molecules [122, 126–127].

Other reported applications include laser cross-linking of coatings, which are conceptually similar to those described in the previous section [128–129]. A similar application, but involving the cure of a continuously reformed liquid surface, is the basis of stereolithography [130–133], a technique which has rapidly become commonplace for prototyping.

6.7   X-RAY AND γ-RAY TREATMENTS

These cannot really be considered surface treatments, since the photon energy is so high that the photon mean free path is quite high [134]. Yet applications similar to UV treatments have been reported, involving mainly cross-linking of polymeric coatings. Furthermore, in the presence of oxygen, high energy photon treatment (e.g. for sterilization purposes) induces the formation of radical sites at surfaces, which then react with atmospheric oxygen forming oxygenated functions.

The apparatus is similar to that used for electron-beam treatments (see Figure 6.6). Similarly to that case, adequate shielding is mandatory. Most applications are reported in the academic literature and involve the study of chemical effects of radiative exposure [135] or radiation-induced curing of coatings [136].

6.8   ELECTRON BEAM TREATMENTS

High energy ($> 50\,keV$, up to a few MeV) electrons have high mean free paths ($> 100\,\mu m$, up to a few mm) in polymeric materials, therefore they are typically used for cross-linking purposes in various applications such as cables, tubing and curing of coatings [36, 137–139]. Applications to the alteration of permeability and diffusion coefficients have also been reported [140–141]. Some literature has also been devoted to the damage of polymer surfaces when exposed to electron beams during electron microscopy examination [142].

The apparatus (Figure 6.6) involves as electron gun and means to drive the sample below it. An adequate shielding is required and it is relatively easy either to raster or to focus the electron beam. For low energy electrons a vacuum chamber is needed, due to the low mean free path of electrons in air.

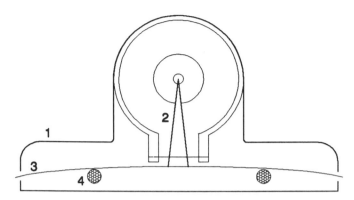

Figure 6.6 Schematic of a system for treating plastic films or coated paper with moderate energy (100–300 keV) electrons. The film or sheet (3), controlled by tensioning devices (4), moves in front of the electron beam (2), which performs a treatment which in most cases involves cross-linking of the coating. The electron beam is generated in a vacuum chamber and exits going through a thin metal film, due to its high energy. All the system is carefully shielded (1) to prevent exposure of workers to X-rays generated by the impact of high energy electrons with the matter.

A few patents and papers suggested the use of low energy electrons (< 25 keV) for surface treatment, aimed at improving adhesion by mechanically strengthening the surface layer and introducing some oxygen at the surface by the formation of radical and consequent reaction with atmospheric oxygen [143–147].

6.9 ION BEAM TREATMENTS

Ion beams can be used for two different purposes, namely to directly alter the surface composition of the material [36, 148] or to sputter off target species which are then deposited on the surface. The latter application will be dealt with in section 6.11.

Ions are energetic species with a high momentum and relatively low mean free path, due to their mass. Therefore they strongly affect surface composition, leading to extensive chemical modification, sometimes resulting in graphitization.

Various types of ion guns do exist; however, the principle always lies in the formation of ions and their acceleration by electrostatic means. In Figure 6.7 the two possible modification methods, ion implantation and ion-assisted deposition, are outlined. The apparatus and the sample in both cases are held under vacuum, unless very high energy is used, again for mean free path considerations.

Patents are very rare and are subordinate to ion implantation on plastics to improve wear resistance or biocompatibility [149]. The literature is mostly

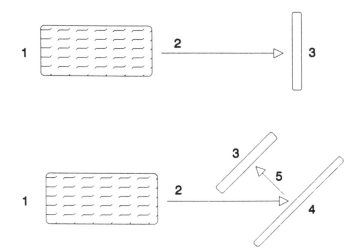

Figure 6.7 Schematic of ion treatment possibilities. At the top ion implantation is shown: the ion gun (1) emits an ion beam (2) which is implanted in the substrate (3). Below an ion-activated deposition is shown. The ion gun (1) emits a beam (92) which hits a target (5) with particular composition, exciting the emission of neutral particles (4) which form a coating on the substrate (4).

academic and is involved with the study of the effect of ion beams on characteristics such as conductivity (by formation of graphitized paths) [150–158], optical properties [159–160] or adhesion/friction (by combined alteration of surface chemistry and morphology) [161–163].

6.10   METALLIZATION

Metallization consists in the coating of polymer surfaces by metal species, typically deposited by evaporation induced by Joule effect or electron-beam excitation. The most widespread application is the aluminium coating of plastic films for electrical (capacitors) or packaging applications (decorative and barrier). As such it is, along with corona treatment, the most frequent industrial high energy density treatment of polymer surfaces.

Metallizers (Figure 6.8) comprise vacuum chambers with systems for metal evaporation, aluminium is particularly convenient since it can be fed as a liquid and easily vaporized at relatively modest vacuum and temperature. Evaporation is directional, therefore simple shapes such as films are preferred. An efficient heat dispersing system is critical for polymers to avoid deformation of the substrate.

The patent literature concerning metallization dates back to the 1960s and 1970s, nowadays metallizers are usually standard equipment. The academic

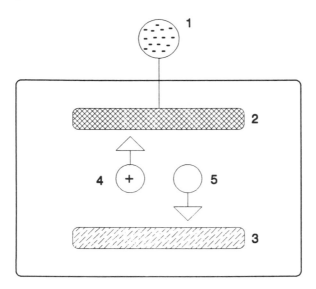

Figure 6.8 Schematic of an apparatus for evaporating aluminium on polymeric films with excitation provided by an electron beam. The polymeric film is wound and unwound on rolls (1) under the control of tensioning devices (5). The electron gun (3) emits a beam which is controlled by a beam scanner (4). The beam, hitting liquid aluminium residing in the crucible (2), induces evaporation and coating deposition. Of course, the whole process involves a vacuum system with efficient pumping systems.

literature is not particularly abundant, although books containing papers from conferences on metallization have recently appeared [164–165], however the most frequently studied topic is the strength of the polymer–metal interface [166]. Further references can be found in the Proceedings of the Society of Vacumm Coaters. Recently some attention has been devoted to the evaporation of oxide material, such as silicon oxides, mainly for packaging applications [167–170].

6.11 SPUTTERING

Sputtering involves creating ions, accelerating them on a target, forming atoms or clusters which are then deposited on the substrate. It is used for producing inorganic coatings when evaporation is not possible. Methods to form ions include ion beams or a discharge (the powered electrode is negatively biased, so ions are accelerated), an extensive review of methods used for sputter deposition is in ref. [33].

A typical apparatus is the one shown in Figure 6.9, it includes an ion source, a target and the sample. The method is used most frequently for discrete objects,

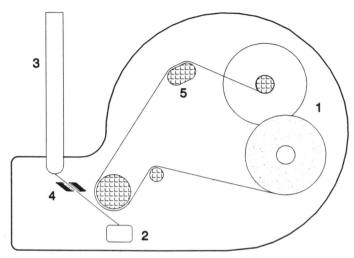

Figure 6.9 Schematic of a sputtering apparatus. In a vacuum chamber (6), fitted with pumping gas introduction systems, positive ions (4) are generated, for example using a plasma. A target of controlled composition is negatively polarized (1) to attract positive ions. Owing to their impact, neutral particles (5) are formed, which then form a coating on the substrate (3).

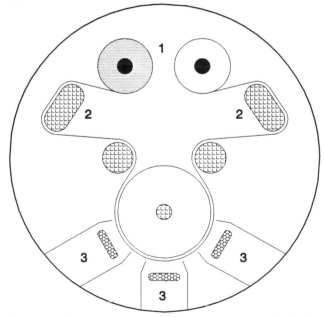

Figure 6.10 Apparatus for continuous sputtering treatment of films. The film is wound and unwound on rolls (1) under the control of tensioning devices (2). Moving in front of sputtering stations (3), it is subsequently coated with layers of controlled thickness and chemistry. Of course, sputtering stations (3) do not interfere reciprocally.

although web applications have been reported recently. A critical feature is the resistance of polymeric material to the heat generated by the process, which could induce deformation. Most sputtering processes are performed on heat resistant polymers such as biaxially oriented PET, PC, polyimides. A system particularly suited for continuous multilayer sputtering on plastic films is represented in Figure 6.10.

Sputtering systems, owing to the presence of energetic species in the system, typically provide good adhesion of the coating to the substrate, with exceptions related to the different dilation coefficient of inorganic coatings and polymeric substrates. Most of the existing literature, both patents and papers, is therefore related to the coating characteristics, aimed mainly at improving electro-optic performance or biocompatibility [171–180].

REFERENCES

[1] S. Wu, *Polymer Interface and Adhesion*, Marcel Dekker, New York (1982).
[2] D. Briggs, D. M. Brewis, and M. B. Konieczko, *J. Mater. Sci.*, **14**, 1344 (1979).
[3] F. Garbassi, E. Occhiello, F. Polato, and A. Brown, *J. Mater. Sci.*, **22**, 207 (1987).
[4] F. Garbassi, E. Occhiello, F. Polato, and A. Brown, *J. Mater. Sci.*, **22**, 1450 (1987).
[5] J. Dillard, F. Cromer, A. Cosentino, T. Rabito, G. Maciver, and J. Saracsan, *Proceedings of ECASIA 87*, Stuttgart, 19–23 October (1987).
[6] D. Y. Wu, E. Papirer, and J. Schultz, *C.R. Acad. Sci. Serie II*, **312**, 197 (1991).
[7] K. Armbruster and M. Osterhold, *Kunstoffe*, **80**, 1241 (1990).
[8] I. Sutherland, D. M. Brewis, R. J. Heath, and E. Sheng, *Surf. Interf. Anal.*, **17**, 507 (1991).
[9] F. K. McTaggart, *Plasma Chemistry in Electrical Discharges*, Elsevier, Amsterdam (1967).
[10] R. S. Sigmond and M. Goldman, in *Electric Breakdown and Discharges in Gases*, E. E. Kunhardt and L. H. Luessen (Eds), *NATO ASI Ser.*, **89b**, Plenum, New York (1983), p. 1–64.
[11] E. Occhiello and F. Garbassi, in *International Encyclopedia of Composites*, S. M. Lee (Ed.), VCH, New York (1991), Vol. 5, p. 390.
[12] M. Kadash and F. C. Schwab, US 4714658 (1987).
[13] J. F. Carley and P. Thomas Kitze, *Polym. Eng. Sci.*, **18**, 326 (1978).
[14] H. L. Spell and C. P. Christensen, *Tappi*, **62**, 77 (1979).
[15] J. F. Carley and P. Thomas Kitze, *Polym. Eng. Sci.*, **20**, 330 (1980).
[16] D. Briggs, D. G. Rance, C. R. Kendall, and A. R. Blythe, *Polymer*, **21**, 895 (1980).
[17] D. Briggs, C. R. Kendall, A. R. Blythe, and A. B. Wootton, *Polymer*, **24**, 47 (1983).
[18] H. Steinhauser and G. Ellinghorst, *Angew. Makromol. Chem.*, **131**, 145 (1985).
[19] F. C. Schwab and M. A. Kadash, *J. Plast. Film Sheeting*, **2**, 119 (1986).
[20] L. J. Gerenser, J. M. Pochan, J. F. Elman, and M. G. Mason, *Langmuir*, **2**, 765 (1987).
[21] H. Iwata, A. Kishida, M. Suzuki, Y. Hata, and Y. Ikada, *J. Polym. Sci. Chem. Ed.*, **26**, 3309 (1988).
[22] M. Strobel, C. Dunatov, J. M. Strobel, C. S. Lyons, S. J. Perron, and M. C. Morgen, *J. Adhes. Sci. Technol.*, **3**, 321 (1989).
[23] C. J. Dias, J. N. Marat-Mendes, and J. A. Giacometti, *J. Phys. D*, **22**, 663 (1989).

[24] Y. Hjertberg, B. A. Sultan, and E. M. Soervik, *J. Appl. Polym. Sci.*, **37**, 1183 (1989).
[25] Y. De Puydt, P. Bertrand, Y. Novis, R. Caudano, G. Feyder, and P. Lutgen, *Br. Polym. J.*, **21**, 141 (1989).
[26] S. Mournet, F. Arefi, P. Montazer-Rahmati, M. Goldman, and J. Amoroux. *Rev. Int. Hautes Temp. Refract.*, **25**, 219 (1989).
[27] W. J. Van Ooij and R. S. Michael, *ACS Symp. Ser.*, **440**, 60 (1990).
[28] Y. De Puydt, P. Bertrand, Y. Novis, M. Chtaib, P. Lutgen, and Feyder, *NATO ASI Ser.*, **176**, 179 (1990).
[29] E. C. Onyiriuka, L. S. Hersh, and W. Hertl, *J. Coll. Interf. Sci.*, **144**, 98 (1991).
[30] R. F. Baddour and R. S. Timmins, *The Applications of Plasmas to Chemical Processing*, MIT Press, Cambridge, MA (1967).
[31] B. Gross, B. Grycz, K. Miklossy, *Plasma Technology,* Iliffe Books, London (1968).
[32] J. R. Hollahan and A. T. Bell (Eds.), *Techniques and Applications of Plasma Chemistry*, Wiley, New York (1974).
[33] J. L. Vossen and W. Kern (Eds.), *Thin Film Processes*, Academic Press, New York (1978).
[34] H. V. Boenig, *Plasma Science and Technology*, Cornell University Press, London (1982).
[35] H. Yasuda, *Plasma Polymerization*, Academic Press, Orlando (1985).
[36] F. Garbassi and E. Occhiello (Eds). *High Energy Density Technologies in Materials Science*, Kluwer, Dordrecht (1990).
[37] D.L. Flamm and V. M. Donnelly, *Plasma Chem. Plasma Process.*, **1**, 317 (1981).
[38] J. W. Coburn, *Plasma Chem. Plasma Process.*, **2**, 1 (1982).
[39] E. Occhiello, F. Garbassi, and J. W. Coburn, *Plasma Chem. Plasma Process.*, **9** (1989) 3.
[40] R. D'Agostino, F. Cramarossa, and F. Illuzzi, *J. Appl. Phys.*, **61**, 2754 (1987).
[41] N. R. Lerner and T. Wydeven, *J. Appl. Polym. Sci.*, **35**, 1903 (1988).
[42] G. Sauvé, M. Moisan, J. Paraszczak, and J. Heidenreich, *Appl. Phys. Lett.*, **53**, 470 (1988).
[43] J. Pelletier, Y. Arnal, and O. Joubert, *Appl. Phys. Lett.*, **53**, 1914 (1988).
[44] A. M. Wrobel, B. Lamontagne, and M. R. Wertheimer, *Plasma Chem. Plasma Process.*, **8**, 315 (1988).
[45] C. A. L. Westerdahl, J. R. Hall, E. C. Schramm, and D. W. Levy, *J. Coll. Interf. Sci.*, **47**, 610 (1974).
[46] R. G. Nuzzo and G. Smolinski, *Macromolecules*, **17**, 1013 (1984).
[47] L. J. Gerenser, *J. Adhes. Sci. Technol.*, **1**, 303 (1987).
[48] M. Hudis, *J. Appl. Polym. Sci.*, **16**, 2397 (1972).
[49] L. Mascia, G. E. Carr, and P. Kember, *Plast. Rubber Proc. Appl.*, **9**, 133 (1988).
[50] B. Westerlind, A. Larsson, and M. Rigdahl, *Int. J. Adhesion and Adhesives*, **7**, 141 (1987).
[51] Y. Nakayama, T. Takahagi, F. Soeda, K. Hatada, S. Nagaoka, J. Suzuki, and A. Ishitani, *J. Polym. Sci. Chem. Ed.,* **26**, 559 (1988).
[52] T. Ogita, A. N. Ponomarev, S. Nishimoto, and T. Kagiya, *J. Macromol. Sci.-Chem.*, **A22**, 1135 (1985).
[53] J. R. G. Evans, R. Bulpett, and M. Ghezel, *J. Adhes. Sci. Technol.*, **2**, 291 (1987).
[54] N. H. Ladizeski and I. M. Ward, *J. Mater. Sci.*, **24**, 3763 (1989).
[55] Y. L. Hsieh and E. Y. Chen, *Ind. Eng. Chem. Prod. Res. Dev.*, **24**, 246 (1985).
[56] P. Jinchang and Z. Baoguan, *Mater. Chem. Phys.*, **20**, 99 (1988).
[57] D. T. Clark and A. Dilks, *J. Polym. Sci. Chem. Ed.*, **17**, 957 (1979).
[58] M. R. Wertheimer and H. P. Schreiber, *J. Appl. Polym. Sci.*, **26**, 2087 (1981).
[59] K. Imada, S. Ueno, Y. Nishina, and H. Nomura, US 4317788 (1982).

[60] H. D. Gesser and R. E. Warriner, US 3925178 (1975).
[61] M. Morra, E. Occhiello, and F. Garbassi, *J. Coll. Interf. Sci.*, **132**, 504 (1989).
[62] E. Occhiello, M. Morra, and F. Garbassi, *Angew. Makromol. Chem.*, **173**, 183 (1989).
[63] M. Morra, E. Occhiello, R. Marola, F. Garbassi, and D. Johnson, *J. Coll. Interf. Sci.*, **137**, 11 (1990).
[64] M. Morra, E. Occhiello, L. Gila, and F. Garbassi, *J. Adhesion*, **33**, 77 (1990).
[65] E. Occhiello, M. Morra, G. Morini, F. Garbassi, and P. Humphrey, *J. Appl. Polym. Sci.*, **42**, 551 (1991).
[66] E. Occhiello, M. Morra, G. Morini, F. Garbassi, and D. Johnson, *J. Appl. Polym. Sci.*, **42**, 2045 (1991).
[67] E. Occhiello, M. Morra, F. Garbassi, D. Johnson, and D. Humphrey, *Appl. Surf. Sci.*, **47**, 235 (1991).
[68] E. Occhiello, M. Morra, P. Cinquina, and F. Garbassi, *Polymer*, **33**, 3007 (1992).
[69] M. Morra, E. Occhiello, and F. Garbassi, *Surf. Interf. Anal.*, **16**, 412 (1990).
[70] G. A. Corbin, R. E. Cohen, and R. F. Baddour, *Polymer*, **23**, 1546 (1982).
[71] I.-H. Loh, R. E. Cohen, and R. F. Baddour, *J. Appl. Polym. Sci.*, **31**, 901 (1986).
[72] T. Yagi, A. E. Pavlath, and A. G. Pittman, *J. Appl. Polym. Sci.*, **27**, 4019 (1982).
[73] T. Yasuda, T. Okuno, K. Yoshida, and H. Yasuda, *J. Polym. Sci. Phys. Ed.*, **26**, 1781 (1988).
[74] M. Strobel, S. Corn, C. S. Lyons, G. A. Korba, *J. Polym. Sci. Chem. Ed.*, **23**, 1125 (1985).
[75] M. Strobel, P. A. Thomas, and C. S. Lyons, *J. Polym. Sci.*, **25**, 3343 (1987).
[76] R. Chasset, G. Legeay, J.-C. Touraine, and B. Azur, *Eur. Polym. J.*, **24**, 1049 (1988).
[77] H. Nomura, P. W. Kramer, and H. Yasuda, *Thin Solid Films*, **118**, 187 (1984).
[78] K. Nakajima, A. T. Bell, M. Shen, and M. M. Millard, *J. Appl. Polym. Sci.*, **23**, 2627 (1979).
[79] C. Arnold, K. W. Bieg, R. E. Cuthrell, and G. C. Nelson, *J. Appl. Polym. Sci.*, **27**, 821 (1982).
[80] G. Smolinski and M. J. vasile, *Eur. Polym. J.*, **15**, 87 (1979).
[81] Y. Haque and B. D. Ratner, *J. Polym. Sci. Phys. Ed.*, **26**, 1237 (1988).
[82] P. J. Astell-Burt, J. A. Cairns, A. K. Cheetham, and R. M. Hazel, *Plasma Chem. Plasma Process.*, **6**, 417 (1986).
[83] J. Sakata, M. Yamamoto, and I. Tajima, *J. Polym. Sci. Chem. Ed.*, **26**, 1721 (1988).
[84] R. A. Assink, A. K. Hays, R. W. Bild, and B. L. Hawkins, *J. Vac. Sci. Technol.*, **A3**, 2629 (1985).
[85] J. Sakata, M. Yamamoto, and M. Hirai, *J. Appl. Polym Sci.*, **31**, 1999 (1986).
[86] K. Okita, S. Toyooka, S. Asako, and K. Yamada, *Sumitomo J. Res. Dev.*, **26**, 194 (1987).
[87] J. Sakata, M. Hirai, and M. Yamamoto, *J. Appl. Polym. Sci.*, **34**, 2701 (1987).
[88] J. Sakata and M. Wada, *J. Appl. Polym. Sci.*, **35**, 875 (1988).
[89] J. F. Evans and G. W. Prohaska, *Thin Solid Films*, **118**, 171 (1984).
[90] V. S. Nguyen, J. Underhill, S. Fridmann, and P. Pan, *J. Electrochem. Soc.*, **132**, 1925 (1985).
[91] N. Inagaki, Y. Ohnishi, and K. S. Chen, *J. Appl. Polym. Sci.*, **28**, 3629 (1983).
[92] Y. Osada, Y. Iriyama, M. Takase, Y. Iino, and M. Ohta, *J. Appl. Polym. Sci. Appl. Polym. Symp.*, **38**, 45 (1984).
[93] C. W. Paul, A. T. Bell, and D. S. Soong, *Macromolecules*, **18**, 2312 (1985).
[94] D. Cohn, I. Tal-Atias, and Y. Avny, *J. Macromol. Sci.-Chem.*, **A25**, 373 (1988).
[95] H. Kita, M. Shigekuni, I. Kawafune, K. Tanaka, and K. Okamoto, *Polym. Bulletin*, **23**, 371 (1989).
[96] L. Martinu and H. Biederman, *Vacuum*, **33**, 253 (1983).

[97] J. Perrin, B. Despax, and E. Kay, *Phys. Rev. B*, **32**, 719 (1985).
[98] L. Martinu, H. Biederman, and J. Zemek, *Vacuum*, **35**, 171 (1985).
[99] N. Morosoff, R. Haque, S. D. Clymer, and A. L. Crumbliss, *J. Vac. Sci. Technol.*, **A3**, 2098 (1985).
[100] R. K. Sadhir and H. E. Saunders, *J. Vac. Sci. Technol.*, **A3**, 2093 (1985).
[101] N. Suzuki, Y. Ogushi, K. Imai, N. Ohta, and S. Ueno, US 4639285 (1987).
[102] S. Ueno, H. Nomura, and K. Imada, US 4429024 (1984).
[103] J. D. Masso, W. D. Brennan, and D. H. Rotenberg, US 4478873 (1984).
[104] S. Ueno, H. Nomura, and K. Imada, US 4548867 (1985).
[105] T. Akagi, I. Sakamoto, and S. Yamaguchi, US 4728564 (1988).
[106] G. A. Senich and R. E. Florin, *J. Macromol. Sci.-Rev.*, **C24**, 240 (1984).
[107] B. Ranby and J. F. Rabek, *Photodegradation, Photo-oxidation and Photostabilization of Polymers*, Wiley, London (1975).
[108] J. F. Rabek, *Intnl. Polym. Sci. Technol.*, **12**, T/91 (1985).
[109] D. J. Carlsson and D. M. Wiles, *Macromolecules*, **2**, 597 (1969).
[110] D. J. Carlsson and D. M. Wiles, *Macromolecules*, **4**, 174 (1971).
[111] D. J. Carlsson and D. M. Wiles, *Macromolecules*, **4**, 179 (1971).
[112] M. Day and D. M. Wiles, *J. Appl. Polym. Sci.*, **16**, 191 (1972).
[113] P. Blais, M. Day, and D. M. Wiles, *J. Appl. Polym. Sci.*, **17**, 1895 (1973).
[114] A. Gupta, A. Rembaum, and J. Moacanian, *Macromolecules*, **11**, 1285 (1978).
[115] J. D. Webb and A. W. Czanderna, *Macromolecules*, **19**, 2810 (1986).
[116] K. Imada, Y. Nishina, and H. Norma, US 4307045 (1981).
[117] H. S. Munro and C. Till, *J. Polym. Sci. Chem. Ed.*, **26**, 2873 (1988).
[118] J. T. C. Yeh, *J. Vac. Sci. Technol.*, **A4**, 653 (1986).
[119] R. Srinivasan, B. Braren, D. E. Seeger, and R. W. Dreyfus, *Macromolecules*, **19**, 916 (1986).
[120] V. Srinivasan, M. A. Smrtic, and S. V. Tabu, *J. Appl. Phys.*, **59**, 3861 (1986).
[121] M. C. Burrell, Y. S. Liu, and H. S. Cole, *J. Vac. Sci. Technol.*, **A4**, 2459 (1986).
[122] S. Lazare and R. Srinivasan, *J. Phys. Chem.*, **90**, 2124 (1986).
[123] P. E. Dyer, S. D. Jenkins, and J. Sidhu, *Appl. Phys. Lett.*, **49**, 453 (1986).
[124] J. E. Andrew, P. E. Dyer, D. Forster, and P. H. Key, *Appl. Phys. Lett.*, **43**, 717 (1983).
[125] B. J. Garrison and R. Srinivasan, *J. Appl. Phys.*, **57**, 2909 (1985).
[126] E. Occhiello, F. Garbassi, and V. Malatesta, *J. Mat. Sci.*, **24**, 569 (1989).
[127] E. Occhiello, F. Garbassi, and V. Malatesta, *Angew. Makromol. Chem.*, **169**, 143 (1989).
[128] R. K. Sadhir, J. D. B. Smith, and P. M. Castle, *J. Polym. Sci. Chem. Ed.*, **23**, 411 (1985)
[129] C. Decker, *J. Polym. Sci. Chem. Ed.*, **21**, 2451 (1983).
[130] C. Hull, US 4575330 (1986).
[131] E. J. Murphy, R. E. Ansel, and J. J. Krajewski, US 4942001 (1990).
[132] G. Sadesh Kumar and D. C. Neckers, *Macromolecules*, **24**, 4322 (1991).
[133] D. C. Neckers, *ChemTech*, **20**, 615 (1990).
[134] H. Wilski, *Radiat. Phys. Chem.*, **29**, 1 (1987).
[135] R. R. Rye, *J. Polym. Sci. Phys. Ed.*, **26**, 2133 (1988).
[136] T. J. Bonk and T. Simpson, US 4563388 (1986).
[137] V. D. McGinniss, in *Encyclopaedia of Polymer Science and Technology*, Vol. 4, p. 418, Wiley (1986).
[138] S. V. Nablo and D. Fussa, US 4100311 (1978).
[139] B. Boothroyd, P. A. Delaney, R. A. Hann, R. A. W. Johnstone, and A. Ledwith, *Brit. Polym. J.*, **17**, 360 (1985).
[140] I. Sobolev, J. A. Meyer, V. Stannett, and M. Szwarc, *J. Polym. Sci.*, **17**, 417 (1955).

[141] H. Kita, M. Muraoka, K. Tanaka, and K. Okamoto, *Polymer J.*, **20**, 485 (1988).
[142] D. Vesely and D. S. Finch, *Ultramicroscopy*, **23**, 329 (1987).
[143] L. Sydney and S. R. Ebner, US 4543268 (1985).
[144] J. L. Evans, K. J. Campbell, C. L. Kreil, and L. Sidney, US 4533566 (1985).
[145] M. Morra, E. Occhiello. and F. Garbassi, EP 423499 (1990).
[146] J. A. Kelber, J. W. Rogers, and S. J. Ward, *J. Mater. Res.*, **1**, 717 (1986).
[147] M. Morra, E. Occhiello, and F. Garbassi, *Angew. Makromol. Chem.*, **180**, 191 (1990).
[148] P. Siohansi, *Thin Solid Films*, **118**, 61 (1984).
[149] P. Siohansi, US 4743493 (1988).
[150] S. Fujimura, H. Yano, and J. Konno, *Nucl. Instr. Meth.*, **B39**, 809 (1989).
[151] S. A. Jenekhe and S. J. Tibbetts, *J. Polym. Sci. Phys. Ed.*, **26**, 201 (1988).
[152] S. Brock, S. P. Hersh, P. L. Grady, and J. J. Wortman, *J. Polym. Sci. Phys. Ed.*, **22**, 1349 (1984).
[153] J. Bartko, B. O. Hall, and K. F. Schoch, *J. Appl. Phys.*, **59**, 1111 (1986).
[154] T. Vendatesan, S. R. Forrest, M. L. Kaplan, C. A. Murray, P. H. Schmidt, and B. J. Wilkens, *J. Appl. Phys.*, **54**, 3150 (1983).
[155] R. M. Faria, B. Gross, and R. G. Filho, *J. Appl. Phys.*, **62**, 1420 (1987).
[156] G. Marletta, C. Oliveri, G. Ferla, and S. Pignataro, *Surf. Interf. Anal.*, **12**, 447 (1988).
[157] G. Marletta, A. Licciardello, L. Calcagno, and G. Foti, *Nucl. Instr. Meth.*, **B37/38**, 712 (1989).
[158] G. Marletta, S. Pignataro, and C. Oliveri, *Nucl. Instr. Meth.*, **B39**, 792 (1989).
[159] M. L. Kaplan, S. R. Forrest, P. H. Schmidt, and T. Venkatesan, *J. Appl. Phys.*, **55**, 732 (1984).
[160] J. R. Kulish, H. Franke, A. Singh, R. A. Lessard and E. J. Knystautas, *J. Appl. Phys.*, **63**, 2517 (1988).
[161] C.-A. Chang, J. E. E. Baglin, A. G. Schrott, and K. C. Lin, *Appl. Phys. Lett.*, **51**, 103 (1987).
[162] P. Bodo and J.-E. Sundgren, *J. Appl. Phys.*, **60**, 1161 (1986).
[163] Z. Bilkadi and W. A. Hendrickson, US 4568598 (1986).
[164] K. L. Mittal (Ed.), *Metallized Plastics 1*, Plenum Press, New York (1990).
[165] K. L. Mittal (Ed.), *Metallized Plastics, Fundamental and Applied Aspects*, Plenum Press, New York (1991).
[166] J. Weiss, C. Leppin, W. Mader, and U. Salzberger, *Thin Solid Films*, **174**, 155 (1989).
[167] J. W. Jones, US 3,442,686 (1969).
[168] E. Jamieson and A. Windle, *J. Mater. Sci.*, **18**, 64 (1983).
[169] A. L. Brody, *Plastics Packaging*, Jan/Feb 1989, p. 22.
[170] T. Sawada, S. Ohashi, and S. Yoshida, EP 311,432 (1989).
[171] A. Anttila, J. Koskinen, R. Lappalainen, J.-P. Hirvonen, D. Stone, and C. Paszkiet, *Appl. Phys. Lett.*, **50**, 132 (1987).
[172] C. Weissmantel, *J. Vac. Sci. Technol.*, **18**, 179 (1981).
[173] G. Biedehoer, *Kunstoffe*, **78**, 763 (1988).
[174] M. J. Brett, R. W. McMahon, J. Affinito, and R. R. Parsons, *J. Vac. Sci. Technol.*, **1**, 352 (1983).
[175] K. Suzuki, K. Matsumoto, and T. Takatsuka, *Thin Solid Films*, **80**, 67 (1981).
[176] R. C. Merrill, G. J. Egan, B. W. Paszek, and A. J. Aronson, *J. Vac. Sci. Technol.*, **9**, 350 (1972).
[177] K. Nakamura, EP 324351 (1989).
[178] M. K. Stern, EP 113555 (1984).
[179] G. J. Hahn, US 4478874 (1984).
[180] K. J. Heimbach, R. Adam, and H. Wenzl, US 4428809 (1984).

Chapter 7

Chemical Modifications

In this chapter those treatments are discussed that modify the chemical composition of polymer surfaces either by direct chemical reaction with a given solution (wet treatments) or by the covalent bonding of suitable macromolecular chains to the sample surface (grafting). Several of these treatments have found a large application in practice for many years, while others are still under trial in research laboratories.

7.1 WET TREATMENTS

Wet treatments were the first surface modification techniques used in order to improve surface properties of polymers. The chemical composition of the solution employed in the treatment was, in general, mutated from general wet chemistry knowledge: thus, for instance, hot chromic acid was used to oxidize polyolefins. In other cases, however, specific solutions were developed in order to exploit specific liquid–polymer interaction.

The comparison with solution organic chemistry is a key issue when discussing wet treatments of organic polymers. The basic question is the following: how is the reactivity of functional groups affected by the reduction of dimensionality, when a reaction occurs at an interface [1, 2]? Another important point involves where the reaction actually occurs. Depending on solvent–polymer interaction one can expect a reaction confined to a nearly geometrical interface or extending far into the substrate.

We will discuss some of these topics while treating surface functionalization of polymers, that is the fine tuning of surface composition and functionalities. The ultimate goal of this approach is to create well defined functional substrates characterized by controlled surface properties or available for further chemistry.

Before this, however, we will describe the most common wet modifications of polymer surfaces, which have already reached the status of commercially exploited treatments, that is the sodium etching of fluoropolymers, the oxidizing treatment, mainly based on chromic acid solutions, of several different polymers and the hydrolysis of polyesters and other polymers. In these cases the emphasis is not only on chemistry. Very often the effect of a given treatment on the surface structure of a polymer is both chemical and physical since, besides the modification of the surface composition, solubilization and etching occur. The effectiveness of a treatment depends strictly on the interplay of the different modification mechanisms. When it comes to adhesion, for instance, pitting of a surface is generally welcomed, since it offers mechanical anchoring sites. Pitting alone, however, is not enough since, as discussed in Chapter 4, cracks and crevices of a low energy surface may not be viable to a wetting liquid (adhesive). Thus, the best results (apart from effects arising from the mechanical coherence of the boundary layer) are obtained by the coupling of pitting and oxidization, the latter effect strongly increasing adhesive penetration into pores.

A last point to note is that, as discussed previously, polymer surfaces are seldom homogeneous. Even when the surface composition is constant throughout the surface, amorphous and crystalline domains can be present on it. A given reactant usually behaves in a different way towards crystalline or amorphous domains, thus, in general, the effect of a wet treatment is not homogeneous on the surface plane. As we will see in the following, this feature is often exploited in commercial treatments.

7.1.1 Etching of Fluoropolymers

Surface properties of fluoropolymers make them very poorly suited to adhesion: due the low surface tension they are very poorly wettable by liquids; a strong attractive field of forces is very difficult to attain, due to the lack of sites able to create chemical bonding and the low value of the Hamaker constant. Thus, in order to bond fluoropolymer, their surfaces must be modified. This task is not very easy, since the inertness towards chemicals is one of the most appreciated features of fluoropolymers. As a consequence, a treatment able to induce major surface modifications on fluoropolymer surfaces must be a rather special one. Since the 1950s, it has been recognized that strong reducing agents readily react with poly(tetrafluoroethylene) (PTFE), leading to a dark-coloured porous surface which can be successfully glued [3, 4]. The most commonly used reducing agent is sodium, either in liquid ammonia or as a complex with naphthalene, solvated by an excess of glycol dialkylether. The last solution, which is more stable and safer to handle, is commercially available under the trade name of Tetra-Etch. These solutions are highly coloured and conductive [5, 6], and the treatment is simply made by dipping the fluoropolymer in the solution for a few minutes [7]. The reaction involves the breaking of the carbon–fluorine bond, that is the defluorination of the polymer and the production of sodium fluoride [8], and the recombination of carbon radicals to yield unsaturated hydrocarbon chains. Some oxygen from the surroundings is also introduced. The treated samples assume a dark colour, due to the great number of conjugated double bonds and to the presence of light scattering pores. The treated depth is about 1 μm [4] and the bond strength in tension using epoxy adhesive is as high as 7–14 MPa.

The sodium etching of fluoropolymers has been a favourite subject in XPS study of polymer surfaces, since the great chemical shift induced in the C1s peak by the fluorine to carbon bond makes it very easy to detect the effects of the treatment on surface groups. Since the early 1970s [9, 10], XPS has shown that the surface of sodium treated polytetrafluoroethylene completely lose the fluorine signal. In a classical paper [10], Dwight and Riggs combined XPS and contact angle measurement to discuss the effect of the reaction and of the post-reaction treatment on the surface composition and peel strength of PTFE and Teflon FEP. Figure 7.1 has been derived from data of that paper, and shows the effect of different treatments on water contact angles. It must be noted that after etching, the C1s peak shifts towards binding energy values typical of hydrocarbons, the fluorine peak is absent and an oxygen peak has appeared. Contact angles show that the surface has been rendered much more hydrophilic. According to the authors [10] (which observed the modification of the surface morphology by SEM), the treatment produces a highly reactive sponge-like carbon network, which reacts with atmospheric oxygen and moisture when removed from the solution. The result is a rough hydrocarbon layer containing unsaturation and carbonyl and carboxyl groups. The modified surface layer is, however, unstable, since several post-reaction treatments induce the recovery of the fluorine signal and the increase of the contact angles, as shown in Figure 7.1 (on the other hand, no contact angle increase is observed when the sample is aged in water, see Chapter 9). This finding was interpreted as a removal of some of the oxidized hydrocarbon layer and consequent exposure of untreated (or less treated) fluoropolymer. No clear cut correlation was observed between either surface elemental composition or wettability and peel strength.

Recently, Rye and coworkers authored an interesting series of papers on XPS studies of sodium-etched PTFE [11–14]. In these papers, the XPS measurements are coupled with thermal desorption mass spectroscopy (TDS), which allows the detection of species that are desorbed from the sample (not only from its surface) as a function of temperature. While XPS results substantially confirm earlier findings, some interesting new features are indicated by TDS. In particular thermal desorption spectra of peak 69 (CF_3^+, typical of low molecular weight fluorocarbon), show a monotonic decrease of the intensity with increasing treatment time, suggesting greater removal of low molecular weight fluorocarbons: after 60s treatment time this contribution is virtually eliminated. This observation concludes that sodium treatment of PTFE leads to a sponge-like carbon-rich network free of low molecular weight species. A further finding, closely related to the one just discussed, is that the recovery of the fluorine signal

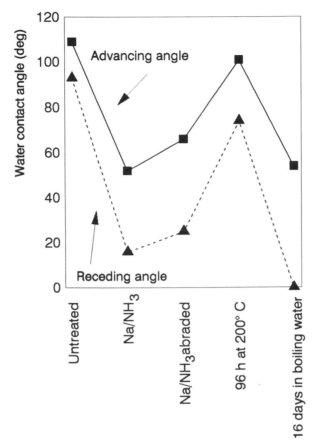

Figure 7.1 Effect of several treatments on advancing and receding water contact angles on 'Teflon FEP'. (Adapted from ref. [10].)

in the XPS spectrum and the decrease of bondability (both promoted by heat exposure) is not caused by the removal of part of the modified surface layer, as suggested by Dwight and Riggs. Rye shows that the desorption of products occurs well before the regeneration of the fluorine signal, so that the heat promoted hydrophobic recovery must be of the same kind as the one that will be discussed for plasma-treated polymers in Chapter 9, that is, it must arise from dynamic processes in the solid phase such as the diffusion of fluorocarbons from the subsurface region (Chapter 2) rather than through the desorption of an overlayer. In this restructuring process the surface area (that Rye and coworkers measured by means of krypton adsorption [12]) shows a factor 2 of loss, going from 1250 cm^2/g after 40 s etching to 610 cm^2/g after heating to 200 °C (the surface area of PTFE itself is, of course, below the measurable limit). This observation suggests two comments: first, returning to adhesion, the high value of the surface area indicates that mechanical interlocking, together with the contribution from oxygen-containing groups, should play a major role in the improvement of the bondability of etched PTFE. An important contribution comes also from the lack of a weak surface layer of low molecular weight chains, as suggested by the quoted TDS data. The second point to note is that, when it comes to contact angles, sodium-etched PTFE surfaces are absolutely non-ideal, as defined in Chapter 4, and the measured contact angles cannot be used to extract surface tension and surface tension components data. Even if this has been done [15], the best way to exploit contact angle measurements on etched PTFE surfaces is to critically read contact angle hysteresis data, as very well done by Dwight and Riggs [10].

An interesting point is that the PTFE surface can be rendered etch-resistant by radiation-induced cross-linking [12–14].

Other papers have discussed the reduction of PTFE by means of benzoin dianion [16] (see the section on functionalization) and Li-ammonia solution [17].

7.1.2   Surface Oxidation and Etching

Surface treatment of various polymers by oxidizing solutions was the answer of the electroplating industry to the low metal-to-polymer adhesion. It was found that a sulphuric acid solution saturated with chromium trioxide at a temperature of about 80 °C can yield easily palatable and strongly adhesive ABS and polypropylene surfaces [18–23]. Early investigations, based mainly on microscopic observations, found that treated surfaces become very rough, full of cracks and crevices that work as mechanical anchoring sites. While in the two phases ABS, chromic acid etching preferentially removes butadiene rubber particles, the rough topography of etched polyolefins surfaces, an example is shown in Figure 7.2, is caused by the different etching rates of the amorphous (more accessible) and crystalline regions [20]. Fitchmun and coworkers presented an accurate analysis of the control of surface morphology of polypropylene (and thus of the etching pattern of chromic acid treated surfaces) produced by compression and injection moulding [20, 21]. It was found that, following the surface morphology imparted by the moulding conditions, radial patterns are developed on spherulithic surfaces, while random patterns are characteristics of transcrystalline ones. Peel adhesion values up to 40 N/cm were recorded, and the highest peel values were associated with moulding-etching conditions that lead to deep and frequent fissuring of the polymer surface and minimum oxidative damage (intended as a lowering of the molecular weight).

While in these early papers the problem was tackled completely from a mechanical point of view, later studies, mainly following the development of surface sensitive analytical techniques, considered also the chemical side of the problem. Kato exploited the derivatization of carbonyl groups by 2, 4-

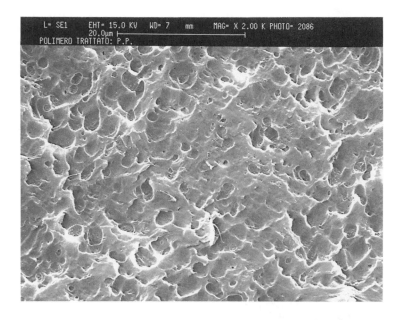

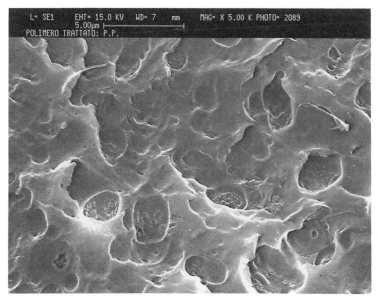

Figure 7.2 Surface morphology of chromic acid-etched polypropylene.

dinitrophenylhydrazine to follow the chemical modification of PP, HDPE and LDPE by UV spectroscopy, coupled with infrared spectroscopy of treated powders, SEM and contact angle measurements [24–27].

Blais and coworkers authored an interesting paper where the effect of a chromic acid solution ($K_2Cr_2O_7:H_2O:H_2SO_4 = 4.4:7.1:88.5$ by weight) on the surface of LDPE, HDPE and polypropylene is assessed by TEM, SEM, contact angles and ATR-IR [28]. The merit of this paper is to clearly highlight the different chemical reactivity towards the etchant of the surfaces of these chemically related polymers. Figure 7.3, which is based on data taken from Table 1 of reference [28], shows that polypropylene is etched more readily than LDPE, which in turn is less resistant than HDPE. Cross-linking polypropylene by

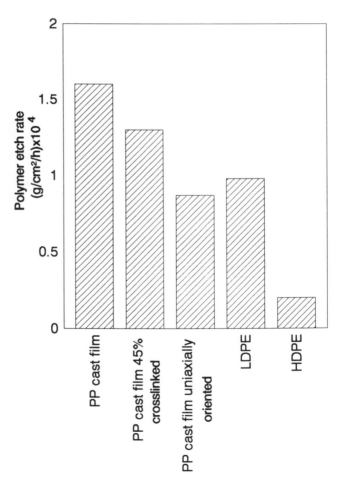

Figure 7.3   Etching rate of polyolefins by chromic acid at 70 °C (from Table I of ref. [28]).

irradiation does not appreciably affect the etch rate. An explanation of the observed behaviour is offered by a comparison with solution chemistry of chromic acid oxidation of low molecular weight alkanes [29]: the relative rate of reaction of $\equiv C - H, - CH_2 -$ and $- CH_3$ is about 4600: 75: 1, thus the rate determining feature is expected to be the number of tertiary carbons on the macromolecular chain (see, however, the section on surface functionalization), according to the measured etch rates and to the finding that cross-linking polypropylene does not reduce its propensity to undergo quick etching. The proposed reaction scheme involves the attack by Cr(VI) in acidic media to give an alcohol through a Cr(IV) ester intermediate [30]. The alcohol undergoes a very rapid reaction to give carbonyl containing scission products, which may be further oxidized to carboxylic acids. The weight loss measured as a function of etching time results from a combination of solubilization of oxidized chain segments and loss of large insoluble polymer particles, which are lost mechanically once relatively few key tie molecules have been cleaved. The latter effect is promoted by the local acidic penetration along amorphous, highly accessible zones and confirms that the interplay of physical (crystallinity) and chemical (reactivity rating) variables control the result of the treatment.

The contact angles reported by Blais and coworkers show that, while etching of LDPE and HDPE causes a continuous decrease of the advancing angle, in the case of PP the contact angle is initially reduced but for longer treatment times it increases again reaching values only slightly lower than those of untreated PP. This behaviour is accounted for by the strong reactivity of PP surfaces, whose closely situated tertiary sites produce a heavily oxidized, completely soluble surface layer, so that the washed surface probed in contact angle measurement is virtually oxidation-free. According to the authors, this suggestion is supported by the lack of appreciable modification of ATR-IR spectra of etched PP. Contact angle data are in disagreement with the findings of Kato [25], who observed a steady decrease, until reaching a plateau value of about 65° of water contact angle on etched polypropylene films as a function of treatment time. These conflicting data deserve some comment: first the chromic acid etching of polypropylene actually introduce oxygen-containing groups on the surface, as readily shown by analytical techniques more surface sensitive than ATR-IR, as discussed below. Thus, from a composition–contact angle relationship one should expect a lowering of the water contact angle. However, a second point is that the rough nature of etched surfaces greatly affects the wetting behaviour, shifting the problem away from a simple composition–contact angle relationship. Since, as shown in references [20, 21], the roughness of etched surfaces strongly depends on the sample preparation, it is not strange that different contact angles were observed on different samples. A last, important, point is that, while contact angles measured by Blais and coworkers are declared as an advancing angle [28, 31], those reported by Kato are unspecified, and likely reflect some intermediate value of the accessible contact angle range (Chapter 4).

As to the modification of the surface composition, ATR-IR detects changes only on LDPE, where it is possible to observe the build-up of hydroxyl, carbonyl and possibly sulphate groups. This finding was attributed to the previously discussed too great reactivity of PP together with the too low reactivity of HDPE, nearly free of tertiary carbons [28]. However, the chemical modification of chromic acid etched polyolefins surfaces can be readily detected by XPS [32, 33]. Reference [32] reports a variable angle XPS analysis of chromic acid etched $(K_2Cr_2O_7: H_2O: H_2SO_4 = 7: 12: 150$ by weight) LDPE and PP. Sulphate, hydroxyl, carbonyl and carboxyl groups are detected on both polymers, while on the most intensely treated polyethylene samples traces of Cr are found. A substantial difference in oxidation rate is highlighted by this study: polypropylene reaches the maximum degree of modification more quickly than polyethylene, but the rate of penetration is higher in the latter case. In fact, variable angle XPS shows that after 1 min of polypropylene etching at 20 °C the oxidation degree is depth dependent, while in the case of polyethylene, as well as in all other etching conditions tested, the samples are homogeneously oxidized within the XPS sampling depth.

The effect of oxidizing solutions other than $K_2Cr_2O_7-H_2SO_4$ on the surface of polyethylene was investigated by Baszkin Ter-Minassian-Saraga and coworkers in an interesting series of papers that, besides trying to assess the surface composition produced by the treatment, describes the temperature–environment effects on surface functional groups [34–39]. The number of oxygen containing functionalities as a function of oxidizing conditions (treatment time and composition of $KClO_3-H_2SO_4$ or $KMnO_4-H_2SO_4$ etching solution) is assessed by the amount of ^{45}Ca adsorbed from an aqueous calcium chloride solution [34], by the use of an adsorbed paramagnetic probe, the ion Mn (II), and ESR spectroscopy [38] or by XPS [39]. Results are always supported by contact angle measurements. $KMnO_4-H_2SO_4$ solutions appear stronger oxidizing agents than either $KClO_3-H_2SO_4$ or $K_2Cr_2O_7-H_2SO_4$. However, the most original result from these papers concern post-treatment effects either in air or in water. It was early recognized that the wettability imparted by the treatment is irreversibily lost if the sample is treated above 80 °C, which is about the temperature at which the melting transition begins [36]. These results were interpreted as an effect of the redistribution of external oxidized groups located at the polymer–air interface that, due to the increase of macromolecular mobility in the melting transition, can reorient or diffuse towards the bulk, thus lowering the polymer surface tension. If contact angle data suffer from some uncertainty, due to the rough nature of etched surfaces (the thermal treatment itself is expected to some how smooth the roughness, thus introducing further variables in the interpretation of data), further arguments to this thesis were offered by the adsorption either of radioactive ^{45}Ca or of paramagnetic Mn (II). In particular, in reference [38] it is shown that the amount of adsorbed Mn (II) is decreased if adsorption is performed after the annealing step, while adsorption followed by annealing seems to block reorientation. These results are confirmed by XPS measurements of $KMnO_4$–H_2SO_4 treated polyethylene surfaces [39]. In particular, it is interesting to note that the S2p peak (arising mainly from $-OSO_3H$ groups) is only slightly reduced if the thermal treatment is performed in water, suggesting that in a high surface energy environment oxidized groups tend to remain at the interface. We will return to this topic when discussing the ageing behaviour of plasma-treated surfaces (Chapter 9).

Finally, McCharty and coworkers exploited chromic acid oxidization to introduce various functional groups on the surface of polyethylene. Since the main point of these works is surface functionalization of polyethylene, rather than simple surface oxidization, this will be discussed in the relevant section .

7.1.3 Hydrolysis

The attack of a nucleophil agent, such as a base, on an electron-deficient carbon atom, has been exploited in several different classes of polymers. Figure 7.4 shows some of the polymer functional groups that have been subjected to hydrolysis in order to improve surface-related properties. Among them, hydrolysis of PET by hot sodium hydroxide attack is probably the oldest and most exploited (Figure 7.5). This reaction clearly increases the number of hydrophilic groups, improving in this way moisture-related properties. As recognized early, the reaction takes

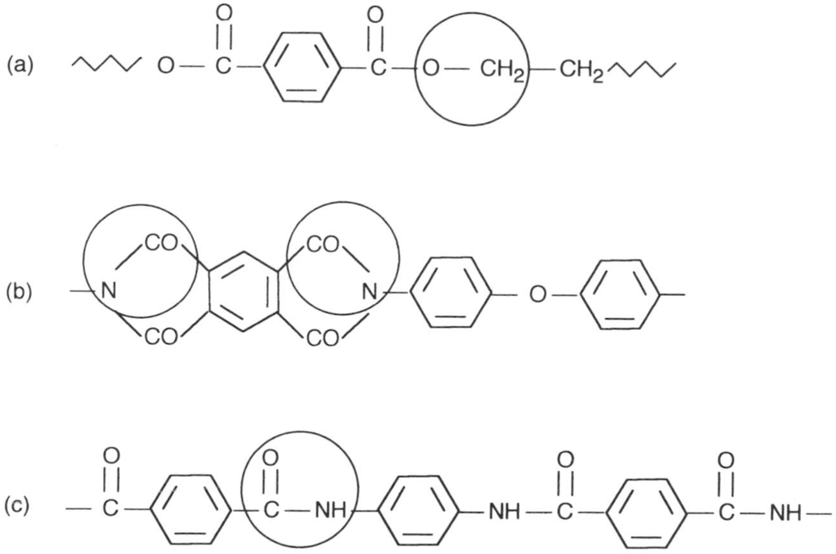

Figure 7.4 Some of the polymers which have been surface modified by hydrolysis. Circles indicate the site of reaction: (a) PET, (b) polyimide, (c) Kevlar.

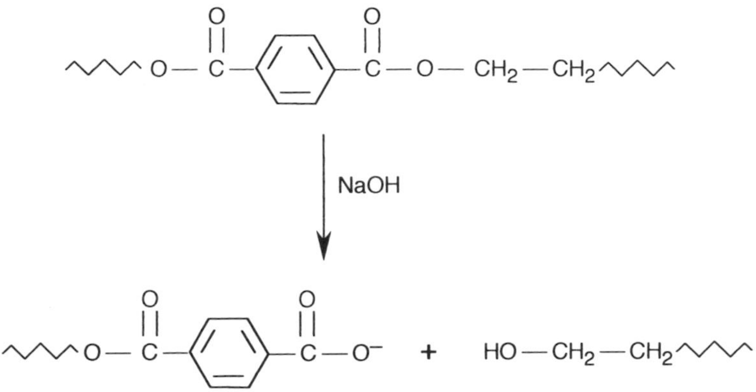

Figure 7.5   NaOH attack on PET.

place in solvents (water or alcohols) that do not have any swelling action on PET, thus the hydrolysis must take place on the fibre surface [40]. The reaction bears a marked weight loss (30% for two hours treatment with 4% solution of caustic soda at 100 °C) [41] and extensive pitting and roughening of the surface of the treated sample. The base randomly attacks the carbonyl groups of the macro-molecules present on the surface and removes them as short chains, which are further hydrolysed in solution. Based on weight loss, the following order of reactivity of several bases was found [40]: hydroxide < *tert*-butoxide < *sec*-propoxide < methoxide < ethoxide. The observed order follows the nuc-leophilicity of the bases, while the relatively low reactivity of *tert*-butoxide and *sec*-propoxide is likely to be related to steric retardation.

The correlation between the modified structure and polyester properties has been extensively investigated [42–51]. Zeronian and coworkers have indicated three reasons for the increased hydrophilicity of NaOH treated polyester [46]: the first one is the increase in the number of surface hydrophilic groups, as shown by the general reaction in Figure 7.5. The second one is the effect on wettability of surface roughness, as discussed in the relevant chapter. Finally, it is possible that the modified surface structure induced by hydrolysis may increase the accessibil-ity of hydrophilic groups present on the surface, that are supposed to be oriented towards the bulk in untreated PET [52]. There has been some controversial views expressed upon the respective merits of hydroxyl and carboxyl groups in promoting the increased hydrophilicity of treated PET surfaces [48, 51], but the preferred opinion is that both of them are equally important.

NaOH treated PET surfaces have been studied by Batich and Wendt by means of chemical derivatization and XPS [53]. Treated surfaces were rinsed with water, a step that removes all Na present on the treated surfaces (that is Na imbibed and not removed by a towel-drying step, besides, of course, that ionically linked to surface salts). After this stage, the sample is derivatized for one minute in 0.01 M NaOH, to promote the $H^+ \rightarrow Na^+$ exchange, so that the number of surface carboxyl groups is detected through the amount of Na. Comparison with bulk end group analysis of the polymer shows that the average acid level is about one-third of the observed surface value.

Polyimides (Figure 7.4) represent another class of polymers whose surface properties can benefit from a treatment with hydroxides. These materials are employed in the microelectronic industry as insulating layers in fabrication of chips and chip carriers. The poor polyimide–polyimide adhesion (3 g/mm, peel strength [54, 55]) is greatly improved if the surface is pretreated with KOH or NaOH, about 1 M. As shown in Figure 7.6, KOH attack on poly(pyromellitic dianhydride-oxydianiline) (PMDA-ODA) yields a surface layer of potassium polyamate, which can be protonated to give the corresponding polyamic acid [56]. Upon thermal curing the polyamic acid is converted back to polyimide, which shows a peel strength with a spin-coated polyimide layer of up to 126 g/mm, after 10 min reaction with KOH at 22 °C. The analysis of the Cls peak detected by XPS shows the changes in the chemical environment of surface carbon atoms [54]: the polyimide carbonyls have the same environment and, accordingly, only one high binding energy component is observed in the Cls peak of untreated polyimide. After KOH treatment the high energy component is split, since the binding energies of carboxylate carbon and amide carbon are different. Of course, protonation of potassium polyamate does not change the shape of the Cls peak, while after curing the original appearance is restored. According to the authors, who also followed the reaction by contact angle measurement, external reflection IR spectroscopy and ellipsometry, the modification method is surface selective. The calculated modification depth for 10 min treatment with 1 M KOH

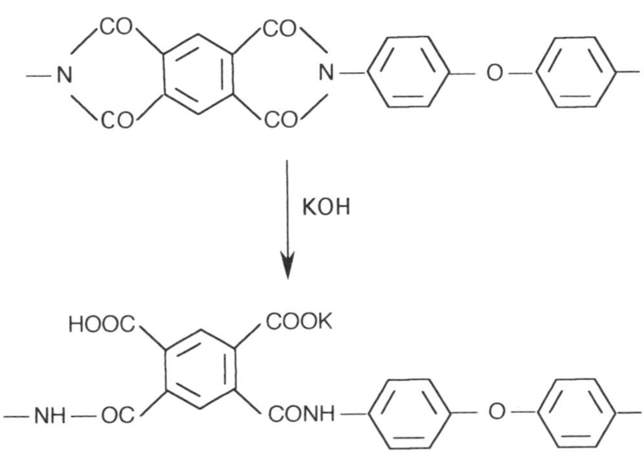

Figure 7.6 KOH attack on polyimide.

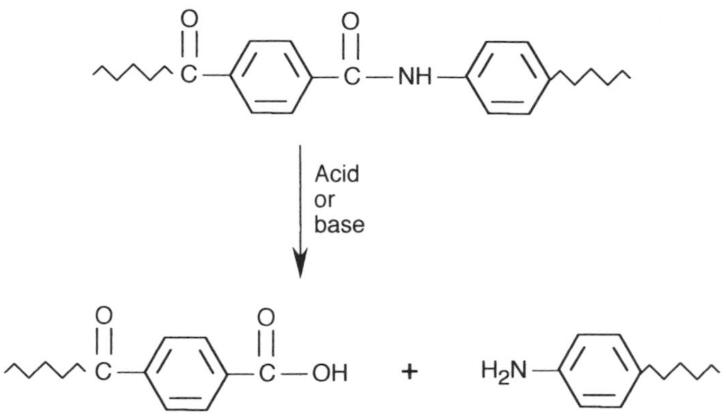

Figure 7.7   Hydrolysis of Kevlar^R.

solution at 22 °C was about 24 nm [54, 55]. The increased adhesion is attributed to the quick diffusion of adherend polymer on the amorphous layer of the modified region. Subsequent curing induces interlocking of the polymer chains and, possibly, transamidization, that is covalent bonding between adherend and adherate.

Finally, poly(aramid) surface functional groups have been hydrolysed either by acid or by bases, according to the reaction scheme shown in Figure 7.7 [57].

The goal of this reaction was the creation of reactive amino groups on Kevlar^R fibre surfaces, for subsequent chemical attachment of a bifunctional epoxide, which could improve the strength of the Kevlar^R fibre–epoxy matrix interface in composites. The reaction is not surface-specific, and if hydrolysis is allowed to proceed for too long, mechanical damage outweighs advantages gained by improved adhesion.

7.1.4   Functionalization

The goal of the functionalization of polymer surfaces is to create a surface layer of well-defined functional groups. In this way one can create polymeric surfaces of controlled properties (providing, of course, a well defined functional group–properties relationship) or a rigid substrate of controlled chemical reactivity. It is clear that the ultimate goal of a two-dimensional array of organic functional groups requires two basic conditions: the first one is a surface modification reaction ideally limited to the topmost atomic layer (surface-specific reaction), while the second is the precise knowledge, already referred to in the introduction, of how far it is possible to extend the established body of knowledge of organic solution chemistry to reactions involving constrained functional groups, whose environment is a solid in one direction and a solution in the other [58].

Surface functionalization of fluoropolymers and polypropylene has been extensively studied by McCarthy and coworkers [16, 58–63]. The modification of poly(vinylidene fluoride) (PVDF) by tetrabutylammonium bromide (TBAB) containing solutions of NaOH is an example of autoinhibitive or surface selective reaction [58]. This reaction introduces double bonds by elimination (as shown by XPS analysis of the treated surfaces derivatized by Br_2), as well as some oxygen likely by the oxidation of the conjugated surface, on top of PVDF. According to the authors the reaction proceeds by phase-transfer catalysis: the impervious interface between hydrophobic PVDF and the NaOH solution requires TBAB as a carrier of hydroxide ions from the aqueous to the organic (PVDF) phase. Using the macroscopic language of wetting, TBAB can be considered as a wetting agent, without which, reactant cannot comes into contact with the substrate. The eliminated surface is, however, impenetrable to TBAB, thus the reaction is autoinhibiting. The authors calculated a maximum reaction depth of about 1 nm, thus it can be concluded that the eliminated surface closely approaches the two-dimensional array of reactive groups that represents the ultimate goal of surface functionalization. Further chemistry, leading to highly wettable surfaces, has been performed on unsaturated bonds [58].

The same authors investigated the surface functionalization of PTFE [16, 61] and poly(chlorotrifluoroethylene) (PCTFE) [59, 60, 63]. In the first case PTFE is modified by benzoin dianion, obtaining a complex carbonaceous surface film that contains carbon–carbon double and triple bonds. In this case the reaction is not surface specific, but, being rather mild, can be controlled to as low as about 15 nm. The reacted surface has been subjected to further chemistry, which allowed the introduction of halogen, hydroxyl, amino, or carboxylic acid functionalities. Surfaces of PCTFE have been reacted with protected alcohol-containing lithium reagents, followed by deprotection [61]. A number of reactions on surface-hydroxylated PCTFE have been carried out [63].

An important outcome of these studies is the temperature and solvent dependence of reaction depth. The depth of surface modifications can, in general, be increased by increasing temperature, probably owing to an increase in the mobility of surface chains. Higher temperatures favour surface mobility, so that more potential reaction sites are exposed to reagents in solution. As to the solvent, solvents which wet or swell the polymer to a greater extent lead to thicker modified layers.

The hydroxylation of polypropylene (PP) allows comment on the relationship between solution versus surface organic chemistry [62]. Lee and McCarthy [63] compared the selective (in solution) oxidation of hydrocarbon tertiary carbon to tertiary alcohol by a solution of chromium (VI) oxide, acetic acid and acetic anhydride with the same reaction on PP surfaces. While this reaction involves retention of configuration in solution [64], owing to the previously discussed sensitivity of tertiary carbons to oxidization [29], it produces a number of different functionalities on solid PP surfaces (with a reaction depth of about

10 nm). By ATR-IR, XPS and UV-vis spectra, the authors detected the presence of hydroxyls, carbonyls and olefins, conjugated to varying extents. The latter groups probably arise from acid-catalysed dehydration of alcohols, while carbonyls are probably introduced by the oxidation of unsaturations and esterifications of hydroxyls groups. The yellow tint of treated films (arising from conjugated unsaturation) readily disappears when the samples are treated in a solution of borane, which reduces carbonyl functionality, a portion of olefins and breaks up the conjugation, nearly doubling the density of hydroxyl groups. Clearly, the oxidation reaction follows different paths with respect to solution chemistry when tertiary carbons are fixed to a solid substrate, but the exact feature that determines the course of the reaction, apart from some intuitive statement about physical constraints, is largely unknown.

Another extensive work on surface functionalization has been carried out by Whitesides and coworkers [65–69]. In this case the studies have concentrated on one substrate, that is LDPE lightly oxidized (1 min at 72 °C) by chromic acid, as previously described in the section on oxidization. The oxidizing mixture is chosen so as to minimize non-carbonyl IR adsorption bands ($CrO_3: H_2O: H_2SO_4 = 29: 42: 29$ by weight) and the sample is further treated by nitric acid in order to solubilize inorganic residues. According to the authors. LDPE surfaces treated in this way contain primarily carbonyl derivatives, with approximately 60% of these present as carboxylic acid groups and 40% as ketones or aldehydes [65]. Even if suffering from etherogeneity (both morphological, as previously discussed, and chemical, since the density and the distribution of functional groups is expected to be a function of the crystalline or amorphous nature of a given spot of the surface) the oxidized LDPE, called PE–CO_2H, works rather well as a functionalized substrate. Figure 7.8 shows the procedures used to couple organic groups to PE–CO_2H. An intriguing observation is that reaction with $SOCl_2$ is not completely 'clean', in the sense that some $SOCl_2$ is dissolved in LDPE (as indicated in Figure 7.8 by the square bracket) and affects successive reactions. This does not happen with hydrazine hydrated that, unlike $SOCl_2$, is immiscible with hydrocarbon solvents and does not swell or dissolve in polyethylene films. This observation points, of course, to the very core of the problem, i.e., what actually is an interfacial reaction? What looks like an impervious separation between two phases to a given reactant may be only a more or less smooth transition from a different environment to another one. The outcome of a broadly defined 'interfacial reaction' can be completely different in the two cases.

The quoted papers present an in-depth investigation of the chemical environment of functional groups, probed by fluorescence, electron spin resonance spectroscopy and contact angle titration, besides ATR-IR and XPS. Fluorescence measurements demonstrate that surface functional groups of films in liquids are solvated and show high rotational mobility [66]. According to the previously discussed findings [36, 38, 39], the functionalized surfaces lose their hydrophilicity if heated in air, while remain hydrophilic when heated in water.

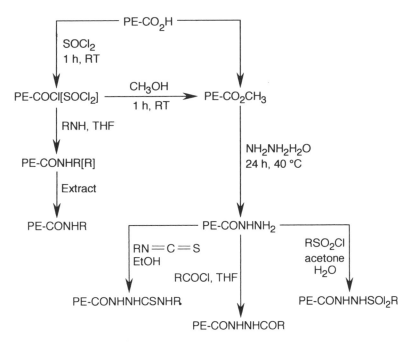

Figure 7.8 Scheme of reactions on surface functionalized PE (see text). (Adapted from ref. [66].)

Evidence of surface (wettability determining) and subsurface functional groups was gathered comparing results of different surface sensitive techniques [69]. A comparison with organic solution chemistry shows that carboxylic acids or ester moieties are both less reactive in hydrolysis and formation reactions than are these groups in organic molecules in solution. The conclusion is that the reactivities of interfacial functional groups depend on structure in ways that have no analogy in reactions in solution [69]. For instance, the rate of base catalyzed hydrolysis of esters of $PE-CO_2H$ is affected by the length of the alcohol component in a way that is absolutely unexpected, if compared with solution chemistry. A possible rationale for the difference in reactivity between short and long chain $PE-CO_2H$ esters is that the longer the alcohol component chain, the greater the ability of the esters to form a hydrophobic, low dielectric constant layer at the polymer–water interphase.

7.2 SURFACE GRAFTING

Contrary to the techniques described in the previous section, where the chemical modification of the surface was obtained by modifying existing chains, in surface grafting this aim is achieved by the covalent bonding of new macromolecules on top of the substrate. As discussed in Chapter 1, only a few polymers bear surface functional groups able to engage direct chemical bonding with surroundings (macro) molecules. Thus a fundamental step in grafting is the creation of reactive groups on the substrate surface. As we will discuss later, this can be done either chemically or (more often) by irradiation (ionizing radiation, UV light and glow discharge). Once reactive sites are created one can, in principle, couple to them a preformed macromolecular chain (bearing, of course, a reactive end group). Even if this graft-coupling technique is sometimes followed, it is rather more common to contact the activated surface with a suitable monomer, so that a growing chain starts from the activated site. Thus, the process of grafting macromolecular chains to a substrate generally involves both chemical linking to the substrate and polymerization of the chains.

It must be noted that grafting techniques are commonly used in 'bulk' macromolecular chemistry: graft copolymers are produced by coupling two different macromolecular segments [70]. Of course, this application of grafting is outside the scope of this text and will not be discussed here. Also, we will not discuss principles and methods of grafting inside the substrate. Many papers have tackled the problem of grafting from the point of view of increasing the grafting yield (that is the amount of grafted polymer), mainly promoting the diffusion of the monomer into the substrate. It is clear that in these cases the process cannot be considered a genuine surface modification technique.

Ideally a grafted surface should look like the one depicted in Figure 7.9(a). In this case a homogeneous covering of the substrate is obtained. Surface properties are controlled by the grafted chains. In solvents these chains are free to follow the behaviour outlined in Chapter 1, expanding or shrinking as the properties of the solvent are changed. In real situations, however, other surface structures are possible: in Figure 7.9(b), cross-linking between growing chains (for instance due to contamination of the monomer by a cross-linking agent) impairs chain mobility [71]. In Figure 7.9(c), an uneven distribution of active sites on the substrate surface leads to a heterogeneous grafted surface, whose properties are, in part, those of the substrate. In Figure 7.9(d), the surface is composed of both grafted and substrate chains, possibly a consequence of a swelling of the substrate by the monomer.

In the following sections we will briefly review the methods commonly employed in surface grafting. In general, the choice of the more suitable grafting technique involves, first, the recognition of the nature of the substrate surface. If it bears suitable functional groups, one can either perform a graft-coupling of a preformed chain or exploit the functional groups as sites whence polymerization of a monomer can start. This step requires an initiation reaction. If, as in the great majority or cases, no chemically reactive functional groups are present on the substrate surface, the only choice is to create them: possible options are chemical treatments followed by an initiation reaction, or irradiation, which can generate radicals as surface sites for graft polymerization (a further way is copolymeriz-

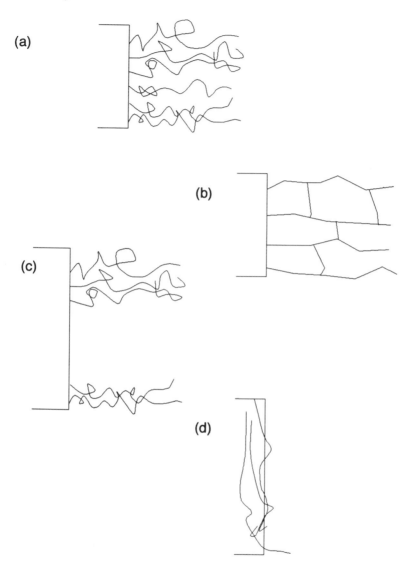

Figure 7.9 Possible structure of grafted surfaces: (a) homogeneous coverage of mobile chains, (b) mobility is constrained by cross-linking among grafted chains, (c) non-homogeneous coverage, the substrate surface is still exposed, (d) the surface layer contains macromolecular chains both grafted and belonging to the substrate.

ation with a monomer bearing reactive groups. In this case it is important to assess if this second component is effectively present on the sample surface). When using some kind of irradiation, two options are possible: irradiation can be either performed in the presence of the monomer (simultaneous irradiation) or before exposure to monomer (pre-irradiation). The involvement of radicals means that the irradiation atmosphere must be taken into account: if it is performed in air, or if the sample is exposed to air after irradiation, oxygen will scavenge radicals. It is often possible to exploit peroxides produced by the reaction between oxygen and radicals as grafting polymerization initiators, by thermally or chemically inducing their decomposition.

The previous discussion was a general, simplified picture of grafting processes. In the following section we will take a closer look at the various techniques.

7.2.1 Activation and Initiation Reactions

The great majority of grafting processes involves a radical mechanism of polymerization of vinyl monomers. Thus, the aim of the surface activation step is to create sites that can yield radicals in the successive initiation reaction. In the case of the simultaneous of mutual irradiation techniques the two steps coincide, while no activation is required when, after pre-irradiation, the substrate is contacted with the monomer without exposure to air.

Graft polymerization on polymers bearing surface functional groups can be induced by a great number of chemical treatments and is the most direct and favourable case. The requirement of suitable functional groups, however, limits this technique mostly to natural polymers and in only a few cases to synthetic macromolecules.

A great deal of work has been done on chemical grafting, both bulk and surface, on wool, where one can take advantage of the reactivity of functional groups of wool proteins. A non-exhaustive list of chemicals and references is shown in Table 7.1. It has often been found that active sites are created by omolytic cleavage of –SS– linkages of cystine [73, 85].

Collagen is another natural polymer whose proteinaceous nature makes it very suitable for chemically induced grafting. Cerium ions, potassium persulphate and similar redox systems have been extensively employed in graft copolymerization of collagen and vinyl monomers [86–93].

Also the other class of natural polymers, that is polysaccharides like cellulose, starch and lignin, bears functional groups which can initiate grafting using redox systems, as discussed in several papers [94–98]. Even if in all these papers the emphasis is more on bulk rather than on surface-related aspects, they can be used as a useful guideline for more surface-specific studies.

When it comes to synthetic polymers, the opportunities for chemically promoted initiation reaction are limited to a few substrates. Hydroxyl containing polymers, such as poly(vinylalcohol) (PVA), poly(2-hydroxyethylmethacrylate)

Table 7.1 Chemicals used for grafting vinyl monomers on wool

System	Reference
Ceric ions	76, 78
Periodate ions	80
Acetonylacetonato-Cu (II) – trichloroacetic acid complexes	72, 77, 79
Potassium permanganate	81, 83
Benzoyl peroxide	75
Azobisisobutyronitrile	82
Dimethylaniline/Copper(II)	84
Tertiary butyl hydroperoxide/	85
ferrous ammonium sulphate	85

(PHEMA) and so on, can be treated with cerium ions, according to the reaction:

$$S\text{-}CH_2OH + Ce(IV) = S\text{-}CH_2O^{\cdot} + Ce(III) + H^+$$

where S stands for the polymer substrate.

A different approach is to use the substrate hydroxyl groups to link polymerization initiators. Ikada and coworkers coupled 4,4′-azobis-4-cyanovaloyl chloride (AIVC), which is capable of initiating radical polymerization, with surface hydroxyl groups of PVA and ethylene-vinyl alcohol copolymer (EVA) [99] (Figure 7.10). Upon heating at 50 °C, AIVC decomposes, producing radicals which initiate the polymerization of acrylic monomers.

When one has to deal with polymers that do not possess hydroxyls or other reactive groups, the only way left in order to perform grafting is to create them. In principle, every one of the previously discussed surface modification techniques can work as the activation step. For instance, the functionalization of PP discussed before [62] (oxidation by a solution of chromium (VI) oxide, acetic acid and acetic anhydride, followed by reduction with borane) yields a highly hydroxylated surface, which can be treated with Ce(IV) to initiate radical polymerization. Even if such a chemical route is, in principle, practicable, it is far more common to create active sites by some kind of radiation. Both ionizing radiation and UV light can create radicals on polymers, as we will briefly discuss.

Ionizing radiations owe their name to the effects they produce on matter. Both electromagnetic and particulate radiation are commonly indicated with this name. In the case of the treatment of polymers the most commonly used ionizing radiation are gamma rays and high energy electrons. When dealing with radiation the basic variables are the energy, the amount of energy imparted to matter and the rate at which it is delivered. While the 'physical' unit is the electron volt which is used preferentially for the energy instead of the SI unit Joule, the adsorbed dose is expressed in gray (Gy), that is:

$$1\,Gy = 1\,J/kg = 100\,rad$$

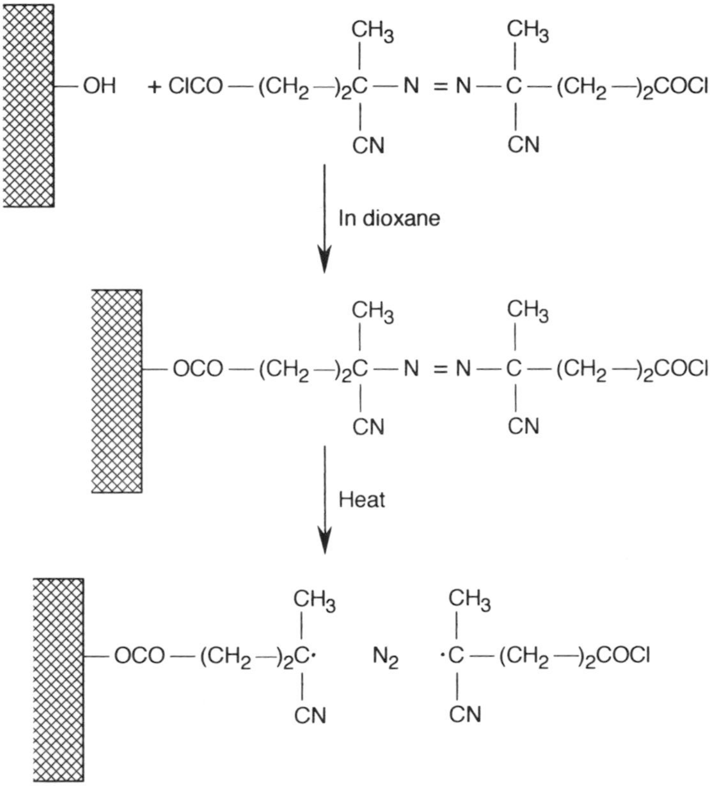

Figure 7.10 Surface grafting by 4, 4'-azobis-4-cyanovaloyl chloride (AIVC).

and the dose rate is expressed in grays (or rad) per unit time. Roughly, the dose needed to modify polymer surfaces ranges from 1 to 10 Mrad.

Table 7.2 compares energy and dose rate of gamma-rays and electron source. ^{60}Co is the most common gamma-ray source (1.17–1.33 MeV), while Van de Graaff accelerators, linear accelerators and dynamitrons are the most common sources of high energy electrons [100–103]. Gamma-rays are more penetrating than electrons: even if the penetration depth can be considered unimportant, as long as one is concerned only with the modification of the surface, it must be noted that care must be taken in order to avoid side-effects resulting from unintentional bulk modification (grafting, cross-linking, chain scission) of the substrate.

A look at Table 7.2 shows that both electrons and gamma-rays bear enough energy to break chemical bonds. The goal of radiation chemistry (of which radiation-grafting is a part) is to exploit the transfer of energy from radiation to matter to destroy some of the existing chemical bonds and to create new ones.

Table 7.2 Energy and dose rate of common gamma-
rays and electron sources

Source	Energy (MeV)	Dose rate (rad/s)
Gamma-ray	0.1–1.5	10^{-1}–10^2
Electrons	0.1–10	10^5–10^9

The interaction between ionizing radiation and matter has been treated in a number of books [104–109]. Here we will provide a brief examination of some basic aspects.

Electrons travelling inside matter lose their energy in several different ways, of which the most important are emission of electromagnetic radiation and inelastic or elastic collisions. The first effect arises when an electron is decelerated when passing close by a nucleus. The lost energy appears as electromagenetic radiation (bremsstrahlung radiation), and this mechanism is the dominant mode of energy loss between 10 and 100 MeV electron energy. Below these values, the dominant process whereby electrons are slowed down by a material (stopping material) is inelastic collision, where the energy lost from the incoming electrons produce excitation and (if high enough) ionization of the stopping material. Thus, an inelastic collision leaves an excited state or a secondary electron which again will interact with surrounding matter, as much as is allowed by its energy, producing a path of excited or ionized species along the track of the primary electron. In the MeV range of energies, a useful relationship for the ratio of the energy loss by radiation (EL^r) to the loss by collision (EL^c), is given by [106]:

$$EL^r/EL^c = EZ/(1600\,m_0 c^2) \tag{7.1}$$

where E is the energy of the incoming electron, Z is the atomic number of the stopping material, m_0 is the rest mass of the electron and c is the speed of light. Finally the Coulomb field of an atomic nucleus may deflect an electron by elastic (without energy loss) scattering, a favoured mechanism in the case of low energy electrons and for high atomic number materials.

As to electromagnetic radiations, the photoelectric effect, the Compton effect and pair-production are the dominant mechanisms of radiation-matter energy transfer. The photoelectric effect has been discussed in the chapter on XPS. The Compton effect is the elastic collision between a photon and an electron, the former transferring part of its energy to the latter. Thus, the result of a Compton encounter is a photon of reduced energy and an emitted or excited electron. Finally, at high energy, a photon can interact with a nucleus and disappear, with the production of two particles, a positron and an electron. This pair-production effect requires, of course, an incoming photon energy greater than twice the rest mass of the electron, i.e. greater than 1.02 MeV.

The three mechanisms quoted are function of the photon energy, and the atomic number of the adsorber. Evans [110] calculated the energy region over which the three processes predominate, as shown in Figure 7.11. It is clear that, in the case of irradiation of polymers by gamma-rays, the Compton effect is the main mechanism of energy transfer.

Thus, it turns out that when a polymer is treated by ionizing radiation, the primary effect is the creation of ions and excited electronic states. Radiation chemistry of polymers is largely the output of secondary events that arise from the as-bombarded surface: coulombic interaction causes the ions and the electrons to recombine in highly excited electronic states. The excited-state macromolecules may lose their excess energy in several different ways: they can undergo radiationless decay, producing heat, or can emit radiation (phosphorescence or fluorescence). A further possibility is to undergo chemical decay, via heterolytic or homolytic bond cleavage. In the former case ions are produced, while the latter case leads to the production of free radicals, which are the main subject of radiation chemistry of polymers and the basis of radiation-induced grafting.

A further way of producing grafting on polymer surfaces involves electromagnetic radiation of far greater wavelength, i.e. UV light, than those we have just discussed. The most commonly employed wavelengths range from 250 to 360 nm, which means that their energy is not enough for homolytically cleaving the strong C–C or C–H bonds of polymers. Using the shorter wavelength, some low strength bonds (either a part of the repeating unit, or present as an impurity) may be cleaved, but the low yield of this reaction makes this approach unsuitable for

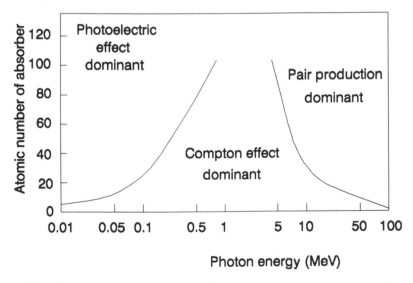

Figure 7.11 Energy region over which the three primary processes predominate. (Adapted from ref. [110].)

practical purposes with most common polymers (although it must be noted that prolonged exposure to the source of irradiation, which means a very high radiation dose, can lead to major effects, as demonstrated by the photodegradation of polymers exposed to sunlight). A way to overcome the problem and practically exploit UV irradiation for grafting, is to add a photo-sensible molecule to the polymer. A kind of photo-sensible molecule is constituted by the so called photoinitiators, which are molecules engineered to homolytically cleave and initiate radical polymerization when struck by light of a given wavelength [111, 112]. Photoinitiators are widely used in UV curing of coatings or UV initiated polymerization [113]. As to grafting, they can be used with substrates containing unsaturated groups, available for co-polymerization. Grafting on saturated polymers requires the creation of radical sites on the substrate surface, and this can be done with the class of photo-sensible molecules called photosensitizers: in this case, the radiation promotes the molecule to an excited state, which can abstract a hydrogen atom from the polymer and produce a macroradical on the substrate. It is this radical site on the substrate that works as the initiator of polymerization. The mechanism of action of the most commonly used photosensitizer, that is benzophenone (BP), is displayed in Figure 7.12: UV light of wavelength between 300 and 400 nm is adsorbed, and excites BP to a singlet state which, by intersystem crossing (ISC), reverts to a reactive triplet state which can abstract a hydrogen atom from the polymer substrate and create a macroradical. If vinyl monomers are present, the macroradical can become the anchoring site of a growing grafted chain.

Photoinitiators and photosensitizers can be added to the substrate in several different ways. In the pioneering work of Oster and coworkers [114], BP was blended with unvulcanized natural rubber, by casting a film from a cyclohexane solution of rubber and BP. Polyethylene films were sensitized by introducing BP prior to extrusion, by soaking for a few minutes in a solution of BP or by placing the films in the presence of BP for two days in a desiccator [115]. Needles performed photografting of methyl acrylate vapours on textile fabrics (both natural and synthetic) in the presence of biacetyl vapours [116] or after treatment with aqueous dispersions of photosensitive metal oxides [117]. Acrylamide (AAm) was photografted on oriented polypropylene films by Tazuke and Kimura by irradiation (through the polymer film) of an AAm-acetone solution containing BP or other sensitizers (methyl-2-benzoylbenzoate, 9-fluorenone 4-bromobenzophenone) [118, 119], a technique used also for the surface modification of poly(ethyleneterephthalate) (PET) [120]. Another approach involves the coating of the substrate by a sensitizer-containing film. Ogiwara and coworkers dipped polyolefins, PET and polyamide films in an acetone solution containing the photosensitizer (BP, anthraquinone, benzoilperoxide) and a film-forming agent (polyvinylacetate, typical concentration 1 wt%). The sensitized substrate was later irradiated either in a solution of the monomer or in the presence of vapour monomers [121, 122]. The substrate can also be sensitized by pre-swelling in a

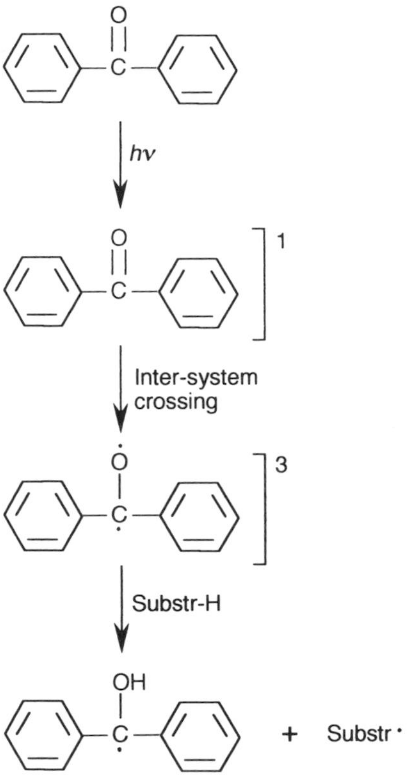

Figure 7.12   The effect of UV light on benzophenone. The excited triplet state can extract a hydrogen from the substrate (Substr-H), creating in this way a macroradical where grafting occurs.

suitable solution, which allows the diffusion of the photosensitizer in the substrate surface region [123]. After this step, the sensitized substrate can be irradiated, either in solution or in an atmosphere of acrylic monomers. Rånby and coworkers performed vapour phase photografting on polyolefins, polystyrene and PET, irradiating the substrates in a thermostated chamber equipped with a quartz window, where sensitizer (BP) and monomer evaporate from a solution of a volatile solvent in an open bucket which is shielded from UV by an aluminium foil [124–128]. They also developed a continuous process, where the substrate is sensitized by pre-soaking in a sensitizer solution before reaching the reaction chamber [129–130]. A further approach involves the synthesis of polymers containing a photosensitive substituent: Inoue and Kohama synthesised diethyl-dithiocarbamated poly (dimethylsiloxane), and irradiated it in solution of hydrophilic acrylic monomers [131]. The activation step is displayed in Figure 7.13.

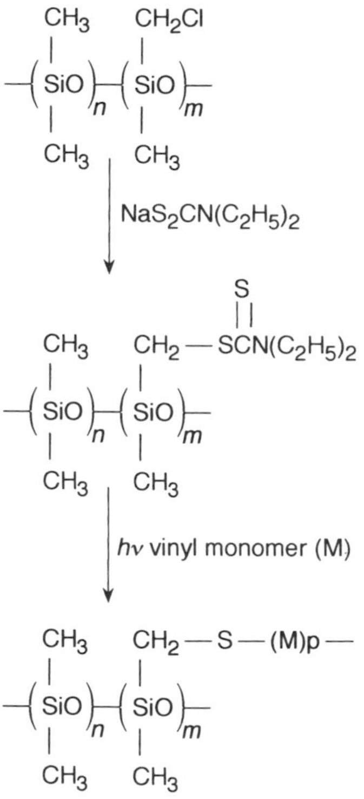

Figure 7.13 Surface photografting on diethyldithiocarbamated poly(dimethylsiloxane). (Adapted from ref. [131].)

The other combination is to make the monomer itself work as a photosensitizer. This can be done, as extensively studied by Bellobono and coworkers [132–138], with some azo dyes, shown in Figure 7.14, which contain polymerizable acryloxy groups and which possess photochemically induced excited states that can abstract hydrogen from polymeric substrates.

Finally, it is possible to promote grafting on polymer surfaces by glow discharge. As discussed in the previous chapter, glow discharges contain lots of species that can create radicals on polymeric substrates, like energetic electrons, excited atoms, molecular fragments and molecules, and UV light. Treated substrate can be exposed to monomer vapours after the plasma treatment, avoiding air contact, such as carried out by Yasuda and coworkers in the grafting of 2-vinyl-pyridine on polypropylene [139], and as commercially exploited in the grafting of acrylic acid on PET fibres [140]. Treated samples can otherwise be exposed to air, so that peroxides are created by the reaction of atmospheric oxygen with surface radicals. If treated samples are heated in a vinyl monomer solution, peroxides decompose homolytically, initiating graft copolymerization.

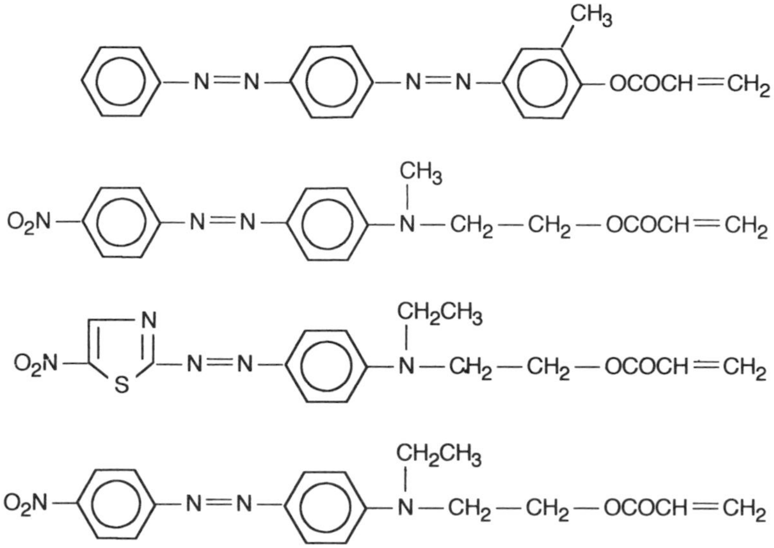

Figure 7.14 Several acryloxy-substituted aromatic diazenes used in photografting. (Adapted from ref. [137].)

This technique was used by Ikada and coworkers for grafting acrylamide on polypropylene [141]. Figure 7.15 (from ref. [141]), shows that the peroxide concentration is very sensitive to plasma treatment time. The grafting yield closely parallels the detected peroxide concentration.

To compare the respective merits of the different techniques, let us remember that, when dealing with surface grafting, one has to take into account several competing processes. The most important of these are surface grafting versus bulk grafting and graft copolymerization versus homopolymerization. When using ionizing radiations, cross-linking and chain scission of the macromolecules of the substrate are other important side reactions.

The non-specific interaction and the high (as compared with the thickness affecting surface properties) penetration depth of ionizing radiation makes them less surface specific than UV light, where initiation occurs only if there is photosensitizer molecules and thus is more tunable. On the other hand, ionizing radiations allow additive-free processing [142], while UV treatment requires addition of photosensitizers that, if remaining on the sample after the treatment, can promote light-induced degradation or can be leached out in the surrounding environment. In general, some control of the depth of modification can be obtained by a proper choice of the grafting parameters. For instance, in the mutual grafting by ionizing radiation in vapour phase, the monomer–substrate interaction plays the most important role. If the monomer is soluble or can swell the substrate, it is expected that grafting will proceed inside the substrate, the

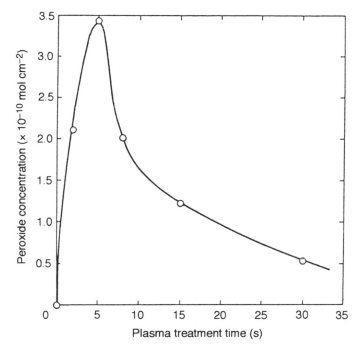

Figure 7.15 Formation of peroxides on a PE film exposed to Ar plasma [141]. (Reprinted with permission from Suzuki *et al.*, *Macromolecules*, **19**, 1804. Copyright (1986) American Chemical Society.)

depth of modification being limited by the irradiation penetration depth and (generally the limiting step) the speed of diffusion of the monomer into the substrate [143, 144]. The last point can be improved using a higher temperature or a liquid phase grafting with a solvent that can swell the substrate and promote monomer diffusion. The effect of the nature of the solvent on the grafting yield (measured by the weight increase) has been a favourite subject on grafting studies but, since the yield increase is obtained by promoting bulk grafting, it is not of direct relevance here. On the contrary, from a surface modification standpoint, indiffusion of monomer should be avoided. Apart from using non-substrate-soluble monomers or non-substrate-swelling solvents, a favourable case is when the polymer is not soluble in its monomer: in this case the grafted layer works as a barrier to indiffusion of the monomer and grafting is confined on the substrate surface.

REFERENCES

[1] P. Pfeifer and D. Avnir, *J. Chem. Phys.*, **79**, 3558 (1983).
[2] R. D. Astumian and Z. A. Schelly, *J. Am. Chem. Soc.*, **106**, 304 (1984).

[3] E. R. Nelson, T. J. Kilduff, and A. A. Benderley, *Ind. Eng. Chem.*, **50**, 329 (1958).
[4] A. A. Benderly, *J. Appl. Polym. Sci.*, **20**, 221 (1962).
[5] G. Lepoutre and M. J. Sienko (Eds), *Metal-Ammonia Solutions: Physicochemicals Properties*, W. A. Benjamin, New York (1964).
[6] J. J. Lagowsky and M. J. Sienko (Eds), *Metal-Ammonia Solutions*, Butterworths, London (1970).
[7] S. Wu, *Polymer Interface and Adhesion*, Marcel Dekker, New York (1982), p. 280
[8] J. Lukas, L. Lochmann, and J. Kalal, *Angew. Makromol. Chem.*, **181**, 183 (1990).
[9] H. Brecht, F. Mayer, and H. Binder, *Angew. Makromol. Chem.*, **33**, 89 (1973).
[10] D. W. Dwight and W. M. Riggs *J. Colloid Interface Sci.*, **47**, 650 (1974).
[11] R. R. Rye and J. A. Kelber, *Appl. Surf. Sci.*, **29**, 397 (1987).
[12] R. R. Rye, *Polym. Phys. Ed.*, **26**, 2133 (1988).
[13] R. R. Rye and G. W. Arnold, *Langmuir*, **5**, 1331 (1989).
[14] R. R. Rye and R. J. Martinez, *J. Appl. Polym. Sci.*, **37**, 2529 (1989).
[15] D. H. Kaelble and E. H. Cirlin, *J. Polym. Sci. A-2*, **9**, 363 (1971).
[16] C. A. Costello and T. J. McCarthy, *Macromolecules*, **17**, 2940 (1984).
[17] N. Chakrabarti and J. Jacobus, *Macromolecules*, **21**, 3011 (1988).
[18] K. Kato, *Polymer*, **8**, 33 (1967).
[19] K. Kato, *Polymer*, **9**, 419 (1968).
[20] D. R. Fitchmun, S. Newman, and R. Wiggle, *J. Appl. Polym. Sci.*, **14**, 2441 (1970).
[21] D. R. Fitchmun, S. Newman, and R. Wiggle, *J. Appl. Polym. Sci.*, **14**, 2457 (1970).
[22] D. M. Brewis, *J. Mater. Sci.*, **3**, 262 (1968).
[23] V. J. Armond and J. R. Atkinson, *J. Mater. Sci.*, **3**, 332 (1968)
[24] K. Kato, *J. Appl. Polym. Sci.*, **18**, 3087 (1974).
[25] K. Kato, *J. Appl. Polym. Sci.*, **19**, 1593 (1975).
[26] K. Kato, *J. Appl. Polym. Sci.*, **20**, 2451 (1976).
[27] K. Kato, *J. Appl. Polym. Sci.*, **21**, 2735 (1977).
[28] P. Blais, D. J. Carlsson, G. W. Csullog, and D. M. Wiles, *J. Colloid Interface Sci.*, **47**, 636 (1974).
[29] F. Mares and J. Rocek, *Coll. Czech. Chem. Comm.*, **26**, 2370 (1961).
[30] K. B. Wiberg and R. Eisenthal, *Tetrahedron*, **20**, 1151 (1964).
[31] P. Blais, D. J. Carlsson, and D. M. Wiles, *J. Appl. Polym. Sci.*, **15**, 129 (1971).
[32] D. Briggs, D. M. Brewis, and M. B. Konieczo, *J. Mater. Sci.*, **11**, 1270 (1976).
[33] D. Briggs, V. J. I. Zichy, D. M. Brewis, J. Comyn, R. H. Dahm, M. A. Green, and M. B. Konieczo, *Surface Interface Anal.*, **2**, 107 (1980).
[34] A. Baszkin and L. Ter-Minassian-Saraga, *J. Polym. Sci.*, **C34**, 243 (1971).
[35] A. Baszkin and L. Ter-Minassian-Saraga, *J. Colloid Interface Sci.*, **43**, 190 (1973).
[36] A. Baszkin and L. Ter-Minassian-Saraga, *Polymer*, **15**, 759 (1974).
[37] A. Baszkin, M. Nishino, and L. Ter-Minassian-Saraga, *J. Colloid Interface Sci.*, **54**, 317 (1976).
[38] B. Catoire, P. Bouriot, A. Baszkin, L. Ter-Minassian-Saraga, and M. M. Boissonnade, *J. Colloid Interface Sci.*, **79**, 143 (1981).
[39] J. C. Eriksson, C. G. Golander, A. Baszkin, and L. Ter-Minassian-Saraga, *J. Colloid Interface Sci.*, **100**, 381 (1984).
[40] C. G. G. Namboori and M. S. Haith, *J. Appl. Polym. Sci.*, **12**, 1999 (1968).
[41] V. V. Korshak and S. V. Vinogradova, *Polyesters*, Pergamon Press, New York (1965), p. 407.
[42] H. D. Weingmann, M. G. Scott, A. S. Ribnick, and L. Rebenfeld, *Textile Res. J.*, **46**, 574 (1976).
[43] H. D. Weingmann, M. G. Scott, A. S. Ribnick, and R. D. Matkowsky, *Textile Res. J.*, **47**, 745 (1977).

[44] W. T. Holfeld and M. S. Shepard, *Textile Chem. and Colorist*, **10**, 26 (1978).
[45] H. L. Needles, P. Vinther, K. Beardsley, J. Chandler, M. Drubach, L. Eggert, L. Fisher, P. Montagne, R. O'Connell, D. Sheriff, P. Spencer, A. Tai, and S. H. Zeronian, *Textile Chem. and Colorist*, **11**, 35 (1979).
[46] E. M. Sanders and S. H. Zeronian, *J. Appl. Polym. Sci.*, **27**, 4477 (1982).
[47] H. L. Needles, A. Tai, and H. Bluen, *Textile Chem. and Colorist*, **15**, 205 (1983).
[48] B. M. Latta, *Text. Res. J.*, **54**, 766 (1984).
[49] M. V. S. Rao and N. E. Dweltz, *J. Appl. Polym. Sci.*, **33**, 835 (1987).
[50] H. L. Needles, C. Walker, and Q. Xie, *Polymer*, **31**, 336 (1990).
[51] S. H. Zeronian, H. Z. Wang, and K. W. Alger, *J. Appl. Polym. Sci.*, **41**, 527 (1990).
[52] B. M. Latta, in *Clothing Comfort-Interaction of Thermal, Ventilation, Construction and Assessment Factors*, N. R. S. Hollies, and R. F. Goldman (Eds) Ann Arbor Science Publishers, Ann Arbor, Mich. (1977), p. 33.
[53] C. D. Batich and R. C. Wendt, *Ann. Chem. Soc., Adv. Chemical Series*, **162**, 221 (1990).
[54] K. W. Lee, S P. Kowalczyk, and J. M. Shaw, *Polymer Preprints*, **31**, 712, (1990).
[55] K. W. Lee, S. P. Kowalczyk, and J. M. Shaw, *Macromolecules*, **23**, 2097 (1990).
[56] H. J. Leary and D. S. Campbell, *Am. Chem. Soc., Adv. Chemical Series*, **162**, 419 (1981).
[57] T. S. Keller, A. S. Hoffman, B. D. Ratner, and B. J. McElroy, in *Physicochemical Aspects of Polymer Surfaces*, K. L. Mittal (Ed.), Plenum Press, New York (1983), p. 861.
[58] A. J. Dias and T. J. McCarthy, *Macromolecules*, **17**, 2529 (1984).
[59] A. J. Dias and T. J. McCarthy, *Macromolecules*, **18**, 1826 (1985).
[60] A. J. Dias and T. J. McCarthy, *Macromolecules*, **20**, 2068 (1987).
[61] C. A. Costello and T. J. McCarthy, *Macromolecules*, **20**, 2819 (1987).
[62] K. W. Lee and T. J. McCarthy, *Macromolecules*, **21**, 309 (1988).
[63] K. W. Lee and T. J. McCarthy, *Macromolecules*, **21**, 2318 (1988).
[64] J. Rocek, *Tetrahedron Lett.*, 135 (1962).
[65] J. R. Rasmussen, E. R. Stedronsky, and G. M. Whitesides, *J. Am. Chem. Soc.*, **99**, 4736 (1977).
[66] J. R. Rasmussen, D. E. Bergbreiter, and G. M. Whitesides, *J. Am. Chem. Soc.*, **99**, 4746 (1977).
[67] S. R. Holmes-Farley, R. H. Reamey, T. J. McCarthy, J. Deutch, and G. M. Whitesides, *Langmuir*, **1**, 725 (1985).
[68] S. R. Holmes-Farley and G. M. Whitesides, *Langmuir*, **2**, 266 (1986).
[69] S. R. Holmes-Farley and G. M. Whitesides, *Langmuir*, **3**, 62 (1987).
[70] P. Dreyfuss and R. P. Quirk, in *Encyclopedia of Polymer Science and Technology*, Wiley, New York (1986), vol. 7, p. 551.
[71] Y. Ikada, *Adv. Polym. Sci.*, **57**, 103 (1984).
[72] W. S. Simpson and W. van Pelt, *J. Text. Inst.*, **58**, T 316 (1967).
[73] M. Negishi, K. Arai, and S. Okada, *J. Appl. Polym. Sci.*, **11**, 115 (1967).
[74] A. J. McKinon, *J. Appl. Polym. Sci.*, **14**, 3033 (1970).
[75] K. Arai, S. Komine, and M. Negishi, *J. Polym. Sci.*, *A-1*, **8**, 917 (1970).
[76] A. Kantouch, A. Hebeish, and A. Bendak, *Eur. Polym. J.*, **7**, 153 (1971).
[77] A. Hebeish, A. Bendak, and A. Kantouch, *J. Appl. Polym. Sci.*, **15**, 2733 (1971).
[78] A. Bendak, A. Kantouch, and A. Hebcish, *Kolor. Ert.*, **13**, 106 (1971).
[79] W. S. Simpson, *J. Appl. Polym. Sci.*, **15**, 867 (1971).
[80] A. Kantouch, A. Hebeish, and A. Bendak, *Text. Res.. J.*, **42**, 7 (1972).
[81] A. Kantouch, S. Abdel-Fattah, and A. Hebeish, *Polym. J.*, **3**, 375 (1972).
[82] A. Bendak and A. Hebeish, *J. Appl. Polym. Sci.*, **17**, 1953 (1973).
[83] S. Abdel-Fattah, A. Kantouch, and A. Hebeish, *J. Chem.*, **17**, 311 (1974).

[84] M. I. Khalil, M. H. El-Rafie, A. Bendak, and A. Hebeish, *J. Appl. Polym. Sci.*, **27**, 519 (1982).
[85] B. N. Misra and D. S. Sood, in *Physicochemical Aspects of Polymer Surfaces*, p. 881, Vol. 2, K. L. Mittal (Ed.), Plenum Press, New York (1983).
[86] K. P. Rao, K. T. Joseph, and Y. Nayudamma, *J. Sci. Ind. Res.*, **29**, 559 (1970),
[87] K. P. Rao, K. T. Joseph, and Y. Nayudamma, *J. Polym. Sci.*, *A-1*, **9**, 3199 (1971).
[88] K. P. Rao, K. T. Joseph, and Y. Nayudamma, *J. Appl. Polym. Sci.*, **16**, 975 (1971).
[89] K. Kojima, S. Iguchi, Y. Kajima, and M. Yoshikuni, *J. Appl. Polym. Sci.*, **28**, 87 (1983).
[90] S. Amudeswari, C. R. Reddy, and K. T. Joseph, *Eur. Polym. J.*, **20**, 91 (1984).
[91] S. Amudeswari, C. R. Reddy, and K. T. Joseph, *J. Macromol. Sci. Chem.*, **A23**, 805 (1986).
[92] A. Klasek, A. Kaszonyiova, and F. Pavelka, *J. Appl. Polym. Sci.*, **31**, 2007 (1986).
[93] D. R. Patil and D. J. Smith, *J. Appl. Polym. Sci.*, **40**, 1541 (1990).
[94] J. J. Hermans, *Pure Appl. Chem.*, **5**, 147 (1962).
[95] J. J. Meister, D. R. Patil, L. R. Field, and J. C. Nicholson, *J. Polym. Sci. Polym. Chem. Ed.*, **22**, 1963 (1984).
[96] J. L. Garnett, *Cellulose: Structure, Modification and Hydrolysis*, R. A. Young and R. M. Rowell (Eds), Wiley, New York (1986).
[97] J. J. Meister, D. R. Patil, M. C. Jewell, and K. Krohn, *J. Appl. Polym. Sci.*, **33**, 1887 (1987).
[98] J. C. Chen, Y. X. Zhang, D. Patil, G. Butler, and T. Hogen-Esch, *J. Macromol. Sci. Chem.*, **A25**, 971 (1988).
[99] R. K. Samal, H. Iwata, and Y. Ikada, *Physicochemical Aspects of Polymer Surfaces*, p. 801, Vol. 2, K. L. Mittal (Ed.), Plenum Press, New York (1983).
[100] W. J. Ramler, *J. Radiat. Curing.*, **1**, 34 (1974).
[101] W. J. Ramler, *Radiat. Phys. Chem.*, **9**, 69 (1977).
[102] R. E. Schuler, *Radiat. Phys. Chem.*, **14**, 171 (1979).
[103] S. Schiller, U. Heisig, and S. Panzer, *Electron Beam Technology*, Wiley–Interscience, New York (1982).
[104] A. Berthelot, *Radiation and Matter*, Leonard Hills, London (1958).
[105] F. A. Bowey, *The Effects of Ionizing Radiation on Natural and Synthetic High Polymers*, Interscience Publishers, New York (1958).
[106] J. W. T. Spinks and R. J. Woods, *An Introduction to Radiation Chemistry*, Wiley, New York (1964).
[107] A. Charlesby, *Atomic Radiation and Polymers*, Pergamon Press, Oxford (1960).
[108] A. Chapiro, *Radiation Chemistry of Polymeric Systems*, Wiley–Interscience, New York (1962).
[109] *Effect of Radiation on Materials and Components*, J. F. Kircher, and R. E. Bowman (Eds), Reinhold Publishing Corporation, New York (1964).
[110] R. D. Evans, *The Atomic Nucleus,* McGraw-Hill, New York (1955).
[111] C. G. Roffey, *Photopolymerization of Surface Coating*, Wiley, New York (1982).
[112] J. P. Fouassier, *Photopolymerization Science and Technology*, N. S. Allen (Ed.), Elsevier, London (1989).
[113] *UV Curing: Science and Technology*, A. Pappas (Ed.) Technology Marketing Corp., Stamford, CT (1978).
[114] G. Oster and O. Shibata, *J. Polym. Sci.*, **26**, 233 (1957).
[115] G. Oster, G. K. Oster, and H. Moroson, *J. Polym. Sci.*, **34**, 671 (1959).
[116] H. L. Needles and K. W. Alger, *J. Appl. Polym. Sci.*, **19**, 2187 (1975).
[117] H. L. Needles and K. W. Alger, *J. Appl. Polym. Sci.*, **19**, 2207 (1975).
[118] S. Tazuke and H. Kimura, *J. Polym. Sci., Polym. Lett. Ed.*, **16**, 497 (1978).

[119] S. Tazuke and H. Kimura, *Makromol. Chem.*, **179**, 2603 (1978).
[120] R. D. Goldblatt, J. M. Park, R. C. White, L. J. Matienzo, S. J. Huang, and J. F. Johnson, *J. Appl. Polym. Sci.*, **37**, 335 (1989).
[121] Y. Ogiwara, M. Kanda, M. Takumi, and H. Kubota, *J. Polym. Sci., Polym. Lett. Ed.*, **19**, 457 (1981).
[122] H. Kubota, N. Yoshino, and Y. Ogiwara, *J. Polym. Sci., Polym. Lett. Ed.*, **21**, 367 (1983).
[123] M. Morra, E. Occhiello, and F. Garbassi, *J. Colloid Interface Sci.*, **149**, 290 (1992).
[124] K. Allmer, A. Hult, and B. Rånby, *J. Polym. Sci. Polym. Chem. Ed.*, **26**, 2099 (1988).
[125] K. Allmer, A. Hult, and B. Rånby, *J. Polym. Sci. Polym. Chem. Ed.*, **27**, 1641 (1989).
[126] K. Allmer, A. Hult, and B. Rånby, *J. Polym. Sci. Polym. Chem. Ed.*, **27**, 3405 (1989).
[127] K. Allmer, A. Hult, and B. Rånby, *J. Polym. Sci. Polym. Chem. Ed.*, **27**, 3419 (1989).
[128] K. Allmer, J. Hilborn, P. H. Larsson, A. Hult, and B. Rånby, *J. Polym. Sci. Polym. Chem. Ed.*, **28**, 173 (1990).
[129] P. Y. Zhang and B. Rånby, *J. Appl. Polym. Sci.*, **41**, 1459 (1990).
[130] P. Y. Zhang and B. Rånby, *J. Appl. Polym. Sci.*, **41**, 1469 (1990).
[131] H. Inoue and S. Kohama, *J. Appl. Polym. sci.*, **29**, 877 (1984).
[132] I. R. Bellobono, F. Tolusso, E. Selli, S. Calgari, and A. Berlin, *J. Appl. Polym. Sci.*, **26**, 619 (1981).
[133] E. Selli, I. R. Bellobono. F. Tolusso, and S. Calgari, *Ann. Chim. (Rome)*, **71**, 147 (1981).
[134] E. Selli, I. R. Bellobono, S. Calgari, and A. Berlin, *J. Soc. Dyers Colour.*, **97**, 438 (1981).
[135] I. R. Bellobono, S. Calgari, M. C. Leonardi, E. Selli, and E. D. Paglia, *Angew. Makromol. Chem.*, **100**, 135 (1981).
[136] I. R. Bellobono, S. Calgari, and E. Selli, *Org. Coat. Appl. Polym. Sci. Proc.*, **46**, 472 (1981).
[137] S. Calgari, E. Selli, and I. R. Bellobono, *J. Appl. Polym. Sci.*, **27**, 527 (1982).
[138] I. R. Bellobono, E. Selli, and B. Marcandalli, *J. Appl. Polym. Sci.*, **32**, 4323 (1986).
[139] H. Yasuda, B. Sherry, M. A. El-Nokaly, and S. E. Friberg, *J. Appl. Polym. Sci.*, **27**, 1735 (1982).
[140] J. D. Fales, A. Bradley, and R. E. Howe, *Res./Dev.*, **27**, 53 (1976).
[141] M. Suzuki. A. Kishida, H. Iwata, and I. Ikada, *Macromolecules*, **19**, 1804 (1986).
[142] A. S. Hoffman, *Adv. Polym. Sci.*, **57**, 142 (1984).
[143] I. Ishigaky. T. Sugo, K. Senoo, T. Okada, J. Okamoto, and S. Machi, *J. Appl. Polym. Sci.*, **27**, 1033 (1982).
[144] I. Ishigaky, T. Sugo, T. Takayama, T. Okada, J. Okamoto, and S. Machi, *J. Appl. Polym. Sci.*, **27**, 1043 (1982).

Chapter 8

Bulk Modifications

In this final chapter on the surface modification of polymers, we will discuss the surface structures arising in multicomponent polymeric systems. The emphasis will therefore be on the relationship between the bulk and the surface structure of heterogeneous, macromolecular systems. Contrary to the techniques described in the previous chapters, where the surface composition was modified and adapted to a given application by some external treatment (plasma, grafting, oxidation), here the surface composition is the result of the effect of the presence of a free surface on the components of the system. The ultimate goal of 'bulk modification' is, thus, the design of a polymer of specific surface composition from the knowledge of the effects of surface-related phenomena on a multicomponent polymeric system.

The meaning of the term 'multicomponent polymeric system' requires some clarification, because, strictly speaking, almost all polymers are multicomponent: even a well prepared, pure homopolymer is composed of chains of different length, whose distribution affects the properties of the system. In this chapter, by 'multicomponent polymeric system' we mean those materials made by macro-molecular chains whose repeating unit is different (blends) or by macromolecules containing blocks of different repeating units in the same chain (block copolymers) (Figure 8.1).

Blending of polymers is a very interesting way of producing materials of improved bulk properties. Many polymeric materials of great technological and commerical importance owe their superior performances to the contribution of two or more cleverly blended parent polymers [1,2]. The thermodynamics of mixing, the occurrence of phase separation and the use of specific (macro) molecules to control the phase-separated structure have been described in many papers and textbooks [1–11]. As to the surfaces, the body of knowledge on the surface properties of polymer blends is much less extensive. Only in the last few years studies on well defined model systems have been published, which evaluate

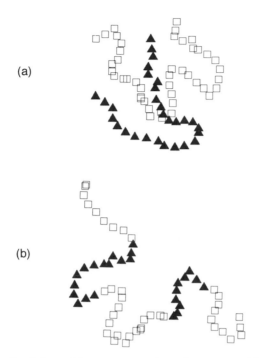

Figure 8.1 Schematic of the multicomponent polymeric systems which will be discussed in this chapter: (a) Blend of two homopolymers. (b) Block copolymer.

the effects on the surface structure of the several variables which affect the nature of the blend. These papers will be reviewed in the following section.

Block copolymers are also generally well understood with respect to their bulk properties [12, 13]. The surface structure is much less known, even if in this case, also, recent years have witnessed a steady increase in the degree of comprehension of these complex systems. Current views on block copolymer surfaces will be discussed in the last section of this chapter.

8.1 POLYMER BLEND SURFACES

Before discussing the basic aspects of polymer blend surfaces, it is necessary to recall briefly a notion which belongs to the thermodynamic bulk treatment of blends. Here as in the following, it is assumed that we are dealing with homopolymer blends, which are the subject of most surface studies. As is well known, the free energy of mixing contains an enthalpic and an entropic component, as follows:

$$\Delta G_{mix} = \Delta H_{mix} - T\Delta S_{mix} \tag{8.1}$$

In order to understand the phase behaviour of the mixture of different components, the enthalpic and entropic contributions must be written in terms related to the molecular characteristics of the components of the system. The simplest model is described in the well-known Flory-Huggins theory [14]. It assumes that only combinatorial entropy contributes to the entropy of mixing. While the entropic contribution always favours mixing, its explicit formula shows a decreasing trend as the polymer molecular weight increases [1, 14]. On the other hand, the enthalpy of mixing is not affected by the polymer molecular weight, and can be favourable or unfavourable according to the value of the Flory's parameter of interaction. Thus, in the case of unfavourable mixing enthalpy (endothermic mixing), the favourable entropic contribution must be large enough to yield a negative free energy of mixing. However, as the molecular weights are increased, a point is reached where the value of the combinatorial entropy becomes so low that the enthalpic contribution prevails, and phase separation occurs. On the other hand, in the case of a favourable enthalpy of mixing (exothermic mixing), the conditions for miscibility are always satisfied, no matter how large the molecular weights are. Thus, using as a first approximation this simplest approach, a blend can be miscible (compatible) for entropic reasons, but in this case there will be a limiting molecular weight (whose value depends on how much is unfavourable to the enthalpic interaction) where phase separation occurs. Otherwise, it can be miscible for enthalpic reasons, and in this case no upper limit to the compatibility is imposed by the molecular weight of chains.

The study of blend surfaces is the study of the effect of the presence of a surface on the rules which control the bulk phase structure. Intuitively, one can expect that the lower surface free energy component will dominate the surface, that is, the surface composition will as a result be enriched, with respect to the bulk, in the lower surface free energy component. Surface enrichment, however, means de-mixing, that is the loss of the combinatorial entropy and, in the case of exothermic mixing, the loss of the mixing enthalpy. Thus, it is clear that the combined effect of the surface free energy and of the free energy of mixing will yield different results in incompatible, entropically compatible and enthalpically compatible blends. Moreover, given the previously discussed dependence of the entropic term and the known dependence of the surface energy on the polymer molecular weight [15], within each quoted class, the polymer molecular weight will, in turn, affect the final result. Of course, in addition to these thermodynamic considerations, kinetics also play an important role: the rate and the degree of attainment of equilibrium, the effect of solvents and crystallinity on the same, are all important variables which makes the study of this problem theoretically and experimentally very demanding.

Despite these difficulties, some brilliant studies on blend surfaces have recently appeared. The effect of the surface energy on surface composition has recently been highlighted by Jones, Kramer and coworkers [16]. The system they studied is an isotopic blend of poly(styrene) (PS) and deuterated PS (d-Ps). Despite the very close similarity of these two polymers, the Flory's interaction parameter for this system is weakly unfavourable, probably owing to the difference in polarizability between C–H and C–D bonds. As to the surface energy, according to the authors' calculation [16], that of d-PS is 7.8×10^{-2} mJ/m^2 lower. The molecular weights they used are 1.8×10^6 and 1.03×10^6 respectively: given the molecular weight dependence of the surface energy [15], this difference in molecular weight implies that the surface free energy of the d-PS used is lower than that of PS, irrespective of the isotopic effect. However, as stated by the authors, the molecular weight effect is too small to account for the observed surface effects, described in the following. Moreover, as stated in note 20 of ref. [16], the molecular weight of PS was varied in a $6.7 \times 10^5 \div 1.8 \times 10^6$ range, without significant difference in the results. Several samples of the increasing bulk volume fraction of d-PS were prepared and sealed samples were annealed at 184°C which, for this system, is a one-phase region of the phase diagram. Thus, according to the previous discussion, this system can be classified as an entropically compatible blend. The distribution of d-PS in the samples was measured by forward-recoil spectrometry (FRES). The results show, after annealing, a definite enriched layer of d-PS on the surface of the blend, whose thickness is some tens of nanometres. In a further paper [17], the authors used secondary ion mass spectrometry (SIMS) and time-of-flight forward recoil elastic scattering (TOF-FRES) to discuss the kinetics of formation and the fine structure of the enriched layer. Figure 8.2, reprinted from ref. [17], shows the evolution of the surface concentration profile as a function of annealing time for an isotopic blend containing 0.33 bulk volume of d-PS.

Considering the surface fraction of d-PS, the authors found an almost linear increase with the bulk volume fraction [16]. Calculations show that the surface enrichment of d-PS is consistent with a surface free difference-driven process of isotopic origin. As to the kinetic side, examined in ref. [17], it is shown that the rate of formation of the enriched profile is controlled by the bulk mutual diffusion coefficient.

The previous results document nicely the interplay of mixing and surface effects. Figure 8.2 is a series of snapshots of surface effects at work. The presence of a polymer–air interface perturbs the system and challenges the results of the free energy of mixing, which, in the region far away from the surface, imposes a homogeneous system. The weakly unfavourable nature of the interaction, coupled with the small combinatorial entropic gain caused by the large chain length, make surface free energy-driven effects win over mixing, and surface segregation occurs.

Surface enrichment of the lower surface energy component was also observed in another homopolymer blend by Pan and Prest [18]. The system under study is

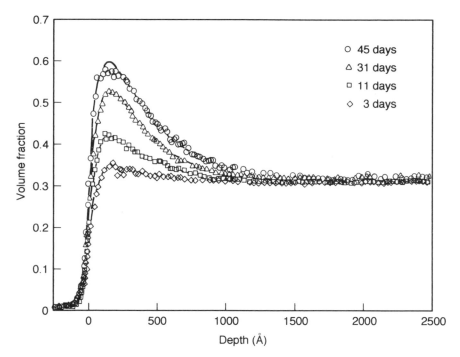

Figure 8.2 Evolution of the surface concentration profile as a function of annealing time for an isotopic PS/d-PS blend containing 0.33 bulk volume fraction d-PS [17]. (Reprinted with permission from Zhao *et al.*, *Macromolecules*, **24**, 5991. Copyright (1991) American Chemical Society.)

PS/poly(vinylmethylether) (PVME), whose behaviour makes it a good candidate for surface–bulk structure relationship studies. In fact, PS/PVME mixtures are miscible in all proportions when they are cast from toluene, but are phase separated when cast from trichloroethylene [19–23]. Moreover, the miscible PS/PVME system can be phase separated by thermal treatment above its lower critical solution temperature [19]. Contrary to the previous example, the surface free energy difference (expressed as critical surface tension) of the two components is rather high: PS = 36 mJ/m^2, PVME = 29 mJ/m^2. Taking into account miscible PS/PVME blends, all the samples (from 5 to 95% by weight of PS) show surface enrichment of PVME when analysed by XPS. According to the typical behaviour of surfactants [24], the relative enrichment of the surface is found to be nearly independent of the bulk composition from 25 to 75 weight per cent of PVME. The 50/50 blend was studied by angular dependent XPS, and it was found that at the lowest sampling depth the PVME concentration is nearly 95% by weight. As to the details of the PVME enriched layer, a simple parallel layer model was incapable of fitting the experimental data and it was impossible to conclude if the surface structure is laterally inhomogeneous.

As to the phase separated PS/PVME blends, it was observed that phase separation increases the degree of surface enrichment by PVME. Specifically, the surfaces of thermally induced phase-separated blend (which underwent a heat treatment at 154 °C for 10 min) are more enriched in PVME than solvent-cast phase-separated blends, which, in turn, are more enriched than miscible blends. This trend apparently correlates inversely the degree of surface enrichment and the degree of bulk mixing [18], but the possible effect of the heat treatment on the surface composition of thermally induced phase-separated blends should also be taken into account. In other parts of this book it was shown (Chapters 2, 7, and 9) that the enhanced macromolecular mobility granted by the heat treatment can deeply affect the surface composition.

Koberstein and coworkers nicely completed the work on miscible PS/PVME blends [25]. In this case, blends of PVME ($M_w = 99\,000$) and PS (M_w from 517 to 127 000) were cast from toluene. The surface energy (by the pendant drop method [15]) and the surface composition (by XPS) were measured. It must be noted that the range of PS molecular weight used guarantees several mJ/m^2 of difference between the surface energy of the heaviest and the lightest PS, according to the LeGrand and Gaines empirical expression [15, 26]:

$$\gamma = \gamma_\infty - K_e/M_n^{2/3} \qquad (8.2)$$

where γ is the surface energy (Chapter 4), γ_∞ is the surface energy at infinite molecular weight, K_e is an empirical constant and M_n is the number average molecular weight. Given the value of the empirical constants for PS [15, 25] a difference of something less of 6 mJ/m^2 results between the surface energy of PS$_{127\,000}$ and PS$_{517}$. If the lowering of the surface free energy of the blend is the driving force for surface segregation, a correlation between the surface amount of PS and its molecular weight is expected. As shown by Figure 8.3, this is indeed the case: the surface weight per cent of PVME (data refer to the 50/50 blend) follows the same trend as the difference between the surface energy of PVME and PS. In other words, as the PS molecular weight and the surface energy of PS increase, the blend surface is more and more enriched by the low surface energy PVME. The reported data could also imply a kinetic contribution to the observed effect: in fact, one could suggest that the lower molecular weight PSs can diffuse more readily, so that the increased surface enrichment of PVME as the PS molecular weight increases could be more the result of a slower diffusion of bulky PS chains than the effect of surface energy differences. However, Koberstein and coworkers were able to find a linear correlation between the surface weight per cent of PVME (calculated from XPS data) and $M_w^{2/3}$ of PS [25], a relationship which clearly reflects the molecular weight dependence of the surface energy of PS and strongly suggests the surface energy difference-driven nature of process.

While the reported data clearly indicate a non-homogeneous distribution of PS and PVME in the vertical dimension (that is the bulk-surface direction), it was impossible to obtain data on the lateral distribution of PS and PVME

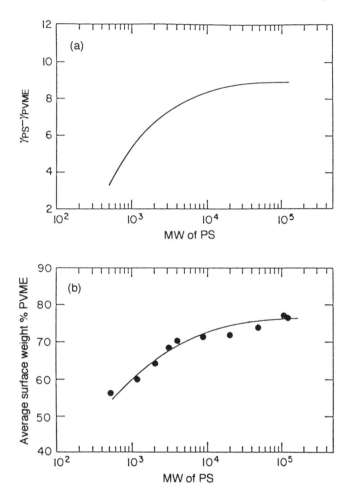

Figure 8.3 Effect of the molecular weight of PS on the surface composition of PS/PVME blends: (a) Surface tension difference between the PS and PVME homopolymers versus molecular weight. (b) Average surface weight per cent of PVME obtained from XPS versus molecular weight of PS [25]. (Reprinted with permission from Bhatia *et al.*, *Macromolecules*, **21**, 2166. Copyright (1988) American Chemical Society.)

components. It is not known if the presence of a polymer/air interface can induce the formation of top layers which are one-phase but more diluted in PS with respect to the bulk, or directly promote phase separation and create PVME rich domains.

Another series of outstanding works on the surface structure of blends has been performed by Gardella and coworkers. An interesting paper deals with the poly-(methylmethacrylate)/poly(vinylchloride) (PMMA/PVC) system [27]. These blends are reported as not compatible when cast from tetrahydrofuran (THF) [27] and compatible when cast from methylethyl ketone (MEK) [28]. As to the surface energy, the reported values for PMMA and PVC are very close: 41.2 and 41.9 mJ/m^2, respectively, using the harmonic mean method (Chapter 4) [15].

Gardella and coworkers prepared both THF and MEK cast blends and observed (by angle dependent XPS and ATR-IR and transmission FTIR) the relationship between the surface and bulk structure of the blends. In general, THF cast blends yielded microdomain structures and XPS showed a definite enrichment of the lower surface energy component (PMMA) for every blend composition. On the other hand, MEK cast blends were homogeneous and only a very slight surface segregation of PMMA was detected. The surface excess of PMMA is so small that, according to the authors, the surface amount of PMMA is within error limits equivalent to the bulk composition. However, the PMMA surface excess, albeit very small, was systematically observed for every blend composition so that it is very likely real and not the result of an experimental artefact. However, it is noteworthy that the combined effect of a favourable free energy of mixing (as shown by the blend compatibility) and a very small surface energy difference between the couple of homopolymers, can yield a surface composition which is very close to the bulk one.

This same point is underlined in another series of papers of Gardella and coworkers, where they studied the system poly (ε-caprolactone)/PVC(PCL/PVC) [29, 30]. The PCL/PVC system is miscible in the amorphous and molten state throughout the entire composition range of 10–90% PCL. At a weight per cent greater than 50%, PCL crystallizes in solid state blends and two phases, crystalline PCL and compatible PVC/PCL blend, are formed [31]. A series of studies, discussed, for instance, in ref. [31], indicate that there are specific interactions between PCL and PVC which are responsible for the compatibility of these blends.

Gardella and coworkers evaluated by angle dependent XPS and transmission and ATR–IR the relationship between the bulk and surface structure of (PCL) ($M_w = 33\,000$) and PVC ($M_w = 190\,000$) blends, in the whole range of weight per cent. Further information on the bulk nature of the blends was collected by differential scanning calorimetry (DSC). The solid surface energy data are reported as 42.9 mJ/m^2 for PCL and 44.0 mJ/m^2 for PVC (this last value is more than 2 mJ/m^2 higher than the one quoted in the discussion on the PMMA/PVC system, a very significant difference. Assuming the previously quoted value of 41.9 mJ/m^2, which is reported by Wu at page 180 of ref. [15], PVC would be the lower surface energy component in the PVC/PCL blend. The reason for the uncertainty of these data must be found in the experimental and theoretical problems involved in the measurement of contact angles and the calculation of the solid surface energy from contact angle data, discussed at length in Chapter 4).

The main results obtained on the relationship between bulk and surface composition of PVC/PCL blends are shown in Figure 8.4, which is reprinted from ref. [29]. It is possible to observe that, for bulk weight percent of PCL lower or equal to 48%, where the blends are found compatible, the surface composition is equivalent to the bulk composition. This is the first, unambiguous example of a blend where the thermodynamics of mixing overcome surface energetics driving forces (as previously discussed, in the quoted study on MEK cast PMMA/PVC blends there is some hint of a small surface excess of PMMA).

As the bulk per cent of PCL increases, however, a pronounced surface excess of PCL is observed (40 < % PCL < 90). As previously discussed, at a weight per cent greater than 50%, PCL crystallizes in solid state blends and, in the simplest model, the overall structure can be represented by a two-phase system, composed by crystalline PCL and compatible PVC/PCL blend [31]. Accordingly, the blends made in this range of composition are opaque, and a melting transition whose value indicates the presence of crystalline PCL is observed by DSC [29]. Thus it is possible to hypothesize that the observed surface excess of PCL is due to the surface segregation of semicrystalline PCL, driven by the previously quoted surface free energy difference (see, however, the previous short discussion on this topic). To make the picture more complicated, the surface of the 90% bulk weight per cent PCL blend contains a higher concentration of the blended phase than the bulk (Figure 8.4).

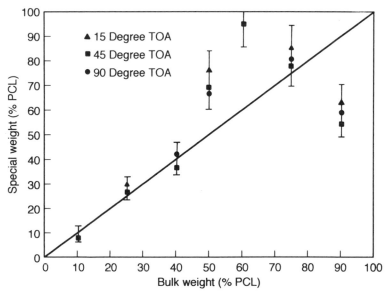

Figure 8.4 Plot of angle-dependent XPS results of surface composition versus bulk composition of PCL/190 PVC blends [29]. (Reprinted with permission from Clark *et al.*, *Macromolecules*, **24**, 799. Copyright (1991) American Chemical Society.)

To get more information, Gardella and coworkers studied the effect of PVC molecular weight in the PVC/PCL system [30]. In addition to the previously quoted PVC of $M_w = 190\,000$ ($PVC_{190\,000}$), blends containing $PVC_{77\,000}$ and $PVC_{275\,000}$ were also analysed. Interestingly, it was observed that, when using $PVC_{77\,000}$, only the 10% PCL was homogeneous, and in this case the surface composition was found equivalent to the bulk one. In general, it was found that the surface composition of this blend is governed by a combination of the degree of crystallinity of PCL and of the molecular weight of PVC. When using relatively short-chain PVC ($PVC_{77\,000}$), the latter is less effective at inhibiting the PCL–PCL interactions which lead to crystallinity. Thus, more crystalline PCL domains are formed and more PCL surface segregation occurs. Accordingly, the data show that for $PVC_{77\,000}$/PCL blends containing 40–60% bulk PCL, the surface composition is about 100% PCL [30]. When using longer chain PVC, the surface excess of PCL is generally lower.

The blends with a bulk weight per cent of PCL higher than 80 generally show a surface excess of the blended PVC/PCL phase [30]. The reason for this behaviour is not completely clear.

The previous discussion was mainly based on compatible blends. Since, in general, the perturbation induced by the surface results in marked effects, it can be expected that in incompatible blends, where the effect of surface energy adds to the unfavourable free energy of mixing, the more surface active component should completely dominate the surface composition. This is indeed found to be the case in several instances, especially when siloxanes are involved. The latter have a very low surface energy, are scarcely miscible with other polymers and readily diffuse. Blends containing silicone homopolymers display silicone-like surface properties [32]. The diffusion towards the surface is so pronounced that it leads to their exudation from the system over a period of time, an occurrence which make them suitable for only temporary surface modifications. For this reason, the exploitation of silicones in the surface modification of polymers by blending, requires the synthesis of properly engineered copolymers, where the surface activity of the siloxane component is coupled to a block or a graft segment which provides miscibility, or at least, some kind of interaction, with the base polymer. Several interesting examples of this form of macromolecular engineering are described in the works of Yilgor and McGrath [33, 34].

Several examples from the literature, however, suggest that even in the case of siloxanes a complete, homogeneous overlayer should not be taken for granted. Gardella and coworkers using XPS and ion scattering spectrosocopy (ISS) studied the surface composition and structure of bisphenol A poly(carbonate)/poly(dimethylsiloxane) blends (BPAC/PDMS) [35]. For PDMS bulk concentration greater than 20%, macroscopic effects of phase separation were observed, and only PDMS was detected on the surface of the blends. When the PDMS concentration was decreased to $1.4 \div 10.8\%$ however, both BPAC and PDMS were observed on the sample surface. According to the relevant values of surface energy (BPAC = 34 mJ/m^2, PDMS = 24 mJ/m^2), all compositions showed a remarkable surface enrichment of PDMS, whose concentration, as detected by XPS, was approxiamtely 87.5%. Thus, despite the great unbalance in surface energy and the well documented, surface active behaviour of PDMS, a certain degree of mixing occurs, resulting either in a two-phase surface structure containing small BPAC domains in PDMS layers or a single phase homogeneous PDMS/BPAC blend [35]. The authors suggest that while surface energy values are useful for predicting which species should show a surface preference, they alone are not enough for predicting the degree of surface enrichment expected.

Another interesting example of surface studies of incompatible homopolymer blends is given by the work of Thomas and O'Malley (here, as in the previous case, a further reason for interest is given by the comparison they made with a block copolymer made by the same components). These aspects will be discussed later [36]. In the PS/PEO system, the surface energy difference between the two polymers is similar to that of the BPAC/PDMS system (PS = 36 mJ/m^2, PEO = 44 mJ/m^2 [35]). Samples were analysed by angle dependent XPS and, in every case, a surface enrichment of the PS was observed. However, the measured surface excess of PS was lower than the excess measured in diblock copolymers of comparable bulk composition by the same authors [37, 38] (these results will be discussed in the next section). This result contrasts with the intuitive feeling that the greater freedom enjoyed by the homopolymer chains in blends with respect to blocks linked in the same chain, should grant to the former a greater chance to follow the thrust of the surface energy difference. Modelling the angle dependent XPS data, the authors found that the surface was laterally inhomogeneous, with PS domains raised somewhat above the PEO domains. Even when a PEO/PS bilayer was made and heated at 130 °C for several hours it was impossible to obtain a pure PS surface. Thus, again, even if, as a general rule, the surface is enriched in the lower surface energy component with respect to the bulk, the quoted results suggest that some other variables come into play to determine the final surface structure. In the present case, for instance, the propensity of PEO to crystallize could affect the observed results, just as was previously discussed for PCL/PVC blend surfaces ([29, 30], Figure 8.4).

This brief review of the most important published works on the surface characterization of homopolymer blends allows some conclusions to be made. Undoubtedly, the previous discussion clearly shows the complexity of the surface phenomena observed even in the comparatively simple systems composed by homopolymer blends (as compared to blends made by more complex polymeric chains, such as block, graft, etc.). However, owing to the growing body of detailed work which is published on this subject, some general conclusions can be made: in general, the surface of homopolymer blends, either compatible or incompatible, is enriched in the lower surface energy component with respect to the bulk. As a general trend, it is often concluded that the degree of bulk mixing affects the degree of surface segregation, the lower the former the higher the latter. In some systems, such as the PS/d-PS and the PVME/PS the surface energy difference between the blend components seems the main (and possibly, the only) variable affecting the surface composition. It is noteworthy to observe that surface energy has an important effect on both systems, despite the surface energy difference between the components it is about $7\,mJ/m^2$ in the latter case and $7.8 \times 10^{-2}\,mJ/m^2$ in the former. Figure 8.2, which shows the build-up of a d-PS rich layer in PS/d-PS blends, also underlines the crucial role played by the experimental conditions and by the design of the experiment itself: time and temperature deeply affect the amount of the surface excess of the low surface energy component. It must be noted, that PS/d-PS blends were annealed at a temperature higher than the T_g of the polymers, where bulk mobility allows macromolecules a greater degree of diffusional freedom, so that the thermodynamic thrust can be followed. When bulky chains are blended at room temperature, without annealing, a different picture can result. In these cases, the plasticizing effect of the solvent can play an important role.

Comparing the PS/d-PS and the PS/PEO data, some conflicting observations can be made: in the quoted PS/PEO bilayers, despite the 130 °C annealing [36], the experimental data show that the extrapolation in time to where a surface of only PS exists yields, for all practical purposes, infinity. Yet the molecular weight of the polymers is lower than that of the PS/d-PS pair, the surface energy difference is more than two orders of magnitude larger and free energy of mixing is not favourable (as previously noted, contrary to the PS/d-PS pair, this blend is incompatible). The propensity of PEO to crystallize is very welcome, since it provides a possible explanation to this disagreement.

However, the most evident conclusion is that the coincidence of the surface and the bulk composition of homopolymer blends is a rare exception, rather than a rule. It was necessary to work with selected compositions of a system characterized by specific interactions between the components (which guarantee a favourable mixing enthalpy) to find no difference between the surface and the bulk composition of a homopolymer blend [29, 30]. This observation clearly underlines the need for a complete understanding of the variables which control the surface segregation phenomena in blends and the potentiality of blending as a surface modification technique.

8.2 BLOCK COPOLYMER SURFACES

In the previous section it was shown, by experimental evidence, that the presence of a surface deeply affects the composition of the outermost layers of a multicomponent polymeric system. The systems under study were composed of homopolymer chains of a different repeating unit. The present section will deal with surface studies on another very important class of multicomponent polymeric materials, that is block copolymers. The problem is, in its general tenet, almost the same: the presence of a free surface affects the global interaction among the different components of the system, so that the surface composition is, in general, different from the bulk one. A fundamental difference, however, exists: in copolymers the constituents are connected, therefore the details of phase separation and the nature of the phase separated structure is affected by the connectivity of the segments. From bulk studies, it is well known that the domain size for block copolymers is generally smaller than that of polymer blends, because the different blocks in a chain of block copolymers are connected by covalent bonds. Connectivity affects all the aspects relevant to the formation of a phase structure: from a thermodynamic point of view, the interplay of surface tension and thermodynamics of mixing must cope with the block length, which limits the domain dimensions. From a kinetic point of view, the time scale required to build up a given phase structure is affected by the mobility of the segments, which, in turn, is affected by the constraints at the block ends.

As in the case of polymer blends, the bulk morphological structures of block copolymers have been widely investigated [12]. It is now widely accepted that the bulk morphology (in equilibrium or near equilibrium conditions) is dictated by the bulk volume fraction of each component. Figure 8.5 shows the evolution of the bulk morphology of a diblock copolymer (A–B) as a function of the volume fraction of the components. When the volume fraction of one of the components is 0.2 or less, this minor component forms spherical domains in a matrix of the major component. If the volume fraction is increased (between 0.2 and 0.4) cylindrical microdomains are formed, which are arranged in a hexagonally close-packed structure. If the average volume fraction of one of the blocks in the copolymer is between 0.4 and 0.6, then the structure is comprised of alternating lamellae (in the narrow regime between those which dictate the formation of cylindrical and lamellar microdomains, the cylindrically shaped microdomains arrange themselves in a structure which exhibits diamond-like symmetry [39]). This structure is not shown in the figure.

The understanding of the copolymer surface structure is far less complete than the bulk one. As in the case of blends, it is possible to suggest that the domain structures shown in Figure 8.5 are modified by the effect of the free surface, and that the lower surface energy component exhibits a preferential affinity for the free surface. A more quantitative and useful knowledge of the problem requires, however, the evaluation of the effect of the block's length and architecture on the surface energy-induced segregation effects, as well as the understanding of the role played by the preparation routine and the processing conditions. As in the previous section, the experimental studies on block copolymer surfaces will now be reviewed. The experimental findings will be completed by the discussion of some recent theoretical suggestions.

In his pioneering work on the application of XPS to polymers, Clark studied the surface composition of PDMS/PS diblock copolymers [40]. The two copolymers studied contained 23 and 59 wt% PS, and the number of average

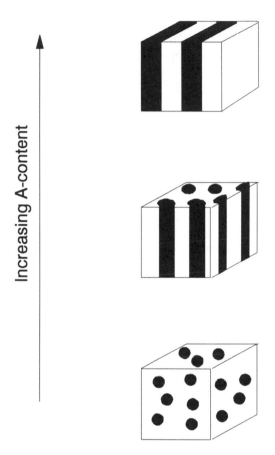

Figure 8.5 Schematic illustration of the various bulk phase structures obtained as the content of the A component (black) is increased.

molecular weights was slightly more than 120 000. While contact angle measurements evidenced a continuous overlayer of PDMS, XPS spectra clearly showed the typical shake-up peak of the aromatic component of the PS unit (see Chapter 3). This result was attributed to the different depth sampled by the two techniques: the less surface sensitive XPS measurement actually also showed the sub-surface region, below the PDMS overlayer. The thickness of the latter was calculated by the XPS spectra: depending on the casting solvent used, it ranged from 1.3 to more than 4 nm. In general, the occurrence of a pure PDMS overlayer shows that, in this extremely incompatible system [41, 42], the surface energy difference between the blocks control the surface structure. If the PDMS block length and the overall content of PDMS are reduced (PDMS < 20%), however, a different picture arises [43]. From contact angle measurement, a pure overlayer of PDMS is no longer found. The analysis of the angle-resolved XPS data suggests that the PDMS blocks form cylindrical domains, whose radii are of the order of the dimensions of the PDMS blocks, orientated normal to the sample surface and slightly projecting above the PS matrix.

While in the previous system PS was the higher surface energy component, in PS/PEO block copolymers it is the lower one. Thomas and O'Malley evaluated the surface structure of diblock and PEO/PS/PEO triblock copolymers by an angle resolved XPS [37, 38] (the results obtained on blends of analogous composition have been described in the previous section). Three PS/PEO diblocks were studied, ranging from 19.6 to 70 wt% PS, with an average number molecular weight of the order of several tens of thousands [37]. As a general result, the surface is not made by a pure overlayer and it is significantly enriched in PS. Angle resolved XPS data suggest morphology similar to the previously described one: the copolymer surfaces are laterally inhomogeneous with cylindrical domains orientated normal to the sample surface. The PS component is often raised above the PEO component.

Another interesting finding involves the role of the solvent on the surface composition: the degree of surface enrichment by PS was found to be deeply affected by the nature of the solvent used in the casting step. The surface amount of PS increases as the solvent becomes more preferential for PS. As an extreme case, when ethylbenzene was used, something less than 70% PS was found on the surface of a copolymer characterized by a bulk composition of 21.4% PS (Figure 8.6, [37]). This dependence on the nature of the solvent was observed for all the three polymers studied.

The work on PEO/PS/PEO triblocks (cast from chloroform) shows the same general trends observed on diblocks [38]. There is a significant surface excess of PS at all copolymer compositions. From a quantitative point of view, a remarkable similarity in the surface behaviour of diblocks and triblocks is observed. Moreover, from the analysis of the XPS peaks, the authors suggest that some limited degree of phase mixing occurs on the sample surfaces, a finding which cannot be explained by bulk results (according to which, PS and PEO are absolutely incompatible [44, 45]). The suggestion that some degree of phase mixing occurs on the surface implies that the local surface interaction parameter is different from the bulk one (or, in other words, that the presence of a free surface affects the value of the intraction parameter). As will be discussed later in this section, recent views on the variables which affect the surface composition and structure suggest that this is indeed the case.

The previously quoted copolymers were characterized by a large difference between the surface energy of the components. Hasegawa and Hashimoto have studied by transmission electron microscopy (TEM) thin sections the free surface of PS-poly(isoprene) (PI) diblocks [46]. In this case, the essentially hydrocarbon-like repeating unit of the two blocks yields very similar surface energy values: 30–32 mJ/m^2 for PI and 33–36 mJ/m^2 for PS [46]. The PS/PI system studied has

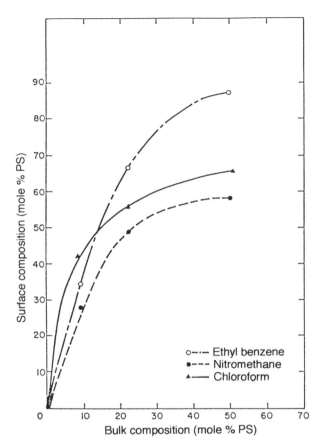

Figure 8.6 Effect of the solvent on the surface composition of PS/PEO diblocks [37]. (Reprinted with permission from Thomas and O'Malley, *Macromolecules*, **12**, 323. Copyright (1979) American Chemical Society.)

a weight fraction PS of 52% and an average number molecular weight of about 500 000. Accordingly, a bulk lamellar structure is observed and a PI overlayer at the free surface is apparently detected. Even when, occasionally, lamellae are oriented with their interfaces normal to the free surface, the PI overlayer maintains its orientation parallel to the free surface. In a beautifully illustrated paper concerning a larger set of polymers, the same authors show that even when the lower surface energy component forms a dispersed phase (spheres or cylinders) and the other one forms the matrix, the surface is made by a continuous overlayer of the former [47]. Contrary to the previously reported findings concerning PS/PEO blocks (Figure 8.6), the nature of the solvent has no effect on the surface structure. However, it must be remembered that this conclusion is based on TEM observation of stained layers and it is not supported by more quantitative surface analysis (XPS analysis of these systems is, however, rather ambiguous, since both components are hydrocarbon in nature, and differ only by the broad shake-up peak of PS). Further information from Hasegawa and Hashimoto show that the free surfaces of block copolymers with crystalline components are governed by crystallization kinetics. Thus, as in the previously reported studies on polymer blends, the formation of crystalline domains is shown to deeply interfere with the push of surface energy.

The effect of the interplay of surface energy and crystallinity on the surface composition of block copolymers is described in another interesting paper of Gardella and coworkers [48]. This time, the copolymers studied are made by poly(tetramethyl-*p*-silphenylenesiloxane)/PDMS (PTM*p*S/PDMS) blocks, in the whole range from 100 to 0%. The PTM*p*S/PDMS system is a very interesting one, as described in a series of extensive studies [49–51]. According to the accepted model, the bulk of PTM*p*S/PDMS copolymers consists of crystalline regions of PTM*p*S interspersed within disordered or amorphous regions of PDMS and PTM*p*S. The degree of crystallinity, the melting temperature and the glass transition temperature exhibit large changes with composition. The analysis of the copolymer surface by ATR/IR and XPS reveals some interesting features: in samples with a high degree of crystallinity and a low PDMS concentration a highly crystalline core is formed. This structure forces the segregation of amorphous PDMS towards the surface, and, accordingly, a greater than bulk surface concentration of PDMS is detected. On the other hand, when the overall crystallinity is lower and the PDMS content is greater, the surface enrichment of PDMS is much less dramatic. While these results underline once again the strong interplay between the push of lowering the surface energy and the energetics of crystallization, it is interesting to compare them with the results obtained by the same group on other block copolymers containing PDMS, that is PDMS/bisphenol A poly(carbonate) (BPAC) [52]. Contrary to PDMS/PTM*p*S diblocks, PDMS/BPAC diblocks are completely amorphous, so that the surface structure is not affected by crystallization. Besides XPS, samples were analysed by the more surface sensitive ion scattering spectroscopy (ISS) (as previously discussed, the same studies were performed on PDMS/BPAC blends [35]). The copolymers studied range from about 22 to about 63 wt% PDMS, with an average PDMS block length of 20 units [52]. Analysis shows that the surface of all copolymers is enriched in PDMS, the lower surface energy, component. ISS shows a much greater degree of surface enrichment than XPS and the comparison of angle resolved XPS and ISS data suggests a model for the surface topography which is similar to that proposed for PS/PEO and PDMS/PS copolymer, previously discussed [37, 38, 43]. In the present case, the surface morphology is thought to be made by microdomains of each component oriented perpendicularly to the surface, with a depth greater than the sampling depth of XPS and with the PDMS microdomains slightly raised above the BPAC regions [52]. Another possible model involves the formation of a very thin homogeneous PDMS overlayer (too shallow to be detected by XPS).

Comparing the results of the surface analysis of the PDMS/BPAC and PTMpS/PDMS diblocks, a much greater degree of surface segregation is observed in the former. This result is in part due to differences in block length (as discussed below) and surface energy difference between the components of the blocks, but the different nature of the copolymers also plays a role: the former is completely amorphous, so that the segregation of PDMS blocks is controlled by thermodynamic incompatibility and by the lowering of the surface free energy of the system. In the case of PTMpS/PDMS diblocks, crystallinity can promote a more vertically homogeneous structure, resulting in a lower degree of surface segregation of PDMS.

The class of block copolymers which has been subjected to the larger number of surface studies is probably that of segmented polyurethanes (SPU) [53–59]. This analytical effort has been prompted mainly by biomedical materials researchers, in the quest for an answer to the unusually mild behaviour of SPUs towards blood, a feature which makes SPUs the materials of choice in many blood contacting applications (Chapter 12). While it is generally accepted that the phase separated structure of SPUs is at the origin of this property, it is not yet completely clear which are the most important details and how it is possible to engineer a SPU in order to obtain a given surface structure (the latter point is, of course, at the very core of the problem discussed in this chapter).

The inherent complexity of the SPU's chemistry makes the discussion of these studies poorly suited for a chapter like this, where, through the analysis of experiments performed on well defined, model systems, some general relationships between bulk and surface structures are sought. Moreover, many of these studies were performed on commercial polymers, where the presence to additives and the possibly broad molecular weight distribution can affect the results. The interested reader is referred to the quoted papers, and to the references they contain. Several aspects of phase-separated polymers for blood contacting applications will be briefly reviewed in Chapter 12. As an example, however, the results of an interesting paper by Yoon and Ratner will be quoted here [60]. Several poly(etherurethanes) (PEU) and poly(etherurethane urea) (PEUU), based on 4,4'-methylenebis (phenylene isocyanate) (MDI) were synthesized. The soft segment consisted of poly(tetramethyleneglycol) (PTMO) of molecular weight 1000 or 2000. Fluorinated chain extenders were used as a 'tag' for XPS analysis. The latter technique avoids the inaccurate results which can be obtained using (as generally done) the N1s signal as a marker of the hard segment. The low intensity of the latter naturally give rise to inaccurate results in the evaluation of angle dependent XPS data. The prepared samples were purified to remove all low molecular weight contaminants. The analysis of the sample's surface by angle dependent XPS allows some interesting results to be obtained on the surface–bulk structure relationship. In particular, as already found for other systems, the greater the degree of bulk incompatibility the greater the degree of surface phase separation. Bulk incompatibility was enhanced by increasing the PTMO molecular weight or by the so called 'odd diol' effect, according to which, chain extenders containing an even number of carbons produce polymers that are more phase separated than those containing chain extenders with an odd number of carbons [61–63]. Polymers characterized by a high degree of incompatibility (high molecular weight soft segment, even number of carbon in the chain extender) showed a definite surface enrichment of the soft segment (no data on the surface energy of the fluorinated hard segments are available, thus it is not possible to state if this effect is driven by the surface energy difference or by, for instance, the effect of PTMO crystallinity). Even in the case of samples highly enriched in PTMO, however, the surface is probably not made by a complete overlayer of the polyether, as shown by the results of surface analysis obtained by a more surface sensitive technique (static secondary ion mass spectroscopy, SSIMS [64]).

Clearly, the suggestion of the effect of the PTMO block weight and of the number of carbon atoms of the chain extender on the surface structure strongly point to a true bulk molecular 'engineering' of the polymer to obtain a given surface structure. This last remark introduces the key point of this section, that is our present understanding of the mechanisms which control the surface structure and are responsible for the effects observed in the above described studies.

As previously discussed, the difference in the surface energy of the different blocks was indicated early as the origin of the observed surface composition and a surface enrichment by the lower free energy component is generally observed. However, as shown already by one of the first studies (that of O'Malley and coworkers on PDMS/PEO diblocks [43]), the surface energy difference alone cannot explain the coexistence of domains of different composition on the polymer surface.

In 1981 Gaines proposed that the experimental observations could be explained by the spreading coefficient criterion [65]. The latter was borrowed from the thermodynamics for liquid spreading: in order for liquid b to spread over the surface of liquid a, it is necessary that [66]:

$$S_{b/a} = \gamma_a - \gamma_b - \gamma_{ab} > 0 \tag{8.3}$$

where S is the spreading coefficient of b on a, γ_i is the surface tension of liquid i and γ_{ij} is the interfacial tension between liquid i and liquid j (equation (8.3) basically states that spreading occurs when the energetic cost to be paid for the building up of a free surface of the liquid b and an interface between a and b is lower than the cost paid for the maintenance of a free surface of the liquid a). According to this criterion, block copolymer surfaces are made by a complete overlayer when $S_{b/a}$ (where a and b are the two blocks) is greater than zero. When the criterion is not satisfied, spreading of one of the components does not occur and isolated domain morphology is to be expected. The spreading criterion can explain the early results on PDMS/PS blocks [40], where surface/interfacial energetics induce the spreading of PDMS and on PS/PEO [37, 38]. In the latter case, the surface energy difference between the components is lower than in the PDMS/PS case and a large interfacial tension is measured between low molecular weight analogues (Gaines reports a value of $10.6 \, mJ/m^2$ for the interfacial tension between ethylene glycol and ethylbenzene [65]). Thus, the criterion for spreading is not satisfied and an isolated domain morphology is observed.

Albeit attractive for its immediateness, the theory of Gaines fails to explain, for instance, the results observed in PDMS/PS diblocks with short PDMS blocks and low PDMS concentration [43].

The suggestion that the component with the lowest surface energy is preferentially segregated at the free surface has been recently resumed [67]. The observation by TEM of PS/poly(butadiene) (PB) diblocks in a large composition range, concludes that, regardless of the microdomain type (Figure 8.5), all samples showed a surface region where the formation of the normal microdomain morphology was inhibited. In every case, evidence was found that the preferential segregation of the block component with the lowest surface energy takes place. While there are no doubts that this attempt to relationalize the surface behaviour of block copolymers can work with the PS/PB system, the problem is how far it is possible to generalize it. As previously discussed, much experimental evidence is found that leave little room to doubt the existence of multiphase surfaces. Even if, in some instances, it is possible to suggest that the preparation routine yields samples far from equilibrium or that the crystallinity of one of the components affects the surface structure, the surface energy difference does not seem enough to explain all of the observed data.

Recently, the theoretical description of block copolymer surfaces has received new impetus, mainly starting from the work of Fredrickson on surface ordering phenomena in block copolymer melts [68]. The theory describes the behaviour of almost symmetrical diblocks near the microphase separation transition in the weak segregation limit. The presence of the surface modifies the Flory interaction parameter and the chemical potential in the adjacent copolymer layer. Fredrickson, Russell and coworkers applied the previous theory to the study, by XPS, of the surface composition of a series of symmetric diblocks of PS and poly(methylmethacrylate) (PMMA) [69]. This system is particularly suitable for the evaluation of the merits of the different theories: the oxygen containing carboxyl group of PMMA makes the surface energy of the latter slightly higher than that of PS. Thus, according to the surface energy criterion, the surface of the diblocks should be made by a pure PS overlayer covering the lamellar bulk morphology of the symmetric copolymers (Figure 8.5). On the other hand, if the Gaines's spreading coefficient criterion is taken into account (equation (8.3)), isolated domain morphology is to be expected, since the spreading coefficient of PS on PMMA is lower than zero [70]. The results of Russell and coworkers confirm the limits of the two approaches: for low N values, where N is the number of statistical segments that compose copolymer chain, both PMMA and PS coexist at the

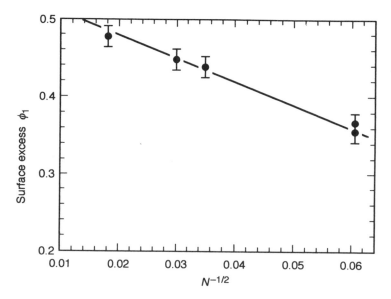

Figure 8.7 Surface excess of PS in PS/PMMA diblocks annealed at 180 °C as a function of $N^{1/2}$ [71]. (Reprinted with permission from Green *et al.*, *Macromolecules*, **24**, 252. Copyright (1991) American Chemical Society.)

surface. At large N, the surface is composed only of PS. Thus, a first important result of this work is that it underlines the need to taking into account the important size scale parameter N. At high xN, where x is the Flory interaction parameter [14], the system is in the strong segregation limit and only PS is observed on the copolymer surfaces. At low xN, where the system is in the weak segregation limit, both components coexist on the surface. In this regime, the surface excess of PS (Φ_{PS}) shows a linear dependence on $N^{-1/2}$, as follows [69]:

$$\Phi_{PS} \approx \alpha - \beta N^{-1/2} \qquad (8.4)$$

The dependence of the surface excess of PS on N is shown in Figure 8.7, reprinted from ref. [71], which also describes the effect of temperature on the surface excess.

The quoted work of Fredrickson [68] supports this dependence of the surface composition on N. Russell and coworkers illustrate that the relationship described in equation (8.4) can be obtained by the application of the Fredrickson theory of block copolymers in the weak segregation limit. The constants α and β contain the two phenomenological parameters that, according to the Fredrickson theory, describe the effect of the surface potential on the bulk potential: the first one is related to the surface energy difference between the components of the copolymer. The second reflects the ability of the surface potential to alter the

Flory interaction parameter in the vicinity of the surface. Explicit relationships between α and β and the quoted parameters can be found in the original papers. Here, it is important to emphasize that the quoted works represent a leap forward in the understanding of the equilibrium composition of block copolymer surfaces. Several empirical observations made in previous works can be explained: for instance, the free surface can effect the Flory interaction parameter, so that miscibility relationships on the surface can differ from that on the bulk, as observed by Thomas and O'Malley [37, 38]. Then, it was frequently observed that the greater the bulk incompatibility the greater the surface segregation. The previous results show that, as the increase of the $\mathbf{x}N$ scale parameter promotes incompatibility, the surface excess of the lower surface energy component increases. Moreover, a given system, such as the previously discussed PDMS/PS diblocks [40, 43], can show a complete overlayer of the lower surface energy component in a given regime and separated microdomains in the weak segregation limit. Finally, it is confirmed that the surface energy difference between the component is not enough to explain the surface structure and composition. Instead, a number of a factors contribute:

(1) The size scale N;
(2) The difference between the surface energy of both components;
(3) The way in which the interaction at the surface differs from that in the bulk.

Several other important hints on the surface composition of PS/PMMA symmetric diblock copolymers can be found in the recent works of Russell and coworkers. In the previously quoted paper [69] it is confirmed that annealing has a profound effect on the surface excess of PS. The equilibrium concentration of PS on surfaces of higher chain length samples is reached only after 130 hours annealing at 170 °C. In another paper [72], it is shown that the rate of the casting solvent evaporation deeply affects the surface excess of PS.

Gardella and coworkers have shown that the dependence of the fractional surface excess of the lower free energy component of N described by equation (8.4) can be extended to PS/PDMS di- and tri-blocks [73, 74]. Moreover, contrary to the copolymers studied in the previous papers [68–72], the PDMS/PS copolymers used by Gardella and coworkers were asymmetric. These papers show that, besides the chain length, the block architecture also has an important effect on the surface morphology of the block copolymer. The comparison of the surface excess of the diblocks with the PDMS–PS–PDMS and PS–PDMS–PS triblocks shows that the PDMS segments of the diblock are the easiest to segregate in the free surface region. When the two triblocks are compared, it is found that the PDMS segments of the PS–PDMS–PS copolymer are the least probable to segregate. This important observation adds to the relationship between the surface excess and the block length. As in the previous case [69–72], it is possible to observe a transition from a weak segregation limit, where the two components coexist on the surface, to a strong segregation limit, where a PDMS overlayer is detected (note that, in the case of triblocks, N is taken as the number of repeat units in one of the PDMS blocks of PDMS–PS–PDMS copolymers, and half of the repeat units of the PDMS block for the PS–PDMS–PS copolymers).

The last results reviewed show that a strong acceleration in the degree of understanding of the relation between surface morphology and molecular composition is presently going on, making more realistic the goal of the molecular design of a polymer of specific surface composition.

REFERENCES

[1] D. R. Paul, J. W. Barlow, and H. Keskkula, in *Encyclopaedia of Polymer Science and Technology*, Wiley, New York (1986), Vol. 12, p. 399.

[2] *Polymer Blends*, D. R. Paul and S. Newman (Eds), Vols I and II, Academic Press, New York (1979).

[3] O. Olabisi, L. M. Robenson, and T. M. Shaw, *Polymer-Polymer Miscibility*, Academic Press, New York (1978).

[4] I. C. Sanchez, *Ann. Rev. Mater. Sci.*, **13**, 387 (1983).

[5] P. G. DeGennes, *J. Chem. Phys.*, **72**, 4756 (1980).

[6] P. Pincus, *J. Chem. Phys.*, **75**, 1996 (1981).

[7] J. Gilmer, N. Goldstein, and R. S. Stein, *J. Polym. Sci. Polym. Phys. Ed.*, **20**, 2219 (1982).

[8] J. Noolandi, *Polym. Eng. Sci.*, **24**, 70 (1984).

[9] H. W. Kammer and J. Kressler, *Int. Polym. Sci. Tech.*, **16**, 83 (1989).

[10] S. Wu, *Polym. Eng. Sci.*, **27**, 335 (1987).

[11] M. Xanthos, *Polym. Eng. Sci.*, **28**, 1392 (1988).

[12] S. L. Aggarwal, *Block Polymers*, Plenum Press, New York (1970).

[13] A. Noshay and J. E. McGrath, *Block Copolymers—Overview and Critical Survey*, Academic Press, New York (1977).

[14] P. J. Flory, *Principles of Polymer Chemistry*, Cornell University Press, Ithaca, N.Y. (1953).

[15] S. Wu, *Polymer Interface and Adhesion*, Marcel Dekker, New York (1982), Ch. 5.

[16] R. A. L. Jones, E. J. Kramer, M. H. Rafailovich, J. Skolov, and S. A. Schwarz, *Phys. Rev. Lett.*, **62**, 280 (1989).

[17] X. Zhao, W. Zhao, J. Skolov, M. H. Rafailovich, S. A. Schwarz, B. J. Wilkens, R. A. L. Jones, and E. J. Kramer, *Macromolecules*, **24**, 5991 (1991).

[18] D. H. K. Pan and W. M. Prest, Jr., *J. Appl. Phys.*, **58**, 2861 (1985).

[19] M. Bank, J. Leffingwell, and G. Thies, *J. Polym. Sci.*, **A-2**, 1097 (1972).

[20] T. Nishi, T. T. Wang, and T. K. Kwei, *Macromolecules*, **8**, 227 (1975).

[21] D. D. Davis and T. K. Kwei, *J. Polym. Sci. Polym. Phys. Ed.*, **18**, 2337 (1980).

[22] A. Robard and D. Patterson, *Macromolecules*, **10**, 1021 (1977).

[23] T. Shiomi, K. Kohno, K. Yoneda, T. Tomita, M. Miya, and K. Imai, *Macromolecules*, **18**, 414 (1985).

[24] A. W. Adamson, *Physical Chemistry of Surfaces*, 5th edn, Wiley, New York (1990).

[25] Q. S. Bhatia, D. H. Pan, and J. T. Koberstein, *Macromolecules*, **21**, 2166 (1988).

[26] D. G. Le Grand and G. L. Gaines, Jr., *J. Colloid Interface Sci.*, **31**, 162 (1969).

[27] J. J. Schmidt, J. A. Gardella, Jr., and L. Salvati, Jr., *Macromolecules*, **22**, 4489 (1989).

[28] D. J. Walsch and J. G. McKeown, *Polymer*, **21**, 1330 (1980).

[29] M. B. Clark, Jr., C. A. Burkhardt, and J. A. Gardella, Jr., *Macromolecules*, **22**, 4495 (1989).
[30] M. B. Clark, Jr., C. A. Burkhardt, and J. A. Gardella, Jr., *Macromolecules*, **24**, 799 (1991).
[31] M. M. Coleman and J. Zarian, *J. Polym. Sci. Polym. Phis. Ed.*, **17**, 837 (1979).
[32] M. J. Owen and T. C. Kendrick, *Macromolecules*, **3**, 458 (1970).
[33] I. Yilgor and J. E. McGrath, *Adv. Polym. Sci.*, **86**, 1 (1988).
[34] I. Yilgor, W. P. Steckle, Jr., E. Yilgor, R. G. Freelin, and J. S. Riffle, *J. Polym. Sci., Polym. Chem.*, **27**, 3673 (1989).
[35] R. L. Schmitt, J. A. Gardella, Jr., and L. Salvati, Jr., *Macromolecules*, **19**, 648 (1986).
[36] H. R. Thomas and J. J. O'Malley, *Macromolecules*, **14**, 1316 (1981).
[37] H. R. Thomas and J. J. O'Malley, *Macromolecules*, **12**, 323 (1979).
[38] J. J. O'Malley and H. R. Thomas, *Macromolecules*, **12**, 996 (1979).
[39] E. L. Thomas, D. B. Alward, D. J. Kinning, D. C. Martin, D. L. Handlin, and L. J. Fetters, *Macromolecules*, **19**, 2197 (1986).
[40] D. T. Clark, J. Peeling, and J. M. O'Malley, *J. Polym. Sci. Polym. Chem. Ed.*, **14**, 543 (1976).
[41] J. C. Saam, D. J. Gordon, and S. Lindsey, *Macromolecules*, **3**, 1 (1970).
[42] J. C. Saam and F. W. G. Featon, *Ind. Eng. Chem. Prod. Res. Develop.*, **10**, 10 (1971).
[43] D. Shuttleworth, J. G. VanDusen, J. J. O'Malley, and H. R. Thomas, Polymer. *Prep. Amer. Chem. Soc. Div. Polym. Chem.*, **20**, 499 (1979).
[44] R. G. Crystal, in *The Colloidal and Morphological Properties of Block and Graft Copolymers*, G. Molav (Ed.), Plenum Press, New York (1971), pp. 279–293.
[45] J. M. Pochan and R. G. Crystal, in *Dielectric Properties of Polymers*, F. E. Karasz (Ed.), Plenum Press, New York (1972), pp. 313–327.
[46] H. Hasegawa and T. Hashimoto, *Macromolecules*, **18**, 589 (1985).
[47] H. Hasegawa and T. Hashimoto, *Polymer*, **33**, 485 (1992).
[48] R. L. Schmitt, J. A. Gardella, Jr., J. H. Magill, and R. L. Chin, *Polymer*, **28**, 1462 (1987).
[49] N. Okui and J. H. Magill, *Polymer*, **18**, 845 (1977).
[50] N. Okui and J. H. Magill, *Polymer*, **18**, 1152 (1977).
[51] H. M. Li and J. H. Magill, *Polymer*, **19**, 416 (1978).
[52] R. L. Schmitt, J. A. Gardella, Jr., J. H. Magill, L. Salvati, Jr., and R. L. Chin, *Macromolecules*, **18**, 2675 (1985).
[53] B. D. Ranter and R. W. Paynter, in *Polyurethanes in Biomedical Engineering*, H. Planck, G. Egbers, and R. Syre (Eds), Elsevier, Amsterdam (1984).
[54] S. C. Yoon and B. D. Ranter, in *Polymer Surface Dynamics*, J. D. Andrade (Ed.), Plenum Press, New York (1988).
[55] S. L. Goodman, C. Li, J. B. Pawley, S. L. Cooper, and R. M. Albrecht, in *Surface Characterization of Biomaterials*, B. D. Ratner (Ed.), Elsevier, Amsterdam (1988), p. 281.
[56] S. L. Goodman, S. R. Simmons, S. L. Cooper, and R. M. Albrecht, *J. Colloid Interface Sci.*, **139**, 561 (1990).
[57] K. G. Tingey, J. D. Andrade, R. J. Zdrahala, K. K. Chittur, and R. M. Gendrau, in *Surface Characterization of Biomaterials*, B. D. Ratner (Ed.), Elsevier, Amsterdam (1988), p. 255.
[58] S. R. Hanson, L. A. Harker, B. D. Ratner, and A. S. Hoffman, *J. Lab. Clin. Med.*, **95**, 289 (1980).
[59] V. Sa Da Costa, D. Brier-Russell, E. W. Salzman, and E. W. Merrill, *J. Colloid Interface Sci.*, **80**, 445 (1981).
[60] S. C. Yoon and B. D. Ratner, *Macromolecules*, **19**, 1068 (1986).

[61] J. Blackwell and M. R. Nagarajan, *Polymer*, **22**, 202 (1981).

[62] J. Blackwell and M. R. Nagarajan, and T. B. Hoitink, *Polymer*, **23**, 950 (1982)

[63] J. Blackwell, J. R. Quay, M. R. Nagarajan, L. Born, and H. Hespe, *J. Polym. Sci. Polym. Phys. Ed.*, **22**, 1247 (1984).

[64] B. D. Ratner, D. Briggs, M. J. Hearn, S. C. Yoon, and P. G. Eldman, in *Surface Characterization of Biomaterials*, B. D. Ratner (Ed.), Elsevier, Amsterdam (1988) p. 317.

[65] G. L. Gaines, Jr., *Macromolecules*, **14**, 208 (1981).

[66] A. W. Adamson, *Physical Chemistry of Surfaces*, 5th edn, Wiley, New York (1990), p. 110.

[67] C. S. Henkee, E. L. Thomas, and L. J. Fetters, *J. Mat. Sci.*, **23**, 1685 (1988).

[68] G. H. Fredrickson, *Macromolecules*, **20**, 2535 (1987).

[69] P. F. Green, G. H. Fredrickson, T. P. Russell, and R. Jerome, *J. Chem. Phys.*, **92**, 1478 (1990).

[70] S. Wu, *Polymer Interface and Adhesion*, Marcel Dekker, New York (1982), p. 111.

[71] P. F. Green, T. M. Christiensen, and T. P. Russell, *Macromolecules*, **24**, 252 (1991).

[72] P. F. Green, T. M. Christiensen, T. P. Russell, and R. Jerome, *Macromolecules*, **22**, 2189 (1989).

[73] X. Chen, J. A. Gardella, Jr., and P. L. Kumler, *Macromolecules*, **25**, 6621 (1992).

[74] X. Chen, J. A. Gardella, Jr., and P. L. Kumler, *Macromolecules*, **25**, 6631 (1992).

Part IV
APPLICATIONS

Chapter 9

Wettability

Wetting is a key phenomenon in many technological fields. Gluing, painting, inking and washing are only a few examples of practical situations where a good contact between a liquid and a solid is sought. On the other hand, a further large number of technical operations require some kind of protection from too strong a liquid–solid interaction: waterproofing and anti-sticking or anti-adherent coatings are probably the best known examples. As often happens when the same topic is shared by many, apparently different and scarcely intercommunicating fields, every sector has developed its own terminology, so that different definitions of what is intended by 'wetting' exist [1]. Adamson [2] states that wetting means that the contact angle between a liquid and a solid is zero, or so close to zero that the liquid spreads easily over the solid surface, while non-wetting means that the angle is greater than 90°, so that the liquid tends to ball-up and run off the surface easily. From a surface energetics point of view, the problem is usually discussed starting from the Young equation (Chapter 4), with the usual conclusion that, in order to promote wetting, one must try to lower both the liquid surface tension and the solid–liquid interfacial tension. The best way to do that is to add a surfactant (commonly called wetting agent) to the liquid phase. While wetting agent studies are a healthy and extremely intriguing section of applied surface science, here we are more concerned with modifications of the solid to be wetted by a liquid of fixed surface tension. Stated otherwise, the variables that come into play are the solid surface tension and the solid liquid interfacial tension. A point to note, however, is that both the quoted definition of wetting and the reasoning based on interfacial energetics, suffer the same problem that we met in the chapter on contact angle measurement, that is contact angle hysteresis. Thus the knowledge that a contact angle is greater than 90° is of little help in understanding if a drop of liquid will run off a surface, as long as we do not know if it is an advancing, a receding or an intermediate contact angle. As shown in Figure 9.1(a), and according to the discussion in Chapter 4, a low receding angle can strongly pin a drop of liquid on a solid surface [3–5]. The peculiar behaviour of drops of water on composite surfaces, which, as also discussed later, is the structure of choice when it comes to waterproofing, is chiefly a consequence of the increase in the receding angle (the advancing angle is usually also very high immediately before the transition to composite surfaces, yet no freely rolling droplets are observed). On the other hand, a receding angle very close to zero does not mean that the liquid will spread on the solid: as shown in Figure 9.1(b) in this case it is the advancing angle that limits spreading. Thus, each practical situation involving wetting must be correctly evaluated and characterized by properly measured contact angles.

On a general ground, wetting can be divided into contact angle and capillary action phenomena [2]. The former involve smooth or moderately rough (that is not dominated by capillary action) surfaces. In this case wetting must be treated on the basis of interfacial energetics and contact angle hysteresis theories, as discussed before. On the other hand, on highly porous or woven surfaces, capillary action sums to contact angles in the mechanism of wetting. As discussed in basic textbooks of the physical chemistry of surfaces, a pressure difference

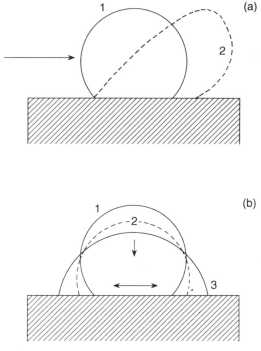

Figure 9.1 The effect of advancing and receding angles on drop motion: (a) A drop of liquid on a surface pinned by a low receding angle. Arrow indicates the direction of the applied force. (b) Spreading of a droplet is controlled by the advancing angle. Arrows indicate the direction of motion of the droplet. Numbers refer to successive time steps.

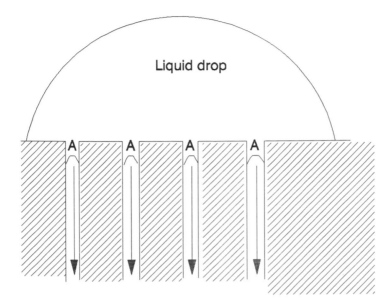

Figure 9.2 A drop of liquid on a porous surface. The liquid in A is at a lower pressure than bulk liquid (which is in equilibrium with the surrounding atmosphere) due to the curved interface caused by the acute contact angle and the small diameter of pores. Arrows indicate the direction of flow of the liquid produced by the difference of pressure.

exists across the curved surfaces of a meniscus, as described by the Laplace equation, the convex side being at a lower pressure than the concave side. Thus, as shown in Figure 9.2, the bulk of the liquid, which is in equilibrium with the surrounding atmosphere, is at a higher pressure than liquid close to the curved interface imposed by an acute contact angle (point A), giving the driving force for capillary penetration. On the contrary, if the contact angle is obtuse, the pressure difference does not allow the liquid to penetrate pores. As discussed below the shape of the pores also plays a major role in the phenomenon.

Thus, there are basically two routes towards the modification of the wetting behaviour of a given material. The first involves the modification of the chemical composition of the material surface, in order to best exploit contact angle phenomena. The second, which can, of course, be coupled with the former, involves the macroscopic modification of the physical structure of the surface. We will discuss examples of both kinds of modification below. In particular, we will discuss the interaction of polymer surfaces with water or, following the general usage, the hydrophilization or hydrophobization of polymer surfaces (other chapters will implicitly discuss wetting of solids by other fluids, for instance adhesives in the chapter on adhesion or biologic fluids in the chapter on biocompatibility). Water is surely the most widely diffuse solvent or liquid medium in technological operation, so that a great deal of work has been done on this topic. The need for surface treatment of water-contacting polymers is shown in Figure 9.3, where the calculated surface tensions of most common polymers, taken from ref. [6], are plotted as well as the measured surface tension of water. It is clear that most polymers fall in the region between the strongly hydrophobic poly(tetrafluoroethylene) (PTFE) and poly(dimethylsiloxane) (PDMS) and water, whose surface tension, as a first order approximation, ion a critical surface tension perspective (Chapter 4), can be considered to be the minimum required solid surface tension for the spreading of water. Taking into account contact angle hysteresis phenomena worsen the picture, since it adds the problem of a lower than expected (from a surface tension point of view) receding angle. The result is that most common polymers are neither hydrophobic nor hydrophilic enough for the huge number of applications involving either the quick rolling off or the spreading of water droplets. The objective of surface modification techniques in this context is, therefore, to shift polymers from the middle towards the extremes of the graph of Figure 9.3. The best way to improve hydrophilicity is to allow water to engage its preferred interactions with the solid, i.e. the short-range hydration, acid-base and hydrogen bonding interactions discussed in Chapter 1 [7]. This can be done with surface modification techniques that introduce hydrophilic groups on the substrate surface. When it comes to hydrophobicity, not only must these interactions be prevented, but also the residual van der

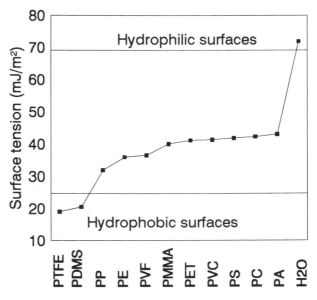

Figure 9.3 Surface tension of common polymers and water.

Waals' interaction must be minimized. As Figure 9.3 shows, this is best done with the introduction of fully fluorinated or methyl groups.

9.1 HYDROPHILIC SURFACES

9.1.1 General Comments

The principle underlying the hydrophilization of polymer surfaces has been described in Chapters 1 and 4, while the modification techniques which can render a polymer surface more hydrophilic have been described in the chapters of Part III. Here, we are more concerned with applications: instead of reviewing the details of the different fields where polymeric hydrophilic surfaces are employed (a tremendously long and difficult task) we will discuss some general problems and aspects of hydrophilized polymer surfaces. Hopefully, this approach can give some help in the difficult choice of the more suited surface modification techniques for a given problem.

In general, all techniques of hydrophilization of polymer surfaces involve an increase of the surface amount of the groups which are generally called polar or, using the more modern approaches discussed in Chapter 4, of the Fowkes acid–base or the Good–Van Oss–Chaudhury electron donor–electron acceptor groups [8–13]. This topic can be discussed from several different points of view:

- From a technological point of view, all of the surface modification techniques described in Part III can be and have been used to render the surfaces of polymers more hydrophilic: each of them has its own merits and shortcomings, in terms of efficiency, costs and range of applicability;
- From a macroscopic physico-chemical point of view, the goal is to exploit as much as possible of the polar (or acid–base) component of the surface free energy of water. As shown in Table 9.1, water, because of its molecular structure, is a peculiar liquid, since the non-dispersive component of its surface tension is more than twice the dispersive one, unlike other common liquids.
- From a microscopic point of view, the basis of surface hydrophilization is to maximize hydration and hydrogen bonding interactions (Chapter 1).

Table 9.1 Surface tension component (mJ/m^2) of some common liquids (data from ref. [6])

Liquid	γ_p	γ_d	γ_p/γ_d
Water	50.2	22.6	2.22
Glycerol	22.8	40.6	0.56
Formamide	22.2	36.0	0.62
Methylene iodide	1.8	49.0	0.04

Hydroxyl, carbonyl, carboxyl, and carboxylate groups contain lone pairs, unshared electrons and asymmetric charge distributions. All sorts of oxygen, nitrogen or sulphur containing organic functional can interact with water more effectively than common carbon-based repeating units. This trend is clearly shown in Figure 9.3, which, at the same time, shows, as already discussed, that even common oxygen containing polymers do not have a surface concentration of hydrophilic groups high enough to promote complete wetting by water. This point highlights the importance and the need of surface modification techniques.

The previous discussion underlines a fundamental aspect of surface hydrophilization: the introduction of surface hydrophilic groups (or the maximization of the water–surface interaction) leads necessarily to an increase of the surface free energy (or the polymer–air interfacial free energy) over that of the parent polymer

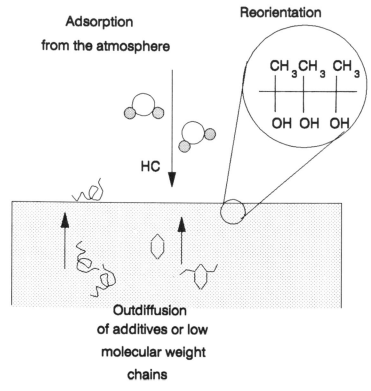

Figure 9.4 Schematics of several of the mechanisms which can reduce the solid surface tension (or, better, the interfacial tension against the surroundings).

(Figure 9.3). This means that, when air is the interfacing medium, a drawing force exists to restore the original surface composition or, in general, to lower the surface tension of the treated material. This can be accomplished by a variety of means, as shown in Figure 9.4: adsorption from the atmosphere of low energy contaminants, out-diffusion of oligomers of hydrophobic finishing agents and additives, reorientation of macromolecule side chains or pendant groups. The effect of these different mechanisms, which leads to a lowering of the surface tension and a decrease of the hydrophilicity imparted by a surface treatment, is collectively known as ageing or hydrophobic recovery. Ageing of hydrophilized surfaces is a matter of great concern in technological applications of polymers. Several different reactions make this class of materials particularly subject to effective ageing:

- Common polymers usually contain additives and finishing aids, often designed to bloom to the surface.
- All common polymers contain a range of molecular weights, including a fraction of low molecular weight oligomers, which can easily diffuse. This aspect involves both the diffusion from one polymer to another (contamination of a surface treated item from packaging material) or out-diffusion from bulk towards the surface. The last point is especially important since, as previously discussed, many surface modification techniques affect only a very thin layer (fractions of μm). Thus, the modified layer to be crossed is thin enough to show ageing effects owing to rapid diffusion.
- The last points are connected to the dynamics of polymer surfaces (Chapter 2). Many polymers operate above their T_g, or above the critical temperature of some side-group or lateral chain relaxation [14]. Moreover, as discussed in Chapter 2, there is now considerable evidence that surfaces are regions of enhanced mobility as compared to bulk. This means that kinetics accompanies the thermodynamic thrust to move towards a more stable state. Surfaces of more rigid materials, such as ceramics and metals, have much less freedom in normal time–temperature conditions, and the only way they can lower their surface energy is through the adsorption of low surface tension molecules from the atmosphere [15].

Owing to the importance of ageing effects in the practical application of surface modification techniques, the next section will be devoted to this topic. In particular, recent studies on hydrophobic recovery of treated surfaces will be reviewed. The discussion will be devoted to plasma treated surfaces since, due to the high efficiency of the treatment and the small modification depth, these systems are examples of extreme surface unbalance. Anyway, the same general reasoning can be applied to systems produced by other surface modification techniques, as shown, for instance, by the effect of heat treatment on sodium-etched fluorocarbon surfaces described in Chapter 7.

9.1.2   Hydrophobic Recovery of Plasma Treated Surfaces

The dynamic behaviour of macromolecules differs from that of more rigid
materials, such as ceramics and metals: the basic components, that is the
macromolecule backbone, side chains and lateral groups, have enough mobility,
even at room temperature, to yield to external stresses, as documented by the
extensive literature on the dynamic properties of macromolecules. As discussed in
Chapter 2, this is true also for surfaces, even if the application of the dynamic
approach to polymer surface science is much more recent [16, 17].

In the present section we are concerned with the response of the polymer
surface to the stress which arises from the peculiar condition of the outermost
layers of the material, which are subjected to an asymmetric intermolecular field
of forces, quantitatively expressed by the surface (interfacial) tension. This
problem is particularly important in macromolecules containing both hydrophilic and hydrophobic groups (amphipathic polymers) since the contemporary
presence of moieties of widely different properties magnifies the possibilities of
thermodynamic unbalance. Polymers subjected to hydrophilizing surface treatments are an important class of amphipathic materials. For instance, oxidizing
plasma treatments can introduce more than 20% atomic concentration of oxygen
on the first few nanometres of a hydrophobic polymer such as polyethylene (PE).
The surface of the just treated PE is completely wettable by water, which means a
more than twice increase in the surface tension of the solid (or, more properly, of
its interfacial tension against air). If the solid is interfaced with air, a thermodynamic push towards the burying of the hydrophilic groups introduced by
the treatment exists. The effect of this driving force depends on material related
properties, such as the nature of the parent polymer and the modified surface or
the kind and concentration of additives and external conditions, such as time and
temperature. Figure 9.5 shows the typical behaviour of advancing water contact
angles of an oxygen plasma treated polypropylene (PP) aged in air [18]: the
hydrophilizing effect of the plasma treatment is lost in a few days ageing in air at
room temperature, while it is maintained if liquid N_2 temperature ageing slows
down macromolecular dynamics (incidentally, this observation rules out the
hypothesis that ageing is due to the adsorption of ubiquitous hydrocarbons from
the atmosphere, a mechanism which would yield an opposite dependence on
ageing temperature).

The complex nature of the phenomena involved in ageing of plasma treated
polymers can be appreciated by the comparison of the ageing behaviour of three
hydrocarbon-based polymers: PP, PE and poly(styrene) PS. As shown in Figure
9.5, oxygen plasma treated PP aged in air at room temperature shows a complete
recovery of the advancing angle (the receding angle, however, does not fully
recover, suggesting that oxygen containing polar groups remain on the restruc-
tured surface [18]). Fixed angle XPS does not detect any modification of the
surface composition upon ageing [18], showing that recorganization goes on in a

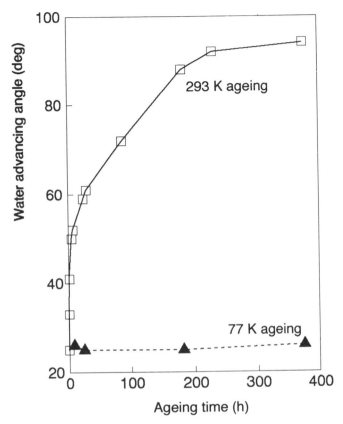

Figure 9.5 The effect of ageing at two different temperatures on the water advancing angle of oxygen plasma treated poly(propylene).

very thin layer, thinner than XPS sampling depth (about 6 nm). On the other hand, static SIMS (SSIMS, Chapter 3), whose sampling depth is comparable with that of contact angle measurement, reveals a modification of the surface upon ageing, as shown in Figure 9.6 [19]. To make the analysis more treatment-specific, a $^{18}O_2$ plasma was used. Figure 9.6 shows the plot of the ratio of the peak at 13 amu (CH^-) to that at 18 amu ($^{18}O^-$). As the ageing time–temperature increases, the ratio of hydrocarbon fragments (CH^-) to hydrophilic fragments ($^{18}O^-$) increases.

What kind of mechanism is responsible for the observed ageing behaviour? The experimental findings previously discussed suggest that hydrophobic recovery of oxygen plasma treated PP goes on through rearrangement within the modified layer, that is through short-range reorientation of side groups. If the time needed to reach the advancing angle typical of untreated PP is taken as a

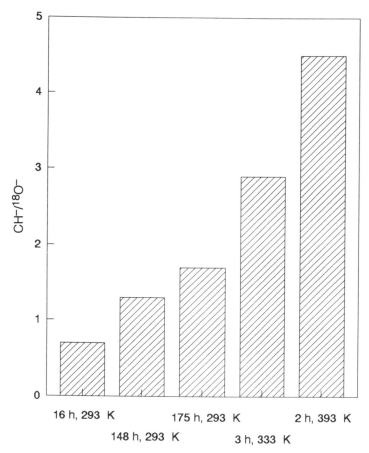

Figure 9.6 The effect of ageing on the ratio of the intensity of $CH^-/{}^{18}O^-$ peaks detected by SSIMS.

reference time, it is possible to calculate an apparent activation energy by an Arrhenius-like equation through the study of the temperature dependence of recovery. Even if the assumptions involved in such a calculation must be taken into account (the relationship between the advancing angle and the surface composition of a non-ideal surface is still an open question, as discussed in Chapter 4), it is satisfactory to notice that the calculated value of 58.1 kJ/mol [19] is of the same order of magnitude as the activation energy of the bulk β relaxation of an oxygen containing polymer(poly(vinylacetate)), as calculated by Pennings and Bosman [20].

Another interesting observation is that the final stage of recovery is not a pure PP surface, but one which still bears oxygen containing moieties, as shown by the value of the receding angle of water [18] and by the SSIMS results shown in

Figure 9.6 [19]. This observation suggests that two opposite pushes are operating: on one side, minimization of the surface free energy tries to build up a homogeneous hydrocarbon-like surface. On the other, this mechanism means the disruption of hydrogen bonding interactions between oxygen containing surface groups, with the loss of interaction energy. The final reorganized surface is the best compromise between these two opposite mechanisms, and it is interesting to note that the value of the fully recovered receding angle tends to increase as the ageing temperature increases [18], suggesting that, with a greater thermal energy input, more hydrogen bonds can be broken. This hypothesis was theoretically tested with good results, by modelling the surface layer with a random copolymer of PP and PP with hydroxymethyl groups [21].

Contrary to PP, high density PE (HDPE) treated in the same conditions and in the same reactor, shows a very small hydrophobic recovery [22], even at 393 K ageing temperature. It must be noted that other authors observed hydrophobic recovery also on plasma treated PE [23]. This is not unexpected, since, beside materials related differences, the effect of plasma is very sensitive to the treatment conditions and reactor design (Chapter 6). Thus, for the sake of comparison between different materials subjected to the same treatment in the same apparatus, we will discuss our results.

A possible explanation of the striking difference between the closely related PP and HDPE arises from their completely different behaviour when subjected to irradiation [24, 25]: the former preferentially undergoes chain scission, while the latter is cross-linked. However, even if extensive cross-linking can be associated with low mobility, literature data suggest that the behaviour of PE cannot be accounted for by cross-linking alone, as discussed in the papers on hydrophobic recovery of oxygen plasma treated, plasma deposited polymeric films [26–29]. It is then important to note that a major difference exists between the amount of oxygen introduced by O_2 plasma treatment, in the conditions of the quoted works [18, 22], in PP or HDPE surfaces: XPS analysis shows that the former has an O/C ratio of 0.19, the latter of 0.13. Thus, it seems very likely that the coupling of a high degree of cross-linking with a large amount of oxygen-containing groups (interacting by hydrogen bonding) is necessary to effectively hinder recovery. We will return to this point after the discussion of the ageing behaviour of the third hydrocarbons-based-O_2 plasma-treated polymer, that is PS.

PS is an interesting subject for recovery studies: it is readily available as standard with a sharp distribution of molecular weights: and the shake-up peak associated with the aromatic ring of the repeating unit allows an easy interpretation of the C1s peak obtained by XPS analysis.

The effect of the polymer molecular weight on the ageing behaviour of five different fractions of PS, O_2-plasma treated in the same conditions of the previously described HDPE and PP, is shown in Figure 9.7 [30–32]. At low ageing temperature, all the samples reach the same limiting value of the advancing angle, irrespective of the specific molecular weight. Also the kinetics of the

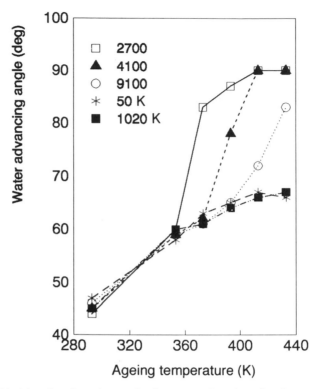

Figure 9.7   Limiting the advancing angle of water as a function of ageing temperature and molecular weight of oxygen plasma treated PS.

process, that is the time taken to reach the limiting value, is unaffected by the molecular weight. Starting from 373 K ageing temperature, however, low molecular weights exhibit a distinct behaviour and the limiting value of the advancing angle, as well as the time course of the process, become molecular weight dependent. At 413 K ageing temperature, the advancing angle of PS_{2700} and PS_{4100} is equal to the advancing angle of untreated PS, while the receding angle, not shown, is lower. At the same temperature, PS_{9100} is beginning to deviate from the behaviour of the two heavy weights, which also follows a common path at higher temperatures. Table 9.2 shows the XPS detected surface composition of the samples: at low ageing temperature, the increase of contact angles occurs without modification of the surface composition, as in the previously discussed case of PP. At higher temperatures, the O/C ratio starts to decrease for lighter samples. Accordingly, the deconvolution of the C1s peak shows a decrease of higher binding energy components, while the shake-up peak, which disappears after treatment and low temperature ageing, becomes observable again.

Table 9.2 XPS surface composition (at. %) of recovered, oxygen plasma treated PS as a function of molecular weight and ageing temperatures

M_w	Element	Ageing temperature (K)					
		293	353	373	393	413	433
2 700	C	79.5	80.3	84.8	86.0	93.8	93.6
	O	20.5	19.7	15.2	14.0	6.2	6.4
4 100	C	80.2		80.5		89.0	90.4
	O	19.8		19.5		11.0	9.6
9 100	C	78.7		79.8		84.9	85.4
	O	21.3		20.2		15.1	14.6
50 000	C	80.8				80.5	81.1
	O	19.2				19.5	18.9
1 020 000	C	80.4	78.6	78.9	80.0	81.0	80.8
	O	19.6	21.4	21.1	20.0	19.0	19.2

The previous results point to two different recovery mechanisms: the first one, as already observed on PP, involves short-range segmental motion within the modified layer, leading to burial of polar groups away from the surface. Since this mechanism involves only short-range reorientation of side groups, it should not be affected by the molelcular weight of the treated polymer, as actually observed. The second mechanism involves long-range motion and out-diffusion of un-treated molecules, which is much more demanding in terms of activation energy. Accordingly it occurs only above 373 K and exhibits a strong dependence on the molecular weight. Another interesting point is the dependence on cross-linking of the two different mechanisms. By a solubilization test in tetrahydrofuran (THF) the insoluble fraction was found to increase with increasing discharge power (from 20 to 150 W) [31]. Contact angle measurements showed that, while only minor effects were observed in the short-range recovery mechanism, the onset of recovery by long-range mobility was lowered, both in terms of temperature and molecular weight, the lower the cross-linking. This demonstrates that, as pointed out by Tead and coworkers [33, 34], cross-linked layers are effective obstacles to macromolecular diffusion.

An interesting observation involves the XPS surface concentration of oxygen and the degree of recovery (expressed as the ratio between the advancing angle after full recovery minus the advancing angle immediately after treatment and the advancing angle of the untreated sample minus the advancing angle immediately after treatment). In Figure 9.8, the degree of recovery is plotted as a function of the XPS surface composition in the case of ageing of PP, HDPE and PS (data are taken from the previously quoted papers). Taking into account that the advancing angle is a crude indication of recovery (as discussed in Chapter 4, it is sensitive to the most hydrophobic part of the surface and in the case of PP, the receding

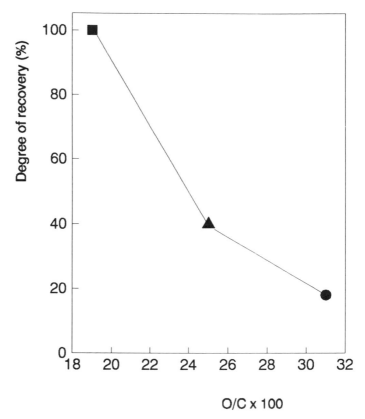

Figure 9.8   Degree of recovery as a function of the surface O to C ratio (from XPS) for PP (square), PS (triangle) and HDPE (circle).

angle shows that recovery is not 100%), it is undeniable that a definite relationship exists between the surface concentration of oxygen and the degree of overturning allowed to the hydrophilic groups introduced. These results confirm that the interactions between surface polar groups play a role in the control of short-range reorientation at least as important as cross-linking.

Besides hydrocarbon based polymers, studies on the mechanism of hydrophobic recovery have been performed on siloxanes [35],. fluoropolymers [36–38], polycarbonate [39], poly(etheretherketone) (PEEK) [40, 41], poly (ethyleneterephthalate) [42].

All of these studies underline the complexity of ageing phenomena. For instance, in an interesting paper Griesser and coworkers [36] suggest that recovery of plasma treated fluorinated ethylene-propylene copolymers (FEP) involves, besides short-range reorientation, also subsurface oxidation processes, starting from a radical entrapped in the subsurface zone after plasma treatment.

According to the principles underlying ageing, polymer surfaces hydrophilized by oxidizing treatments should not decay and air equilibrated surfaces should become hydrophilic again when stored in water or high energy media. The technological recognition of this phenomenon is described in some recent Japanese patents [43, 44] while experimental observations have been performed, besides plasma treated polymers [19, 28, 29, 35], also on oxidized PTFE obtained by sputtering from a PTFE target [45] and on chemically oxidized PE [46]. On the other hand in the quoted paper of Munro and McBriar, hydrophobic recovery was also observed when oxygen plasma treated samples were stored under water [40].

The reason for this different behaviour is not completely clear. In principle, there is no reason why a hydrophobic interface should be created in a purely physical process where the only variable is the interfacial free energy between the solid and the liquid phase. But, as shown by the quoted studies, events in plasma treated polymers are controlled by a large number of different interactions, both chemical and physical, involving modified chains and the environment, surface and subsurface molecules and interactions among the treatment-introduced functional groups.

Finally, it must be noted that the previous studies were performed on laboratory grade materials. With real polymers, additives and finishing agents can create further degrees of freedom to the evolution of the surface composition and properties of hydrophilized surfaces.

9.1.3 Hydrophilic Surfaces and Very Low Contact Angle Surfaces

The previous discussion raises some interesting points. The most fundamental of which concerns the definition itself of hydrophilic surface: the O_2 plasma-treated PP surfaces whose advancing water contact angles are shown in Figure 9.5 have the same overall composition, as detected by fixed angle XPS (about 16% O and 74% C) [18]. Yet, as judged by contact angle measurement, the as-treated or liquid N_2 temperature aged surfaces would be termed hydrophilic, while the room temperature aged surface would be called hydrophobic.

The same problems of definition are met when one considers the poly-(hydroxyethylmethacrylate) (PHEMA) surface, whose behaviour was discussed in Chapter 4. Despite the approximate 40% equilibrium water content [47], which indicates the hydrophilic nature of this polymer, the advancing water contact angle is something less than 80°, due to the preferential exposure of the low energy methyl group at the solid–air interface [48].

The most elegant example of such a dilemma is probably given in the work of Bain and Whitesides on self-assembled organic monolayers [49–51]. These authors have used monolayers of long chain thiols adsorbed on gold to mimic organic surfaces. One of their findings, adapted from ref. [51], is shown in Figure 9.9: pure monolayers of $HS(CH_2)_{11}$ OH and $HS(CH_2)_{19}OH$ show an

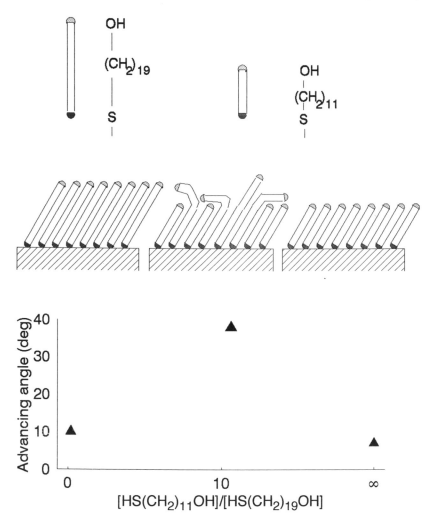

Figure 9.9 Schematic illustration of monolayers of hydroxy-terminated thiols on gold. In the mixed monolayer shielding of hydroxyls by methylene groups occurs. In the graph the advancing contact of water for the three different situations is reported. The abscissa represents the ratio of the concentrations of the thiols in solution. (Adapted from ref. [51].)

advancing water contact angle lower than 15°, as expected from a surface exposing hydroxyl groups. On the other hand, in a mixed monolayer, the last eight carbons of the longer chains can form a disordered liquid-like layer on top of the densely-packed lower phase of the monolayer. This disordered region exposes polymethylene chains to the water droplet, as shown in Figure 9.9, and the advancing water contact angle shows a sharp increase.

The three quoted examples underline basic aspects of the wettability/ hydrophilicity relationship: due to the nature of the forces involved in the process (Chapters 1 and 4), the shape of a drop of water on a surface is controlled by the very top of the surface of the material. In order to obtain a very low ($< 15°$) contact angle, not only must hydrophilic groups be present, but also the details of the molecular conformation on the surface must be favourable: this is shown by the strong dependence of the contact angle on aged, plasma-treated surfaces on short-range recovery mechanisms and by the increase on the water contact angle on the mixed hydrophilic monolayers of Figure 9.9.

The consequence of this attitude is that an important difference exists between hydrophilic surfaces and very low contact angle surfaces: the former can adsorb some water or moisture from the surroundings. Their surface and subsurface zone can be extensively hydrated, yet the water contact angle can be similar to that of hydrophobic polymers. A typical example is given by the previously quoted PHEMA. These surfaces can be fully composed by molecules exposing hydroxyl groups, yet a moderate shielding effect, as shown in Figure 9.9, can produce a higher than expected contact angle.

On the other hand, very low contact angle surfaces require not only surface groups which can interact with water, but also all shielding or short-range hydrophobizing effects should be prevented.

From a practical point of view, the above difference is extremely important and should always be considered when selecting the most proper surface modification technique. For instance, static charge accumulation on polymer surfaces can be reduced by hydrophilization of the surface, since the adsorbed moisture can carry away surface charges [52]. Thus, the control of antistatic behaviour by surface modification requires a hydrophilic and not necessarily a very low contact angle surface. Figure 9.10 shows the comparison between surface composition, advancing water contact angle and antistatic behaviour of poly(methylmethacrylate) (PMMA) and PMMA coated by 200 nm of plasma deposited PHEMA [53]. The samples were triboelectrified by rubbing against a wool cloth. The great improvement in charge dissipation is unexpected from contact angles alone: the former is controlled by the surface and subsurface hydrophilicity, which allows moisture adsorption and improves electrical conduction. The latter by the short-range effects of the interaction between the drop of water and the macromolecular moieties (incidentally, receding angles of PMMA and PHEMA-coated PMMA are very different. As discussed in Chapter 4, the receding angle is much more sensitive to hydrophilic groups. This points again to the need for measuring both advancing and receding angles [54]).

On the other hand, several applications definitely require a very low advancing water contact angle, such as anticlouding or antifogging of transparent plastics [39]. Dropwise condensed water gives a cloudy appearance to transparent polymeric sheets, due to light diffusion by water droplets. The amount of transmitted light is severely reduced, a problem of great concern, for instance, in

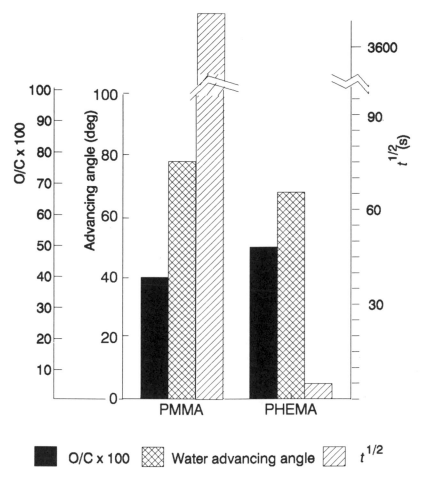

Figure 9.10 Comparison of the surface O to C ratio (from XPS), the advancing water contact angle and the time taken to reduce to one half the surface potential induced by triboelectrification ($t^{1/2}$) for PMMA and PMMA coated with 200 nm of PHEMA deposited from plasma.

greenhouses. In these cases short-range mobility can have a detrimental effect on properties, as displayed in Figure 9.11, where the effect is shown of the previously discussed ageing behaviour of plasma-treated PS on clouding resistance [39]. Here the problem is completely controlled by the structure of the topmost molecular layer.

The previous examples clearly show that no general recipe exists on the best way to hydrophilize polymer surfaces. Depending on the peculiar application, each of the surface modification techniques described in Part III can be suitable or not. Practical applications involve many more variables than the few we

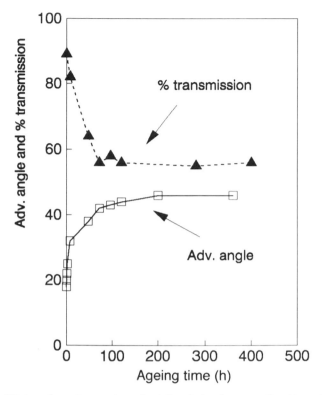

Figure 9.11 Water advancing angle and misting behaviour as a function of ageing time for oxygen plasma treated PS.

quoted. For instance, the hydrophilized surface could also be subjected to wear, soil, and contact with other solids. In some instances ageing can be a problem, while in other cases either the lifetime of the recently treated surface structure is long enough or the fully recovered surface maintains the amout of hydrophilic groups needed.

9.2 HYDROPHOBIC SURFACES

9.2.1 General Comments

As shown by Figure 9.3 and by everyday experience, all common synthetic polymers are, in a general sense, hydrophobic, since a drop of water does not spread over their surfaces. However, in many practical applications, it is important not only to avoid spreading, but also to promote a ready flowing away of water. Thinking again of everyday experience, this condition is fulfilled by a far lower number of polymers.

As discussed in the introduction to this chapter, many definitions of what is intended by hydrophobic surfaces exist, the most commonly quoted being that of a greater than 90° contact angle [2]. In this respect, it is important to remember that definitions based on an equilibrium contact angle are often of little help in practical applications of polymer surface science, since practical applications must necessarily face the widespread occurrence of contact angle hysteresis. This problem is particularly important in water repellency since, as shown by Figure 9.1 (a) and by the discussion in Chapter 4, it is the receding angle which limits the flowing away of a drop from a surface. Thus, rather than giving general and vague definitions, it is better to underline the need to evaluate carefully the role played by the advancing and receding angle in a given water–surface interaction.

From a basic point of view, the first requirement for hydrophobization is to prevent water from establishing its peculiar interactions. This is, of course, the opposite of what was discussed in the previous section: thus, groups able to engage hydrogen bonding with water should not be present on hydrophobic surfaces. A less obvious requirement is to minimize the other component of the interaction, that is the Lifshitz–van der Waals or dispersive one (Chapters 1 and 4). As discussed in Chapter 1, the extent of the interaction force arising from electrodynamic phenomena (Debye + Keesom + London) can be evaluated by the Hamaker constant, which arises from the details of the atomic structure of interacting bodies. Thinking of hydrophobicity, the problem can be stated as the minimization of the Hamaker constant between water (w) and the material (m) interacting in air (a). Using the combining relation, the latter can be written as [55]:

$$A_{wma} = (\sqrt{A_w} - \sqrt{A_a})(\sqrt{A_m} - \sqrt{A_a}) \tag{9.1}$$

where A_i are the individual Hamaker constants. It is clear that, since the Hamaker constant of air is zero, in order to minimize A_{wma} the Hamaker constant of the polymer should be as small as possible. Polytetrafluoroethylene (PTFE) is, of course, the synthetic polymer with the smallest Hamaker constant (3.8×10^{-20} J [55]), followed by hydrocarbon-based polymers. The minimization of Lifshitz–van der Waals' forces and the lack of specific water–polymer interactions leads to an advancing water contact angle of about 114° on PTFE [2] (the receding angle, on carefully prepared surfaces, is no more than five degrees lower).

Besides material related effects, hydrophobization of surfaces can also take advantage of morphological effects. As discussed in Chapter 4, when roughness exceed a given limit, water can no longer penetrate the cracks and crevices, owing to the constraints imposed by the local contact angle, and composite surfaces are created. In this case, the drop of water must largely face air, both in the upper and in the lower side of the drop. It follows that the apparent contact angle against the solid must approach 180°, with a very small difference between the advancing and the receding angle (Figure 4.8 and 4.9). Such very high contact angles make drops of water roll on the surface as freely as mercury drops on glass.

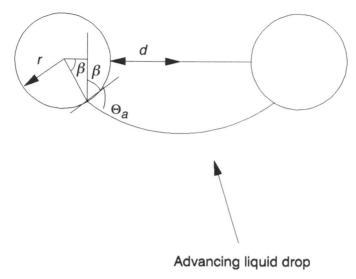

Advancing liquid drop

Figure 9.12 Penetration of liquid into pores (explanation in the text).

The last observation is also of great importance from a practical point of view: porous, open structures are very effective in increasing the contact angles (both advancing and receding) and imparting water repellency. On the other hand, the open structures allow gaseous exchange (transpiration, breathing etc.) through the pores. Thus, it is not strange that porous or textured or woven surfaces are commonly found, both in nature and in manmade articles, when it comes to hydrophobization.

The importance of the design of porous structures can be understood by the following example [57]: Figure 9.12 shows a section across a small ring made of a yarn of circular section (radius r), bent into a circle and joined at the ends. The radius of the central hole is d at its narrowest point. The advancing angle of the liquid with the yarn is $< 90°$. In this case, the liquid penetrates beyond the narrowest point of the pore, but there is a point where the combined effect of the angle of curvature of the yarn (β) and the liquid–solid contact angle give rise to a meniscus concave upward. According to the Laplace equation ([2], Figure 9.2), in this case an over-pressure is required in order to drive the liquid through the hole. The over-pressure $(P - P_0)$ required to drive the liquid through the hole of the ring is given by [57]:

$$P - P_0 = (2\gamma\cos(\Theta_a + \beta))/(r + d - r\cos\beta) \qquad (9.2)$$

Only if Θ_a is zero will the liquid go right through the hole with no external pressure to drive it. Thus, in principle, a proper design of the texture prevents, even with a small contact angle, the spontaneous passing of the liquid through the holes in a woven fabric.

More in general, equation (9.2) shows that resistance to penetration increases for small values of r and d, that is fine yarns and close weaves. It increases also for high contact angles, which again underlines the importance of the combined effect of materials related and structure related aspects.

Referring to the previous example, it is also important to note that the advancing angle is important for the initial entry in the pores. If a sudden, high pressure allows water penetration, a channel of water will be established, which will remain unless the receding angle is finite and as large as possible.

It must be noted that the terms water repellent and waterproof are used loosely, often to describe the same effect. Strictly speaking, waterproofing means to render a fabric completely impervious to water, either in the liquid or vapour phase. Thus, it depends on the application of a continuous film and film continuity provides water resistance. The coating materials used for waterproofing are applied to fabrics as hot melts (waxes or synthetic resins), solvent solutions or aqueous latexes. On the other hand, a water repellent body allows the gaseous exchange of air or water vapour, but does not allow water penetration. Water repellent fabrics are composed of open, porous structures, which are made by hydrophobic or hydrophobized fibres or materials.

In summary, the hydrophobization of polymer surfaces can exploit, from a strict surface physico-chemical point of view, two different structures: the first one is based on fluorinated groups, while the second one involves hydrocarbon groups. The latter can be provided by hydrocarbon polymers (natural or synthetic), by organometallic complexes of metal ions and long-chain carboxylic acids, by long-chain hydrocarbon compounds bearing a polar head or by siloxane backbones containing pendant hydrocarbon groups. The need for a molecular structure more complex than the simple hydrocarbon chain, arises from the search for hydrophobizing agents with improved interaction with the substrate, resulting in a better and more stable effect. In fact, an effective agent must always cope with two interfaces, one with the surrounding medium and the other with the substrate. While fluorinated or hydrocarbon groups can satisfactorily give an answer to the first requirement, the second one must be addressed by other functional groups present in the molecule and on the surface or by a proper surface modification technique. In general, the stronger the interaction with the substrate, the more lasting will be the effect. This aspect is, of course, shared by many other technological fields involving surface modification. However, in surface hydrophobization it plays a very important role, since, by definition, water-proofed or water repellent objects must come into contact with an aqueous phase, which may wash off the surface layer and reduce or destroy its hydrophobicity. Another critical stage is that of cleaning, which can also require high temperature treatments. The search for durable hydrophobizing agents is an interesting topic of polymer surface science. Several aspects of the materials involved in this field will be discussed in the next sections.

The main object of the discussion, as already evident from the previous description, will be repellents used on textiles. The main reason for this choice is that we are concerned with hydrophobization of polymer surfaces, and polymeric fibres employed in clothes, tents, carpets and so on are by far the greatest application of hydrophobization of polymers. However, it must be noted that polymeric repellents are used on a wide range of other organic or inorganic materials and these applications constitute a very large market. Besides waterproofing of paper (which can be somehow assimilated to the waterproofing of textile cellulosic fibres) the hydrophobization of soils and building materials is probably the best example of this large and intriguing field. The importance of repellent in this area can be immediately realized with the knowledge that inorganic materials have, in general, high energy surfaces [15] and thus are hydrophilic. Moreover, they are often porous and the combined effect of these two features makes them vulnerable to capillary penetration. Everyone has faced some adverse effect of the water sensitivity of common building materials.

Another interesting application of waterproofing to inorganic materials is that of dropwise condensation of steam, a problem of great relevance in heat transfer applications.

Returning to polymer surfaces, it must be noted that several thousands of documents on textile water repellents are reported in the *Chemical Abstracts* from 1970 to 1990. Most of them are patents, whose main goal is, obviously, to obtain or avoid a monopoly rather than to inform competitors of the details of one's own discovery and advancements. A thorough discussion of the theoretical background and patent literature on water repellency to 1963 is given in reference [58].

9.2.2 Hydrocarbons

Naturally occurring hydrocarbon based materials such as waxes, oils and rubber latex were probably the first water repellent materials used by man [59]. The discovery that insoluble metallic salts could improve the effect led to more refined processes. The combined effect of aluminium salts and waxes has been largely-exploited in water repellent treatments of cellulosic fabrics. A one bath wax/aluminium salt, combined in a stable emulsion, gives very good water repellency, but the product is easily removed either by washing or drycleaning. A certain improvement is given by the replacement of aluminium with zirconium salts [60]: the salts of the latter with fatty acids are more hydrophobic than the corresponding aluminium salts. Moreover, zirconium does not form compounds with alkalis and provides an improved wash fastness. Anyway, in both cases no direct bond between the hydrophobizing agent and the substrate exists, so that fastness is not completely satisfactory. Once widely used owing to their low price and ease of application, their share of the market is today decreasing [61].

Academic work on cellulose chemistry was probably at the origin of the idea that cellulosic fabrics might be made water repellent by chemical reaction of hydrophobic compounds with the surface groups of cellulose. The most interesting expected side-effect is of course durability to washing and dry cleaning, brought about by the covalent bond between the water repellent and the substrate. This kind of work must face several of the aspects described in Chapter 7 on the chemical modification of surfaces: on one hand, surface modification is made easier by the presence of reactive groups on the substrate. In this respect cellulosic or protein based natural fibres can offer many different reactive surface groups. On the other hand, reactions are planned following the general rules of solution organic chemistry, and no knowledge exists of how far these rules can be extended to reactions involving a liquid and a surface phase.

One of the first water repellents developed to produce a chemical bond with a specific group of the substrate (cellulosic fabrics) was the stearamidomethyl-pyridinium chloride. This compound is applied to cotton or other cellulosic fibres, and cured at about 150 °C [59]. The curing stage brings about the decomposition of the quaternary compound, and the main end product is probably the methylol derivative of methylene distearamide. Probably, chemical combination takes place between the cured compound and cellulose through an ether linkage. It is, however, likely that the major portion of the end product after curing is not covalently bonded, but rather forms a strong physical attachment which gives a very high resistance to washing and dry-cleaning.

Today, most pyridinium compounds have been removed from the market, because of the toxicities of the intermediates used and the coming of age of better repellents (fluorocarbons). Anyway, the quoted compounds (marketed as Velan PF by ICI and later as Zelan A by Du Pont or Cerol WB by Sandoz) are an important milestone in surface modification techniques. They stimulated research work in the coupling of naturally occurring surface groups and manmade compounds. A product of this research effort was an aqueous dispersion of octadecylethylene urea, whose suggested mechanism of reaction is shown in Figure 9.13. The behaviour of this compound is not completely understood.

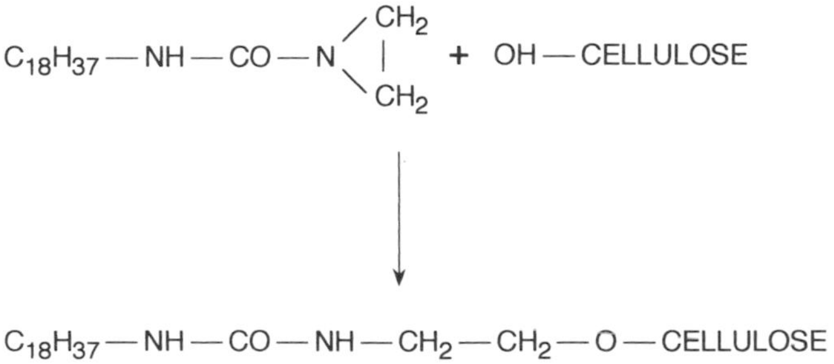

$$C_{18}H_{37}-NH-CO-N\begin{array}{c}\diagup CH_2 \\ | \\ \diagdown CH_2\end{array} \quad + \quad OH-CELLULOSE$$

$$\downarrow$$

$$C_{18}H_{37}-NH-CO-NH-CH_2-CH_2-O-CELLULOSE$$

Figure 9.13 Suggested mechanism of reaction of a repellent based on octadecylene urea with cellulose.

Interestingly, both the quoted agents are less effective (either in immediate hydrophobicity or fastness) on synthetic fibres, showing that surface chemical reaction provides at least a contribution to the overall result [62].

Another interesting product is octadecyl isocyanate, which couples the reactive isocyanate group with the hydrophobic long-chain hydrocarbon [63]. In general, these products were used as part of more complex formulations in a durable system. The role played by the long-chain isocyanate was the increase of the standard of water repellency, as well as durability to washing, dry-cleaning and weathering [59].

9.2.3 Silicones

Silicone-based water repellents were introduced on the market in the 1950s. Today, they are second only to fluorochemicals in terms of the volume used as textile repellents.

From a wettability point of view, the use of polymers composed of a siloxane skeleton and pendant hydrocarbon groups (for instance methyl groups in the common poly(dimethylsiloxane) (PDMS)) is a very effective way of creating a closely packed hydrocarbon surface. In fact, the very flexible siloxane skeleton allows great orientational freedom to pendant groups, which, in this way, can be exposed towards the interfacing environment to their best effect [64]. The advancing water contact angle on PDMS (about 112° [65]) indicates a fully methylated surface, effectively shielding the high energy inorganic-like backbone (see also Chapter 4).

As previously discussed, another important requirement of water repellents is the interaction with the substrate, which can impart fastness to the treatment. Durability of silicone-treated textiles to washing and drycleaning is good, and several mechanisms of repellent–substrate interactions have been suggested in the past. Among these, suggestions were made of hydrogen bonding between surface hydroxyls and siloxane oxygen, chemical reactions between Si–H in the siloxane and hydroxyls in the substrate, adsorption to the substrate via the polar siloxane oxygen [66]. It seems, however, likely that the permanency of silicone repellents, at least for the reactive type, which, as discussed below, is by far the most used, stems from the very good film forming properties and excellent flowability of liquid siloxanes in the uncured state. Liquid silicones can cover very thoroughly the details of the surface morphology of the fibres. After curing, a tough, flexible, insoluble film around the fibre is formed. This mechanism was demonstrated by Fortess in a classical paper [67], where cellulose acetate treated with silicone was dissolved by acetone, leaving cylindrical skins on the undissolved residue.

Silicone repellents can loosely be classified as reactive or non-reactive. The former are by far the most used on textiles and are in general composed of reactive polymethylsiloxane (PMS), where reactivity is caused by the presence of some

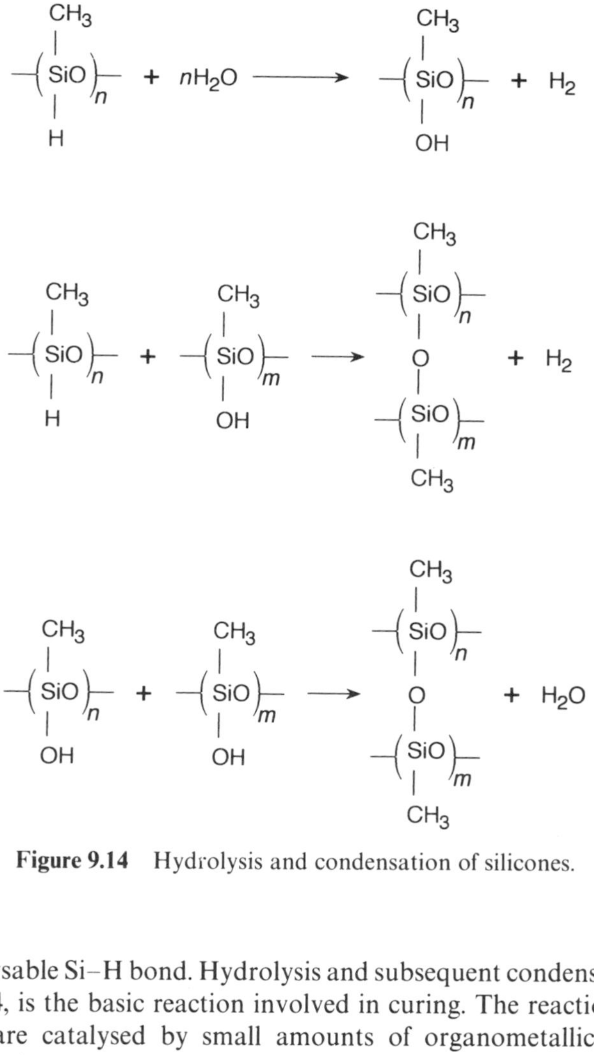

$$\underset{\underset{H}{|}}{\overset{\overset{CH_3}{|}}{-(SiO)_n-}} + nH_2O \longrightarrow \underset{\underset{OH}{|}}{\overset{\overset{CH_3}{|}}{-(SiO)_n-}} + H_2$$

$$\underset{\underset{H}{|}}{\overset{\overset{CH_3}{|}}{-(SiO)_n-}} + \underset{\underset{OH}{|}}{\overset{\overset{CH_3}{|}}{-(SiO)_m-}} \longrightarrow \begin{array}{c} \overset{CH_3}{|} \\ -(SiO)_n- \\ | \\ O \\ | \\ -(SiO)_m- \\ | \\ CH_3 \end{array} + H_2$$

$$\underset{\underset{OH}{|}}{\overset{\overset{CH_3}{|}}{-(SiO)_n-}} + \underset{\underset{OH}{|}}{\overset{\overset{CH_3}{|}}{-(SiO)_m-}} \longrightarrow \begin{array}{c} \overset{CH_3}{|} \\ -(SiO)_n- \\ | \\ O \\ | \\ -(SiO)_m- \\ | \\ CH_3 \end{array} + H_2O$$

Figure 9.14 Hydrolysis and condensation of silicones.

easily hydrolysable Si–H bond. Hydrolysis and subsequent condensation, shown in Figure 9.14, is the basic reaction involved in curing. The reactions shown in Figure 9.14 are catalysed by small amounts of organometallic catalysts at temperatures above 100 °C. Typical compounds are organo-tin or zinc, the former being more effective [68]. Another component which is usually added in the formulation is non-reactive PDMS. The latter acts as a softener or plasticizer, since the use of PMS alone leads to a brittle resin film on the textile fibre, and impairs handling of the fabric. The use of PDMS containing some silanol groups can result in direct linking between PMS and the softener, and this can give rise to improved properties. The formulation of active polymer, catalysts and softener can be supplied either in aqueous emulsion or in organic solvent solution.

Non-reactive silicones are mainly based on PDMS. When applied to inorganic substrates such as glass, they require a curing stage above 200 °C. This limits the applicability of these systems to organic fibres and textiles. It was, however, found that the use of non-reactive silicones and titanium or zirconium compounds could reduce the maximum temperature required to obtain good water repellence. Typical compounds are alkyl titanate or zirconate, which readily hydrolyse in thin films by atmospheric moisture. Titanium or zirconium compounds can readily coordinate with several substrates, and titanates are commonly used as primers [69]. Probably a coordination bond between Ti or Zr atoms and the polar Si–O siloxane repeating unit takes place, which imparts a certain degree of fastness to non-reactive silicone repellents. Better results are obtained with a short time high temperature curing (below 200 °C), whose main effect seems to be the opening out of the structure of PDMS, thus reducing the shielding effect of methyl groups and allowing an easier interaction between the inorganic Ti or Zr based substrate and the siloxane skeleton.

Covalent bonding of silicone-based repellents through irradiation has been studied. Repellents used are of the reactive type (PMS), they are applied to polyester fibres and exposed to a radiation dosage of 70–460 rad [70]. The siloxane is linked through the fibre by reactions involving radicals induced by irradiation. High durability of repellency has been reported.

9.2.4 Fluorocarbons

Like silicones, fluorocarbons were first marketed as repellents for textiles in the 1950s. A technical study shows that fluorochemicals from 1975 to 1980 constituted about 50% of the weight volume and 90% of the dollar volume of all repellents [71]. The higher price of these materials is due to their very good performance, both in terms of hydrophobizing power and durability. Moreover, it must be noted that fluorochemicals can impart not only repellency to water, but also to oils. The reason for this property is the low surface tension of fluorocarbons: as previously discussed, and as shown in Figure 9.3, PTFE, whose surface is composed by $-CF_2$ groups, has the lowest surface tension among common polymers, lower than that of hydrocarbons. Thinking of the definition of critical surface tension (Chapter 4), this means that fluorocarbon surfaces are the only polymer surfaces where hydrocarbon based soils will not spread, and where a finite contact angle between a hydrocarbon drop and the surface will be observed. Figure 9.11 and the relevant discussion show that in this case a proper design of the fabric will impart oil proofness. A further point to note is that, as shown by the fundamental work of Zisman [72], the surface tension of solids fluorocarbons can be still decreased by the creation of a surface exposing $-CF_3$ groups. Even if a bulk polymer fully exposing $-CF_3$ does not exist [73], suitable chemicals can be attached to surfaces of conventional polymers.

The first commercial fluorochemicals were based on chromium complexes of perfluoroacids. They were not very durable and, again, covalent linking with the surface was attempted. Among the reactions investigated were the esterification of cellulose with perfluoroalkanoic acid chloride (not resistant to alkaline detergency) [74] and the attack of the glycidyl ethers of the 1,1-dihydrofluoroalkanols [75]. The availability of reactive surface groups has led to many patents on molecules suitable for hydrophobization of wool fibres [76]. Some examples of groups appearing in fluorocarbon repellent are amino, sulphido, amido, ester and thioester [77], sulphonamido [78], carbamyl [79], trialkoxysilyl [80] and oxazolynil [81]. The Ce^{4+} promoted grafting of trifluoroethylmethacrylate to starch surfaces (see Chapter 7) has been claimed to impart good water repellency to water and oil [82].

In all these applications the surface modification technique takes advantage of reactive groups which are naturally present on the fibre surface. These areas are largely covered by patent literature and fundamental studies on the details of thermodynamic and kinetic parameters of supposed chemical reactions involving the repellent and the surface bonded reactive groups are not known.

As discussed in the chapter on grafting, materials that do not contain surface reactive groups must be subjected to some kind of activation step (either by UV light or by ionizing radiation) in order to obtain grafting of acrylate or methacrylate fluoroesters.

In addition to the previously quoted reactive repellents, the most common fluorocarbon repellents are of a polymeric nature and are coated on the substrate. These repellents are used primarily for textiles and are copolymers of the fluoroalkyl acrylate or methacrylate, where the length of the perfluoroalkyl group is at least four, and very often between six and eight. Copolymers are often designated to control the physical properties of the polymeric product (handling of the fabric, softness etc.) and maintain the cost of the product at a low enough level. Typical comonomers are hydrocarbon-based esters of acrylic or methacrylic acid, sometimes containing alkylamide or polyether groups.

Fluorochemical finishes are applied by conventional means, such as spraying or foam finishing or padding. Many of the polymeric fluorocarbon repellents require a high temperature curing (about 150 °C) whose main effect is to allow melt-spreading to ensure effective coating of the substrate.

A peculiar repellent is Gore-Tex[R], or expanded porous PTFE. Usually Gore-Tex[R] films are laminated on supporting fabrics and the final product is called Gore-Tex[R] fabric. Clearly, in Gore-Tex[R], both the chemical composition and the physical nature of the surface contribute to impart superior properties.

Finally, it must be noted that there is some interest in water repellence obtained by the deposition of hydrophobic fluorinated films from plasma (Chapter 6) [83]. Yasuda and coworkers discussed the deposition of fluorocarbon films from plasma on a Nylon fabric [84]. Surfaces of treated fabrics became very hydrophobic and after washing tests durability was improved by treatment with saturated fluorocarbons as compared to unsaturated fluorocarbons. This technique is still in the stage of laboratory research.

REFERENCES

[1] F. J. Holly, in *Physicochemical Aspects of Polymer Surfaces*, K. L. Mittal (Ed.), Plenum Press, New York (1983), Vol. 1, p. 141.

[2] A. W. Adamson, *Physical Chemistry of Surfaces*, 5th edn, Wiley, New York (1990).

[3] J. J. Bikerman, *J. Colloid Sci.*, **5** 349 (1950).

[4] K. Kawasaki, *J. Colloid Interface Sci.*, **15**, 402 (1960).

[5] C. W. Extrand and A. N. Gent, *J. Colloid Interface Sci.*, **138**, 431 (1990).

[6] S. Wu, *Polymer Interface and Adhesion*, Dekker, New York (1982), p. 184.

[7] J. N. Israelachvili, *Intermolecular and Surface Forces*, Academic Press, London (1985).

[8] F. M. Fowkes, *Physicochemical Aspects of Polymer Surfaces*, Vol. 2, K. L. Mittal (Ed), Plenum Press, New York (1983), p. 583.

[9] F. M. Fowkes, *Surface and Interfacial Aspects of Biomedical Polymers*, Vol. 1, J. D. Andrade (Ed), Plenum Press, New York (1985), Ch. 9.

[10] F. M. Fowkes, *J. Adhesion Sci. Tech.*, **1**, 7 (1987).

[11] C. J. van Oss, R. J. Good, and M. K. Chaudhury, *J. Colloid Interface Sci.*, **111**, 378 (1986).

[12] C. J. van Oss, M. K. Chaudhury, and R. J. Good, *Adv. Colloid Interface Sci.*, **28**, 35 (1987).

[13] C. J. van Oss, R. J. Good, and M. K. Chaudhury, *Separ. Sci. Technol.*, **22**, 1 (1987).

[14] J. D. Andrade, D. E. Gregonis, and L. M. Smith, in *Surface and Interfacial Aspects of Biomedical Polymers*, J. D. Andrade (Ed.), Plenum Press, New York (1985), Vol. 1, Ch. 2.

[15] S. Wu, *Polymer Interface and Adhesion*, Dekker, New York (1982), Ch. 6.

[16] *Surface and Interfacial Aspects of Biomedical Polymers*, J. D. Andrade (Ed.), Plenum Press, New York (1985), Vol. 1.

[17] *Polymer Surface Dynamics*, J. D. Andrade (Ed.), Plenum Press, New York (1988).

[18] M. Morra, E. Occhiello, and F. Garbassi, *J. Colloid Interface Sci.*, **132**, 504 (1989).

[19] E. Occhiello, M. Morra, G. Morini, F. Garbassi, and P. Humphrey, *J. Appl. Polym. Sci.*, **42**, 551 (1991).

[20] J. F. M. Pennings and B. Bosman, *Colloid Polym Sci.*, **257**, 720 (1979).

[21] F. Garbassi, E. Occhiello, M. Morra, L. Barino, and R. Scordamaglia, *Surf. Interf. Anal.*, **14**, 595 (1989).

[22] M. Morra, E. Occhiello, L. Gila, and F. Garbassi, *J. Adhesion*, **33**, 77 (1990).

[23] M. Suzuki, A. Kishida, H. Iwata, and I. Ikada, *Macromolecules*, **19**, 1804 (1986).

[24] H. V. Boenig, *Plasma Science and Technology*, Cornell University Press, Ithaca (1982).

[25] V. D. McGinnis, in *Encyclopaedia of Polymer Science and Technology*, Wiley, New York (1986), Vol. 4, p. 432.

[26] H. Yasuda, A. S. Sharma, and T. Yasuda, *J. Polym. Sci., Polym. Phys. Ed.*, **19**, 1285 (1981).

[27] A. K. Sharma, F. Millich, and E. W. Hellmuth, *J. Appl. Polym. Sci.*, **26**, 2205 (1981).

[28] T. Yasuda, K. Yoshida, T. Okuno, and H. Yasuda, *J. Polym. Sci., Polym. Phys. Ed.*, **26**, 2061 (1988).

[29] T. Yasuda, T. Okuno, K. Yoshida, and H. Yasuda, *J. Polym. Sci., Polym. Phys. Ed.*, **26**, 1781 (1988).

[30] E. Occhiello, M. Morra, P. Cinquina, and F. Garbassi, *ACS Polym. Preprints*, **31**, 308 (1991).

[31] E. Occhiello, M. Morra, and F. Garbassi, *Polymer*, **33**, 3007 (1992).

[32] E. Occhiello, M. Morra, F. Garbassi, D. Johnson, and P. Humphrey, *Appl. Surf. Sci.*, **47**, 235 (1991).
[33] S. F. Tead, W. E. Wanderlinde, A. L. Ruolf, and E. J. Kramer, *J. Appl. Phys. Lett.*, 52, 101 (1988).
[34] S. F. Tead, W. E. Wanderlinde, G. Marra, A. L. Ruolf, E. J. Kramer, and F. D. Egitto, *J. Appl. Phys.*, **68**, 2972 (1990).
[35] M. Morra, E. Occhiello, R. Marola, F. Garbassi, P. Humphrey, and D. Johnson, *J. Colloid Interface Sci.*, **137**, 11 (1990).
[36] H. J. Griesser, J. H. Hodgkin, and R. Schmidt, in *Progress in Biomedical Polymers*, C. B. Gebelein and R. L Dunn (Eds), Plenum Press, New York (1990), pp. 205–15.
[37] M. Morra, E. Occhiello, and F. Garbassi, *Surf. Interface Anal.*, **16**, 142 (1990).
[38] H. J. Griesser, D. Youxian, A. E. Hughes, T. R. Gegenbach, and A. W. H. Mau, *Langmuir*, 7, 2484 (1991).
[39] M. Morra, E. Occhiello, and F. Garbassi, *Angew. Makromol. Chem.*, **189**, 125 (1991).
[40] H. S. Munro and D. I. McBriar, *J. Coatings Technol.*, **60**, 41 (1988).
[41] W. J. Brennan, W. J. Feast, H. S. Munro, and S. A. Walker, *Polymer*, 32, 527 (1991).
[42] Y. L. Hsieh and E. Y. Chen, *Ind. Eng. Chem. Prod. Res. Dev.*, **24**, 246 (1985).
[43] Japan Kokay Tokkyo Koho, JP 62103140, to Hiraoka Shokusen KK (1987).
[44] Japan Kokay Tokkyo Koho, JP 62110973, to Hiraoka Shokusen KK (1987).
[45] E. Ruckenstein and S. V. Gourisankar, *J. Colloid Interface Sci.*, **107**, 488 (1988).
[46] J. C. Eriksson, G. C. Gölander, A. Baszkin, and L. Ter Minassian-Saraga, *J. Colloid Interface Sci.*, **100**, 381 (1984).
[47] B. D. Ratner, in *Surface and Interfacial Aspects of Biomedical Polymers*, J. D. Andrade (Ed), Plenum Press, New York (1985), Vol. 1, Ch. 10.
[48] F. J. Holly and M. F. Refojo, *J. Biomed. Mater, Res.*, **9**, 315 (1975).
[49] C. D. Bain and G. M. Whitesides, *Science*, **240**, 62 (1988).
[50] C. D. Bain and G. M. Whitesides, *J. Am. Chem. Soc.*, **110**, 3665 (1988).
[51] C. D. Bain, and G. M. Whitesides, *Angew. Chem. Int. Ed. Engl.* **28**, 506 (1989).
[52] R. A. Reck, in *Encyclopaedia of Polymer Science and Engineering*, Wiley, New York (1985), Vol. 2., p. 99.
[53] M. Morra, E. Occhiello, and F. Garbassi, Unpublished results.
[54] R. J. Good, *J. Colloid Interface Sci.*, **59**, 398 (1977).
[55] J. N. Israelachvili, *Proc. Roy. Soc. Lond. A*, **331**, 39 (1972).
[56] D. B. Hough and L. R. White, *Adv. Colloid Interface Sci.*, **14**, 3 (1980).
[57] N. K. Adam, in *Water Proofing and Water Repellency*, J. L. Moillet (Ed.), Elsevier, Amsterdam (1963), p. 15.
[58] *Water Proofing and Water Repellency*, J. L. Moillet (Ed.), Elsevier, Amsterdam (1963).
[59] E. B. Higgins, in *Water Proofing and Water Repellency*, J. L. Moillet (Ed.), Elsevier, Amsterdam (1963), p. 188.
[60] W. B. Blumental, *Ind. Eng. Chem.*, **42**, 640 (1950).
[61] M. Hayek, in *Kirk-Othmer, Encyclopedia of Chemical Technology*, Wiley, New York (1984), Vol. 24, p. 440.
[62] J. W. Bell and C. S. Whelwell, in *Water Proofing and Water Repellency*, J. Moillet (Ed.), Elsevier, Amsterdam (1963), p. 250.
[63] C. Hamalainen, J. D. Reid, and W. B. Berard, *Am. Dyes. Rep.*, **43**, 453 (1954).
[64] J. Owen, *Ind. Eng. Chem. Prod. Res. Dev.*, **19**, 97 (1980).
[65] M. Morra, E. Occhiello, R. Marola, and P. Humphrey, *J. Colloid Interface Sci.*, **137**, 11 (1990).
[66] R. L. Bass and M. R. Porter, in *Water Proofing and Water Repellency*, J. L. Moillet (Ed.), Elsevier, Amsterdam (1963), p. 142.

[67] F. Fortess, *Ind. Eng. Chem.*, **46**, 2325 (1954).
[68] J. A. C. Watt, *J. Textile Inst.*, **51**, Tl (1960).
[69] S. Wu, *Polymer Interface and Adhesion*, Dekker, New York (1982), p. 430.
[70] N. Nishide and H. Shimizu, *Text. Res. J.*, **45**, 591 (1975).
[71] E. G. Hochberg and Associates, *Textile Finishing Agents, A Marketing Technical Study*, Chester, New York (1976).
[72] W. A. Zisman, *Am. Chem. Soc. Adv. Chem. Ser.*, **43**, 1 (1964).
[73] M. J. Owen, *J. Appl. Polym. Sci.*, **35**, 895 (1988).
[74] R. J. Berni, R. R. Benerito, and F. J. Philips, *Textile Research J.*, **30**, 393 (1960).
[75] R. J. Berni, R. R. Benerito, and F. J. Philips, *Textile Research J.*, **30**, 576 (1960).
[76] B. M. Lichstein, in *Surface Characteristics of Fibres and Textiles*, M. J. Schick (Ed.), Marcel Dekker, New York, 1977, p. 514.
[77] U.S. Pat. 3,655,732 C. S. Rondestvedt, Jr.
[78] Brit. Pat. 1,283,266 P. C. Bouvet, C. M. H. E. Brouard, and J. P. C. Lalu.
[79] U.S. Pat. 3,679,634 E. Schuirer, W. Renz, and H. Sommer.
[80] Brit. Pat. 1,267,224 to Nalco Chemical Ltd.
[81] U.S. Pat. 3,677,812 D. G. Gagliardi.
[82] Jpn. Kokai, 01,203,411 T. Ueda, S. Takagi, H. Fukuda, N. Hishiki, K. Nagatsuka (to Nippon Starch Chemical Co. Ltd).
[83] H. Yasuda, *Plasma Polymerization*, Academic Press, Orlando, (1985).
[84] Y. Iriyama, T. Yasuda, D. L. Cho, and H. Yasuda, *J. Appl. Polym. Sci.*, **39**, 249 (1990).

Chapter 10

Adhesion

Adhesion is a surface property relevant in many applications of polymeric materials. For instance, it is necessary to obtain the permanent adhesion of ink on shopping bags; the adhesion of paint is important on bumpers and other body car parts like hoods; the stable adhesion of a thin layer of metal (mainly aluminium) is a must in the decoration of objects; in engineering plastics applications structures can be joined together using an adhesive; on the other hand there are applications where a low or even a null 'adhesion' degree is requested, like in antiblocking or in solid lubricants. As a definition, adhesion is the joining together of two dissimilar materials, while 'cohesion' is the joining together of different portions of the same material; the opposite of adhesion is often referred as 'abhesion'.

To obtain adhesion, a necessary condition is the tight contact between the two moieties. As a consequence, if one of them is a liquid (at least during the joining operation), wettability of the other one is a strictly correlated property. In other words, the first step of gluing an adherend is to wet it efficiently with the adhesive (generally a more or less viscous liquid). Wettability of polymer surfaces is treated in Chapter 9.

In this chapter, theories of adhesion are discussed, with a special emphasis towards polymer specificity. Methods for practical adhesion measurements are then briefly reviewed, followed by methods developed for varying the adhesion level.

10.1 THEORIES OF ADHESION

As reported by K. W. Allen [1], Director of Adhesion Studies at The City University, London and widely known editor of the volumes titled *Adhesion*, early in the 1920s the need for a proper understanding of the fundamental basis of adhesion for relevant scientific and applicative reasons was recognized. Several theories have been proposed, often with the aim to offer a universal solution.

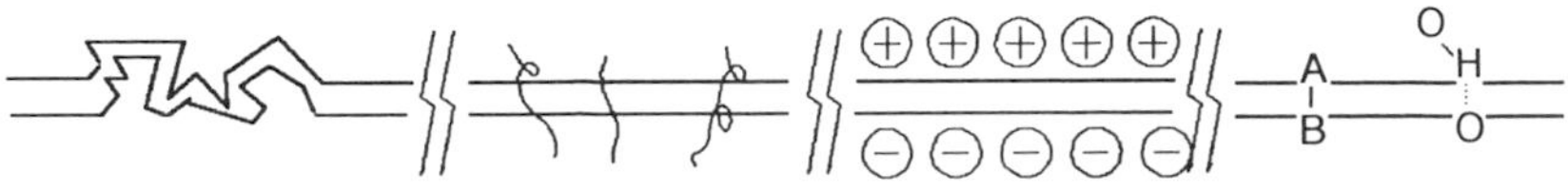

(a)   (b)   (c)   (d)

Figure 10.1 Contributions to adhesion: (a) mechanical interlocking; (b) interdiffusion of chains; (c) electrical interactions; (d) chemical interactions.

Today, it is generally acknowledged that different mechanisms work, alone or sometimes together, in every particular system. The subject is treated in several books on adhesion and adhesives, to which one can refer for probing [2–6]. Presently, four main mechanisms survive, involving physical or chemical forces: mechanical interlocking, interdiffusion, electrostatic attraction and chemical interactions. Their action at the interface between two different materials is sketched in Figure 10.1. It is immediately evident that the above mechanisms work at different scales of distance between the two materials: from the atomic scale for chemical bonds to an undetermined scale (but reasonably in the range of micrometres) in the case of mechanical interlocking. A complete discussion of such mechanisms, their importance and weaknesses is reported in ref. [3].

10.1.1 Mechanical Interlocking

Rather than from theory, mechanical interlocking originates from rough practice observations. In fact, it is rather usual that the concept of gluing is more successful on rough and irregular surfaces. The use of grinding paper on adherends before gluing, often recommended by adhesives manufacturers, is considered a confirmation of such a circumstance.

However, evidence can be found that good adhesion can also occur on smooth surfaces: the strong adhesion between mica sheets is an example coming from the natural world, while a common technical practice is to obtain good joints by gluing flat optical glasses. However, some experimental work demonstrated the fallacy of the assumption: in the 1940s, Maxwell [7] prepared maple wood specimens of different roughness and measured their strength after bonding with an urea/formaldehyde resin. He observed a decrease of strength by increasing roughness (Figure 10.2). Even if the above experiment retains some qualitative validity, it also failed to establish a quantitative relationship, because the roughness level was not translated in figures.

Another difficulty is that wood surfaces are not well characterized and reproducible, because every specimen can have different defects and treatments made were able to introduce even more defects, all influencing the final bonding level.

Similar considerations can be raised on other systems where mechanical interpenetration seems to play a role, like rubber to textiles from natural fibres

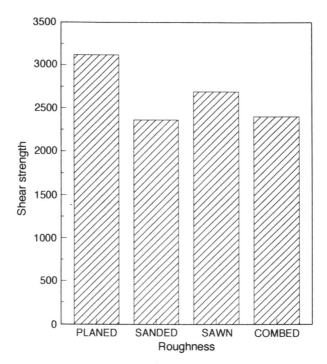

Figure 10.2 Shear strength (pounds/inch^2) versus roughness for maple wood samples. (From. ref. [7].)

[8] and adhesive film to leather [9]. It is not by chance that wood, textiles, or paper etc. are considered to offer examples of mechanical interlocking: in fact, the above materials have a finely divided or porous nature, where the adhesive flows before solidification, thus realizing a sort of anchoring. A similar case is that of the presence of protrusions, around which the adhesive can solidify.

The simple fact of putting mechanical interlocking together with the other adhesion theories is controversial: Lee, for instance, considers such a mechanism only as a technological means to achieving a good level of adhesive bonding [5]. In fact, a major success of this mechanism was achieved in structural adhesive bonding of aluminium or titanium alloys. This work was pioneered by Venables and coworkers at Martin Marietta Laboratories and is described in a comprehensive review article by Venables [10]. Making a comparison with the result of etching methods of Al before bonding, both chemical and electrochemical, the authors found evidence, mainly by STEM, of the role of the morphology of the oxide created by the etching process in determining the strength of metal–polymer structures and their durability as well. A series of high resolution stereo micrographs showed the formation of a hollow hexagonal cell structure

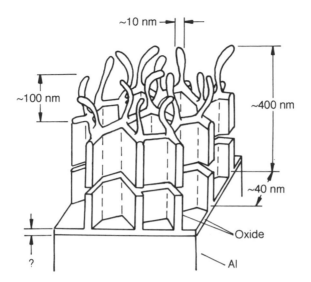

~10 nm

~100 nm

~400 nm

~40 nm

Oxide

Al

?

Figure 10.3 Oxide morphology developed on Al by phosphatic acid anodization (PPA) treatment. (From. ref. [10].)

with whisker-like protrusions (Figure 10.3), thus realizing in the same structure the presence of pores and protrusions. Even if this constitutes a nice example of mechanical interlocking, it is difficult to use it to assess the importance of the mechanism. Once again, it is strictly connected to a particular material, furthermore the main question remains open: does producing (or taking advantage of, if already present) a particular morphology activate a new, independent adhesion mechanism, or does it simply increase the effectiveness of other adhesion mechanisms, owing to the increased area involved in adhesion and/or the removal of dirty and weak layers from the etched surface? By using chemical or physical methods, highly porous surfaces can also be produced on polymers, but it remains to be demonstrated that high strength is obtained also in the presence of a weak intrinsic interaction between the two solids. An area where mechanical interlocking can play a relevant role in adhesion is that of metal plating on polymer surfaces, where chemical pretreatment of the latter is generally carried out. Examining the adhesion of electroplated Cu on PP, Perrins and Pettett [11] concluded that adhesion was controlled by mechanical interlocking, basing their conclusion on the comparative results obtained on surfaces having a different degree of roughness of the substrate surfaces. The superior control on the morphological state of the surface obtainable by chemical treatments with respect to mechanical treatments allowed the comparison of peel energies of surfaces of increasing roughness, ranging from $0.7 \, kJ/m^2$ in the case of flat surfaces to $2.4 \, kJ/m^2$ when high angle pyramids and dendrites are contemporarily present.

10.1.2 Interdiffusion

Interdiffusion means that if two polymers are put in close contact at a temperature above their T_g, some long-chain molecules or at least some segments reciprocally diffuse in the other moiety. This concept has been suggested by Voyutskii [12] and its effectiveness strictly depends on the degree of compatibility between the two polymers, reaching a maximum degree when they are equal. In this particular case, it is worthwhile to speak about 'autohesion' and large values of interdiffused layer thicknesses in the order of some micrometres were observed [13], while incompatible or scarcely compatible polymers give rise to layers as thin as 10 nm or less.

A quantifiable theory of interdiffusion in polymers has been developed by Vasenin [14], in analogy with mixing and interdiffusion in liquids. In the case of autohesion, an equation was proposed where the peeling force F is directly proportional to the rate of separation r and to the 4th root of the contact time t, and inversely proportional to the 2/3rd root of the molecular weight. As discussed by Allen [1], several experimental results were found to fit with the Vasenin equation, however the approach was later criticized, attributing the peeling strength increase to an increase of the contact surface due to local rheological processes [15].

Evidence was provided by experiments on the occurrence of interdiffusion in several situations, for instance by radiometric studies [16] and optical microscopy [17]. It remains that this mechanism has only a limited relevance, when both moieties are polymers and their compatibility is high enough. Attempts to extend the theory to polymer/metal systems were not successful [2].

10.1.3 Electrostatic Attraction

A theory was developed by Derjaguin on the occurrence of an electrical double layer at the interface between two solids, so justifying the adhesion force with the establishment of an electrical attraction [18]. The original observations concerned a pressure sensitive tape, where crackling noise and light emission occurred by fast peeling in dry conditions. Subsequently, a quantitative theory was developed, attaining satisfactory agreement with experiment. The majority of experiments concerned the adhesion degree of vacuum-deposited thin metal films on polymer substrates. It was found [19] that the force necessary to remove the film by scratching with a stylus increases with time. Since the effect occurs also with gold, slow oxidation processes at the metal–polymer interface, allowing the formation of chemical bonds, must be discounted. Furthermore, if the specimen is exposed to a corona discharge after deposition and ageing, when the maximum adhesion strength is reached, the strength value drops to zero, supporting the electrostatic origin of adhesion.

The electrostatic attraction theory was subjected to three major controversies [9]:

(1) The electrostatic double layer can be identified only after breaking the adhesive bond.
(2) Its effect on the adhesion force is overconsidered.
(3) Some overlapping was made between electrostatic and acceptor–donor interactions [20].

About the first point, a SEM investigation was recently carried out that confirmed the existence of the electrostatic double layer, determining the potential distribution at the interface [21]. The actual charge density and the work associated with the electrostatic double layer were also determined, with values of $2.7 \times 10^{-2} C/m^2$ and $1.71 \times 10^{-3} J/m^2$, respectively [22, 23].

Since the peel work of a LDPE film from an Al foil is $1 J/m^2$, that is 600 times more, it is suggested that the electrostatic component is a very minor fraction of the adhesive strength. Also, similar conclusions were reached in rubber adhesion studies [24].

The role of the electrostatic component in adhesion has recently been discussed by Hays [25]. Starting from the consideration that electrostatic effects occur when one or both materials acquire a net charge through an imbalance of charges (carried by ions or electrons), the various possible situations were examined. Regarding polymers, the case of planar, insulating materials is the most interesting. The electrical component of the adhesive pressure was calculated, depending on the distance of charge carriers, showing that the electrostatic component can be comparable or larger than the van der Waals' component, depending on density of charge carriers. The concentration at which it appears relevant is around 1% of monolayer. This result is in agreement with the viewpoint of Derjaguin.

10.1.4 Chemical Interactions

Chemical interactions between two bodies embrace a variety of forces, like non-polar dispersion forces (London interactions), dipole–dipole interactions (Keesom interactions), dipole–induced dipole interactions (Debye interactions), all generally considered as van der Waals' forces. Then, the adsorption forces must be considered, not forgetting the establishment of true chemical bonds. The contribution of the various types of interactions to the adhesion force has been recently reviewed by Gutowski [26] and Good [27, 28].

The above forces differ very much in strength and range, as shown in Figure 10.4, from 2 to $1000 kJ/mol$ and from 0.1 to $1.0 nm$, respectively. While the establishment of true chemical bonds depends on the nature of the bodies under consideration, van der Waals' forces are universally present. They also operate in a large range, and their decrease with increasing distance is less. It is worth noting that an essential condition for establishing the above interactions is to put the two parts in close and intimate contact, for this reason wetting properties are so relevant in adhesion affairs. Wetting properties of polymer surfaces are considered

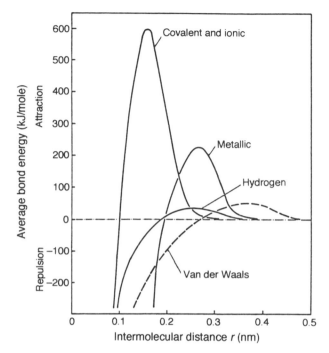

Figure 10.4 Strength and range of typical intermolecular and interatomic forces. (From ref. [26].)

in Chapter 9. Theoretically, as can be shown in experiments when trying to separate mica lamellae, two perfectly plane and parallel plates can develop a huge attractive force per surface unit when the centres of molecules of the two surface layers are separated by 1 nm (10^7 N/m^2, in the case of dielectrics), much more than is practically achieved in adhesive joints.

To take into account the different forces above, Fowkes [29] proposed an additive contribution to the work of adhesion W_A:

$$W_A = W_A^d + W_A^h + W_A^k + W_A^{ab} + \cdots$$

including dispersion forces (d), hydrogen bonds (h), Keeson and Debye interactions (k), acid/base interactions (ab) and, in principle, also covalent bonds. The relationship is sometimes simplified introducing just two terms, representing dispersive and polar, non-dispersive forces:

$$W_A = W_A^d + W_A^p$$

Because a work on the enthalpies of acid/base reactions [30, 31] demonstrated that dipole interactions do not contribute measurably to the enthalpies of molecular interactions, the work of adhesion is now considered from two components, dispersive and acid/base. To confirm this result, Fowkes and coworkers [32] prepared a series of copolymers of ethylene with either acrylic acid or vinyl acetate, thus controlling their acidity or basicity. The acid-base contribution to the work of adhesion was determined by measuring the contact angles of a series of liquids of known acidity, and was found in agreement with theory. To show the effect directly, the strong adhesion of a cast PMMA film (basic) on an acidic glass, dropped to a low value in the case of a basic glass [33].

This occurrence has an important fall-out in the choice of the right adhesive for many materials, also taking into account the possibility to regulate the acid/base degree of resins constituting most of the adhesives, by the introduction of appropriate chemical groups; for instance, adhesives containing carboxylic acids are suitable for Fe substrates (amphiphilic), while adhesives with amino groups are appropriate for silica-glass substrates (basic), and those based on phenols are appropriate for acidic substrates like MgO [34].

Good and coworkers recently made a complete treatment of adhesive forces across interfaces on the basis of Lifshitz–van der Waals (LW) [27] and acid-base components [28]. In this view, PTFE has a purely apolar surface free energy, that is only the LW component gives a contribution. On the other hand, PMMA and several other polymers (polyoxyethylene, cellulose acetate, etc.) are monopolar solids, always behaving like a Lewis base. For PMMA, the Lewis base character is given by the ester group, while for PEG it is given by the ether group. On the other hand, the classification of cellulose nitrate like a monopolar solid is surprising, since on the basis of the structure it should be bipolar. For this and other polymers whose acid-base character is unexpected, like PVC, which does not show a Lewis acid character, the Lewis neutralization of the hydroxyls by hydrogen bonding, or of H atoms by Cl atoms acting like electron donors, was respectively invoked.

10.1.5 Conclusions

The examination of the various theories made on the basis of adhesion, and specifically referred to polymers, has shown that only those based on chemical interactions have a general character and a wide application. Since various types of interactions and bonds can be established, the occurrence of them strictly depends on the chemical nature of the adherends and the conditions where the adhesion is established.

Electrostatic interactions can play a role in adhesion, but their importance and the strength that they contribute is generally less important than chemical interactions.

Interdiffusion is surely important in some specific systems but cannot be applied in the generality of cases. Finally, mechanical interlocking seems to have only a peripheral role.

In conclusion, the best way to increase adhesion properties of polymer surfaces seems to involve their chemical properties. The next section is devoted to the methods to measure adhesion, while methods to change adhesion properties are presented in the following section.

10.2   MEASUREMENT OF ADHESION

Methods for measuring adhesion properties strongly depend on the nature of the system under examination. Dealing with polymers, three main cases will be discussed, i.e. polymer/metal systems, polymer/polymer joints polymer/fibre systems in composite materials.

10.2.1   Polymer/Metal Systems

Polymer/metal systems are present in several market segments like packaging materials, electronic devices, appliances, toys, etc. This subject has developed great importance in recent years, as testified by the arrangement of several symposia on the field, from which a series of proceedings books was originated [35–37]. In most cases a thin metal layer is deposited on a polymer substrate, with the aim to change substantially a specific application property, like impermeability to gases, electrical conductivity, electromagnetic shielding, decorative purposes, etc. In a more limited number of applications, such as laminates or in metallic inserts both moieties (polymer and metal) have a bulky consistency. In the latter case, adhesion measurements can be carried out by adopting some of the methods developed for adhesive joints.

For metal deposits on polymeric substrates, some tests have been elaborated similar to those used for paints. The tape test [38] offers a simple way to measure the adhesion degree of a coating on a substrate, but is hardly quantitative. On a proper coating deposited on a flat surface, a series of cuts is traced with a sharp razor-blade, using sufficient pressure to have the cutting edge reach the substrate. Depending on the thickness of the coating, normal cross-cuts can be traced, or a single cross-cut. An example with six cuts for each direction is reported in Figure 10.5. A piece of semi-transparent pressure-sensitive tape is placed on the centre of the grid and pressed on the specimen, then removing the tape by pulling it off after 90–120 s. By visual inspection, the adhesion is rated from OB (no adhesion) to 5B (no damage) as reported in Figure 10.5. Obviously, the operator's judgement plays an important role in determining the adhesion degree. In a simplified version, a tape is pressed on to the surface and then peeled off. The analysis of both sides by a surface sensitive technique like XPS (see Chapter 3) can give some information on the locus of failure.

In order to obtain a direct measurement of adhesion strength, commercial devices are available to carry out pull tests. A pull stud pre-coated with an epoxy adhesive is positioned over the coating to be tested, secured with a spring clip,

Adhesion level	Visual inspection
5B	None
4B	
3B	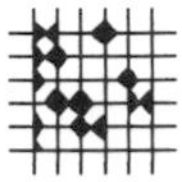
2B	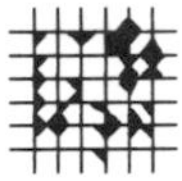
1B	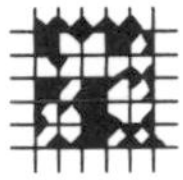
0B	Greater than 65%

Figure 10.5 Classification of results of adhesion cross test (ASTM D3359).

then cured in an oven at suitable temperature and time. After cooling at room temperature, the test stud assembly is put into the pulling grip. A progressive load is applied concentrically through the grip–sample–platen system (Figure 10.6(a)), until the sample fails. The load is continuously monitored by a strain gauge load cell and the maximum load. In all cases when the strength of the stud-sample assembly is larger than the metal–substrate adhesion, such a load corresponds to the adhesion strength between metal and plastics. It is essential that the joint is properly made, with the axis of the stud perfectly normal with respect to the specimen surface; also the force must be exerted without significant deviations from that direction. Otherwise, the strength measured is a relevant, not determinable fraction of pull strength, hindering the direct comparison with other specimens.

If the coating has a sufficient consistency to be handled without damage, a peeling test can be adopted (see below). It must be provided that the cohesive strength of both moieties is larger than their adhesive strength. In order to avoid dependence on the deformation behaviour of the film and the substrate, a stretch-deformation method has been developed [39]. Using the sample geometry of Figure 10.6(b), a force is applied in the direction of the arrows,

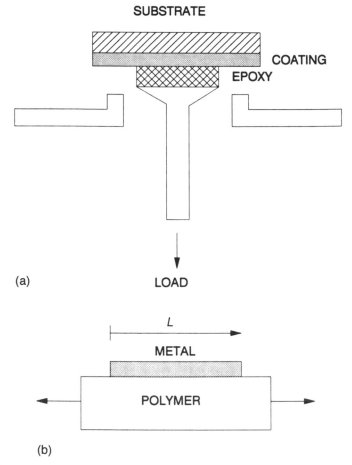

Figure 10.6   (a) Geometry of *z*-axis pull-out test; (b) geometry of stretch deformation test.

detecting the load versus elongation curve and following the formation of cracks in the metal film by means of *in-situ* optical microscopy.

10.2.2   Polymer/Polymer Systems

Polymer/polymer adhesion can be met in several practical situations. In this chapter, two main typical situations are treated, i.e. the direct adhesion between two polymers (for instance, two contiguous layers of a multilayer film) or their joining by means of an adhesive, the second one constituting the most common case. Since the question to be treated here is to control the adhesion properties of the substrate(s), the discussion is limited to them, neglecting the role of the adhesive, even if most of them—after curing—have a polymeric nature. A numerous literature is dedicated to this subject [2, 3, 40–44].

Strength analysis of adhesive joints can be made using several mechanical tests, widely described in the above literature [2, 3, 42, 44]. They are all destructive. For the purpose of this book, single lap shear tests and peeling tests can be sufficient to characterize the adhesive properties of a substrate. They are described in standardization procedures [3]. Common geometrical arrangements are reported in Figure 10.7. Peeling tests can be used also for multilayer films.

During the fracture of an adhesive joint, the adhesive layer and the substrates are deformed in a way that depends on the joint geometry. Because substrate and

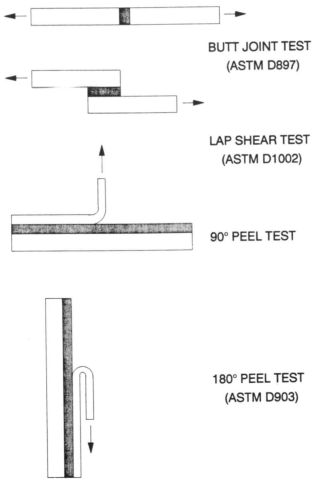

Figure 10.7 Geometry of some standard tests for adhesive joints.

adhesive have an imperfect elasticity, a certain amount of energy is dissipated, so obtaining a final strength value that could be higher than the true value coming from equilibrium considerations. Roughly, the adhesive, which is generally a thermoset, is more regid than thermoplastic polymers. Different situations can be found when both adherends are polymers (equal or different), or if one of them is a metal, like steel or aluminium. Owing to the variety of situations, absolute measurements of adhesion strength imply a very accurate analysis of the experimental conditions, and often need the evaluation of the work of adhesion. When it is important to check the effectiveness of treatments to change the adhesion level, comparative tests in the same conditions can be sufficient.

10.2.3   Composite Materials

In an accepted definition, a 'composite' is a multiphase material resulting from a combination of materials, differing in composition or form, which remain bonded together, but retain their identities and properties [45]. In order to limit such a wide definition, only the combinations of a polymeric matrix with a reinforcing agent constituted by fibres will be considered here. Matrices can be both thermoplastic or thermoset polymers, while reinforcements are continuous strands, unidirectional fabrics, mats, chopped strands, etc. In high performance composites, carbon or Kevlar[R] fibres are often used, while in low performance segments glass fibres are preferred.

In composite materials, the interfacial behaviour between the matrix and the reinforcement is a key factor influencing their mechanical properties. As in the previously treated cases, it is essential that fibres be wetted by the resin during the manufacturing process. The main contribution to the strength is then constituted by chemical interactions, while mechanical interlocking and electrostatic attraction play generally a marginal role. When glass fibres are used, silane based coupling agents are used, which couples the resin to the matrix via one or more reactive groups [46]. The general formula of silanes is X_3Si-R, where X is a hydrolysable group, like Cl-, methoxy- or ethoxy-, and R is an organic group, like vinyl, methacryloxypropyl or aminopropyl; such a group should be compatible with the polymer matrix. Silanols are formed by hydrolysis in water solution, which are attached to the glass surface by hydrogen bonds. Subsequent drying induces the formation of a polysiloxane layer onto the glass, covered by a layer rich in R groups.

Unlike glass fibres, carbon fibres are quite inert, so they must be activated to favour adhesion to the matrix. Oxidative treatments, both by gases or by liquids, are commonly used for this purpose. Two main effects have been observed, that is the removal of a weak, defective outer layer initially present on the fibre, and the formation of oxygen-containing surface groups able to interact with the matrix resin [47].

Characterization of the interface in composite materials is made in three different ways, that is physico-chemical methods described in Chapter 3, bulk mechanical tests or single-fibre tests. Bulk mechanical tests give a measure of the strength of the whole specimen, from which the mechanical properties of the interface *per se* must be inferred. Tests for shear strength, tensile strength, compressive strength and flexural strength are carried out [48], with special attention to the transverse 90° tensile strength (ASTM D 3039), that has been proposed as a primary meaure of interfacial bond strength [49]. A good indication of the adhesion degree can be reached observing by scanning electron microscopy (SEM) the fracture surface. Bare fibres and those completely covered by the matrix correspond to the two limit situations of no adhesion and good adhesion, respectively (Figure 10.8).

Because bulk mechanical tests cannot directly determine the interfacial

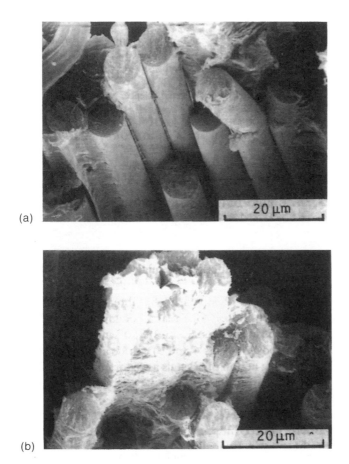

(a)

(b)

Figure 10.8   Examples of poor (a) and good (b) adhesion between fibres and a matrix.

strength, several methods have been developed to measure the adhesion of a single fibre to the surrounding matrix. The compressing microbonding test [50] is the only single-fibre test able to analyse actual composite specimens. It consists of a stylus which applies a compressive loading to an individual fibre end oriented normal to a polished surface (Figure 10.9(a)). The load necessary to cause debonding is measured by optical microscopy or by an automated system. Specially prepared specimens are required for other tests: single-fibre pull-out, microdroplet pull-off and embedded fibre critical length test.

In the single-fibre pull-out test a very thin resin disc is cast around a fibre, then a force is applied to initiate pull-out [51] [Figure 10.9(b)). In the microdroplet pull-off test, instead of embedding the fibre, a droplet of the resin is deposited on to the fibre. After curing, the fibre is placed in a microdevice mounted on the

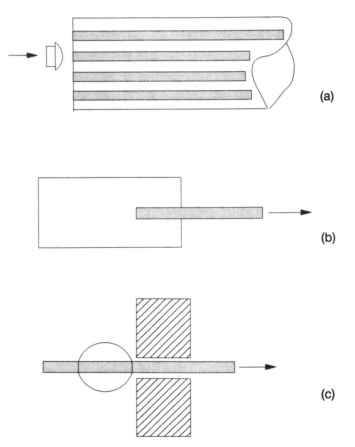

Figure 10.9   Evaluation of fibre/matrix strength in composite materials: (a) compressive microdebond test; (b) single fibre pull-out test; (c) microdroplet pull-off test.

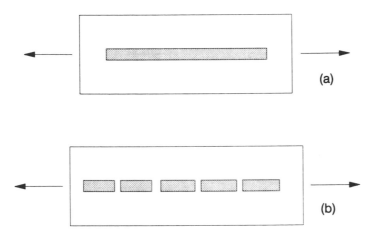

Figure 10.10 Evaluation of fibre/matrix strength in composite materials; embedded fibre critical length test; (a) before test; (b) after test.

cross-head of a tensile testing apparatus (Figure 10.9(c)), where the droplet is sheared from the fibre [52]. The average interfacial shear stress is derived from the slope of the resulting line yield.

In the critical length test, a single fibre is embedded longitudinally in the resin, from which a tensile strength specimen is prepared as shown in Figure 10.10(a). applying a strain in the direction of the arrows, the fibre will break into many small fragments (Figure 10.10(b)). The latter are observed by using a polarized optical microscope, and their size distribution is calculated. Various methods have been elaborated to obtain the interfacial shear strength from this experiment [53, 54]. This test has been widely used in fibre/matrix adhesion studies, although several objections have been raised to the corresponding equations and the extrapolation to real composites [48].

10.2.4 Conclusions

Several methods are available for measuring the adhesion level in systems like polymer/metal, polymer/polymer and polymer/fibre. The values obtained must be taken with care, because they represent the sum of a large number of single events, which can have a different nature. Also the experimental conditions strongly influence the result, and tests of a different nature do not necessarily give the same scale for a group of specimens. Thus, in the presence of different adhesion levels, comparisons can be made only when standard conditions are rigorously maintained. A bibliography limited to the field of films and coatings, but extended to all types of materials on adhesion measurement has been compiled by K. L. Mittal [55].

10.3   METHODS FOR MODIFYING ADHESION

Polymer surfaces generally exhibit low surface tension values, ranging, for commodity polymers, from 19 mJ/m^2 of PTFE to 43 mJ/m^2 of polyamide 6, 6. The surface tensions reported in Figure 9.3, except the cases of PTFE and PDMS, do not consider polymer surfaces as completely hydrophobic; however, they are quite removed from materials with high energy surfaces, such as oxides and metals (Figure 10.11).

Considering adhesion theories as briefly discussed in Section 10.1, in order to obtain adhesion in polymer-based systems it is necessary to achieve an intimate-molecular contact between the two moieties involved in the adhesion process. This fact introduces the role of surface contaminants, the deleterious effect of

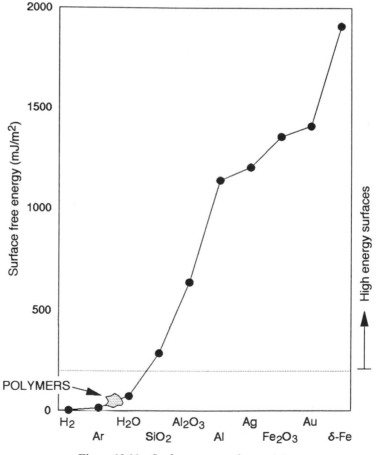

Figure 10.11   Surface energy of materials.

which must be removed. The reason is that molecular energies across an interface strongly decrease with distance: both polar interaction (Keesom) and non-polar dispersion forces (London) decrease greatly with the intermolecular distance, as discussed in Chapter [34]. Since clean surfaces having a surface tension such as those reported in Figure 9.3 are hardly wetted by liquids or melts, many treatment methods have been elaborated to modify their properties.

Treatment of polymers with the aim to change adhesion properties can be classified in two main groups, chemical surface modifications and physical surface modifications. In the first case, the polymer surface is placed in contact with a reactive species, like a liquid, a solution or sometimes a gas. In the second case, the surface is irradiated by a high energy medium. A third group, of minor importance in the present context, considers the obtaining of better surface properties through modification of bulk properties like crystallinity or even chemical composition. The majority of methods have been extensively treated in Part III of this book, therefore only aspects strictly connected to adhesion will be considered here.

In Table 10.1, treatment methods are summarized following the above scheme. They tend to substantially introduce high energy functional groups at the surface, mostly containing oxygen.

In the description of the effects of the various methods on the adhesion properties of polymers, rather than follow the above scheme, the subject is divided into three main parts concerning the chemistry: polyolefins, fluoropolymers and 'reactive' polymers. In this way, it will be possible to describe many practical

Table 10.1 Methods to modify polymer surfaces

Chemical
    Wet etching
    Chemical oxidation
    Radical activation
    Grafting
    Plasma polymerization

Physical
    Flame
    Corona
    Plasma
    Ion beams
    Electron beams
    Photon beams (UV, Laser, X-ray, etc.)

Bulk
    Blending
    Additivation
    Recrystallization

examples on how adhesion problems of polymers have been dealt with and solved. A final section is devoted to fibres.

10.3.1　Polyolefins

Polyolefins include several important classes of polymers, like poly(ethylene) (PE) and poly(propylene) (PP), as well as their copolymers. Due to their structure that is totally built by hydrocarbon units, their surfaces are quite non-polar and their adhesion properties are scarce. For this reason a lot of work has been carried out in order to improve the adhesion properties, using most of the methods of Table 10.1. Chemical methods have been assessed in the 1970s, while physical methods became more popular later.

10.3.1.1　Chemical methods

Chemical methods to oxidize the surface of polyolefins include the use of strong oxidants like fuming sulphuric acid, nitric acid and chromic acid/sulphuric acid mixtures [56]. Chain scission and etching are obtained, together with the introduction of oxygen-containing functional groups. Chromic acid action on poly(ethylene) has been extensively studied, due to its commercial importance [57–60]. In a solution of chromic trioxide in water and in the presence of sulphuric acid, the following species exist, for the establishment of several chemical equilibria: $HCrO_4^-$, $Cr_2O_7^{2-}$ and $O_3CrOSO_3^{2-}$ [60], all causing oxidation of the polymer surface. It is a difficult task to determine which of the functional groups introduced assumes a major role in increasing the adhesion properties. A study has been carried out in order to clarify this point [61], by introduction of functional groups, one at a time on to poly(ethylene) (both HDPE and LDPE), through bromination of the surface followed by specific reactions. The following functions have been introduced, as checked by XPS and multiple internal reflectance IRS: brominated, unsaturated by dehydrobromination,

Table 10.2　Joint strengths of poly(ethylene) after functional-ization (from ref. [61])

Functionality introduced	Joint strength (N/mm²)	
	LDPE	HDPE
Untreated	1.4	0.7
Brominated	6.7	5.4
Unsaturated	3.7	2.9
Hydroxyl	4.6	—
Hydroxyl (vicinal)	4.9	—
Carboxyl	11.3	7.1
Carbonyl	15.7	9.3

hydroxylated (single or vicinal –OH groups), carboxylated and carbonylated. Adhesion strengths, reported in Table 10.2, were measured by butt tests using an epoxy adhesive. Two main pieces of evidence have been provided, i.e. that the presence of functional groups is essential to increase adhesion, and that LDPE and HDPE behave in a different way. At least in the limits of the experiments, carboxylic and even better carbonyl groups were found to be most effective in promoting adhesion [61].

Returning to chromic acid treatments, the reagent can selectively etch the regions of low or zero crystallinity [62]. This action produces a rough surface, the occurrence of which can contribute to adhesion for the increased surface area or mechanical interlocking. The latter mechanism, however, cannot be considered the principal one, because strong joint strengths were also observed in the case of relatively smooth surfaces. For PE, by increasing the etching time, both etching depth and degree of surface oxidation did increase [57], as well as peel strength after bonding with a poly (acrylic acid) adhesive [62]. There was a long debate about the role of wettability, surface polarity, degree of surface oxidation and acid–base interactions in determining the adhesive properties, and establishing various correlations. Briggs and coworkers [57, 58] observed a correlation between adhesion and surface oxidation, while Ranhut [63] found a correlation between shear strengths and critical surface tension, which reflects surface polarity. Also the presence of a weak boundary layer (WBL) can be important on joint strength. For this purpose, it was argued that etching can efficiently remove the WBL, in case it covered the polyolefin surface [62, 64]. However, this conclusion is seriously weakened by the fact that no trace of PE was found on the epoxy side of a fractured joint glued with an epoxy adhesive [57]. Furthermore, if the oxidized layer is removed, the adhesion again becomes poor, so indicating the role of chemical function with respect to the WBL, which, if present, has been removed as well [65].

In addition to chromic acid, other chemical treatments have been proposed, like ammonium peroxidisulphate [66]. Starting from a shear strength of 3 MPa for untreated PE, over 30 MPa have been achieved after a treatment for 30 min or more. The presence of a small insoluble gel fraction after treatment suggested the formation of a cross-linked surface layer. However, the lack of surface analysis (only ATR–IRS was performed) does not allow such a result to be considered as conclusive.

Sulphonation of PE with fuming sulphuric acid leads to surfaces containing $-SO_3H$ groups [67]. Short treatment times are necessary in order to avoid char formation. Likewise $-SO_2Cl$ groups are introduced using chlorosulphonic acid as reagent. A review of many other reagents useful to increase the polyolefin adhesion properties has been compiled taking into special account their use with cyanoacrylate adhesives [68]. Many active reagents diluted in organic solvents have been developed as 'primers' and often patented, mainly by Toa Gosei [69], Three Bond Co. [70] and Loctite [71]. Generally, they are complex mixtures of

Table 10.3   Effect of fluorination on PE films [75].

Film	Butt joint strength		XPS at conc. (%)		
	(MPa)	(mJ/m^2)	C	O	F
LDPE					
Untreated	0.6	31.0	100	—	—
Treated	17.3	39.6	52	7	41
HDPE					
Untreated	0.6	29.9	100	—	—
Treated	23.5	27.8	48	2	50

substances like organometallic compounds, amines, phosphines, guanidines and fluoro-containing compounds. By the use of such primers, an increase in the tensile shear strength of PE from 0 to 6–9 N/mm^2 has been reported [68]. Even if free radical halogenation can be used to modify polyolefin surfaces [56], it is difficult to avoid diffusion into the bulk; for this reason, some procedures have been developed to maintain the surface specificity of the treatment. Dilution of the reagents, to control the reaction time or the use of a separate phase gas have been proposed [72–74]. Fluorination is an example of the latter process, which has gained a special place in this type of treatment and is commercially available. Some fluorination results are collected in Table 10.3 [75], showing that the increase of surface free energy is not necessary to reach a good adhesion with an epoxy adhesive, while the presence of oxygen-containing groups is essential. The role of fluorine is in activating the PE surface for subsequent oxidation.

More sophisticated chemical processes to oxidize polyolefin surfaces have been mutated from studies on the function of hydrocarbons [56]. Such studies include the use of metal catalysts like FeCl$_3$, which demonstrated an ability to introduce carbonyl groups at the surface of PE, in the presence of H$_2$O$_2$ as an overall oxidant. This process has the advantage that etching is not a side-effect, however the concentration of function introduced is no more than 20% of a conventional oxidative etching.

Chemical methods for increasing the adhesion properties of PP do not differ substantially from those used for PE, due to the similar chemical composition of the two polymers. A huge increase of strength of PP joints has been reported by simple trichloroethylene solvent vapour exposure [64]. The effect was an increase form 2 MPa to 12 MPa for 1 second exposure, followed by a progressive decrease to the initial value if exposure is prolonged up to 30 s. Since a dull white finish is imparted to the surface, which also assumes a highly porous structure, the removal of a WBL and the possible formation of a semi-interpenetrating network structure after curing of the liquid adhesive have been indicated as responsible for the results. On the other hand, contact angle measurements did not indicate any increase of the surface free energy. The subsequent loss of strength was attributed to a rapid degradation and weakening of the surface layer.

Even if other treatments like chromium acid etching show similar effects on PP and PE [57, 58], there are some indications that in the case of PP the effect of modification is lower [6.2]. Such differences between PP and PE can be attributed to the different crystallinity of samples (isotactic PP have been used on most experiments) and the relative reactivity of crystalline and amorphous regions. Furthermore, the presence of more reactive tertiary–CH bonds can favour chain scission, producing low molecular weight chains. In this sense, etching can be more destructive for PP and contribute to the formation of a WBL.

10.3.1.2 Physical methods

Flame and plasma treatments are the most popular physical mehtods adopted for the improvement of the adhesion properties of polyolefin surfaces. Plasma treatments can be carried out both in vacuum and in air, using in the latter case a corona apparatus.

Flame treatments have been performed for many years by practitioners, but were scarcely considered from a scientific point of view. Even if flame has been used for films, it is preferably used for thick objects, such as blow-moulded containers or car bumpers [76]. The aim is to improve the adhesion performance of labels or paints, respectively. The main parameters of flame treatment are the geometry of the burner, the composition of the burning gas (generally a mixture of a light hydrocarbon like methane and air), the distance between the burner and the sample surface and the exposure time [77]. Since high temperatures, over 1000 °C, can be experienced by the polymer, the experimental assessment is critical in order to avoid excessive exposure, up to complete combustion. Exposure times of 1 second or less are typical. After exposure to flame, an oxidized layer of 4–9 nm has been observed on PE, where the oxygen concentration is remarkably increased, as well as the lap shear strength of joints [78]. Data reported in Table 10.4 do not show any effect of additives such antioxidants, which are supposed to volatilize during the process. It is also apparent that adhesion reaches a good level after a short treatment and increasing the treat-

Table 10.4   Effects of flame treatments on PE [78]

Sample	Treatment time (s)	O/C (XPS) (at.%)	Lap shear strength (MPa)
PE	0	0.2	0.5
—no additives	1.2	16.9	6.6
	4.8	31.0	7.2
PE	0	<0.2	0.4
—contains an	1.2	˙20.5	5.6
antioxidant	4.8	33.4	7.2

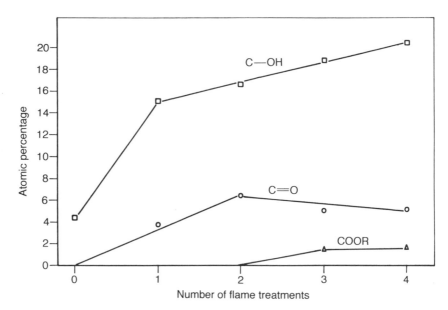

Figure 10.12 Atomic percentages of C_{1s} components on PP surface created by successive flame treatments. (From ref. [79].)

ment time does not give a substantial advantage. On PP, similar effects have been obtained, observing by XPS and SIMS surface analysis a progressive oxidation of the $-CH_3$ pendant group, with the progressive formation of hydroxyl, carbonyl and carboxylic functions [79, 80]. This behaviour has been in evidence performing several treatments in standard conditions on the same specimen (Figure 10.12). Again, only minor variations have been observed after the first treatment [79]. Comparing the adhesion strength of a paint and the variation of surface tension, a parallel behaviour has been found (Figure 10.13), suggesting that any oxygen-containing function has the same effect on adhesion and a limited concentration of them is sufficient to reach a good adhesion level [80].

Further SIMS studies suggested that some elimination of methyl groups can occur, together with the elimination of low molecular weight fragments [81]. The latter behaviour is much more effective in PP than in PE, where repeated flame treatments, studied by contact angle measurements, showed a layer-by-layer progressive degradation followed by oxidative decomposition [82]. Ethylene–propylene copolymers exhibit substantially similar results, reaching under selected flame conditions joint strengths to a polyurethane paint as high as 19 MPa, compared to 3 MPa obtained after solvent wiping and 13.4 MPa using a chlorinated primer specific for polyolefins [83]. The fracture loci were complex, so was the correlation between adhesion levels and chemistry at the surface.

Among the modification techniques of polymer surfaces, plasma is quite

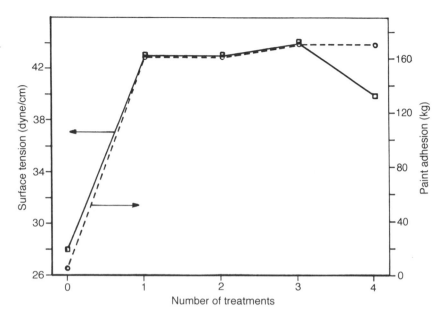

Figure 10.13 Paint adhesion (dotted line) and surface tension (continuous line) as a function of the number of surface treatments. (From ref. [80].)

popular because, depending on the adopted experimental conditions and the nature of the polymeric substrate, different effects can be obtained, several of them occurring simultaneously:

1. Plasma cleaning, due to the removal of low molecular weight contamination, like additives, processing aids and adsorbed species.
2. Plasma etching, by degradation and ablation of polymer chains on the top layer of samples.
3. Cross-linking, derived from the introduction of radicals along the polymer chains and the subsequent formation of chain-to-chain bonds by radical–radical interaction.
4. Introduction of functional groups, by reaction of the activated surface with radicals, ions and excited molecules present in the plasma region.
5. Polymerization or grafting on to the substrate surface, when a suitable monomer (reactive or polymerizable, respectively) is introduced into the plasma reactor.

All the above modification effects can influence the adhesion properties of the surface, since their effects are one or more of the following: removal of a low surface energy layer or a WBL, strengthening of the top layer, introduction of reactive groups, creation of a high energy coating tightly bonded to the substrate.

Plasma treatments with the aim to improve adhesive bonding were first used by Schonhorn and Hensen in 1966 [84]. A radiofrequency excited plasma of He applied for 10 s was sufficient to increase the double lap shear strength of a LDPE/epoxy/Al joint up to 19 MPa, starting from a value as low as 1 MPa. In the following years, many experiments were carried out on polyolefins, fluoropolymers and other polymers, examining the effect of variables [85–88]. Commercial equipment is now available from many manufacturers, suitable for treatment of articles in batch (for instance, golf balls, bottles, car bumpers), as well as roll-to-roll arrangements for continuous treatment of films or fibres. A critical review of the use of plasma treatments with the aim to improve adhesion properties of polymers is in publication [89].

Early studies by Schonhorn and coworkers [90–93] on the mechanism by which plasma enhances adhesion reported that such treatments did not change the wettability of PE. On the contrary, an amount of cross-linked polymer was found after treatment by gel permeation chromatography. Since an inert gas was used for plasma experiments, the method was often referred to as CASING (Crosslinking by Activated Species of Inert Gases). The lack of sensitive surface analysis techniques delayed the consideration of the importance of the chemical modification of surfaces as successive studies clarified [94–96]. In fact, also with inert gas plasma, the interaction of the activated surface with residual oxygen atoms present in the reactor atmosphere caused the formation of oxygenated function at the surface of treated polyolefins. With respect to PE, polypropylene has an opposite behaviour in plasma: in fact, while PE tends to cross-link, in the same conditions PP underwent chain scission [96]. This occurrence, as Table 10.5 shows, has no relevant effects on the adhesion level, even if the strength of PE joints is always higher. Results reported in Table 10.5 also show that the complexity of phenomena involved in the adhesion mechanism does not allow any provision on the strength attainable for a particular system or test from figures obtained for different systems or tests.

Plasma polymerization techniques have not been widely applied to improve

Table 10.5 Adhesion properties of oxygen plasma treated polyolefins

Sample	Shear strength (epoxy) N/mm^2	Pull test (epoxy) kg/cm^2	Pull test (Al) N/cm^2	Peel test (Al) N/cm
PE				
Untreated	0.3	—	—	—
Just treated	3.4	180	48	—
PE				
Untreated	0.2	—	—	—
Just treated	1.4	40	37	4.5
	ref. [96]	ref. [96]	ref. [97]	ref. [98]

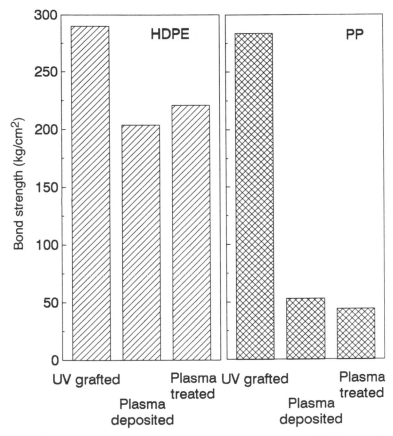

Figure 10.14 Shear strengths as a function of treatment for HDPE and PP.

the adhesion properties of polyolefins, perhaps because of the rather long time (many minutes) necessary for treatments. Among other polymers, PE and PP have been treated in a plasma of various polymerizable gases, like trimethylsilyldimethylamine, hexamethyldisilazane [99] and hydroxyethyl methacrylate (HEMA) [100]. In the latter case, a good adhesion has been obtained in pull tests on PE, compared with other modification methods like oxygen plasma and HEMA grafting by UV irradiation. Conversely, when the substrate is PP, the probable occurrence of a WBL due to the energy involved in the technique considerably lowers the adhesion strength (Figure 10.14). A somewhat different approach has been adopted by Yamakawa [101], who performed the vapour phase polymerization and grafting of methyl acrylate by γ-irradiation on to PE, followed by hydrolysis. In this way an external layer of hydrolysed poly(methyl acrylate) is obtained on to an inner grafted copolymer layer. Subsequent bonding

Table 10.6   Peel strength of PE/epoxy joints [101]

Treatment	Strength (kJ/M^2)
No treatment	< 0.04
Chromic acid etch	0.82
O$_2$ plasma	2.34
Hydrolysed PMA* graft	> 11.7[†]

* Poly(methyl acrylate)
[†] Cohesive failure

with an epoxy adhesive resulted in very high joint strengths with respect to other pretreatment methods, as reported in Table 10.6.

Corona-discharge treatments are largely applied on polymer films, in order to achieve a good performance in printing with inks, lamination, vacuum metalliz-ation, etc. [76]. Regarding the mechanism whereby corona-discharge improves the adhesion of PE, the formation of carbonyl groups and surface insaturations have been initially proposed [103]. XPS studies later confirmed the occurrence of surface oxidation, that was considered to proceed through a free radical mechanism [104–106]. Corona treatments have also been applied to PP, with less success than to PE, even if adhesion is increased to some extent [107–108]. In fact, lap shear tests on PP joints gave results rather lower than conventional

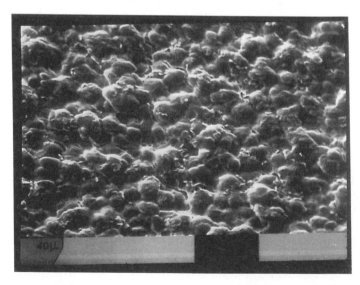

Figure 10.15   SEM micrograph of a PP surface after 30 min photografting with hydroxyethyl acrylate.

chemical treatments like chromic acid etching and the use of chlorinated primers [108].

Grafting operates on polyolefin surfaces via radical chemistry [56]. The first step consists in creating radicals at the surface, and can be performed in many ways, from hydroperoxide chemistry to plasma activation. A comfortable option to promote grafting is ultraviolet (UV) light irradiation, adopting the radical chemistry of methacrylate monomers [109–111]. Grafting on semicrystalline polymers yields a bumpy surface, probably due to the difference in the activation reaction of amorphous and crystalline domain [112, 113]. Figure 10.15 shows the surface morphology of PP after grafting under UV exposure of HEMA, carried out after preliminary activation of the polymer in benzophenone solution [100]. Results of pull tests reported in Figure 10.14 suggest that UV grafting of HEMA is quite an effective method, both for PE and PP. The reason why adhesion is higher than that achieved with other techniques is not yet clear, but can consist in a combination of contributions from the bumpy surface texture (other modified surfaces are smooth), higher oxygen concentration and cross-linking induced by UV irradiation [100].

10.3.1.3   Bulk modification methods

Bulk modification methods are meant as those methods that, even if not directly intended to modify surface properties, assume a secondary effect on it. Typical examples are blending, grafting or copolymerization of a polypolefin with another moiety containing reactive functionalities. For thermodynamic or kinetic reasons, such functionalities can be located preferentially at the surface of the material, so contributing to an increase its adhesion properties. Although widely used, these methods are far from the main purpose of this book.

Other bulk methods involve recrystallization of the polymer. The background of such methods consists in the observation that melting a polymer on to a high-energy surface like a metal or an oxide can generate a good joint strength, while the same polymer, for instance, behaves poorly in film lamination. It has been proposed, in the case of PE, that the strength observed is the result of extensive heterogeneous nucleation [114]. Even if transcrystallinity has been observed on the interfacial region of PE/metal systems, the interpretation of such results must be completed taking into account the possibility that some surface oxidation can occur during the melting process, favoured by the oxidized metal surface.

10.3.2   Fluoropolymers

The most important fluoropolymers are poly(tetrafluoroethylene) (PTFE), poly(vinyl fluoride) (PVF), poly(vinylidene fluoride) (PVDF) and various copolymers like fluorinated poly(ethylene-propylene) (FEP). Because all of them exhibit a very low surface energy (see Figure 9.3), and are also chemically inert, the task of increasing their adhesion properties is a demanding one and began to be pursued very early. A review has been published recently on the specific argument [115].

The first observation about the possibility of chemical modification was made on a PTFE stirrer that became darker if immersed in a solution of sodium and naphthalene in tetrahydrofuran [116]. In such conditions the PTFE surface was reduced to a graphite-like material, also showing some appreciable strength in epoxy joints. Patents were released on the argument [117, 118] and, when surface analysis techniques became available, the action mechanism was deepened. Using XPS and contact angle measurements on PTFE and FEP, it was found that the treatment caused depletion of fluorine, introduction of oxygen due to subsequent reaction in air of the unsaturated surface and increased wettability [119]. Etching with a solution of Na in liquid ammonia gave similar results [119]. Commercial etching systems are available based on such chemistry [120].

Reduction of fluoropolymers was also performed using other reagents like benzoin dianion $K_2[PhC(O)C(O)Ph]$ [121], obtaining modified layers of relevant depth (up to $2\,\mu m$) and reactivity. The introduction of carboxylic acid equivalents or hydroxylic groups was obtained using nucleophilic organolithium reagents [122, 123].

Etching treatment of PTFE have found important industrial applications, such as bonding of bushes to metals, bonding of insulator components to electrical appliances and improving potting seals. In order to avoid premature failure of the joints, the presence of a WBL must be avoided. In fact, if the etching time is too long, a WBL forms on fluoropolymers as an effect of degradation, as observed in a study on FEP substrates and a cross-linked styrene–butadiene–rubber adhesive [124]. Figure 10.16 illustrates how the intrinsic adhesive fracture energy increases for an etching time shorter than $500\,s$, while for longer times G_0 begins to decrease. Because UV radiation is also reactive towards etched surfaces, leading to joint failure, UV absorbers are often incorporated into the substrate and/or the adhesive.

In addition to chemical methods, electrochemical treatments have been studied, producing similar effects, i.e. a dark colour, reduced wetting and a good adhesion [125–128]. Comparative tests with Al/epoxy/PTFE/epoxy/Al joints gave good results when PTFE is reduced in the presence of an electrolyte (tetrabutylammonium tetrafluoroborate) and naphthalene in dimethylformamide as a solvent [128].

Energetic beams have been widely experimented with to modify the surface properties of fluoropolymers [115]. Ions, atoms, electrons and X-rays have been used, obtaining important morphological variations. Splitting of PTFE into filaments have been observed after heavy atom bombardment in the low keV range [129], as well as the development of microcracks and surface splitting [130]. Texturing of fluoropolymer surfaces also occurred after ion bombard-

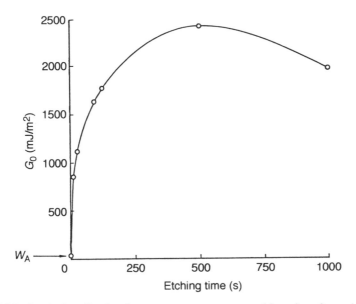

Figure 10.16 Intrinsic adhesive fracture energy versus etching time for rubber/FEP joints. (From ref. [124].)

ment, where cone-like or grass-like microstructures were observed [131,132]. Besides morphology changes, electrons [133,134] and X-ray [135] irradiation causes cross-linking. Improvement of bond strength has often been observed after such type of treatments. Metal–PTFE adhesion was found to increase after sputtering [136], ion bombardment [133,134] and electron irradiation [135]. Analogous improvements were observed in bonding with epoxy adhesives [131,137,138]. Interestingly ion bombardment was found more effective in improving peel strength than Na/naphthalene etching in the case of FEP, but less effective for PTFE (Table 10.7) [131]. Since cohesive failure was involved, the occurrence of cross-linking or WBL could have influenced such results.

Irradiation of PTFE surfaces can be performed in the presence of reactive monomers like methyl acrylate. In such conditions the monomer is grafted to the surface and can be subsequently hydrolysed [137]. Also in this case peel strengths

Table 10.7 Effects of treatment on fluoropolymer/epoxy bond strengths [131]

Polymer	Ar ion bombardment (N/cm)	Na/naphthalene etch (N/cm)
PTFE	166	244
FEP	270	30

appear higher than those obtained by etching. The mechanical strength attained by the surface layers can explain this result.

Plasma treatments have been largely applied in order to improve the adhesion properties of fluoropolymers [91, 136, 139–142]. Depending on conditions, etching of surfaces [143] and cross-linking [144] have been observed, together with changes of the surface chemistry. A deep roughening of the PTFE surface under the action of oxygen plasma has been observed (Figure 4.9) [143]. In order to obtain the best results in terms of adhesion, plasma conditions have to be adjusted in order to compensate chain breakings. The use of a mixture of a noble gas and oxygen during the discharge maximize the adhesion improvement, because reactive fragments are removed as volatiles for reaction with oxygen, thus avoiding the formation of a WBL [140]. Also the use of a hydrogen plasma gives good results [91]. Comparing the effects of O_2 or $(SF_6 + O_2)$ plasmas, where the latter gave shear strengths only 15% higher than the former, the importance of having both fluorine and oxygen radicals in the plasma atmosphere was assessed [145]. Plasma pretreatment with similar gas mixtures followed by plasma deposition of a thin metallic layer from suitable complex precursors (that is, plasma enhanced chemical vapour deposition) was used before electroless and electrochemical Cu plating of PTFE [146], obtaining excellent results in terms of peel strength.

Bulk methods for fluoropolymers have also been experimented. Melting a fluoropolymer against a high energy substrate has been found to produce a surface suitable for adhesive bonding [147, 148]. It shows high mechanical strength and increased surface free energy, attributed to transcrystallization [148] or, alternatively, to the formation of a thin oxygenated layer [119]. In such processes, the presence of an evaporated metal (Au and Al were the most frequently used) can also change the chemistry at the surface. Defluorination of the polymer, formation of organometallic species and metal fluorides [149, 150] and cross-linking [151] were observed. The complexity of the effects is also proved by the fact that different substrates gave different relative results as regards shear strength. By treatment of FEP by gold evaporation, values similar to those obtained for Na/naphthalene etching were observed [148]; in the case of

Table 10.8 Shear strengths of Al/epoxy/PTFE/epoxy/Al joints after different PTFE pretreatments [139]

Pretreatment	Shear strength (kg/cm^2)
None	7–14
Na/naphthalene etching	162–176
Recrystallization in the presence of Au	85–106
CASING with He	49–70

PTFE, this was not true, because strengths after recrystallization in the presence of Au remained well lower with respect to etching (Table 10.8) [139]. Apart from other differences in experimental conditions of treatments and joint formation, the role of WBL cannot be excluded.

10.3.3 "Reactive" Polymers

Polymers other than polyolefins and fluoropolymers generally do not present adhesion problems at the level described in the previous sections. The reasons are several, first, due to the presence of polar groups, they have a rather higher surface free energy (Figure 9.3). Furthermore, reactive pendant groups can directly participate in establishing an adhesive bond. Nevertheless, in order to maximize the adhesion strength, pretreatments have been studied and often carried out in practice. In this way, reactive groups can be used for grafting or as a site for subsequent transformations, aromatic rings can be modified by electrophilic substitution, and relatively weak bonds, like those of polyesters, can be exploited in order to change the backbone composition of the polymer.

Apart from the necessary adjustments required by the nature of the substrate and the particular adhesion problem (metallization, adhesive bondings, coatings, etc.), modification methods used for reactive polymers are substantially similar to those already described. Looking at the recent literature, chemical treatment with liquid or gaseous reagents and cold plasma treatment are the methods more studied. The considerations orienting the utilizer towards a specific treatment are not only technical, but take into account important economic and environmental issues. Chemical reagents are normally cheap and also easy to use for objects of complex shape, but the elimination of exhausts could constitute a danger or a pollution problem. On the other hand, plasma treatments are clean, safe and equally isotropic, but their cost, essentially due to the vacuum pump/plasma reactor system and the relative slowness, could discourage many applications. In fact, more sophisticated techniques, like ion bombardment, only accepted in fields where they are already present for other tasks, as in the electronics industry.

In the following, rather than giving a complete view of the subject, too extensive in this context, and available in books [152] and reviews [89], some examples are given concerning several polymers of relevant industrial importance, mainly polyesters and polyimides.

10.3.3.1 *Polyesters*

The more representative polyester from an industrial point of view is poly(ethylene terephthalate) (PET), which is used in the manufacture of fibres, containers and films. Even if PET is a polar polymer, for the presence of ester groups in the backbone, it poses adhesion problems, connected for instance to the autoadhesion of films, the metallization of the same and the bonding of labels on

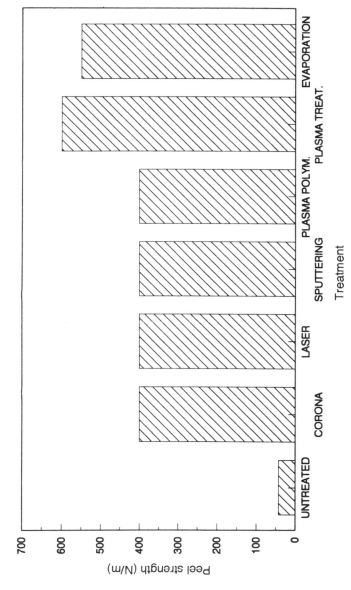

Figure 10.17 Comparison of peel strengths for Al/PET films after different pretreatments. (From ref. [159].)

bottles. In order to improve adhesion of PET, many methods have been experimented, from chemical etching with NaOH [153] to grafting [154], to discharge methods like corona [155, 156] and plasma [153, 157, 158].

After treatment, chemical and morphological modifications are observed on PET surfaces. Oxygen plasma treatment can favour many reactions, such as hydrolytic ester scission with formation of hydroxy and carboxylic end-groups, destruction of aromatic rings, attack of hydroperoxy radicals to aromatic rings or to other reactive sites, formation and polymerization of vinyl groups, etc. [159]. NaOH solutions give porous and rough surfaces for ester hydrolysis and subsequent etching [153], the adhesive properties of which are comparable or superior to those obtained by CF_4/O_2 plasma treatment. The comparison of several treatment methods in enhancing the peel strength of Al on PET showed (Figure 10.17) that most of them are effective and reach a good adhesion level, suggesting that the mechanism is almost the same in all cases [159]. Assuming that the highest level of adhesion is reached by formation of a chemical bond between metal and polymer through oxygen atoms present as functionalities in the polymer structure and as surface oxide on metals [160], a correlation was found between adhesion strength and the heat of formation of the oxide (Table 10.9). This is true for oxides having only one oxygen atom in their structure, like MgO and CuO, while in the case of Al_2O_3 geometrical consideration suggests that the formation of bonds is allowed if polymer chains are free to accommodate owing to chain breaking or fragmentation. This explains the lower adhesion strength of Al with respect to Mg, in spite of the larger heat of formation of the oxide. Cohesive failure instead of adhesive failure suggests in fact the formation of a WBL due to the above degradation phenomena [160].

Adhesion strengths of copper deposited on PET by sputter-deposition/electroplating or electron beam evaporation were improved by previous photografting on the surface of several acrylic or allylic monomers [154]. Starting from a residual adhesion on virgin PET due to physical interactions only, peel strengths considerably increased if an additional annealing of the PET/Cu samples was performed for 18 h in air. From XPS analysis, it was argued that annealing furnished the activation energy necessary to promote chemical interactions

Table 10.9 Correlation between peel strength and heat of formation of the oxide $\Delta(H)$ for evaporated metal/PET systems [60]

Metal	Oxide	ΔH (kcal/mole)	Peel strength (g/inch)
Al	Al_2O_3	− 399	400
Mg	MgO	− 144	1300
Ag	Ag_2O	− 39	140
Cu	CuO	− 7	40

between polymer and metal. It must be remarked that the annealing temperature was chosen above the glass transition temperature of PET, so greatly increased the mobility of chains.

10.3.3.2 Polyimides

Polyimides (PI) are a family of high performance polymers of increasing importance as insulators in the electronics industry, provided that they exhibit excellent

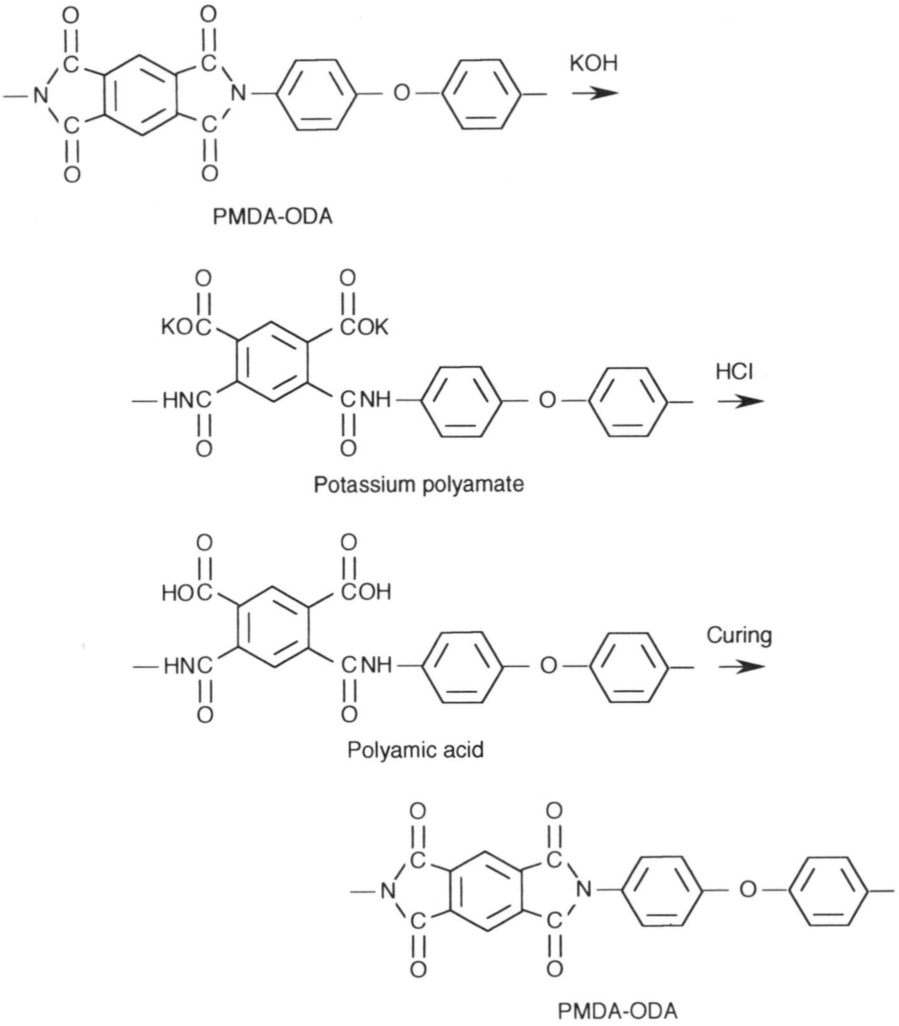

PMDA-ODA

KOH

Potassium polyamate

HCI

Polyamic acid

Curing

PMDA-ODA

SCHEME I

thermal and chemical stability, as well as very good electrical properties. In such applications, metallization with Cu or other metals is very important [161]. A typical PI, such as poly(pyromellitic dianhydride-oxydianiline) (PMDA–ODA) has in its structural formula several C=O groups, an ether bridge and two tertiary amine groups. As is shown in Scheme I, treatment with alkali (KOH or NaOH) gives the corresponding polyamate, which can be protonated to polyamic acid and finally converted back to polyimide by heating [162]. This chemistry has been exploited to increase adhesion, gaining at least one order of magnitude in peel strength in polyimide-to-polyimide systems [163]. The same approach can be applied with similar results to other polyimides [164].

Plasma treatments have been carried out in order to increase PI adhesion using several gases in the discharge, like oxygen [157, 165], water vapour [166], argon, ammonia or CF_4 [165]. A strong dependence of peel strength of vacuum deposited thin films has been observed on the radiofrequency power of oxygen plasma (Figure 10.18). Adhesion is less sensitive to treatment time [157]. Water vapour plasma caused a significant increase of oxygen concentration on the PI surface, primarily in the form of ketone, carboxyl and hydroxyl groups. Hydroxylation was considered to be the predominant effect, producing phenol and catechol groups. The effect on chromium-to-polymer interfacial adhesion was also dramatic, starting from a value of 2 g/mm to values around 40–50 g/mm, always on fully cured polyimide [166]. Considering treatments with different gases on the same plasma reactor, ammonia was found to give the best adhesion

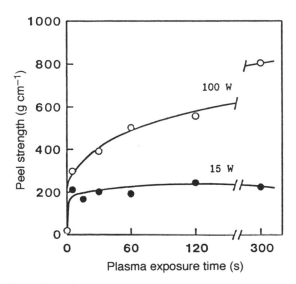

Figure 10.18 Effect of oxygen plasma exposure time on peel strength of Fe/polyimide systems at different radiofrequency power. (From ref. [157].)

of evaporated Al films (Scotch-tape peel test) with respect to CF_4/O_2 and argon in a decreasing order [164]. In another study on the adhesion of chromium after sputtering of the PI surface with Ar ions, peel strengths varied in a similar way (from 1 g/mm to 100 g/mm) [167]. The lack of morphological effects of the treatment allowed the exclusion of the role of material intermixing and mechanical interlocking on enhancing adhesion, that appears to be governed by the degree of the chemical bonding at the interface. This occurs through breaking of C=O and C—N bonds, O and N depletion and formation of a carbon-rich surface induced by ion bombardment [168, 169]. Carbide-like Cr—C bonds are supposed to form [167, 170].

The interaction of ion beams with polymer surfaces induces a wide number of chemical and physical modifications, which can be tuned to some extent by changing the type, energy and fluence of the ions. At the typical fluence of ion bombardment experiments, i.e. 10^{12}–10^{16} ions/cm^2, and starting from lower values, cross-linking is obtained, followed by loss of hydrogen atoms and heteroatoms, ejection of stable functional groups, polymer backbone rearrangements, amorphization and finally graphitization. Adhesion properties are influenced at intermediate fluence values, where some loss of heteroatoms or small molecules occur. Irradiation of a polyimide with Ar$^+$ atoms with an energy of 150 keV causes variations mainly of the concentration of imidic nitrogen and carbonyl groups (Figure 10.19) [171]. Carbonyl depletion has been explained

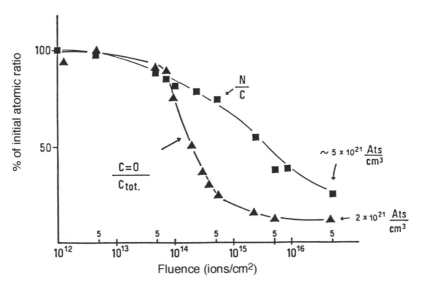

Figure 10.19   Normalized variation of XPS C=O/C$_{tot}$ and N/C ratios on the surface of BTDA–ODA polyimide bombarded with Ar$^+$ ions at 150 keV as a function of ion fluence. (From ref. [171].)

in terms of imide ring isomerization to an isoimidic form and subsequent elimination of CO_2 [172]. Such a process shows a sharp threshold in terms of fluence and goes to completion in a relatively narrow fluence range $(5 \times 10^{13} - 5 \times 10^{14}\,\text{ions/cm}^2)$. Conversely, the depletion trend of nitrogen atoms is a progressive process without a precise saturation point. The reason for the different behaviour of the two processes lies in the fact that the second process is essentially collisional, while the first one is caused by electronic interactions [171]. Besides the complex and expensive instrumentation necessary for experiments, ion beam bombardment offers the possibility to realize modifications resolved in space. The size and shape of these modified regions can be tailored by taking advantage of the focusing capability of ion beams. Coupling this possibility with fluence variations, which confers different properties to the treated polymer (i.e. adhesion, conductivity, optical properties, etc.), makes it possible to build specific devices *in situ*.

10.3.4 Fibres

The necessity to treat fibres in order to improve their adhesion properties comes mainly from their use as reinforcements in composite materials. Besides glass fibres and carbon fibres, which are largely the most used in the field, some polymeric fibres have become increasingly important materials, owing to their remarkable mechanical properties, like tensile modulus and strength: aramid-based fibres and, more recently, ultra–high–molecular weight polyethylene fibres (UHMWPE).

Owing to the nature of continuous filaments or yarns, pretreatment methods offering an isotropic result have been reserved for fibres, mainly chemical treatment and plasma treatment.

High performance PE fibres have extremely high strength (4 GPa) and modulus (120 GPa). However, they cannot be wetted by the matrix nor bonded to it, making a surface treatment necessary. In fact chromic acid etching [173], corona treatment [174] and oxygen plasma treatment [173–177], all oxidative, have been experimented. Chromic acid treatment induces a threefold increase of pull-out adhesion test on monofilaments [178], while oxygen plasma treatment resulted in a tenfold improvement [179]. This difference was attributed mainly to morphological factors. In fact, the wet treatment does not alter dramatically the surface roughness, as does the plasma treatment. It can be argued that two different mechanisms operate, i.e. chemical interactions of the matrix with functional groups formed at the surface of the fibres and mechanical interlocking. As regards the former, the nature of the matrix can have a remarkable influence. For epoxy resins, where curing agents are often amines, the introduction of amine groups on the fibre surface can usefully contribute to promoting covalent bonding. Ammonia plasma treatment has been found suitable for this aim [80]. In Table 10.10, the interlaminar shear strength of UHWMPE/epoxy composites

Table 10.10   Interlaminar shear strength of UHMWPE/epoxy
composites after surface treatment of fibres [180]

Fibre treatment	Shear strength (MPa)
Untreated	5.7
NH$_3$ plasma, 1 min	11.1
NH$_3$ plasma, 2 min	11.8
NH$_3$ plasma, 10 min	11.8
Corona discharge	7.0
O$_2$ plasma, 2 min	6.6

after various fibre treatments is reported. Ammonia plasma treatment appears more effective than corona or oxygen plasma, even at short treatment times. Longer times do not significantly increase the shear strength value, although the degree of surface amination detected by XPS continued to increase for up to 20 min and more [180]. This was attributed to the fact that increasing the interfacial shear strength through the surface amine concentration allowed failure to pass to fibre failure. From a chemical point of view, primary amine groups and probably secondary amine groups are introduced at the polymer surface [181]. Various combinations of plasma treatments with other chemical treatments have also been experimented, taking into account chromic acid etch, sulphuric acid etch and glycerol [182]. There are indications that further improvements can be achieved using a combination of treatments.

Apart from high performance fibres, applications have been carried out in other areas such as paper and concrete–polyolefin composites, operating with fibrillated PE films [183].

Aramid fibres, in particular poly(p–phenylene terephthalamide) sold by Du Pont under the trade name of Kevlar^R, are frequently used as reinforcements in composite materials, particularly in tyre cords. Owing to the high crystallinity and low availability of functional groups, Kevlar^R surfaces are rather inactive, resulting in poor adhesion to the matrix. Experimental techniques to improve such circumstances include grafting [184], substitution reactions in amide groups [185, 186], application of sizings [187], plasma treatment with various gases [188–190], also in combination with coupling agents [191].

Surface modification by chemical reaction can be obtained following Scheme II. The first step is metallation of the polymer in a solution of sodium methylsulphinylcarbanion in dimethyl sulphoxide, followed by a specific reaction. Experiments have been carried out with bromoacetic acid, acrylonitrile and n–octadecyl bromide [192]. Rough surfaces were obtained, and the depth of the modified layer was estimated at about 1 μm. Improvements of stress–strain

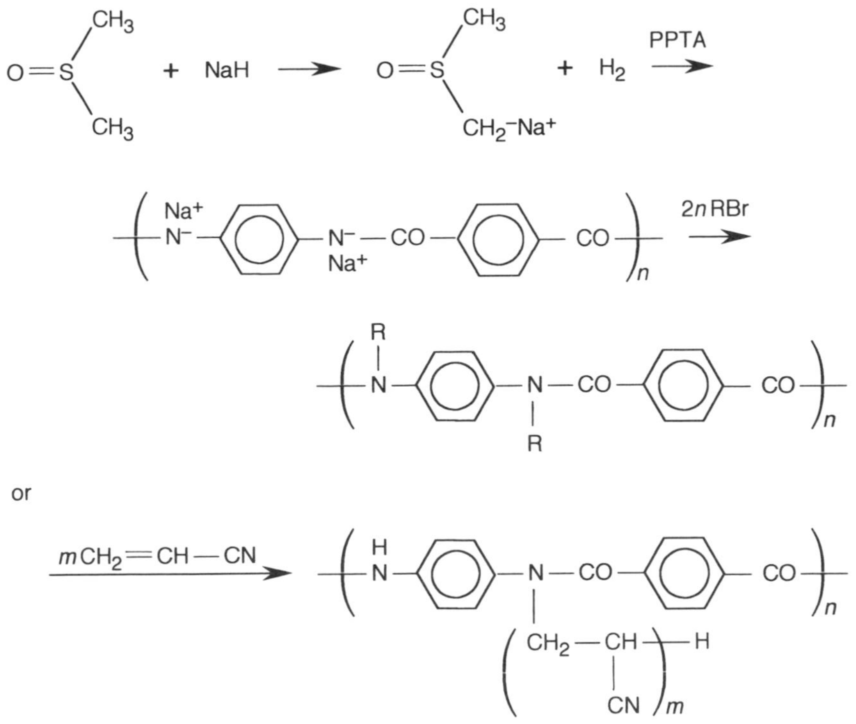

Where R = — C$_{18}$H$_{37}$, — CH$_2$COOH.

SCHEME II

curves were observed for composites made with modified fibres with respect to unmodified ones [192].

Cold plasma treatment resulted in slightly contradictory results. Treatments with air or nitrogen gave bond strength values double those of untreated fibres [188]. More recent studies with argon, nitrogen and carbon dioxide [189] showed slight improvements only in the case of nitrogen plasma in interfacial shear strength by microbond tests [193]. Two reasons can explain the difference, the difficulty of evaluating the interfacial strength of small diameter fibres and the different set up of plasma experiments. In fact, earlier measurements noted etching at the fibre surface [188], while the more recent did not [189]. The use of a coupling agent was found useful to improve the adhesion (evaluated by a pull-out test) of a Kevlar^R yarn and a silicone rubber (Figure 10.20). The coupling agent was a silicon adhesive, chosen in order to have a chemical structure compatible with that of the rubber [191].

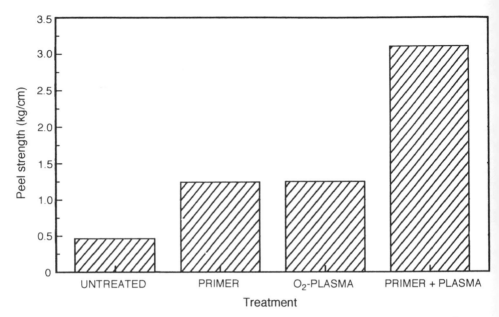

Figure 10.20 Effect of priming and plasma treatments on pull-out force on Kevlar[R] 49. (From ref. [191].)

10.3.5   Ageing Effects

Although the effect of the dynamic behaviour of polymers has been extensively treated in Chapters 2 and 9, it is important to point out here the effects that such behaviour has on adhesion. Surface treated polymers can be considered as amphipathic materials [194]. In fact, with the aim of improving adhesion, the polymer was subjected to some treatment that has often resulted in a remarkable increase in the surface concentration of hydrophilic groups. This circumstance creates a highly asymmetric structure, causing noticeable stress and consequently activating several stress-induced relaxation mechanisms, such as out-diffusion of untreated subsurface molecules through the modified layer, short-range reorientation of chains, chemical reactions between the introduced groups and the surrounding polymer segments. Such mechanisms can have a strong influence on adhesion. In Figure 10.21, the pull strength behaviour of Al coatings on various polymers is reported as a function of ageing time. In the case of PE, no dependence on ageing was observed, while for PP, PET and PC a decrease of adhesion with ageing time occurred. A steep initial decrease was followed by a plateau, which is near 0 in the case of PP. Water contact angle measurements and surface characterization by XPS, SIMS and SEM indicated a chemical mechanism as responsible for the Al/polymer adhesion. In this case, the ability to allow hydrophobic recovery seems the driving force for adhesion variations: PE

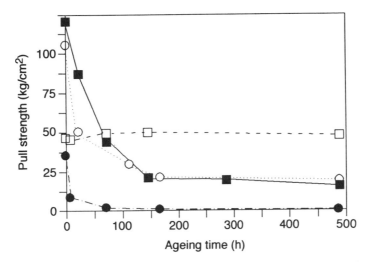

Figure 10.21 Pull strength of Al coatings on plasma treated polymers as a function of ageing at room temperature; PE: (□); PP:(●); PET: (○); PC: (■).

is rather stable in its tendency to cross-linking, forming a less mobile surface layer, while in PP a fast reorientation of oxygenated groups for the topmost layer is allowed. The other polymers also have an intermediate behaviour for the intrinsic polarity of their structure [97].

REFERENCES

[1] K. W. Allen, Proc. 5th Internat. Joint Military/Government–Industry Symp. on Structural Adhesive Bonding, Nov. 1987, Picatinny Arsenal, Dover, p. 1.

[2] W. C. Wake, *Adhesion and the Formulation of Adhesives*, Applied Science Publ., Barking, UK, 2nd Edn (1982).

[3] A. J. Kinloch, *Adhesion and Adhesives*, Chapman & Hall, London (1987).

[4] L.-H. Lee (Ed), *Fundamentals of Adhesion*, Plenum Press, New York (1991).

[5] L.-H. Lee (Ed), *Adhesive Bonding*, Plenum Press, New York (1991).

[6] V. L. Vakula and L. M. Pritykin, *Polymer Adhesion*, Ellis Horwood, New York (1991).

[7] J. W. Maxwell, *Trans. Am. Soc. Mech. Engns*, **67**, (1945) 104.

[8] E. M. Boroff and W. C. Wake, *Trans. Inst. Rubber Industry*, **25** (1949) 199 and 210.

[9] B. M. Haines, in *Aspects of Adhesion 3*, D. J. Alner (Ed.), Chap. 3, p. 40, Univ. of London Press (1967).

[10] J. D. Venables, *J. Mater. Sci.*, **19** (1984) 2431.

[11] L. E. Perrins and K. Pettett, *Plastics and Polymers*, **39**, 391 (1971).

[12] S. S. Voyutskii, in *Autohesion and Adhesion of High Polymers*, Wiley–Interscience, New York (1963).

[13] N. A. Krotova and L. P. Morozova, *Dokl. Akad. Nauk SSSR 127*, 141 (1959).

[14] R. M. Vasenin, *Adhesive Age*, **8**, 21 and 30 (1965).

[15] J. N. Anand, *J. Adhesion*, **1**, 31 (1969); **2**, 23 (1970); **5**, 265 (1973).

[16] S. E. Bresler, G. M. Zacharov, and S. V. Kirillov, *Vysok Soed.*, **3**, 1072 (1961).

[17] N. A. Krotova and L. P. Morozova, *Kokl. Akad. Nauk. SSSR 127*, 141 (1959).

[18] B. V. Derjaguin, *Research*, **8**, 70 and 363 (1955).

[19] C. Weaver, in *Aspects of Adhesion 5*, D. J. Alner (Ed.), Ch. X, p. 262, Univ. of London Press (1969).

[20] B. V. Derjaguin, N. A. Krotova, and V. P. Smilga, *Adhesion of Solids*, Engl. Transl. (by R. K. Johnson), p. 143, Plenum Press, New York (1978).

[21] W. Possart and A. Roder, *Phys. Status Solidi*, **A84**, 319 (1984).

[22] W. Possart and J. Muller, *Phys. Status Solidi*, **A106**, 525 (1988).

[23] W. Possart, *Int. J. Adhes. Adhes.*, **8**, 77 (1988).

[24] A. D. Roberts, *J. Phys.*, **D10**, 1801 (1977).

[25] D. A. Hays, in Ref. [4], Ch. 8, p. 249, Plenum Press, New York (1991).

[26] W. Gutowski, in Ref. [4], Ch. 2, p. 87, Plenum Press, New York (1991).

[27] R. J. Good and M. K. Chaudhury, in Ref. [4], Ch. 3, p. 137, Plenum Press, New York (1991).

[28] R. J. Good, M. K. Chaudhury, and C. J. van Oss, in Ref. [4], Ch. 4, p. 153, Plenum Press, New York (1991).

[29] F. M. Fowkes, *J. Adhesion*, **4**, 155 (1972).

[30] R. S. Drago, G. C. Vogel, and T. E. Needham, *J. Am. Chem. Soc.*, **93**, 6014 (1971).

[31] R. S. Drago, L. B. Parr, and C. S. Chamberlain, *J. Am. Chem. Soc.*, **99**, 3203 (1977).

[32] F. M. Fowkes and S. Maruchi, *Organic Coatings and Plastics Chemistry Preprints*, Am. Chem. Soc., **37**, 605 (1977).

[33] F. M. Fowkes, in *Physicochemical Aspects of Polymer Surfaces*, K. L. Mittal (Ed.), Vol. 2, p. 583, Plenum Press, New York (1985).

[34] S. Wu, *Polymer Interface and Adhesion*, Marcel Dekker, New York (1982).

[35] K. L. Mittal and J. R. Susko (Eds), *Metallized Plastics 1*, Plenum Press, New York (1989).

[36] K. L. Mittal (Ed.), *Metallized Plastics 2*, Plenum Press, New York (1991).

[37] K. L. Mittal (Ed.), *Metallized Plastics 3*, Plenum Press, New York (1992).

[38] ASTM D 3359–83. Standard Method for Measuring Adhesion by Tape Test.

[39] Y. H. Jeng, F. Faupel, S. T. Chen, and P. S. Ho, in Ref. [35], p. 265.

[40] A. J. Kinloch (Ed.), *Structural Adhesives*, Elsevier Applied Science, New York (1986).

[41] A. J. Kinloch (Ed.), *Durability of Structural Adhesives*, Applied Science, New York (1983).

[42] S. R. Hartshorn (Ed.), *Structural Adhesives*, Plenum Press, New York (1986).

[43] J. D. Minford (Ed.), *Treatise on Adhesion and Adhesives*, Vol. 7, Marcel Dekker, New York (1991).

[44] I. Skeist (Ed.), *Handbook of Adhesives*, Van Nostrand Reinhold, New York (1977).

[45] Stuart M. Lee (Ed.), *International Encyclopedia of Composites*, VCH Publ., New York (1990) Vol. 1, p. vii.

[46] E. P. Plueddemann, *Silane Coupling Agents*, Plenum Press, New York (1982).

[47] L. T. Drzal, M. J. Rich, M. F. Koenig, and P. F. Lloyd, *J. Adhes.*, **16**, 133 (1983).

[48] D. L. Caldwell, in Ref. [45], Vol. 2, p. 361.

[49] G. M. Nevaz, *SAMPE Q*, **15**, 20 (1984).

[50] F. Mandell, J. H. Chen, and F. J. McGarry, *Int. J. Adhes. Adhes.*, **1**, 40 (1980).

[51] J. P. Favre and J. Perrin, *J. Mater. Sci.*, **7**, 1113 (1972).

[52] B. Miller, P. Muri, and L. Rebenfeld, *Compos. Sci. Technol.*, **28**, 17 (1986).

[53] L. Ongchin, W. K. Olender, and F. H. Ancker, *Proc. 27th Annual Techn. Conf. Reinforced Plastics/Composites Div.*, SPI, Sect. 11-A (1972).

[54] W. A. Fraser, F. H. Ancker, A. T. DiBenedetto, and B. Elbirli, *Polym. Compos.*, **4**, 234 (1983).

[55] K. L. Mittal, *J. Adhes. Sci. Technol.*, **1**, 247 (1987).
[56] D. E. Bergbreither, in *Chemically Modified Surfaces*, H. A. Mottola and J. R. Steinmetz (Eds), Elsevier, Amsterdam (1992), p.133.
[57] D. Briggs, D. M. Brewis, and M. B. Konieczko, *J. Mater. Sci.*, **11**, 1270 (1976).
[58] D. Briggs, V. J. Zichy, D. M. Brewis, J. Comyn, R. H. Dahms, M. A. Green, and M. B. Konieczko, *Surf. Interf. Anal.*, **2**, 107 (1980).
[59] S. R. Holmes-Farley and G. M. Whitesides, *Langmuir*, **3**, 62 (1987).
[60] L. M. Siperko, *Appl. Spectrosc.*, **43**, 226 (1989).
[61] A. Chew, D. M. Brewis, D. Briggs, and R. H. Dahm, in *Adhesion 8*, K. W. Allen (Ed.), Ch. 6, Elsevier Applied Science (1984).
[62] P. Blais, J. Carlsson, C. W. Csullog, and D. M. Wiles, *J. Coll. Interf. Sci.*, **47**, 636 (1974).
[63] H. W. Ranhut, *Adhesive Age*, **12**, 28 (1969).
[64] E. W. Garnish and G. G. Haskins, in *Aspects of Adhesion 5*, D. J. Alner (Ed.), p. 259, University of London Press, London (1969).
[65] A. Chew, R. H. Dahm, D. M. Brewis, D. Briggs, and D. G. Rance, *J. Coll. Interf. Sci.*, **110**, 88, (1986).
[66] C. E. M. Morris, *J. Appl. Polymer Sci.*, **14**, 2171 (1970) and **15**, 501 (1971).
[67] D. A. Olsen and A. J. Osteraas, *J. Polymer Sci.*, **Part A-1**, 7, 1913, 1921 and 1927 (1969).
[68] P. F. McDonnell, in *Adhesion 15*, K. W. Allen (Ed.), Ch. 5, Elsevier Applied Science (1991).
[69] Toa Gosei Chem. Ind. Co., U.S. Patent 4837260 (1989).
[70] Three Bond Co., Eur. Patent 271675 (1988).
[71] Loctite Co., U.S. Patent 4869772 (1989).
[72] J. F. Elman, L. F. Gerenser, K. E. Goppert-Berarducci, and J. E. Pochan, *Macromolecules*, **23**, 3922 (1990).
[73] E. M. Cross and T. J. McCarthy, *Polymer Prepr.*, **31**, 422 (1990).
[74] J. F. Kinstle and S. L. Watson, Jr., *Polymer Sci. Technol.*, **10**, 461 (1977).
[75] I. Brass, D. M. Brewis, I. Sutherland, and R. Wiktorowicz, *Int. J. Adhes. Adhes.*, **11**, 150 (1991).
[76] D. Briggs, in *Surface Analysis and Pretreatment of Plastics and Metals*, D. M. Brewis (Ed), p. 199, Applied Science Publ., London (1982).
[77] I. Sutherland, D. M. Brewis, R. J. Heath, and E. Sheng, *Surf. Interf. Anal.*, **17**, 507 (1991).
[78] D. Briggs, D. M. Brewis, and M. B. Konieczko, *J. Mater. Sci.*, **14**, 1344 (1979).
[79] F. Garbassi, E. Occhiello, and F. Polato, *J. Mater. Sci.*, **22**, 207 (1987).
[80] F. Garbassi, E. Occhiello, F. Polato, and A. Brown, *J. Mater. Sci.*, **22**, 1450 (1987).
[81] Y. DePuydt, D. Leonard, and P. Bertrand, in Ref. [37] p. 245.
[82] D. Y. Wu, E. Papirer, and J. Schultz, *C. R. Acad. Sci. Paris*, **312**, Sect. II, 197 (1991).
[83] E. Sheng, I. Sutherland, D. M. Brewis, and R. J. Heath, *Surf. Interf. Anal.*, **19**, 151 (1992).
[84] H. Schonhorn and R. H. Hansen, *J. Polymer Sci.*, **B4**, 203 (1966).
[85] J. R. Hall, C. A. Westerdahl, A. T. Devine, and M. J. Bodnar, *J. Appl. Polymer Sci.*, **13**, 2085 (1969).
[86] J. R. Hall, C. A. Westerdahl, M. J. Bodnar, and D. W. Levi, *J. Appl. Polymer Sci.*, **16**, 1465 (1972).
[87] P. W. Rose and E. M. Liston, *Plastics Eng.*, **43**, 41 (1985).
[88] H. Yasuda, C. E. Lamaze, and K. Sakaden, *J. Appl. Polymer Sci.*, **17**, 137 (1973).
[89] E. M. Liston, L. Martinu, and M. R. Wertheimer, *J. Adhes. Sci. Technol.*, in press (1993).
[90] H. Schonhorn, F. W. Ryan, and R. H. Hansen, *J. Adhesion*, **2**, 93 (1970).

[91] H. Schonhorn and R. H. Hansen, *J. Appl. Polymer Sci.*, **11**, 1461 (1967).

[92] H. Schonhorn, in *Adhesion, Fundamentals and Practice* (Ed.), p. 12, McLaren & Son, London (1969).

[93] H. Schonhorn and R. H. Hansen, in *Adhesion, Fundamentals and Practice* (Ed), p. 22. McLaren & Son, London (1969).

[94] S. Wu, *Polymer Interface and Adhesion*, Marcel Dekker, New York (1982).

[95] H. V. Boenig, *Plasma Science and Technology*, Cornell University Press, Ithaca (1982).

[96] M. Morra, E. Occhiello, L. Gila, and F. Garbassi, *J. Adhesion*, **33**, 77 (1990).

[97] M. Morra, E. Occhiello, and F. Garbassi, in Ref. [36], p. 363.

[98] V. André, F. Arefi, J. Amouroux, G. Lorang, Y. DePuydt, and P. Bertrand, in *Polymer–Solid Interfaces*, J. J. Pireaux, P. Bertrand and J. L. Bredas (Eds), p. 269, IOP Publishing Ltd, Bristol (1992).

[99] N. Inagaki, A. Kishi, and K. Katsuura, *Int. J. Adhes. Adhes.*, **2**, 223 (1982).

[100] M. Morra, E. Occhiello, and F. Garbassi, *Proc. 16th Meeting of Adhesion Society*, F. J. Boerio (Ed.), p. 393 (1993).

[101] S. Yamakawa, *Macromolecules*, **12**, 1222 (1979).

[102] S. Yamakawa and F. Yamamoto, *J. Appl. Polymer Sci.*, **25**, 25 and 41 (1980).

[103] K. Rossman, *J. Polymer Sci.*, **19**, 141 (1956).

[104] A. R. Blythe, D. Briggs, C. R. Kendall, D. G. Rance, and V. J. I. Zichy, *Polymer*, **19**, 1273 (1979).

[105] D. Briggs and C. R. Kendall, *Polymer*, **20**, 1053 (1979).

[106] D. Briggs and C. R. Kendall, *Int. J. Adhes. Adhes.*, **2**, 13 (1982).

[107] D. Briggs, C. R. Kendall, A. R. Blythe, and A. B. Wootton, *Polymer*, **24**, 47 (1983).

[108] A. Beevers and T. Norris, in *Adhesion 15*, K. W. Allen (Ed.), Ch. 4, Elsevier Applied Science (1991).

[109] H. Kubota, N. Yoshino, and Y. Ogiwara, *J. Appl. Polymer Sci.*, **39**, 1231 (1981).

[110] Y. Ogiwara, M. Kanda, M. Takumi, and H. Kubota, *J. Polymer Sci., Polymer Lett. Ed.*, **19**, 457 (1981).

[111] K. Allmer, A. Hult, and B. Ranby, *J. Polymer Sci., Polymer Chem. Ed.*, **26**, 2099 (1988) and **27**, 1641 (1989).

[112] B. D. Ratner, in *Surface and Interfacial Aspects of Biomedical Polymers*, J. D. Andrade (Ed.), Plenum Press, New York, p. 373 (1985).

[113] M. Morra, E. Occhiello, and F. Garbassi, *J. Coll. Interf. Sci.*, **149**, 290 (1992).

[114] H. Schonhorn, in *Polymer Surfaces*, D. T. Clark and W. J. Feast (Eds), Ch. 10, John Wiley & Sons, Chichester (1978).

[115] L. M. Siperko and R. R. Thomas, *J. Adhes. Sci. Technol.*, **3**, 157 (1989).

[116] E. R. Nelson, T. J. Kilduff, and A. A. Benderly, *Ind. Eng. Chem.*, **50**, 329 (1958).

[117] R. J. Purvis and W. R. Beck, U. S. Patent 2789063 (to 3M) (1957).

[118] G. Rappaport, U. S. Patent 2809130 (to General Motors Corp.) (1957).

[119] D. W. Dwight and W. M. Riggs, *J. Coll. Interf. Sci.*, **47**, 650 (1974).

[120] W. Goldie, in *Metallic Coatings of Plastics 1*, p. 314, Electrochemical Publications Ltd, London (1988).

[121] C. A. Costello and T. J. McCarthy, *Macromolecules*, **19**, 2819 (1987).

[122] A. J. Dias and T. J. McCarthy, *Macromolecules*, **20**, 2068 (1987).

[123] K.-W. Lee and T. J. McCarthy, *Macromolecules*, **21**, 2318 (1988).

[124] E. H. Andrews and A. J. Kinloch, *Proc. Roy. Soc.*, **A332**, 385 and 401 (1973).

[125] D. M. Brewis, D. J. Barker, R. H. Dahm, and L. R. J. Hoy, *Electrochim. Acta*, **23**, 1107 (1978).

[126] D. M. Brewis, D. J. Barker, R. H. Dahm, and L. R. J. Hoy, *J. Mater. Sci.*, **14**, 749 (1979).

[127] R. H. Dahm, D. J. Barker, D. M. Brewis, and L. R. J. Hoy, in *Adhesion 4*, K. W. Allen (Ed.), p. 215, Applied Science Publ., London (1980).
[128] D. M. Brewis, R. H. Dahm, and M. B. Konieczko, *Makromol. Chem.*, **43**, 191 (1975).
[129] R. Michael and D. Stulik, *Nucl. Instrum. Methods Phys. Res.*, **B14**, 278 (1986).
[130] R. Michael and D. Stulik, *J. Vac. Sci. Technol.*, **A4**, 1861 (1986).
[131] B. A. Banks, J. S. Sovey, T. B. Miller and K. S. Crandall, *NASA Rep. TM-7888* (June 1978).
[132] M. J. Mirtich and J. S. Sovey, *NASA Rep TM-79004* (Dec. 1978).
[133] B. T. Werner, T. Vreeland, Jr., M. H. Mendenhall, Y. Qui, and T. A. Tombrellow, *Thin Solid Films*, **104**, 163 (1983).
[134] G. M. Sessler, J. E. West, F. W. Ryan, and H. Schonhorn, *J. Appl. Polym. Sci.*, **17**, 3199 (1973).
[135] D. R. Wheeler and S. V. Pepper, *NASA Rep. TM-83413* (Aug. 1983).
[136] C.-A. Chang, *Appl. Phys. Lett.*, **51**, 1236 (1987).
[137] S. Yamakawa, *Macromolecules*, **12**, 1222 (1979).
[138] J. A. Kelber, J. W. Rogers, Jr., and S. J. Ward, *J. Mater. Res.*, **1**, 717 (1986).
[139] H. Schonhorn and F. W. Ryan, *J. Adhesion*, **1**, 43 (1969).
[140] A. Moshonov and Y. Avny, *J. Appl. Polym. Sci.*, **25**, 771 (1980).
[141] T. Hirotsu and S. Ohnishi, *J. Adhesion*, **11**, 57 (1980).
[142] E. Ruckenstein and S. V. Gourisankar, *J. Coll. Interf. Sci.*, **107**, 488 (1985).
[143] M. Morra, E. Occhiello, and F. Garbassi, *Langmuir*, **5**, 872 (1989).
[144] W. K. Fisher and J. C. Corelli, *J. Polym. Sci., Polym. Chem. Ed.*, **19**, 2465 (1981).
[145] G. P. Hansen, R. A. Rushing, R. W. Warren, S. L. Kaplan, and O. S. Kolluri, *Int. J. Adhes. Adhes.*, **11**, 247 (1991).
[146] H. Meyer, R. Schulz, H. Suhr, C. Haag, K. Korn, and A. M. Bradshaw, in Ref. [36], p. 121.
[147] H. Schonhorn, *Macromolecules*, **1**, 145 (1968).
[148] H. Schonhorn and F. W. Ryan, *J. Polym. Sci.*, **7**, (A2), 105 (1969).
[149] H. Schonhorn and R. F. Roberts, *Polym. Prepr. ACS Div. Polym. Chem.*, **16**, 146 (1975).
[150] M. S. Toy, *J. Polym. Sci.*, **C34**, 273 (1971).
[151] S. L. Vogel and H. Schonhorn, *J. Appl. Polym Sci.*, **23**, 495 (1979).
[152] A. J. Kinloch, *Adhesion Land Adhesives*, Ch. 4, Chapman Hall, London (1987).
[153] E. Occhiello, F. Garbassi, and G. Morini, *Mat. Res. Soc. Symp. Proc.*, **119**, 235 (1988).
[154] R. D. Goldblatt, J. M. Park, R. C. White, L. J. Matienzo, S. J. Huang, and J. F. Johnson, *J. Appl. Polym. Sci.*, **37**, 335 (1989).
[155] D. K. Owens, *J. Appl. Polym. Sci.*, **19**, 3315 (1975).
[156] B. Leclercq, M. Sotton, A. Baszkin, and L. Ter-Minassian- Saraga, *Polymer*, **18**, 675 (1977).
[157] K. Nakamae, S. Tanigawa, and T. Matsumoto, in Ref. [35], p. 235.
[158] L. Martinu, J. E. Klemberg-Sapieha, H. P. Schreiber, and M. R. Wertheimer, *Le Vide, Suppl.*, **258**, 13 (1991).
[159] J. F. Friedrich, in *Polymer—Solid Interfaces*, J. J. Pireaux, P. Bertrand, and J. L. Bredas (Eds), p. 443, IOP Publishing Ltd, Bristol (1992).
[160] J. F. Silvain, A. Veyrat, and J. J. Erhardt, in *Polymer—Solid Interfaces*, J. J. Pireaux, P. Bertrand and J. L. Bredas (Eds), p. 281, IOP Publishing Ltd, Bristol (1992).
[161] K. L. Mittal (Ed.), *Polyimides: Synthesis, Characterization and Applications*, Vols 1 and 2, Plenum Press, New York, (1984).
[162] N. Nishizaki, *J. Chem. Soc. Japan, Ind. Chem. Sect.*, **69**, 1393 (1966).

[163] K.-W. Lee, S. P. Kowalczyk, and J. M. Shaw, *Macromolecules*, **23**, 2097 (1990).
[164] K.-W. Lee, S. P. Kowalczyk, and J. M. Shaw, *Langmuir*, **7**, 2450 (1991).
[165] J. E. Klemberg-Sapieha, L. Martinu, O. M. Kuttel, and M. R. Wertheimer, in Ref. [36], p. 315.
[166] R. D. Goldblatt, L. M. Ferreiro, S. L. Nunes, R. R. Thomas, N. J. Chou, L. P. Buchwalter, J. E. Heidenreich and T. H. Chao, *J. Appl. Polym. Sci.*, **46**, 2189 (1992).
[167] T. S. Oh, S. P. Kowalczyk, D. J. Hunt, and J. Kim, *J. Adhes. Sci. Technol.*, **4**, 119 (1990).
[168] J. Davenas, G. Boiteux, X. L. Xu, and E. Adem, *Nucl. Instrum. Methods*, **B32**, 136 (1988).
[169] G. Marletta, S. Pignataro, and C. Oliveri, *Nucl. Instrum. Methods*, **B39**, 792 (1989).
[170] F. S. Ohuchi and S. C. Freilich, *J. Vac. Sci. Technol.*, **A4**, 1039 (1986).
[171] S. Pignataro and G. Marletta, in Ref. [36], p. 268.
[172] G. Marletta, C. Oliveri, G. Ferla, and S. Pignataro, *Surf. Interf. Anal.*, **12**, 447 (1988).
[173] M. Nardin and I. M. Ward, *Mater. Sci. Technol.*, **3**, 814 (1987).
[174] S. L. Kaplan, P. W. Rose, H. X. Nguyen, and H. W. Chang, in *Proc. 33rd Internat. SAMPE Symp.*, G. Carrillo, E. D. Newell, W. D. Brown, and P. Phelan (Eds), p. 551, SAMPE, Covina (1988).
[175] H. X. Nguyen, G. Riahi, G. Wood, and A. Poursartip, in *Proc. 33rd Internat. SAMPE Symp.*, G. Carrillo, E. D. Newell, W. D. Brown, and P. Phelan (Eds), p. 1721, SAMPE, Covina (1988).
[176] N. H. Ladizesky and I. M. Ward, *Compos. Sci. Technol.*, **26**, 129 (1986).
[177] N. H. Ladizesky and I. M. Ward, *J. Mater. Sci.*, **18**, 533 (1983).
[178] I. M. Ward and N. H. Ladizesky, in *Composite Interfaces*, H. Ishida and J. L. Koenig (Eds), p. 37, Elsevier, New York (1986).
[179] N. H. Ladizesky, I. M. Ward, and L. N. Phillips, U. S. Patent 4410586 (1983).
[180] P. J. C. Chappell, J. R. Brown, G. A. George, and H. A. Willis, *Surf. Interf. Anal.*, **17**, 143 (1991).
[181] S. Holmes and P. Schwartz, *Compos. Sci. Technol.*, **38**, 1 (1990).
[182] S. Gao and Y. Zeng, *J. Appl. Polym. Sci.*, **47**, 2065 (1993).
[183] B. Westerlind, A. Larsson, and M. Rigdahl, *Int. J. Adhes. Adhes.*, **7**, 141 (1987).
[184] M. Takayanagi, *Pure Appl. Chem.*, **55**, 819 (1983).
[185] M. Takayanagi and T. Katayose, *J. Polym. Sci., Polym. Chem. Ed.*, **19**, 1133 (1981).
[186] M. Takayanagi, S. Ueta, W.-Y. Lei, and K. Koga, *Polym. J.*, **19**, 467 (1987).
[187] D. B. Eagles, B. F. Blumentritt, and S. L. Cooper, *J. Appl. Polym. Sci.*, **20**, 435 (1976).
[188] M. R. Wertheimer and H. P. Schreiber, *J. Appl. Polym. Sci.*, **26**, 2087 (1981).
[189] K. Kupper and P. Schwartz, *J. Adhes. Sci. Technol.*, **5**, 165 (1991).
[190] F. Garbassi, E. Occhiello, L. Nicolais, and R. D'Agostino, *Proc. 9th Int. SAMPE Europ. Conf.*, p. 47, SAMPE (1988).
[191] N. Inagaki, S. Tasaka, and H. Kawai, *J. Adhes. Sci. Technol.*, **6**, 279 (1992).
[192] M. Takayanagi, T. Kajiyama, and T. Katayose, *J. Appl. Polym. Sci.*, **27**, 3903 (1982).
[193] L. S. Penn and S. M. Lee, *Fibre Sci. Technol.*, **17**, 91 (1982).
[194] M. Morra, E. Occhiello, and F. Garbassi, in *Polymer—Solid Interfaces*, J. J. Pireaux, P. Bertrand, and J. L. Bredas (Eds), p. 407, IOP Publishing Ltd, Bristol (1992).

Chapter 11

Barrier Properties

A barrier is of course related to permeability, which in turn is the combination of permeant diffusion and solubility in the polymer. Lowering one or both of them results in an improved barrier. Imparting barrier properties is a paradigm of surface treatments, in fact it involves maintaining bulk related properties, essentially mechanical ones, while getting a barrier approximating that of other (barrier) polymers or even inorganic materials, such as metal and glass.

Improving the barrier properties of polymers is dramatically important in packaging applications (food, pharmaceuticals, cosmetics, fine chemicals, etc.). A barrier is also important whenever a permeant material has to be transported. Tubes and hoses for gasoline, fluorocarbon vapours, etc. (important in automotive and appliance applications), require solutions to improve barrier.

The type of barrier varies depending on the application, yet it is possible to draw a division in three main groups, namely barriers to inert gases (e.g. oxygen, nitrogen), to chemically active gases (e.g. water, carbon dioxide), and finally to liquids and vapours (e.g. aromas, fine chemicals, gasoline). Of course, the extent of the barrier and the shape of the container also depend on the application.

So far three kinds of solutions have been suggested, namely multilayer systems (obtained by coextrusion, coinjection, lamination), blends with lamellar morphology and surface treatments. The latter can be further divided in coating (most frequently from emulsions) with barrier polymers (usually vinylidene chloride copolymers for packaging applications) and 'true' surface treatments (fluorination, sulphonation, metallization, plasma enhanced chemical vapour deposition).

Packaging being of paramount economic importance, the existing literature consists mainly of patents, although contributions from academic groups are also available.

11.1 COATING

The coating of polymer films and containers with barrier polymers has been practised since the 1960s and 1970s and has been reviewed in a few books [1–5].

The typical substrates are commodity polymers such as PE, PP and PET. In the case of films, mainly biaxially oriented PET and PP are treated, although polyethylene has become more important recently. The basic patents date back to the late 1950s [6–7] and the apparatus (schematics in Figure 11.1) is perfectly consistent with that used for the printing and lamination of plastic films [1].

As usual in fields which are already 'mature' little academic literature is available. Patent activity was substantial up to the early 1980s and mostly related to coating formulations and apparatus. Although it may seem controversial, coating with polymeric layers is not really a surface problem. In fact, the typical coating formulations are adjusted to reasonably wet polymer surfaces and the only possible problem lies in adhesion. The latter is favoured by heat and swelling intrinsic in the coating process and can at most be further enhanced by corona treatment of the parent film or priming.

By far the most used system is PVDC [8–26], which should be understood as a family of copolymers with the main component being vinylidene chloride and various comonomers, among which are vinyl chloride, acrylonitrile, methyl methacrylate, etc. Chemical modifications of the polymer are aimed mainly at improving its processing characteristics, namely its ability to form water emulsions of satisfactory properties. Other patents are concerned with coating formulation and machinery to increase the rate and efficiency of coating deposition.

Further patents were devoted to the use of poly(vinyl alcohol) and ethylene-vinyl alcohol copolymers as coating layers [27–34]. These systems are technically much more troublesome, owing to their water affinity. This complicates their deposition, sometimes requiring cross-linking to stabilize them, and worsens the oxygen barrier in humid environments, owing to water-induced swelling. Yet the recent drive to recycling consciousness is pushing aside vinylidene chloride-based systems and giving more momentum to vinyl alcohol based coatings, which are presently intensely experimented with.

In this case some patents have been devoted to the coating formulation itself, to deposition techniques and to the improvement of its interface with the substrate,

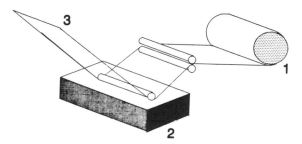

Figure 11.1   Schematics of a coating line, the substrate (1) is drawn through the emulsion of the barrier polymer (2) and emerges coated (3) to undergo further processing, e.g. drying.

for instance claiming the use of primers or high energy density treatments [35–36].

The aforementioned systems are those which actually achieved industrial application. In the patent literature a number of other chemistries has been reported, for applications ranging from food packaging (see the examples in Figure 11.2) to containers or pipes for gasoline or fine chemicals.

Lots of industrial research has been spent in looking for alternative barrier polymers to be deposited from solution or emulsion. Among systems investigated

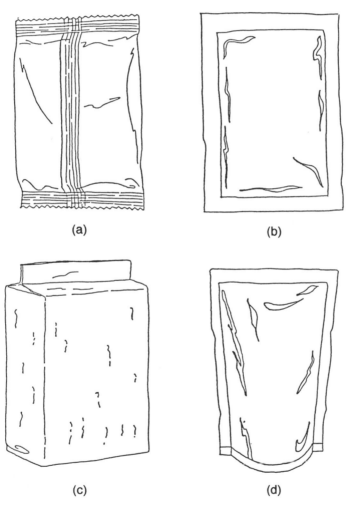

(a)   (b)

(c)   (d)

Figure 11.2 Examples of packages based on film structures including barrier layers: pillow pouch (a), four-side fin (b), vacuum brick (c), delta-pac (d).

it is possible to quote polyesters [37–38], polyamides [39–40], acrylics [41–43], isocyanate polymers [44], siloxanes [45–46] and even biodegradable polymers such as polysaccharides [47–48].

Many of these activities stem from research devoted to scratch-resistant layers (see corresponding chapter). Apart from the chemistry and formulation of coating systems, some patents, for instance those related to siloxane coatings, deal with curing processes, with the use of electron beams [45] or plasma [46]. Others are related to improving the compatibility of these systems with existing production cycles within the film or container industries.

Coating is now a sound industrial activity, particularly in the case of films, although application to containers, mainly bottles, is reported as well. Some concerns are being voiced due to recycling problems. The typical coating thicknesses (at least a few µm) are such that the bulk per cent of the barrier polymer in recyclate is a few per cent, which sometimes, especially in the PVDC case seriously hinders recycling. Furthermore, in many countries chlorine containing polymers are being excluded from use in food packaging.

11.2 SULPHONATION AND FLUORINATION

These techniques were developed based on the need for improving the barrier properties of HDPE fuel tanks and plastic containers for solvents, fine chemicals, etc.

Sulphonation [49–57] was developed based on know-how for the chlorosulphonation of polyethylene. In the most frequent version, the process involves the introduction in a preformed tank or container or sulphur trioxide or chlorine/ sulphur trioxide mixtures. Later on the excess acidity is removed by ammonia treatment. Finally the tank is washed to remove process residues. The process results in the formation of a barrier layer a few micrometres thick and it allows the dramatic improvement of the barrier properties (1–2 orders of magnitude).

The advantages of the treatment consist first of all in the limited investment required to perform it, furthermore the technology is rather straightforward and low-cost. Disadvantages include its slowness (hours per container) and the use of toxic chemicals, which need to be dealt with after being used.

Recently Dow introduced a new version, involving the blow moulding of tanks or containers using a nitrogen/sulphur trioxide blowing gas, with ammonia post-treatment to remove acidity. This latter method is much faster (a few minutes), due to the high container wall temperature during moulding [50–52].

Most applications of sulphonation were for car fuel tanks, since in many countries surface treatments to improve barrier properties are mandatory.

Further to sulphonation, in the early 1970s Dow suggested surface fluorination approaches, without pursuing them commercially. Later on Air Products suggested the so-called Airopak process, involving the use of a nitrogen/fluorine

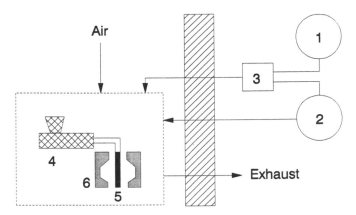

Figure 11.3 Schematic of the fluorination during blow-moulding process. Fluorine (1) and nitrogen (2) are mixed in a blender (3) to provide the blowing mixture and are fed, along with air and nitrogen for purging, to the extrusion blow-moulding equipment. The latter includes an extruder (4), which forms the parison (5), which is clamped by the moulds (6) and blown using the nitrogen–fluorine mixture. Of course, both the fluorine reservoir and blending and the blowing environment are carefully contained to prevent worker exposure to fluorine.

mixture for blow-moulding HDPE fuel tanks [58–63]. A schematic representation of the process is presented in Figure 11.3.

Union Carbide patented an alternative process involving the treatment of formed containers with nitrogen/fluorine gaseous mixtures [64–65]. Finally, in a Battelle patent, fluorination was claimed to avoid plasticizer surface migration and improve surface slipperiness [66].

The Airopak process is conceptually very straightforward, it involves using a nitrogen/fluorine mixture (0.5–5% fluorine) in extrusion blow-moulding processes. A sophisticated treatment of exhaust gases is of course necessary, involving the use of sodium hydroxide traps to absorb unreacted fluorine and hydrogen fluoride reaction products [64–65, 67–68]. Reaction times coincide with blowing times. A surface layer, 10–20 nm thick is formed, involving the presence of –CHF– and –CF$_2$– groups, with the former prevailing [66].

The physics of permeability of these layers was studied by Koros *et al.* and others [69–71]. It was shown that the barrier improvement is dramatic for non-polar compounds (e.g. alkanes, aromatic compounds), close to nil for oxygen and halogen containing molecules. In Table 11.1 permeability data for different solvents in Airopak treated containers are reported. A problem of this treatment is that 'green' fuels can be a problem, leading to differential permeability effects.

The SMP process, involves contacting nitrogen/fluorine mixtures (up to 10% fluorine) with formed containers and tanks. Unreacted fluorine is recycled, while hydrogen fluoride is captured by sodium hydroxide traps. A fluorinated layer

Table 11.1 Percentage weight loss of various solvents in HDPE containers, untreated and Airopak process treated, when allowed to rest at 40 °C for 11 days

Solvent	Weight loss (%)	
	Treated	Untreated
Pentane	0.21	98.10
Hexane	0.19	61.29
Heptane	0.08	24.26
Iso-octane	0.03	4.54
Benzene	3.65	36.68
Toluene	1.80	41.23
p-Xylene	0.54	59.20
Methylene chloride	46.26	50.81
Chloroform	38.17	44.93
Carbon tetrachloride	0.05	28.26
Acetone	2.58	2.51
Ethyl acetate	3.39	3.57
Tetrahydrofuran	45.66	53.93
Dioxane	3.04	4.23
Methanol	0.66	0.75
Propanol	0.35	0.34
Butanol	0.27	0.30

20–50 nm thick is formed, with a chemical composition close to that obtained by the Airopak process. As compared to the latter, a higher number of $-CF_2-$ groups are formed and, including thickness effects, SMP-treated containers prove frequently less permeable than Airopak treated ones. Union Carbide, in its patents claimed improved oxygen barrier properties along with fuel barrier, in agreement with the relative oxygen permeability characteristics of polyethylene and fluorinated polymers (e.g. polyvinylidenefluoride, polytetrafluoroethylene).

Fluorination became very popular owing to its speed (many tanks per hour can be treated), relatively low investment and good barrier. The main disadvantage lies in the use of fluorine, which requires transportation, storing and use precautions, due to its highly corrosive nature.

An alternative to fluorination by fluorine gas, developed by MIT, is the fluorination in a cold plasma, using a fluorine/inert gas mixture. An advantage of this method is a more complete fluorination in shorter times [72–74], yet it still requires the use of fluorine gas, with corresponding environmental constraints. This led to attempts to apply fluorinated layers by plasma decomposition of fluorine-containing molecules, such as CF_4 and SF_6 [75–76].

Further efforts to obtain hydrocarbon barrier layers using vacuum techniques involved the deposition of dense layers from hydrocarbon plasmas [77–79] and the deposition by pyrolysis of poly(2-chloroxylylene) layers [80], see schematics

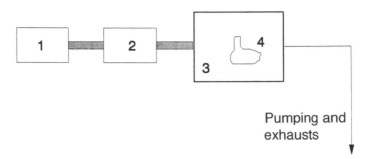

Figure 11.4 Schematic of a poly(2-chloroxylylene) deposition apparatus. The starting compound, 2,2′-dichloroparacyclophane, is vaporized at some 200 °C and 1 torr (1). It is then fed to the pyrolysis region, which occurs just below 700 °C (2). The diradical which is formed is directed into the deposition chamber (3), where it coats the substrate (4).

in Figure 11.4. None of these techniques have so far reached the stage of industrial application. The main alternative to fluorination or sulphonation lies in the production of containers with coextruded structure [81–82] or laminar morphology [83–86].

11.3 EVAPORATION

By far the most commercially successful surface treatment to improve barrier properties (mainly to oxygen, water, carbon dioxide) is aluminium evaporation. It involves the formation of a thin flexible layer (about 50 nm thick) and is typically applied to dimensionally stable polymeric substrates (e.g. biaxially oriented PP and PET films), to prevent barrier loss by formation of cracks in the barrier layer by slackening of the film [1].

The patent literature is mostly relative to instrumental improvements and is therefore not always abstracted in chemical data banks [87–89]. Furthermore there is little published information, with the exception of proceedings of vacuum conferences [94–98].

While metallization can be used for many purposes, e.g. electrical and decorative, barrier requires a dense and pinhole-free layer, therefore it needs a particularly good setting of evaporation machinery and limits attainable speeds. Actually, when using electron beam assisted evaporation with PET films, deposition speeds up to 10 m/s can be obtained, working on film widths even higher than 2 m. Metallization is most efficient on films and docs not perform well on objects, due to its inherent directionality.

As in the case of the deposition of polymeric coatings, surface-related problems are not really important for the success of the treatment. The only precaution which is needed is to obtain a reasonable adhesion of the metallic layer, this is typically obtained by corona treatment of the surface, to form a surface oxidized layer, by avoiding the presence in the film formulation of additives which could worsen adhesion by surface migration and by degassing of the film prior to metallization. In the packaging industry, metallized films are then typically further processed by printing and lamination with sealing layers [2–5], thus the metal layer is protected from environmental exposure and adhesion is much less critical.

The efficiency in improving barrier properties depends solely on the quality of the deposited layer, namely in the number of pinholes. It is known that the measurement of permeability depends on the instrumentation being used and on the method (e.g. manometric versus isostatic), yet the typical result of metallization on a biaxially PET substrate is a more than $100 \times$ reduction in oxygen permeability, as shown in Table 11.2 (measurements performed in the authors' laboratory). To achieve particularly high barriers, metallization on both sides can be performed.

Apart from the accuracy of deposition, barrier can depend very much upon further processing. In fact, due to the thinness of the metal layer and to its mechanical properties, very different from those of the substrate, dimensional variations occurring during processes such as lamination (Figure 11.5) and packaging itself can lead to ruptures in the metal layer and worsening of barrier performance. Typically dimensional variations below 5% are recommended to prevent barrier loss. Given typical stress–curves for polymer films (Figure 11.6), this is easy to achieve for biaxially oriented PET, less so for biaxially oriented PP, and very difficult for unoriented films such as PE.

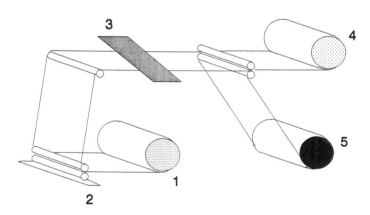

Figure 11.5 Schematic of a dry lamination process. The substrate film (1) is passed through an adhesive (2), which is dried (3). It is then adhered to the other film (4), forming the final product (5).

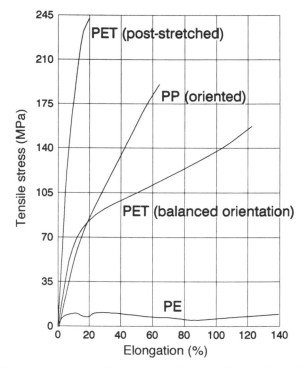

Figure 11.6 Stress–strain curves for some of the most frequently surface metallized plastic films.

The cost of aluminium evaporation depends mainly on investment and on conventional operation of vacuum systems. The alternatives present in the market consist in heat-induced and electron beam-induced evaporation. The former are small-scale and easy to operate systems and are frequently used by converters who work both on barrier and decoration. The latter are much better in terms of productivity, efficiency and control of the quality of the deposited layer, but require more investment and more competence in their use, therefore they are typically used by large-scale converters.

A disadvantage of metallization with aluminium lies in the fact that it necessarily involves the formation of an opaque layer. This is advantageous for foods which can undergo light-induced degradation (e.g. those containing fatty acids), but consumer requirements tend to favour transparent packaging. This led to studies to obtain evaporation of transparent layers, based mainly on silicon oxides.

This process, i.e. evaporating silicon oxides, was originally patented by DuPont [99], and a renewed interest was observed in the 1980s [100–108]. The process is similar to evaporating aluminium, although silicon oxides are fed as granules rather than as a wire and the thickness of the deposited layer is reported to be somewhat higher (80–100 nm). It can anyway be performed on similar reactors, with electron beam operated systems scoring better.

Silicon oxide evaporation has not yet become a viable economic alternative to aluminium metallization, since silicon monoxide has to be of sufficient purity and is much more expensive than aluminium itself. Alternatives such as the *in-situ* formation of silicon monoxide by reaction of silicon and silicon dioxide or silicon and glass have been envisioned [105, 109]. Furthermore, the productivity of equipment evaporating silicon oxide is always more limited, owing to the different feeding mechanism (solid versus liquid in the case of aluminium) and to the necessity of limiting the speed to obtain a uniform layer (about 3 m/s versus 10 m/s for aluminium) because of evaporation inhomogeneities.

So far, the barrier properties obtained by evaporation of silicon oxides are somewhat worse than those reached by aluminium metallization (Table 11.2). Some groups support the view that uneven stoichiometries (namely $SiO_{1.5-1.7}$ compositions) give better results [110], even if a tendency to yellowing of the film develops. Other groups on the other hand support the view that very good barrier properties can be obtained even with SiO_2 compositions, which envisages the evaporation of SiO_2, a rather less expensive starting material [111–113]. The debate is still ongoing, while a commonly accepted view is that there is no real adhesion problem.

The evaporation of silicon monoxide or, worse, silicon dioxide, requires temperatures much higher than for aluminium. Thus the thermal load to the substrate is much higher than for aluminium metallization. This exasperates substrate related problems, all applications reported so far have been performed on biaxially oriented PET (most frequent) and PP films. As for the aluminium metallized counterparts, further processing steps include printing and lamination with heat sealable layers (cast PP or PE), which protect the inorganic barrier layer from environmental exposure. The resistance of the layer to dimensional variation upon further processing is no better than in the aluminium metalliz-ation case.

Table 11.2 Oxygen permeability, as measured according to ASTM D1434, of biaxially oriented PET, as such and with inorganic top layers obtained by evaporation

Sample	Permeability (cm^3/m^2 24 h atm)
PET	160
Al metallized PET	1
SiO treated PET	3

11.4   PECVD AND SPUTTERING

The possibility of using other vacuum treatment, such as plasma [81–82] and sputtering [114], for altering barrier characteristics has long been known. A schematic representation of the mechanism of plasma polymerization processes is reported in Figure 11.7. In the 1980s both processes have been explored as a possible alternative to evaporation for the deposition of transparent inorganic barrier layers.

Sputtering processes have been pursued by various groups, namely BOC [115] and Galileo [111], very good barriers with low thicknesses (about 20 nm) have been obtained, but everybody found that it is too slow (metres per minute as compared to metres per second) for the particular application.

On the other hand, an efficient PECVD process for depositing SiO_2 has recently been developed and the first commercial applications are already under way [116–120]. This process consists in the plasma decomposition of a mixture including a siloxane (hexamethyldisiloxane or tetramethyldisiloxane) in the presence of oxygen and an inert gas, to yield an SiO_2 layer with low thickness (20–30 nm). Deposition rates are not as fast as for evaporation, but they are reported above 1 m/s, barrier properties are reported better than those obtained with silicon oxide evaporation, owing to a more efficient densification of the inorganic layer.

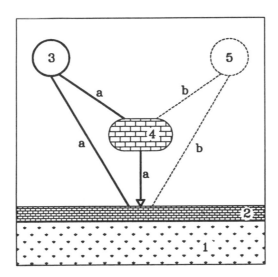

Figure 11.7   Schematics of a plasma polymerization process (see ref. [81]). The monomers (3) are fragmented by the plasma and can (a) polymerize to give the coating (2) directly or through gas-phase oligomers (4). The competitive reaction (b) is depolymerization of the coating to provide volatile compounds (5) eliminated through the pumping system.

As compared to the latter process, plasma is more intricate in terms of process control, since it requires the simultaneous check of pressure, multiple gas injection lines and plasma parameters such as power and tuning of exciting electromagnetic waves. Also it is a tough engineering task to obtain plasma homogeneity over the widths (typically 1.5–2 m) used in the packaging industry.

On the other hand, a definite advantage lies in the low thermal load applied to the substrate, which somewhat relaxes constraints on the film rigidity. Furthermore, plasma is less directional than evaporation, therefore this process is better suited for containers and the application has been thoroughly explored [121].

The origin of the process goes back to attempts to deposit scratch resistant layers on optical plastics [122–124] and to experiments aimed at continuous treatment of large plastic parts [122–126]. A different approach, but again requiring the application of a tightly cross-linked quasi-inorganic layer, suggested the use of amorphous carbon coatings [77].

A rather extensive review of the permeation characteristics of plasma films has been put together by Yasuda [81]. Plasma polymers are characterized by a high degree of cross-linking and low segmental mobility. Therefore permeability, rather than a solution-diffusion mechanism, acts according to a molecular-sieve mechanism. Therefore, plasma polymers are particularly suited to prevent the diffusion of 'big' molecules. As shown in the preceding sections, a nearly inorganic layer need to be put together to block the diffusion of molecules such as oxygen.

This is part of the reason why plasma (mainly using halocarbon gas mixtures) has frequently been suggested for the formation of barrier layers for hydrocarbons and other vapours [127–141]. Another field which has attracted a strong interest is the use of plasma polymers as coating for membranes, for instance to improve selectivity [142–170].

REFERENCES

[1] O. J. Sweeting (Ed.), *The Science and Technology of Polymer Films*, Wiley, New York (1971).
[2] G. L. Robertson, *Food Packaging*, Dekker, New York (1993).
[3] M. Bakker and D. Eckroth (Eds), *The Wiley Encyclopaedia of Packaging Technology*, Wiley, New York (1986).
[4] K. M. Finlayson (Ed.), *Plastic Film Technology*, Technomic, Lancaster (1989).
[5] C. R. Oswin, *Plastic Films and Packaging*, Applied Science Publ., Essex (1975).
[6] A. F. Chapman, US 2824024 (1958).
[7] J. J. Stewart, US 2829068 (1959).
[8] M. Hilger, U. Wallwitz, R. Roesner, and T. Reh, EP 449102 (1991).
[9] R. O. Ranck, US 3975573 (1976).
[10] D. H. Klein and J. Hatziiosifidis, DE 3518875 (1986).
[11] K. Hiyoshi, N. Matsuura, M. Matsuguchi, N. Onofusa, and T. Nishikage, US 4535120 (1985).
[12] W. Quack, DE 3334845 (1985).

[13] K. Hiyoshi, N. Matsuura, M. Matsuguchi, N. Onofusa, and T. Nishikage, US 4446273 (1984).
[14] D. S. Gibbs and W. L. Treptow, EP 68461 (1983).
[15] K. Hiyoshi, N. Matsuura, M. Matsuguchi, N. Onofusa, and T. Nishikage, US 4350622 (1982).
[16] D. S. Gibbs, J. F. Sinacola, and D. E. Ranck, US 4324714 (1982).
[17] W. E. Walles, US 3916048 (1975).
[18] D. G. Grenley and H. J. Townsend, US 3353992 (1967).
[19] C. L. Beeson, R. T. Cole, G. I. Deak, and H. H. Leidorf, WO 92-02573 (1992).
[20] T. Yajima, H. Araki, and E. Ueda, EP 466124 (1992).
[21] T. Min and R. E. Touhsaent, US 4898782 (1990).
[22] D. S. C. Fong, US 4690865 (1987).
[23] W. Quack, US 4710413 (1987).
[24] S. C. Chu, EP 227385 (1987).
[25] L. L. J. Liu, EP 221690 (1987).
[26] D. S. C. Fong, US 4690865 (1987).
[27] S. J. Markiewicz, EP 254468 (1988).
[28] T. J. Min and R. E. Touhsaent, EP 461772 (1991).
[29] G. Grosbard, US 4824701 (1989).
[30] C. E. Gibbons, C. L. Tanner, and A. A. Whillock, EP 293098 (1988).
[31] V. C. Haskell, US 4376183 (1983).
[32] C. E. Gibbons and A. A. Whillock, US 4701360 (1987).
[33] J. Rauser and G. Knudsen, EP 85919 (1983).
[34] G. Roullet and P. Legrand, GB 2014160 (1979).
[35] Y. Maruhashi, I. Tanikawa, S. Hirata, J. Yazaki, and K. Sakano, US 4393106 (1983).
[36] J. F. E. Pocock, C. A. Cole, J. G. Villanueva, J. D. Matlack, and S. E. Lewis, US 4534995 (1985).
[37] M. N. Mang and J. L. Brewbaker, US 5138022 (1992).
[38] G. F. Billovits, M. N. Mang, and J. E. White, US 5134201 (1992).
[39] D. J. Brennan and J. E. White, US 5134218 (1992).
[40] B. Davis and R. B. Barbee, US 4640973 (1987).
[41] A. G. Harrison, D. E. Higgins, and W. N. E. Meredith, EP 498569 (1992).
[42] R. Hinterwaldner and G. Bolte, DE 3814111 (1989).
[43] R. G. Booth and D. L. Bartlett, EP 147984 (1985).
[44] A. Knott and W. Huber, DE 3644239 (1988).
[45] J. E. Wyman, EP 392115 (1990).
[46] G. Boccalon, R. Costantino, and A. Zanaboni, EP 291113 (1988).
[47] O. R. Fennema, S. L. Kamper, and J. J. Kester, WO 87-03453 (1987).
[48] J. Stock, DE 3530366 (1987).
[49] G. G. Cameron and B. R. Main *Polym. Degrad. Stab.*, **11**, 9 (1985).
[50] W. E. Walles, US 4615914 (1986).
[51] J. R. B. Boocock, EP 207204 (1987).
[52] W. E. Walles and D. L. Tomkinson, EP 356966 (1990).
[53] G. Zeitler and E. Zentgraf, DE 3113919 (1982).
[54] R. A. Shefford, EP 35292 (1981).
[55] A. W. Hawkins, M. J. O'Hara, F. P. Gortsema, and E. Hedaya, DE 2644508 (1978).
[56] W. E. Walles, *ACS Symp. Ser.*, **423**, 266 (1990).
[57] W. E. Walles, US 4220739 (1980).
[58] D. M. Dixon, D. G. Manly, and G. W. Rechtenwald, US 3862284 (1975).
[59] H. Fukushima, T. Handa, and K. Kodama, US 4394333 (1983).

[60] J. P. Hobbs, M. Anand, and B. A. Campion, *ACS Symp. Ser.*, **423**, 280 (1990).
[61] K. A. Goebel, V. F. Janas, and A. J. Woytek, *Polym. News*, **8**, 37 (1982).
[62] A. J. Woytek and J. F. Gentilcore, *Plast. Rubb. Process.*, **4**, 10 (1979).
[63] J. F. Gentilcore, M. A. Trialo, and A. J. Waytek, *Plast. Eng.*, **34**, 40 (1978).
[64] P. F. D'Angelo, US 4142032 (1979).
[65] G. Tarancon, US 4467075 (1984).
[66] V. G. McGinniss and F. A. Sliemers, US 4491653 (1985).
[67] D. M. Buck, P. D. Marsh, and K. J. Kallish, *Polym. Plast. Technol. Eng.*, **26**, 71 (1987).
[68] J. P. Hobbs, M. Anand, and B. A. Campion, *ACS Symp. Ser.*, **423**, 280 (1990).
[69] W. J. Koros, V. T. Stannett, and H. B. Hopfenberg, *Polym. Eng. Sci.*, **22**, 738 (1982).
[70] C. L. Kiplinger, D. F. Persico, R. J. Lagow, and D. R. Paul, *J. Appl. Polym. Sci.*, **31**, 2617 (1986).
[71] T. Volkmann and H. Widdecke, *Kunstoffe*, **79**, 743 (1989).
[72] M. Anand, R. F. Baddour, and R. E. Cohen, US 4264750 (1981).
[73] M. Anand, R. F. Baddour, and R. E. Cohen, US 4404256 (1983).
[74] G. A. Corbin, R. E. Cohan, and R. F. Baddour, *J. Appl. Polym. Sci.*, **30**, 1407 (1985).
[75] D. Rebhan and R. Strigl, DE 3739994 (1989).
[76] A. De la Rocheterie, M. H. Legay, and P. Bouard, FR 2570964 (1986).
[77] P. J. J. Kincaid, V. B. Kurfman, and N. M. Sbrockley, US 4756964 (1988).
[78] H. Yasuda, *Plasma Polymerization*, Academic, Orlando (1985).
[79] H. V. Boenig, *Plasma Science and Technology*, Cornell Univ. Press, Ithaca (1982).
[80] A. Cicuta and A. Collina, EP 302457 (1989).
[81] C. Irwin, in *Encyclopaedia of Polymer Science and Technology*, Wiley, New York (1985), p. 447.
[82] I. I. Rubin (Ed.), *Handbook of Plastic Materials and Technology*, Wiley, New York (1990).
[83] P. M. Subramanian, EP 15556 (1980).
[84] J. D. Booze and P. M. Subramanian, EP 90554 (1983).
[85] P. M. Subramanian, EP 211649 (1987).
[86] P. M. Subramanian, EP 238197 (1987).
[87] H. Renke, V. Bauer, A. Feuerstein, and W. Dietrich, US 4724796 (1988).
[88] V. Bauer, A. Feuerstein, and H. Ranke, DE 3420246 (1985).
[89] A. Feuerstein, G. Thorn, and H. Ranke, DE 3330092 (1985).
[90] J. C. Egermeier, EP 345015 (1989).
[91] K. N. Tsujimoto and P. S. McLeod, DE 2815627 (1978).
[92] A. Carletti, EP 266315 (1988).
[93] L. Maugi and R. Cardaio, EP 199122 (1986).
[94] E. Hartwig, J. Meinel, T. Krug, and G. Steininger, *Proc. 35th SVC Ann. Tech. Conf.*, p. 121 (1992).
[95] J. V. Marra, *J. Plast. Film Sheeting*, **4**, 27 (1988).
[96] E. Hartwig, T. Krug, R. Fischer, and J. Meinel, *Proc. 34th SVC Ann. Tech. Conf.*, p. 152 (1991).
[97] A. Feuerstein, E. K. Hartwig, and G. Harwath, *Proc. 31st SVC Ann. Tech. Conf.*, p. 53 (1988).
[98] A. Feuerstein, *Proc. 29th SVC Ann. Tech. Conf.*, p. 118 (1988).
[99] J. W. Jones, US 3442686 (1989).
[100] A. Brody, *Plast. Packaging*, 22 (1989).
[101] R. S. A. Kelly, GB 2210826 (1989).
[102] T. Sawada, EP 311432 (1989).
[103] S. Schiller, M. Neumann, G. Zeissig, and H. Morgner, DE 4113364 (1992).

[104] I. G. Deak, S. C. Jackson, EP 460796 (1991).
[105] A. Hirokawa and K. Ozaki, GB 2211516 (1989).
[106] R. W. Phillips, C. M. Shevlin, and J. S. Matteucci, EP 62334 (1982).
[107] T. Suzuki, M. Sato, K. Tsukagoshi, Y. Takahashi, K. Inagawa, S. Tsukahara, T. Yamamori, and K. Matsumori, *Proc. 34th SVC Ann. Tech. Conf.*, p. 86 (1991).
[108] D. Chahroudi, *Proc. 34th SVC Ann. Tech. Conf.*, p. 79 (1991).
[109] E. Occhiello, M. Morra, and F. Garbassi, EP 492853 (1992).
[110] T. Krug and K. Ruebsam, *Proc. Barrier Pack Conference*, London (1990).
[111] C. Misiano, E. Simonetti, P. Cerolini, M. Carrabino, F. Staffetti, and F. Rimediotti, *Proc. 5th Int. Conf. Vac. Web Coating*, p. 123 (1991).
[112] G. Scateni, F. Grazzini, A. Fusi, C. Misiano, and E. Simonetti, *Proc. 5th Int. Conf. Vac. Web Coating*, p. 104 (1991).
[113] C. Misiano, E. Simonetti, P. Cerolini, F. Staffetti, and A. Fusi, *Proc. 34th SVC Ann. Tech. Conf.*, p. 105 (1991).
[114] C. Boehmler, *Proc. III Conference on Vacuum Web Coating*, (1989).
[115] M. K. Stern, GB 2132229 (1984)
[116] J. Felts and E. Lopata, US 4888199 (1989).
[117] J. Felts, J. Hoffmann, and R. Hoffmann, US 4847469 (1989).
[118] J. Felts, EP 469926 (1992).
[119] J. T. Felts, *Proc. 5th Int. Conf. Vac. Web Coating*, p. 239 (1991).
[120] J. T. Felts and A. D. Grubb, *J. Vac. Sci. Technol.*, **A10**, 1675 (1992).
[121] J. T. Felts, *Proc. 34th SVC Annu. Tech. Conf.*, p. 99 (1991).
[122] G. Menges, W. Michaeli, P. Plein, R. Ludwig, and T. Brinkmann, *Coating*, **22**, 2 (1989).
[123] J. Kieser and M. Neusch, *Thin Solid Films*, **118**, 203 (1984).
[124] J. Kieser and M. Neusch, *Proc. 27th SVC Ann. Tech. Conf.*, p. 47 (1984).
[125] M. R. Wertheimer and H. P. Schreiber, CA 1266591 (1990).
[126] E. Sacher, M. R. Wertheimer, and H. P. Schreiber, US 4557946 (1985).
[127] H. Yasuda and T. Hirotsu, *J. Polym. Sci. Chem. Ed.*, **15**, 2749 (1977).
[128] H. Yasuda and T. Hirotsu, *J. Appl. Polym. Sci.*, **21**, 3167 (1977).
[129] E. Sacher, J. R. Susko, J. E. Klemberg-Sapieha, H. P. Schreiber, and M. R. Wertheimer, *Org. Coat. Appl. Polym. Sci. Proc.* **47**, 439 (1982).
[130] E. Sacher, J. E. Klemberg-Sapieha, H. P. Schreiber, and M. R. Wertheimer, *J. Appl. Polym. Sci. Appl. Polym. Symp.* **38**, 163 (1984).
[131] N. Inagaki, S. Kondo, M. Hirata, and H. Urushibata, *J. Appl. Polym. Sci.*, **30**, 3385 (1985).
[132] E. Sacher, M. R. Wertheimer, and H. P. Schreiber, US 4557946 (1985).
[133] I. Terada, T. Haraguchi, and T. Kajiyama, *Polym. J.*, **18**, 529 (1986).
[134] N. Inagaki and D. Tsutsumi, *Polym. Bull.*, **16**, 131 (1986).
[135] H. Yasuda, US 4692347 (1987).
[136] M. Sanchez Urrutia, H. P. Schreiber, and M. R. Wertheimer, *J. Appl. Polym. Sci. Appl. Polym. Symp.*, **42**, 305 (1988).
[137] N. Inagaki and Y. Kubokawa, *J. Polym. Sci. Chem. Ed.*, **27**, 795 (1989).
[138] A. K. Sharma and H. Yasuda, *J. Appl. Polym. Sci.*, **38**, 741 (1989).
[139] C. P. Ho and H. Yasuda, *J. Appl. Polym. Sci.*, **39**, 1541 (1990).
[140] T. Ihara and H. Yasuda, *Polym. Mater. Sci. Eng.*, **62**, 32 (1990).
[141] T. Masuda, M. Kotoura, K. Tsuchihara, and T. Higashimura, *J. Appl. Polym. Sci.*, **43**, 423 (1991).
[142] O. Koichi, DE 3220037 (1983).
[143] M. Kawakami, Y. Yamashita, M. Iwamoto, and S. Kagawa, *J. Membr. Sci.*, **19**, 249 (1984).

[144] M. Yamamoto, J. Sakata, and M. Hirai, *J. Appl. Polym. Sci.*, **29**, 2981 (1984).
[145] P. Canepa, M. Nicchia, and S. Munari, *Chem. Ind.*, **66**, 604 (1984).
[146] J. Sakata, M. Yamamoto, and M. Hirai, *J. Appl. Polym. Sci.*, **31**, 1999 (1986).
[147] N. Inagaki and J. Ohkubo, *J. Member. Sci.*, **27**, 63 (1986).
[148] N. Inagaki and H. Kawai, *J. Polym. Sci. Chem. Ed.*, **24**, 3381 (1986).
[149] C. P. Ho and H. Yasuda, *Polym. Mater. Sci. Eng.*, **56**, 705 (1987).
[150] T. Hirotsu, *J. Appl. Polym. Sci.*, **34**, 1159 (1987).
[151] N. Inagaki and H. Katsuoka, *J. Membr. Sci.*, **34**, 297 (1987).
[152] J. Sakata, M. Hirai, and M. Yamamoto, *J. Appl. Polym. Sci.*, **34**, 2701 (1987).
[153] T. Hirotsu and S. Nakajima, *J. Appl. Polym. Sci.*, **36**, 177 (1988).
[154] T. Haraguchi, S. Ide, T. Nagamatsu, and T. Kajiyama, *J. Appl. Polym. Sci. Appl. Polym. Symp.*, **42**, 357 (1988).
[155] N. Inagaki, *J. Appl. Polym. Sci. Appl. Polym. Symp.*, **42**, 327 (1988).
[156] T. Kashiwagi, K. Okabe, and K. Okita, *J. Membr. Sci.*, **36**, 353 (1988).
[157] N. Inagaki, S. Tasaka, and Y. Kobayashi, *Desalination*, **70**, 465 (1988).
[158] H. Kita, T. Sakamoto, K. Tanaka, and K. Okamoto, *Polym. Bull.*, **20**, 349 (1988).
[159] M. Karakelle and R. J. Zdrahala, *J. Membr. Sci.*, **41**, 305 (1989).
[160] J. Sakata, M. Yamamoto, and M. Hirai, *J. Appl. Polym. Sci.*, **37**, 2773 (1989).
[161] H. Kita, M. Shigekuni, I. Kawafune, K. Tanaka, and K. Okamoto, *Polym. Bull.*, **21**, 371 (1989).
[162] T. Hirotsu and M. Isayama, *J. Membr. Sci.*, **45**, 137 (1989).
[163] N. Inagaki, S. Tasaka, and T. Murata, *J. Appl. Polym. Sci.*, **38**, 1869 (1989).
[164] N. Inagaki, S. Tasaka, and M. S. Park, *J. Appl. Polym. Sci.*, **40**, 143 (1990).
[165] R. Schallauer, M. Gazicki, F. Kohl, F. Olcaytug, and G. Urban, *Polym. Mater. Sci. Eng.*, **62**, 279 (1990).
[166] H. Iwata, M. Oodate, Y. Uyama, H. Amemiya, Y. Ikada, *J. Member. Sci.*, **55**, 119 (1991).
[167] T. Hirotus and A. Arita, *J. Appl. Polym. Sci.*, **42**, 3255 (1991).
[168] M. Anand, C. A. Costello, and K. D. Campbell, US 5013338 (1991).
[169] T. Yamaguchi, S. Nakao, and S. Kimura, *Macromolecules*, **24**, 5522 (1991).
[170] J. Weichart and J. Muller, *Progr. Coll. Polym. Sci.*, **85**, 111 (1991).

Chapter 12

Biomedical Materials

There are many reasons why a chapter on biomedical materials, i.e. those materials that are used in contact with living tissue and biological fluids for prosthetic, therapeutic and storage applications [1], should be included in a book on polymer surfaces. The first and most obvious is that every biomaterial application involves, by definition, the creation of at least one interface between the material itself and the biological environment [2]. The events going on at the interface play a crucial role in the overall performance of the biomedical device [3], so that surface science knowledge must be applied in the design of every biomaterial [4]. Another reason is that, since it is very difficult to find a single material which possess both the highly specialized surface and bulk properties required for a given application, biomaterials are, in general, of composite nature, and surface modification techniques are often mandatory. Finally, since that of biomaterials is a very high value added market, the creativity of the surface scientist in this particular field is less limited by economic consideration than in other lower value material applications. In other words, the full arsenal of surface modification techniques previously described can, in principle, be exploited.

From a market perspective, a recent survey [5] suggests that biomaterials revenues are expected to grow from $185.1 million in 1986 (data refer to US market) to $623.2 million in 1996, as shown in Figure 12.1. Of that total, the share of polymeric biomaterials (whose applications are outlined in Table 12.1) is about 24%. Another survey suggests [6] that, since 'traditional materials' are continuing to be replaced by thermoplastics, the average annual growth rate over the next five years of medical grade thermoplastics in non-packaging applications is estimated at 6.2%, against a general average annual growth rate in the medical supply industry of 5.6%.

In the present chapter we will review some of the current researches and developments in two of the fields described in Table 12.1, namely blood contacting devices and contact lenses. The reason for this choice is that, while a complete discussion on the surface properties–biomaterials connections would require

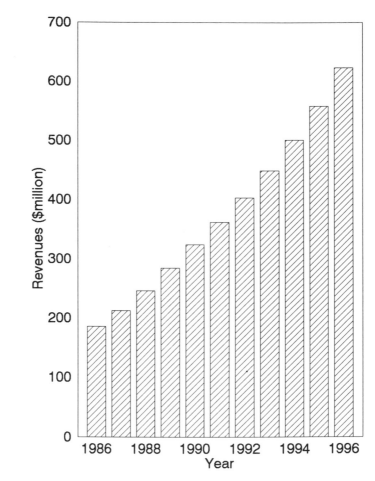

Figure 12.1   Total US biomedical materials market revenue forecasts (1986–1996).

much more than a single chapter of a book, it is the opinion of the authors that a survey of surface-related problems and researches in these two topics can give a good overview of this intriguing field and of how surfaces are currently manipulated in order to obtain specific properties at the synthetic polymer–biological environment interface.

12.1   BLOOD CONTACTING DEVICES

It is generally recognized that one of the greatest problems, in biomaterials, which hinders the full exploitation of synthetic vascular graft and organs, is the

Table 12.1 Examples of application of biomaterials

Intracorporeal materials

Temporary devices
  Surgical dressing
  Sutures
  Adhesives

Simple semipermanent devices
  Tendons
  Joint reconstruction
  Bone cement
  Heart valves
  Tubular devices (vascular grafts, oesophageal prosthetics, etc.)
  Soft-tissue replacement materials for cosmetic reconstruction
  Interocular and contact lenses
  Drug delivery implants

Complex devices simulating physiological processes
  Artificial kidney/blood dialysis
  Artificial lung/blood oxygenator
  Artificial heart
  Artificial pancreas/insulin delivery system

Paracorporeal or extracorporeal materials

  Catheters
  Blood bags
  Pharmaceutical containers
  Tubing
  Syringes
  Surgical instruments
  Sterile and non-sterile packaging materials

---

thrombogenic nature of most materials in blood contacting applications. The quest of a truly haemocompatible synthetic surface is an elusive goal which requires many million dollars research expenses each year. Even if large diameter (> 5 mm) veins are routinely substituted with synthetic materials, as discussed in the following, small diameter vascular grafts are not yet available.

To fully understand the difficulties involved in this problem, it is important to remember that the endothelial cells that line the inner wall of the vascular system are practically the only material 'inert' with respect to platelets and blood-coagulation factors [7]. Any disruption of the cell lining, or the replacement of the natural material with a synthetic surface, such as in prosthetic applications, results in the triggering of the so-called 'thromboembolic cascade', a sequence of chemical reactions which leads to the formation of a coalescent or agglutinated solid mass of blood components in the blood stream [8], which is commonly called thrombus. Of course, thrombus formation produces occlusion of the arteries and prevents blood flow at the site at which the thrombus first deposits. In other cases, a loosened thrombus can travel through the vascular system, until causing obstruction at some distal point. The mass of thrombus that moves is referred to as an 'embolus', and the two phenomena are lumped together under the term 'thromboembolic disease'.

It is important to appreciate that the thromboembolic response to synthetic materials proceeds by an interfacial mechanism [8–13]. When a material contacts blood, blood proteins are immediately adsorbed on its surface (this step begins within the first few seconds of blood contact). The mechanisms of adsorption and the nature of the adsorbed protein layer(s) are of the utmost importance for the following events, which is one of the reasons why so many studies have been devoted to protein adsorption [14–18] on solid surfaces. The second step involves adhesion of platelets on the protein-conditioned surface, and adhered platelets undergo, in the third step, morphological changes, aggregation and release of products which further promote aggregation and adhesion. The blood coagulation factors, i.e. specialized plasma proteins, becomes activated by the contact with the foreign, platelet-covered surface, and start the cascade of reactions which lead to the generation of thrombin and the enzymatic polymerization of fibrinogen, as shown in Figure 12.2. Platelet aggregates are embedded in a network of fibrin, creating a thrombus.

The above description is a highly simplified picture of an enormously complicated phenomenon: plasma proteins are hundreds, ranging in molecular weight from a few thousand to a few million daltons, and in concentration from a few $\mu g/litre$ to several $g/litre$. They can assume very different conformations, denature or reorganize after adsorption. Moreover, the fluido-dynamic conditions of the blood stream, speed, turbulence and so on, can change enormously in different locations of the vascular system, with profound effects on thromboembolic phenomena, with the result that the performance of a potential biomaterial depend also on the details of its geometry and position in the vascular system.

The main conclusions that can be drawn from the previous discussion are the following: thromboembolic phenomena are controlled by interfacial interactions so, in principle, a correct design of the surface structure of a biomaterial should result in haemocompatible synthetic materials. The accomplishment of this task is, however, tremendously difficult: the system is intrinsically very complex and not completely understood. A major difficulty also arises in the characterization steps. Because of the very nature of the problem, researchers must rely on *in vitro* test or on *in vivo* animal models. The former are of limited significance, while the latter, besides ethical and economic considerations, must cope with the fact that none of the animals currently used has blood which is identical to human blood in properties [19], since they differ in their coagulation and fibrinolytic mechanisms [20, 21] and in the degree of platelet adhesion to biomaterials [22, 23]. The great number of currently used test methods, shown in Table 12.2 [24], testify to the complexity of this problem.

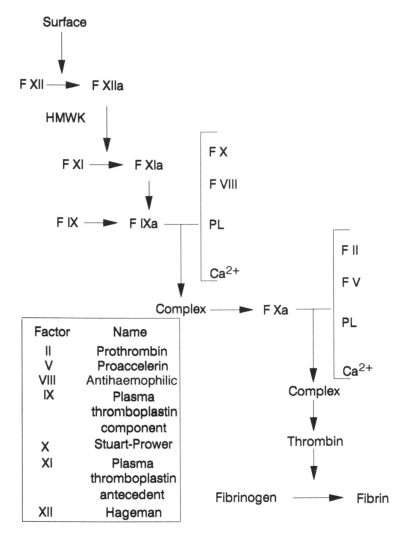

Figure 12.2 Interfacial phenomena in blood clotting. The common name of blood-coagulation factors is shown in the inset. HWMK is a high molecular weight kininogen, PL is a phospholipid. Activated factors (indicated by 'a') trigger the subsequent steps of the cascade.

From a practical point of view, the most commonly employed permanent blood contacting devices are large diameter vascular grafts [4, 25]. These grafts are made either of woven or knitted PET (Dacron[R]) or of porous PTFE (Gore-Tex[R]): as soon as these materials contact blood thrombi form, which reorganize in the porous or textured surface creating what is generally called

Table 12.2   Some of the methods used to assess blood–material interactions
(many of these tests can be performed in different configurations, see ref. [24])

Whole blood clotting time
Partial thromboplastin time (PTT)
Thrombus deposition in flow cells
Thrombus deposition in *in vivo* rings and short tubes
Thrombus deposition in (or on) shunts, patches, flags and catheters
Platelet adhesion
Platelet survival, turnover and consumption
Platelet aggregability and release
Emboli formation
Device performance

'pseudoneointima'. Thus, in working conditions, blood is interfaced with a phase
of natural origin. The same principle cannot be exploited in smaller diameter
vascular grafts, since in this case occlusion of the lumen occurs. The choice is then
to use (when possible) suitable veins of the patient (autologous graft), while, at
present, no synthetic alternative exists. Researches on blood compatible materials of synthetic origin are mainly focused on polymer bearings bioactive
agents such as heparin (anticoagulant), fibrinolytic enzymes, antiaggregation
agents etc. [19] and on intrinsically haemocompatible polymers. While hepa-
rinized surfaces are currently of extreme interest, in the next pages we will discuss
the research on intrinsically haemocompatible surfaces, since they are more in
tune with the scope of this book. In particular, we will review the many theories
that have been proposed on the thrombogenic character–surface properties
relationship. It will appear clear that, even if many property–performance
relationships have been suggested, no theory is completely satisfactory. Moreover, many theories are not only in disagreement, but even antithetic. While these
problems arise because of the extreme complexity of the basic underlying
phenomena and of the correct characterization procedure, it is however extremely useful to review them, in order to best appreciate the difficulties that are met in
this important topic and the efforts that are made to try to overcome them.

12.1.1   Effect of Electrical Properties

The electrical properties of polymer surfaces have been associated with blood
compatibility. In particular, surface charge has been the subject of many speculations. Electrostatic interactions between bodies, such as those discussed in
Chapter 1, are known to deeply affect the stability of colloidal suspension or, in
general, the interaction between suspended particles and vessel walls. In this
connection, it is interesting to find that both vein inner walls and formed bodies of
the blood bear, in physiological conditions, a net negative charge, as discussed,
for instance, by Sawyer and coworkers [8, 26]. In an interesting series of papers

[8–11, 26–29], it was observed that an injury to the blood vessel wall, generally accompanied by thrombus formation, reduces the magnitude of the negative charge density, and very often even causes a reversal in the sign of the surface charge. Streaming potential measurements carried out *in vivo*, before and after the injection of selected drugs, showed that antithrombogenic drugs increase the magnitude of the negative charge density, whereas thrombogenic drugs have the opposite effect (in many cases the charge sign is reversed). With respect to prosthetic materials, it was found that the positively charged ones are thrombogenic, while negatively charged surfaces tend to be non-thrombogenic. As an example, Figure 12.3, reprinted from reference [8] shows the degree of thrombus

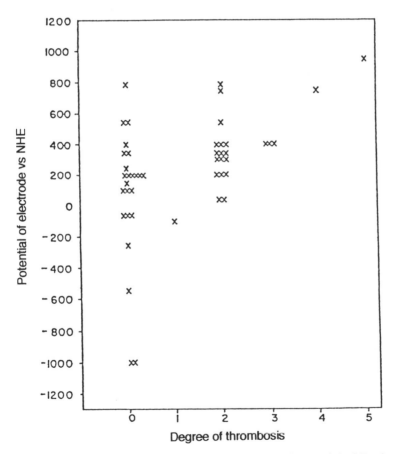

Figure 12.3 Degree of thrombus formation as a function of potential of Pt electrodes inserted into canine carotid and femoral arteries. Degree 0 = no thrombus formation; 1–5 increasing thrombus formation. (Reproduced by permission of Academic Press from ref. [8].)

formation as a function of potential of platinum electrodes inserted into canine carotid and femoral arteries. Clearly, when the potential of the electrodes are on the negative side thrombi formation rarely occurs, while an increasing trend towards thrombogenic character is observed when the potential is on the positive side. These results suggest that the electrostatic interaction between the charged blood vessel walls and the components of blood (a strong electrolyte) controls the interfacial phenomena giving rise to the thromboembolic cascade.

Further studies confirmed the importance of the sign of the surface charge, in particular in connection with the contact activation of the Hageman factor [30, 31]. Boffa *et al.* [32] showed that, *in vitro*, surfaces characterized by a high content of anionic groups (0.8 ÷ 1.2 mmol/g) give low platelet adhesion, while surfaces containing cationic groups readily adsorb platelets. In the latter case, platelet adhesion seems linked to the adsorption of the von Willebrand factor, a plasma protein which mediates platelet adhesion under conditions of high shear rates.

Bruck advanced the hypothesis that another electrical property, namely electrical conduction and semiconduction, may have a relationship to blood compatibility [33]. The basic idea is that collagen, which in normal conditions is covered by the lining endothelium, is a highly ordered structure, with a strong parallel alignment of the basic building block, the dipolar tropocollagen unit. Because of this strong alignment, the entire collagen fibril acquires a permanent dipole moment along its main axis, as shown in Figure 12.4. Electron mobility in the physiological environment may be maintained by the shielding effect of a surrounding 'structured water', that is water molecules that assume a preferred conformation because of hydration effects such as those described in Chapter 1. This 'vicinal' water, which has been postulated for biological macromolecules [34], creates a kind of 'shield', which allows the prevention of the neutralizing effect of the ionic components of the blood on the electrical properties of

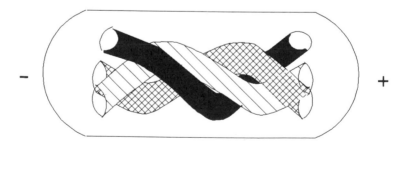

Figure 12.4   Tropocollagen with permanent dipole moment.

biopolymers. Bruck's suggestion is that the electrical properties of collagen may influence the adsorption and conformation of plasma proteins, which in turn could affect the platelets and other blood elements [35, 36]. Thus, when the endothelium lining is disrupted, collagen may be exposed and, if there is a sufficient disturbance of the native structure of collagen, electrical properties may be modified by the inorganic ions and organic polyelectrolytes of the blood. The result is that when the environment of collagen molecules shifts from underneath the normal endothelium to flowing blood, fundamental changes occur on the electrical properties of the collagen fibril. While these changes can result in little effects in collagen itself, they may deeply affect the adsorptive behaviour of collagen, leading to alteration in the protein adsorption step, which, in turn, affects the response of the blood coagulation factors. At the basis of the Bruck hypothesis there are some interesting findings of the same author: Bruck observed that an aromatic polyimide, that is poly[N, N'-(p, p'-oxydiphenilene) pyromellitimide (PPMI), can be thermally converted into electrically conducting and semiconducting condensation products well below graphitization conditions [37–39]. *In vitro* experiments on these materials showed that semiconducting and conducting pyrolitic polymers have clotting times two to three times longer than siliconized glass, while the non-conducting control sample has a clotting time of one-third that of siliconized glass. Moreover, the conducting polymers showed very low or no platelet aggregation, with no evidence of activation, in contrast to the control sample. These results strongly suggest that intrinsic conduction properties of natural and synthetic polymers may be involved in blood compatibility.

12.1.2 Effect of Interfacial Energetics

Many theories have been proposed on the surface–interfacial energetics–blood compatibility relationship. The starting point is, of course, that surface energetics control bioadhesive events at the vein–blood interface, which, in turn, control blood clotting. In this section these theories will be briefly reviewed, and reference will be made to interfacial energetics definitions and theories described in Chapter 4.

The first suggestion on this topic dates back to 1885 [40], when it was observed that blood clotting time is increased if glass is covered by vaseline. The same effect was later obtained when glass was covered with paraffin wax [41]. Extensive experiments were made by Neubauer and Lampert, which reported an inverse relationship between blood clotting time and surface wettability [42] and these observations gave rise to the 'Lampert rule of blood clotting time'. However, as the number of experiments increased, it was clear that the matter was much more complex than expected from the simple Lampert rule. Moreover, as more refined theories on interfacial energetics were developed (Chapter 4), a more precise thermodynamic meaning could be attached to wettability measurements.

Lyman and coworkers first suggested a relationship between surface free energy and blood compatibility [43]. The latter is expressed as blood clotting time, and reasoning is made on *in vitro* data for human blood and *in vitro* and *in vivo* data for canine blood. Surface free energy is expressed following both the Good–Girifalco theory (taking the interaction parameter as unit) and the Zisman critical surface tension. When blood clotting time is plotted as a function of the solid surface tension or the critical surface tension an inverse relationship was observed. Bischoff suggested that a better correlation is obtained if surface energetics are described as the work of adhesion instead of surface free energy, but warned that, since the adsorption of proteins and formed bodies significantly alters the surface parameters, only gross trends may be suggested by correlations with macroscopic surface variables [44].

The next step in the interfacial energetics–blood compatibility relationship is represented by the Baier theory. Basically, Baier stated that blood compatibility can be obtained by the minimization of the adhesion between blood components and foreign surfaces [45, 46]. In this way, the layer of proteins which is readily adsorbed when the material is first exposed to blood [47] does not undergo denaturation, is readily exchanged with fresh proteins from the flowing blood and the mean adhesional time is very low. At the basis of this theory there are some intriguing observations on the correlation between critical surface tension of the substrate and the degree of protein denaturation on adsorption that Baier himself pioneered [48]. If the bioadhesive interaction is low, platelet aggregates which adhere on the adsorbed protein layer cannot grow too much before being detached by the shear force of the flowing blood, where they can plainly be handled, because of their small dimensions, by the normal enzymatic and digestive processes in the flowing stream. In this way, the mechanism by which the solid surface tension controls thromboembolic phenomena is not that it prevents the formation of a thrombus, but rather that it controls adhesive events at the interface and hinders the formation of uncontrollable thrombi and emboli. As Baier stated, the focus of attention has shifted from the substrates themselves as inducers of thrombogenicity, to the substrates as dictators of a special configuration of adsorbed protein molecules that will favour or inhibit the subsequent events [46].

The zone of minimum bioadhesion was found by Baier, after extensive experiments, in the region between 20 and $30\,mJ/m^2$ of critical surface tension (Figure 12.5). According to Baier, this zone of minimum bioadhesion controls interfacial events, besides thromboembolic phenomena, in widely different fields such as biofouling, adhesion and spreading of cells [49]. Dexter and Schrader [50, 51], gave a theoretical explanation of the origin of the 'Baier's window', in the frame of current interfacial energetics theories.

Also the theory of Andrade is based on the minimization of adhesive interaction, but this time the macroscopic variable which is considered is the interfacial free-energy (the theory is known as the minimum interfacial free-energy hypoth-

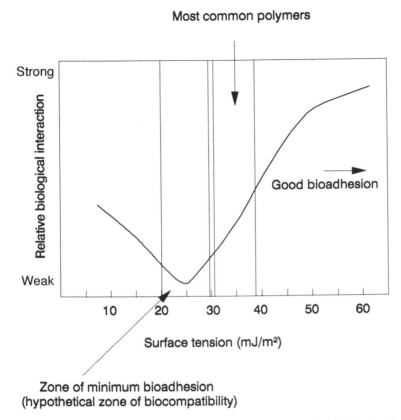

Figure 12.5 Correlation between surface tension of solids and their biological interactions, as suggested by Baier. The zone of minimum bioadhesion is also known as Baier's window.

esis) [52, 53]. According to Andrade, since the working environment of biomaterials is aqueous, it is incorrect to base the hypothesis on the value of the free-energy at the solid–air interface. The interfacial free energy against blood, which can be readily calculated by underwater (or under blood or blood-simulating fluids) contact angle measurements, controls adhesive events: if low enough (ideally equal to zero) no disturbance should be brought to blood and its components. In this respect, hydrogel surfaces, that is the surfaces of hydrophilic, water swollen polymers [54], should be good candidate materials. Later studies by Ratner, Hoffman and coworkers [55], however, did not support the minimum interfacial free energy hypothesis and Andrade *et al.* stressed the need of a multiparameter evaluation of both the materials properties and the biological interactions [56]. From their side, Ratner and Hoffman suggested that an optimum polar/apolar sites ratio is required for biocompatibility [55]. The critical role played by the amount of surface sites of a given nature (polar or apolar, according to the first Fowkes theory, Chapter 4) was suggested by Nyilas in 1975 [57, 58], who observed that, in a series of analogous or homologous polymers and in fixed haemodynamic conditions, the greater the polar force contribution the stronger the interfacial effect on native plasma proteins. From their side, strongly denaturated adsorbed proteins have a greatly increased capability of attracting and activating the formed elements of blood.

The problem of bioadhesion and biocompatibility was also tackled by Kaelble [59, 60]. According to a fracture mechanics approach to bioadhesive phenomena, Kaelble suggested that a strongly adsorbed plasma protein film on the implant surface provides the best blood compatibility and low thrombogenic effects. If adhered proteins are loosely bonded, adsorbed blood components detach from the vein wall and cause emboli. The equation required for the calculation of the Griffith energy for fracture, that is the energy required for the displacement by the blood of the adsorbed protein layer, according to Kaelble's method, are shown in Table 12.3. A positive value means that the process is thermodynamically unfavourable, and the greater the value the more stable the adsorbed layer. The required data are the polar and dispersive surface tension components of the three phases involved, that is the protein, the blood and the implant. Figure 12.6 shows the calculated Griffith energy for fracture for several implant materials (data taken from ref. [59]). The analysis shows that high dispersion–low polar surface free energy for the implants provide surface energetics favouring stable plasma protein film retention.

A detailed analysis of the Kaelble theory was carried out by Sharma and coworkers [61–66]. In particular, Sharma suggested that the excellent haemo-compatibility of some polymeric surfaces could be because of a preferential adsorption of albumin, a plasma protein which seems to increase thrombo-resistance, as discussed in the next section. According to Sharma, preferential adsorption is caused by a proper value of the polar and dispersion component of the surface field of forces.

Table 12.3 Equations relevant to the obtainment of the Griffith surface energy for fracture (G) according to the Kaelble method

---

$G = R^2 - R_0^2$
$R_0^2 = 0.25[(\alpha_1 - \alpha_3)^2 + (\beta_1 - \beta_3)^2]$
$R^2 = (\alpha_2 - H)^2 + (\beta_2 - K)^2$
$H = 0.5(\alpha_1 + \alpha_3)$
$K = 0.5(\beta_1 + \beta_3)$

Notes: Subscript 1 refers to the adhesive layer, 2 to the environment and 3 to the substrate. $\alpha_i = \gamma_i^d$
$\beta_i - \gamma_i^p$

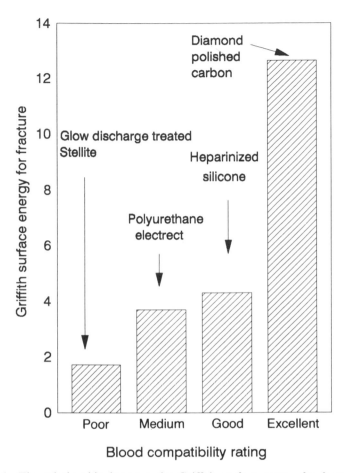

Figure 12.6 The relationship between the Griffith surface energy for fracture at the protein–material interface, as calculated from the Kaelble method and blood compatibility. (From Table 5 of ref. [59].)

The basic assumption of the theory of Ikada [67–69] is that the polymer surface which does not adsorb any plasma protein must be blood compatible. The critical variable is then the underwater work of adhesion, that is the free energy change during the process of protein adhesion to a solid surface in biological (aqueous) media. Underwater work of adhesion ($W_{12, w}$) is expressed by the following equation (12.1), while, as a reminder, the work of adhesion in air or vacuum is expressed in equation (12.2):

$$W_{12, w} = \gamma_{1w} + \gamma_{2w} - [\gamma_{12}]_w \tag{12.1}$$

$$W_{12} = \gamma_1 + \gamma_2 - \gamma_{12} \tag{12.2}$$

where γ_i is the surface free energy of body i and γ_{ij} is the interfacial free energy, while γ_{iw} is the free energy of the interface between body i and water, and $[\gamma_{12}]_w$ is the γ_{12} value in water. According to Ikada calculations, there are two possibilities for non-adhesive polymer surfaces in aqueous media ($W_{12,w} = 0$), that is super-hydrophobic or superhydrophilic surfaces. Since the former cannot be synthesized (no polymer has $\gamma_1 = 0$), superhydrophilic surfaces appear to be more promising for long-term blood compatibility. They can be created by grafting water soluble polymer chains on substrates, producing what Ikada calls a diffuse surface. It must be noted that a diffuse surface greatly differs from hydrogel surfaces, which are cross-linked and have a relatively low water content.

Finally, Ruckenstein and Gourisankar suggest that a compromise between adhesive and non-adhesive properties is the requirement for blood compatibility [70–75]. This hypothesis stems from the consideration that, while the driving force for the adsorption of blood components should be minimized, a certain degree of mechanical stability of the interface is also required. By comparison with the interfacial free-energy between blood and its formed elements, which are of course blood-compatible and whose interface with blood is mechanically stable, it is stated that an interfacial tension of $1–3 \, \mathrm{mJ/m^2}$ satisfy the above criterion.

The above brief review of current theories shows that no general agreement exists on the surface energetics–blood compatibility relationship. Moreover, while some theory suggests that protein adhesion should be minimized, other maintains that it should be maximized, or should involve only selected proteins. This state of the art probably stems from the enormous complexity of the system (blood, its components, the nature of the polymer surfaces, the interfacial events). Even very careful experiments cannot be described by one parameter only (albeit split in two or more components), as discussed by Andrade, who suggests a multiparameter approach to protein–surface interactions [15, 56]. Finally, since the previous theories are based on experimental measurements of interfacial energetics, as calculated from contact angle data, they share all the difficulties involved in the recognition of the correct thermodynamic status of the measured contact angle and the problems connected with the proper evaluation of the solid surface free energy discussed in Chapter 4.

12.1.3 Interaction with Selected Materials

In this section blood compatibility–surface structure relationships are discussed involving selected and promising classes of polymeric materials. In particular, we will discuss plasma deposited polymeric films, poly(ethyleneoxide) (PEO) containing surfaces and microphase separated surfaces.

12.1.3.1 *Plasma deposited polymeric films*

The technique of deposition of thin polymeric films from plasma has been discussed in Chapter 6. Current interest in polymer deposition from plasma for blood-contacting applications is mainly focused on fluoropolymers and siloxanes [24], that is on hydrophobic surfaces. On the other hand, early reports on this topic involved deposition from acrylamide [76] and other oxygen containing monomers besides hydrophobic siloxane and hydro- and fluorocarbons [77]. Siloxane deposition was extensively investigated by Chawla [78—80], who found that these coatings decrease adhesion of canine platelets, while ozdural and coworkers observed that siloxane coatings minimize platelet interactions [81]. Fluorocarbon deposition was extensively studied by Yasuda and coworkers [77, 82–85] and by Ratner, Hoffman and coworkers [24, 86–89].

An intriguing aspect of the quoted studies is the exploitation of the conformal nature of plasma deposited films on textured or porous fabrics to evaluate the

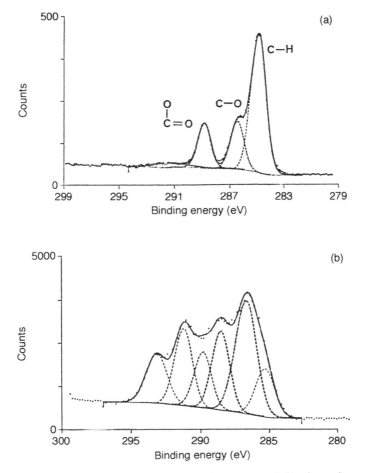

Figure 12.7 (a) High resolution C1s spectrum of Dacron and (b) of tetrafluoroethylene (TFE) treated Dacron. (Reproduced by permission of John Wiley & Sons, Inc. from ref. [89].)

respective effects of surface topography and chemistry. In particular, Figure 12.7 (taken from ref. [89]) shows the C1s peak of Dacron[R] and of the same material after deposition of a fluoropolymer film from tetrafluoroethylene (TFE) plasma. The latter shows the typical structure of plasma deposited polymers, arising from the considerable molecular rearrangement and fragmentation of the TFE molecule that occurs during the plasma discharge reaction (Chapter 6). *In vitro* microembolization studies were carried out on 4 mm internal diameter samples. The experimental procedure involves the recirculation of fresh citrated baboon blood in a closed loop containing TFE treated or untreated Dacron grafts, at a fixed flow rate (100 ml/min). The size and number of aggregates generated is measured by laser light scattering and a computer base optical sizing system. The total emboli volume is calculated from the size and number of aggregate particles. The results of *in vitro* microembolization tests are shown in Figure 12.8, and, clearly, the rate of microemboli productions is greatly reduced after TFE treatment.

A similar investigation was performed by Yasuda and coworkers [89]. This time, fluoropolymer thin films from hexafluoroethane/H_2 were deposited on Gore-Tex[R], that is expanded PTFE. Thus, in this case, two surfaces of identical topography and closely related (fluoropolymer based) chemistry are compared. Again, the results show that plasma-polymer coated surfaces perform much better than untreated ones (the experimental technique used was the measurement of platelet deposition as a function of exposure time in *in vivo* studies on baboons).

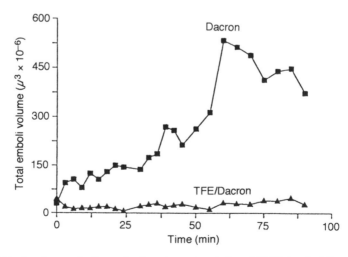

Figure 12.8 *In vitro* emboli production by untreated and TFE treated Dacron grafts. (Reproduced by permission of John Wiley & Sons, Inc. from ref. [89], with permission.)

While the previous results show that deposition of thin polymer films from plasma is a promising technique in blood contacting applications, the reason why this happens is still an open question. According to Ratner and coworkers, the main advantages of plasma deposited films in biomaterials application [24] are the previously quoted conformal nature, the pinhole- free structure, their unique chemistry and good adhesion to the substrate and the low leachability. While these characteristics are surely a good premise for candidate biomaterials, they cannot explain the superior performance of coated Dacron^R and Gore-Tex^R over the uncoated materials. Yasuda and coworkers [89] suggest a possible explanation based on polymer surface dynamics (Chapters 2 and 9): plasma polymer surfaces are highly cross-linked, so their surface mobility is greatly reduced and their surface configuration is not likely to undergo any significant change when the surface is subjected to the interaction forces induced by the protein molecules adsorbed on the polymer surface. The stability of the surface configuration could influence the extent of the initial adsorption of proteins and the conformational change of protein molecules upon adsorption (denaturation), which, in turn, influence the subsequent cellular interactions. The hypothesis that a stable surface configuration is required for good blood compatibility is supported, according to Yasuda, by the observation that those polymers which are rather well behaved towards blood are constituted by rotationally symmetric macromolecules, thus their surface configuration is stable even if their polymer chains are mobile. Another intriguing observation in support of this hypothesis is that of Reichert and coworkers who observed a substantial reduction in thrombus accumulation when side-chain motions at the polymer–blood interface were restricted by irradiation [90].

12.1.3.2 *Polyethyleneoxide (PEO) bearing surfaces*

PEO bearing surfaces are currently one of the most promising approaches to blood compatibility [91, 92]. PEO containing polymers for blood-contacting applications have been obtained by the synthesis of block copolymers [93–95], or, with the goal of creating a surface layer of pendant PEO chains, by grafting methoxy (polyethyleneglycol) monomethacrylates (MnG, Figure 12.9) on poly(vinylchloride) (PVC) photosensitized by dithiocarbamate groups [96], by random copolymerization of MnG and methylmethacrylate (MMA) [97], by coating polyethylene (PE) with PEO containing non-ionic surfactants [98].

The most interesting feature of PEO bearing surfaces is their low tendency to protein adsorption (protein resistant surfaces [98]). A possible explanation of this behaviour is the very low interfacial free energy against water shown by PEO, but this feature is shared by many other natural and synthetic polymers (dextran, agarose, polyacrylamide), which are more interactive with proteins than PEO. Thus, a more likely explanation involves the peculiar solution properties of PEO chains: unlike closely related polymers such as polymethylene oxide and poly-

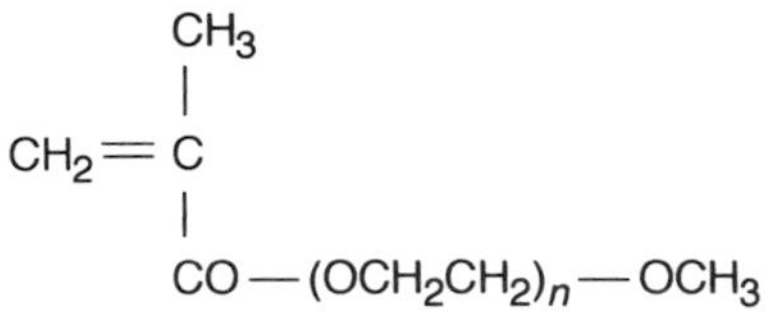

$$CH_2 = \overset{\displaystyle CH_3}{\underset{\displaystyle CO-(OCH_2CH_2)_n-OCH_3}{C}}$$

MnG

Figure 12.9   Methoxypoly(ethyleneglycol) monomethacrylate.

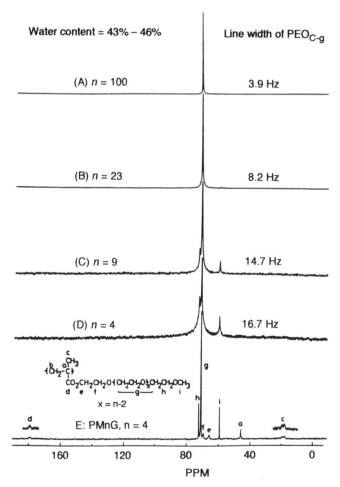

Figure 12.10   256 MHz ^{13}C NMR spectra and line widths of poly (MMA-co-MnG), measured in H$_2$O at 25 °C. The longest PEO chain shows the highest mobility. (Reproduced by permission of J.B. Lippincott Company from ref. [97].)

propylene oxide, which are insoluble in water, PEO is completely miscible with water at room temperature. The strange behaviour of this closely related polymeric chain is explained by the nature of PEO-water interaction [99]: the PEO segments dimension is such that the ordered water structure imposed by hydrogen bonding between water molecules (whose disrupture is the main obstacle to solubility in water) is minimally perturbed. In other words, PEO segments nicely fit the voids in the hydrogen-bond controlled water structure. As a consequence, PEO is very soluble in water, and PEO chains enjoy an extraordinary freedom of motion in water-containing environments. The same is true for hydrated PEO chains linked at one end to a substrate surface, as shown by ^{13}C NMR spectroscopy by Nagaoka and coworkers [97]. They synthetised polymers with PEO chains by random copolymerization of MnG and MMA, with n, the chain length of PEO, ranging from 4 to 100. The line widths of the ^{13}C NMR signal of PEO chains in the hydrated state (Figure 12.10) indicate that the chain mobility increases with increasing chain length, with a correlation time decreasing 10^{-10} s for $n = 100$ (a value close to that of PEO in free motion) to $10^{-9} \div 10^{-8}$ s for the shortest chains. The increased mobility of the longer chains is also indicated by the coalescence of all carbons in a single line, as shown in Figure 12.10. Proteins and platelets adsorption is strongly affected by the PEO chain length, as shown in Figure 12.11, taken from the same reference. There is a significant decrease in the amount of adsorbed blood components with increasing chain length, a finding that, together with the NMR evidence, strongly

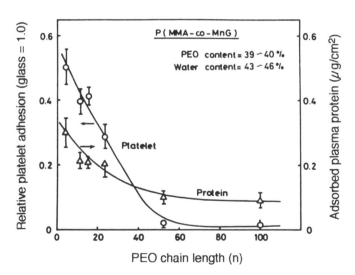

Figure 12.11 Effect of PEO chain length (n) on adhesion of platelets and adsorption of plasma proteins on to poly (MMA-co-MnG). Experiments were conducted in rabbit platelet rich or platelet poor plasma for 3 h. (Reproduced by permission of J.B. Lippincott Company from ref. [97].)

supports the hypothesis that the peculiar behaviour of PEO bearing surfaces is caused by the dynamic nature of its interface with water. Adsorption of proteins means a decrease of the freedom of mobile PEO chains, the loss of many possible configurations, that is an entropically unfavourable process, as described in Chapter 1. Hydrated dynamic surfaces are currently one of the most promising ways towards intrinsically haemocompatible surfaces [92].

A very interesting modelling of the PEO–protein interaction was recently discussed [100, 101]. The free energy of interaction as a function of PEO chain length and surface density is modelled by a steric repulsion, van der Waals' attraction and hydrophobic interaction contribution. Since no theory derived, analytical expression of the latter exists, the hydrophobic interaction free energy function determined experimentally by Pashley and coworkers is used [102]. It was found that the van der Waals' attraction is rather low, since PEO has the lowest refractive index among water soluble polymers, and small as compared to steric repulsion. In general, high surface density and long chain length are desirable for protein resistance, the former having a greater effect on steric repulsion. If the protein dimension is taken into account [101], it turns out that surface density should be properly tuned to specific protein dimensions in order to obtain the best results: a relatively high surface density is best to resist small proteins, while a lower surface density is better for larger ones. Even if these theoretical findings still lack experimental confirmation, and if some unrealistic assumptions are contained in the model, these papers have the extraordinary merit of highlighting the strong effect of the details of the fine structure of PEO bearing surfaces and of the protein solutions on the overall results.

12.1.3.3 *Microphase separated surfaces*

The best known example of polymer surfaces having microphase separated structures is given by segmented polyurethanes (SPU) (Chapter 8). Polyurethane elastomers used in blood contacting applications are segmented block copolymers, composed of alternating soft and hard blocks [103]. The soft segments, usually polyester or polyether have a glass transition temperature well below the temperature of use and are therefore rubbery, while the hard, rigid segments can be either aliphatic or aromatic diisocyanates, chain extended with diols or diamine. The interest in SPU in biomaterials applications was triggered by the finding that a polyether polyurethane, originally developed for elastic thread, was both hydrolytically stable and blood compatible enough for use in several blood contacting devices [104, 105]. Since then an enormous body of literature on bulk and surface properties of SPU has been published and many proprietary SPU specifically intended for blood contacting applications have been developed [103, 106]. Even if these studies succeeded in unravelling many of the secrets of the complex structure of SPU, the basic question, that is the reason why SPU surfaces are blood compatible, is still unanswered. The most often quoted hypotheses are that the surface is completely dominated by the blood compatible, often PEO containing (see the previous section), soft segment [107, 108], that the microphase separated structure gives the correct ratio between polar and apolar sites [58] or that the dimension of surface domains (of the same order of magnitude of individual proteins or cell surface receptors) allows specific interactions with proteins or formed elements of blood [15, 109]. Recently Goodman, Cooper and coworkers reported very interesting findings on protein adsorption on polybutadiene–polyurethane, using colloidal gold of 3 nm diameter to label adsorbed proteins [110]. The results suggest that molecular-sized microdomain structures have significant and direct effects on the adsorption behaviour of individual proteins.

Unfortunately, the already complex problem of the overall interaction is further complicated by the peculiar nature of these composite surfaces. The details of the phase separated surface structure are controlled by the interaction with the surrounding environment, thus they are not directly correlated with bulk structure [109, 111]. Moreover, since the soft segment is, at room temperature, well above its T_g, it is expected to be highly mobile, and to rearrange and reorganize in response to the changing environment [112, 113], introducing in this way a further, poorly controllable, variable. Thus, it is not surprising that three different groups, all of them based on XPS measurements and (different) haemocompatibility tests, reached different conclusions on which segment is better for blood compatibility [114–117].

Even if no general agreement exists on the basic mechanism, many researchers found that, besides SPUs, the blood surface interaction is deeply affected by the degree of phase separation in many other polymeric systems [118–125]. A typical result is shown in Figure 12.12 (adapted from ref. [125]), where the mean in *in vivo* occlusion time obtained using several hemo- and block copolymers is plotted. Clearly, copolymers in every case perform much better than related homopolymers. *In vitro* studies on protein adsorption on these materials (from mono-protein solutions) showed a marked effect of the microdomain structure on the protein adsorption pattern, with albumin selectively adsorbed on the hydrophilic microdomain and gamma-globulin and fibrinogen adsorbed on the hydrophobic microdomain [126]. Organized structures corresponding to the microdomian pattern were observed in competitive adsorption studies. Based on these results, the authors formulated the hypothesis that the organized protein-layer formed on the microdomain surface influences the activation of adhering platelets, for instance by the control of the distribution of the binding sites between platelets and the block copolymer surfaces [125]. As suggested by Andrade [98], howerer, the main limitation to speculations on this topic is the use of surfaces which are not are characterized in environments and under situations related to proteins and cell deposition. In this respect, a major step forward is expected from atomic force microscopy (AFM), whose atomic scale resolution and capability of underwater measurements, perfectly match the analytical

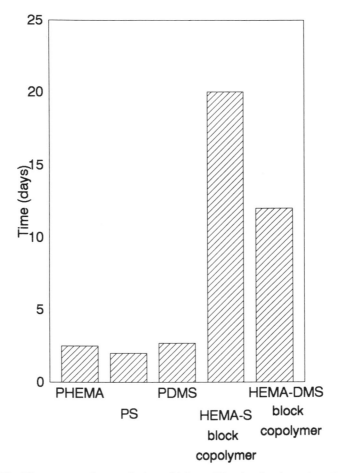

Figure 12.12   Time to complete occlusion of 1.5 mm ID tubes by thrombus. (Reproduced by permission of J. B. Lippincott Company from ref. [125].)

requirements researchers were waiting for. Underwater measurements of SPU by AFM have already been reported [127].

12.2   CONTACT LENSES

The first use of devices contacting ocular tissue dates back to the end of the 18th century. The material of choice was then, of course, glass, but the true development of a contact lens industry had to wait for the coming of age of optically transparent, finishable, light weight plastics. After the Second World War poly-(methylmethacrylate) (PMMA) became easily available and boosted the contact

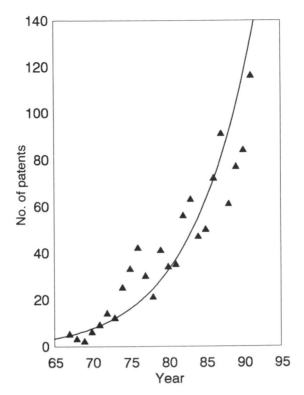

Figure 12.13 Trend of the number of patents on contact lenses (data refer to patents indexed in the *Chemical Abstracts*).

lens industry [128]. Today PMMA lenses are no longer used, as new, specifically designed polymers with improved properties appear. The contact lens industry is undergoing rapid change with new lenses made of a variety of materials appearing with increasing frequency. A research in *Chemical Abstracts* shows that, from 1967 to the end of 1991, 1029 patents on contact lenses appeared, with a yearly trend (Figure 12.13) that underlines the restless status of present day contact lens industry.

The driving force for this research and development effort arises from the need to design a material which can satisfy the large number of requirements imposed by the regulatory laws, the market, the manufacturing step and the in service performance. The latter are particularly critical as buyers' preference shifts towards extended wear (as opposed to daily wear) lenses. Several of the fundamental requirements of contact lenses, taken from ref. [129], are shown in Table 12.4.

Before discussing the role played by interfacial effects in contact lens performances it is necessary to introduce some form of material based classification. It

Table 12.4 Some of the important properties of contact lens materials

Hardness
Elongation
Light transmission, optical clarity, colour
Wettability/lubricity
Stress free, isotropic, stable structure with high T_g
Minimal response to pH, osmolarity, temperature and humidity
Lipid deposition
Disinfection
Monomer cost, availability, reproducibility
Patentability, rapid approval
Reliance on standard manufacturing methods, lens design and fitting
Modulus
Tear strength
Pore size
Toxicity
Handling and optical performance
Oxygen permeability
Refractive index
Protein deposition
Preservative uptake

must be remembered, however, that the following is a fairly general classification, and many variants and sub-divisions exist.

In general, the evolution of contact lens materials can be understood if one thinks that cornea is an avascular tissue, which obtains the oxygen supply it needs from the atmosphere, through the tear film which always covers the cornea. The stability of the tear film is closely controlled by the complicated lacrimal apparatus and by the blinking process. Thus, when cornea is separated from the atmosphere by a foreign material, some way must be found of mantaining an adequate oxygen supply.

The first generation material, PMMA, is practically impermeable to oxygen. Thus, under a PMMA lens ('hard lenses'), oxygen can reach the cornea only through oxygen-rich tear exchange. This is the reason why PMMA lens were so difficult to adjust in a comfortable way, and they have to be specifically fitted to each patient, so that normal blinking will permit continuous and adequate exchange of tears under the lens.

Every time that permeability to oxygen must be improved, one thinks of silicones. In the early 1980s rigid gas permeable (RGP) lenses, basically silicone/acrylate copolymers, entered the market. The permeability increase is most often owing to the incorporation of tris-(trimethylsiloxy)-silylpropylmethacrylate (TRIS, Figure 12.14) as the silicone portion of the copolymer. The amount of TRIS which can enter the composition is, however, limited by its adverse effect on hardness and wettability.

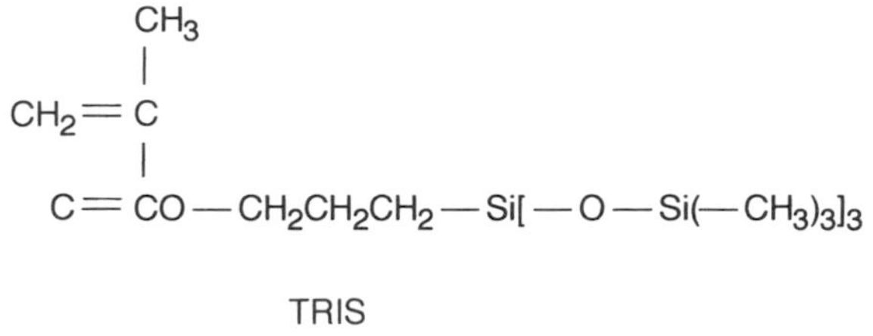

$$CH_2 = \underset{\underset{C = CO - CH_2CH_2CH_2 - Si[- O - Si(- CH_3)_3]_3}{|}}{\overset{\overset{CH_3}{|}}{C}}$$

TRIS

Figure 12.14 Tris-(trimethylsiloxy)-silylpropylmethacrylate (TRIS).

Silicone rubber based materials comprise the class of so-called flexible lenses [128]. They are, of course, very satisfactory with respect to oxygen permeability, but are poorly wettable.

Finally, soft lenses are based on hydrogels, that is cross-linked water swellable or water-soluble polymers. Synthetic hydrogels (based on 2-hydroxyethyl-methacrylate, HEMA) for contact lenses application were first proposed by Wichterle and Lim [130]. In this case, the oxygen supply is due to the oxygen permeation through the aqueous portion of the hydrogel.

Almost all contact lenses on the market today are of the RGP or of the soft type [131].

In order to understand the role played by interfaces, let us consider in some detail the natural tear film and a contact lens in its working environment. It is known that the well-being of cornea require a coating by a continuous fluid film (tear film) [132]. The tear film that continuously covers the cornea has a thickness of about 10 μm in normal conditions [133]. It is composed of two layers, since the aqueous tear layer is covered by a superficial lipid layer which is about two orders of magnitude thinner than the former [134, 135]. The stability of thin films is controlled by interfacial energetics, that is whether it is more convenient to have a system composed of a tear–air and a tear–cornea interface or to give off both of them and create a single air–cornea interface. From a thermodynamic point of view, a convenient way of expressing this balance is through the disjoining pressure, introduced by Derjaguin [136, 137], and defined as the difference between the pressure of the interlayer on the surfaces confining it, and the pressure of the bulk phase in equilibrium with the interlayer.

In normal conditions, the tear film stability is assured by the superficial lipid layer, which lowers the interfacial tension at the tear–air interface, and by the mucus distributed by the lacrimal apparatus and continuously refreshed by the blinking process, which controls the interfacial tension at the tear–cornea interface.

When a contact lens is fitted to the eye, new thin films and interfaces are created, as shown in Figure 12.15: the lens (from 40 to 300 μm thick) is coated by a tear film (pre-lens film) quite similar to the precorneal films of non-contact lens bearing eyes [138–141]. Lens and eye, both mucus coated, are separated by

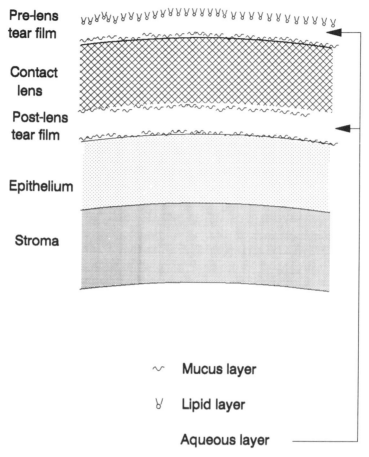

Figure 12.15 Structure of the tear film around a contact lens.

another aqueous film (post-lens film), which lacks the lipid layer, since the lens itself generally forms a barrier to lipids contained in the surface of the pre-lens film. This is an important point, since, while the pre-lens film is continuously refreshed by blinking, the post-lens film is much less effectively replaced. Accumulation of lipids in the post-lens film would create hydrophobic surfaces and high interfacial tensions, seriously threatening the stability of the aqueous layer.

Since a contact lens is completely immersed in an aqueous fluid, whose continuity and stability is fundamental for good performance of the ocular apparatus, a good wettability by water is a fundamental requirement. However, a close examination of the problem shows that this statement is not completely satisfactory: first, a zero water contact angle does not mean that interfacial free energy is zero [53], but rather that water is no longer able to characterize the interfacial field of forces [72]. Thus, the previous statement refers to a necessary but insufficient condition. Then, most important, the working environment of a contact lens is not pure water, but a complex biological fluid, whose components, often of amphiphilic nature, can be adsorbed on the surface and change its properties, much as previously described for blood contacting devices. The adsorbed layer can change the surface properties of the lens, eventually promoting biomolecule deposition, with adverse effects on vision as well as comfort. Protein deposition is the largest single cause of contact lens replacement [129].

Thus, contact lens surfaces must be able to guarantee pre- and post-lens film stability, and minimize the adsorption and deposit of potential properties impairing biomolecules. A possible answer is the periodic installation of properly formulated contact lens wetting solutions. Another is to specifically design or modify contact lens surfaces. This second way is somewhat hindered by the incomplete knowledge of the tear film physiology and by the complex nature of a biological fluid continuously produced by a human being [133], a problem which reaffirms the difficulties that must be faced in the design of blood contacting materials. In the next sections, we will review researchers effort in this second field.

12.2.1 Hard Lenses (HL)

The importance of wettability in contact lens applications can be readily recognized if one thinks of the long predominance of poly(methylmethacrylate) (PMMA) in the field of HL [141]. Many other polymers offer the same or better characteristics than PMMA with respect to workability, optical properties, not to speak of oxygen permeability. The last property, critical as it is in contact lens performance, seems to exclude PMMA from the list of useful polymers since, as already stated, PMMA is almost impermeable to oxygen. As shown in Table 12.5, that compares the oxygen permeability and the advancing water contact angles of PMMA and of some competitors, the (at least) one order of

Table 12.5 Oxygen permeability coefficient (P_g) and advancing water contact angle of some hard contact lens material at room temperature

Material	$P_g \times 10^4$ (μl(STP) cm)/(cm^2 h kPa)	Advancing angle of water (degrees)
PMMA	0.27	72
Poly(4-methyl-1-pentene)	176–64.8	100
Poly(pentamethyldisiloxanyl methylmethacrylate-co-methyl-methacrylate)	7.95	106
Poly(perfluoroalkyl alkyl methacrylate-co-methyl-methacrylate)	*ca* 7.5–27	110

magnitude increase in oxygen permeability of the latter is paralleled by the increase in water contact angle. In the case of poly(4-methyl-1-pentene) (PMP), the whole polymeric chain is hydrophobic, and this is reflected by the, hydrocarbon-like value of the water contact angle (Chatper 9). On the other hand, fluoroacrylate polymers (FAP) contain the same esterified carboxyl groups as PMMA (moreover they are copolymer with PMMA), but are much less wettable. Clearly, surface-active, fluorine containing moieties accumulate at the surface, as described in Chapter 8 on block copolymer surfaces, so that the topmost layer of this acrylic polymer is composed of extremely hydrophobic fluorine containing functions. As a consequence the polymer results are poorly wettable, despite the number of oxygen containing groups that are about the same as in PMMA. Thus, it is the sensitivity of wetting phenomena to the very surface of a material that makes this polymer unsuitable for contact lens applications.

Surface modification of PMMA and candidate HL materials were attempted in order to improve overall performance. Plasma or other electrical discharge treatments have been popular techniques, as shown by the large number of patents on this topic ([142–147] for instance). Many different gases or vapours have been proposed from oxygen, to water, to noble gases, to nitrogen-containing vapours and gases. As discussed in Chapter 6, these treatments introduce a huge number of polar (or acid-base, see Chapter 4) functions to the sample surface, so that the water contact angle is, in general, greatly lowered. For instance, in ref. [144] it is claimed that the contact angle of water on a PMMA-fluoroalkyl-ethylmethacrylate copolymer is reduced from 96–101° to 16.5–21° after 20 cycles of 8–12 kV discharges, with a discharge time of 0.5 seconds and 1.5 seconds between discharges. In the patent of Gesser and Warriner [143], water, hydrazine, acetic or formic acid plasma produce a decrease of the contact angle of water on PMMA from about 60° to about 18°. These values are typical of polymer surfaces in the just plasma treated conditions. However, as previously discussed (Chapter 9), treated surfaces undergo ageing, and the improved wettability is partially or totally lost as time from treatment elapses. To offset this problem, deposition of polymeric films from plasma in the presence of oxygen was attempted [148]. According to Yasuda, the very cross-linked nature of polymers deposited from plasma makes them less prone to ageing [149]. However, none of these treatments appear to be practical, since most HL were manufactured simply from unmodified PMMA. The lack of commercial exploitation of plasma treated PMMA is probably caused by the incomplete effectiveness of plasma treatment, in particular with respect to ageing, and by the shift of the contact lens market towards other materials, as discussed below. The most successful surface modification technique in PMMA HL was probably copolymerization with hydrophilic monomers, namely poly(hydroxyethylmethacrylate) (PHEMA), the basic component of soft contact lenses [128]. In the early days of soft lenses commercialization, PHEMA–PMMA copolymerization was sought as a way of combining the beneficial qualities of both hard and soft type lenses and eliminating all the adverse effects, since the copolymer (generally PHEMA < 10%) is readily wettable and has high dimensional stability [150, 151].

As previously discussed, the long predominance of PMMA in the contact lens industry was caused by the lack of effective alternatives, in particular with respect to surface properties. Its lack of permeability to oxygen, however, put it into a vulnerable position and the continuous development of new and improved materials finally put it out of the market. Surface related aspects of RGP lenses will be discussed in the next section (together with some discussion on the other class of siloxane based lenses, namely flexible lenses, which are not yet important from a market perspective but are very interesting from a surface science point of view). Finally, we will discuss hydrogel based soft lenses the other kind of material which takes the lion's share of the market.

12.2.2 Rigid Gas Permeable (RGP) and Flexible Lenses (FL)

The first gas permeable material to reach the market stage in contact lens application was cellulose acetate butyrate (CAB), whose oxygen permeability is about 100 times better than that of PMMA [152]. Generally speaking, CAB is not a single material, but rather a family of plastics in which the properties of the cellulose derivative can be controlled by the ratio of acetyl and butyryl content. In the contact lens industry, the most often used material is the 13 wt % acetyl, 37% butyryl and 1–2 wt % unsubstituted hydroxyls [138]. The wettability of this CAB in the dry state is about the same as PMMA, but CAB becomes more wettable than PPMA after soaking in tears. Moreover, CAB lenses can be surface treated in order to improve wettability: a chemical treatment of CAB lenses by sodium or ammonium hydroxide is disclosed in a Japanese patent [153]. After soaking in water and a cationic surfactant solution, the wettability of the treated lens results improved.

The main drawback of these RGP lenses are the insufficient mechanical properties, which lead to poor shape memory, insufficient dimensional stability and tendency to warp, deform, scratch and chip. In the early 1980s they were supplanted by TRIS-based RGP, which combine the oxygen permeability of silicones with the chemistry and mechanical properties of acrylates [154]. TRIS can be copolymerized with MMA, giving rise to dimensionally stable, oxygen permeable polymers. The effect of the amount of TRIS on the oxygen permeability of TRIS–MMA copolymers is shown in Figure 12.16. The amount of TRIS that can be incorporated in the copolymer is, however, limited by the adverse effect of TRIS itself on the mechanical properties and wettability. The 'open' structure and the high flexibility of the Si–O bonds, the basic reason of the high permeability to oxygen of siloxanes, also allows the creation of a low energy surface composed mainly of the hydrophobic methyl groups. The same problem is even more pressing in the recently developed RGP lenses based on fluorosiloxane-acrylate polymers [152].

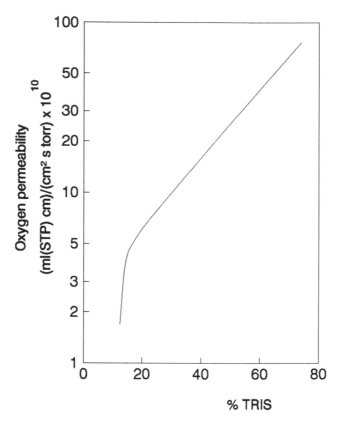

Figure 12.16   Effect of TRIS on the O_2 permeability of TRIS/MMA copolymers.

Two surface modification techniques have been mainly employed in order to render the surfaces of siloxane-acrylate and fluorosiloxane-acrylate contact lenses more wettable, namely copolymerization with hydrophilic acrylate and plasma treatment.

The former technique has some problem when used in TRIS based lenses: the hydrophobic TRIS and the hydrophilic acrylate are incompatible and the resulting copolymer tends to phase-separate, giving rise to opaque materials. Thus, the amount of hydrophilic moieties that can be copolymerized with TRIS is usually small. A possible way out of this problem is to use more hydrophilic siloxane monomers, such as the methyldi(trimethylsiloxy)sylylpropylglycerol methacrylate (SiGMA) or the methyldi(trimethylsiloxy) sylylpropylglycerolethyl methacrylate (SiGEMA) shown in Figure 12.17 [155, 156]. The hydroxyl group and the ether bond in the repeating unit make this monomer compatible with hydrophilic acrylate such as HEMA. The resulting polymer is, of course, more hydrophilic than a comparable TRIS-MMA copolymer. Thus, here is an example

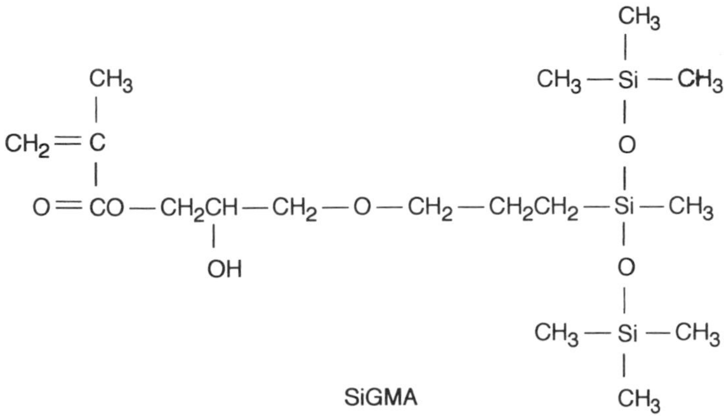

Figure 12.17 Methyldi(-trimethylsiloxy) sylylpropylglycerol methacrylate (SiGMA) and methyldi(-trimethylsiloxy) sylylpropylglycerolethyl methacrylate (SiGEMA).

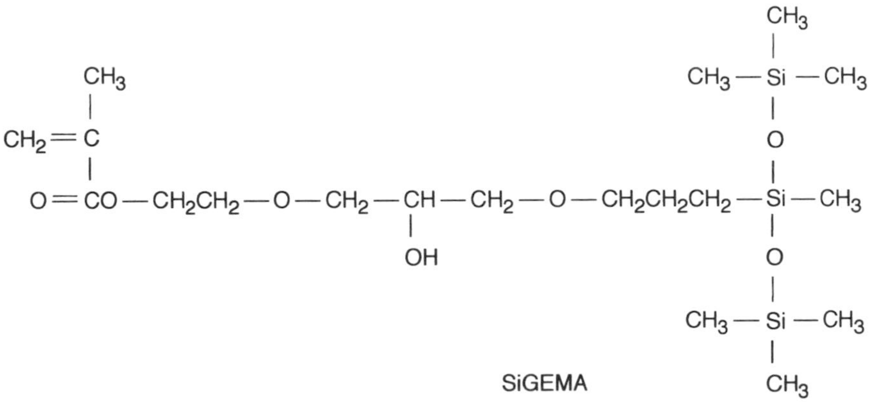

of how technical constraints (the need of a transparent material) make necessary a bulk modification at the molecular level (to switch from TRIS to SiGMA or (SiGEMA) in order to obtain the improvement of a surface property (wettability).

Plasma treatment of RGP is described in several patents. In ref. [157], it is claimed that plasma treatment with inert gases greatly improve the wettability of fluorosiloxane-acrylate RGP lenses. A typical composition of these lenses is about 35 parts of TRIS, 40 parts of trifluoroethyl methacrylate, 5 parts of methylmethacrylate and 6 parts of the cross-linker ethylene glycol dimethacrylate. It is interesting to observe that these basically methacrylate polymers, despite the huge number of oxygen containing functions, have a water contact angle greater than 100°, produced by the orientation of the hydrophobic moieties at the solid–air interface. After several minutes plasma treatment, the contact angle is reduced to about 30°.

The effect of oxygen plasma treatment on the wettability and the surface composition of siloxane-acrylate lenses has been deeply investigated by Fakes and coworkers [158–160]. The organosiloxane unit of the material studied is of the kind shown in Figure 12.17. Thus, it is more hydrophilic than the lenses described in ref. [157] and, accordingly, the advancing water contact angle on the untreated lens is about 78° [159]. The effect of the plasma treatment has been studied, besides contact angle measurement, by X-ray photoelectron spectroscopy (XPS) and static secondary ion mass spectroscopy (SSIMS, Chapter 3). The plasma treatment greatly reduces the advancing contact angle of water (*ca* 30° after 160 seconds treatment. Ageing effects on treated surfaces are not discussed in these papers). Both XPS and SSIMS analysis show convincingly that oxygen from plasma preferentially attacks the C–Si bond, removing the organic moieties from the siloxane-backbone and creating a kind of inorganic silica-like phase (or, more in general, increasing the average number of Si–O–Si bonds). The depletion of organic moieties and the increase of the surface concentration of oxygen produce the improvement of wettability. These results are in agreement with other XPS and SSIMS studies on the effect of oxygen plasma on silicones, namely poly(dimethylsiloxane) (PDMS) [161, 162].

It is interesting to note that the highly cross-linked, inorganic-like, Si–O–Si structure produced by the plasma treatment is potentially very deleterious to oxygen permeability and thus to lens performance (inorganic silicates, such as glass, have for centuries been exploited for their low oxygen permeability, for instance in food packaging). Fortunately, as discussed in Chapter 6, the effect of plasma is confined to a very thin surface layer (approximately 20 nm for 160 s treatment at 50 W [158]), so that oxygen permeability is not greatly affected. On the other hand, the thickness of the plasma modified layer in PDMS in greater, probably owing to its very high permeability to gases, as confirmed by the fact that, unlike 'conventional' polymers [162], the effect of plasma treatment on PDMS is readily observed by ATR–IR [161, 164]. In this case, a decreased permeability is measured after plasma treatment [165].

Plasma-induced grafting of acrylamide on siloxane-acrylate lenses is disclosed in ref. [166]. The main advantage of this treatment over plasma is that no ageing of treated surfaces occurs, but it is clearly more expensive and time consuming than a simple plasma treatment.

The other class of contact lenses which exploits the siloxane chemistry is that of FL [167, 168]. As discussed previously, silicone rubber is highly permeable to oxygen and with a proper tuning of substitution of the organic groups, branching cross-linking and molecular weight, a large range of mechanical properties can be obtained. The main drawback of silicone rubber as a contact lens material is its strong hydrophobicity: despite several years of research efforts, surface modification techniques still cannot bridge the gap between silicone rubber based FL and siloxane-acrylate RGP or soft contact lenses.

The most commonly suggested treatments are plasma [169, 170], grafting

[171–173], and coating or copolymerization [174, 175]. As previously discussed, plasma treatment of silicones produces highly wettable surfaces, but the effect is only temporary [161]. A more durable hydrophilization is obtained, according to Yasuda, by oxygen plasma treatment of plasma deposited thin polymeric films [176, 177].

The effect of plasma treatment on protein and lipid adsorption by silicones is discussed by Holly and Owen [170]. In an *in vitro* study, they observed that, wettability increases, protein adsorption increases while lipid adsorption diminishes. Moreover, more proteins are retained on the surface of the treated samples after cleaning with a surfactant solution. Thus, surface hydrophilization alone, even if necessary, is not the definite solution to all contact lenses problems, since the lens substrate must face all the different interactions forces arising from the complex molecular architecture of proteins. This problem, which reflects the discussion on blood contacting devices of the first part of this chapter, will be further discussed in the next section on soft (hydrogel based) contact lenses.

12.2.3 Soft Lenses (SL)

Soft contact lenses (SL) were the first alternative class of materials to PMMA that reached commercial success in the contact lens market. As discussed in the introduction, today they share with RGP lenses almost all of the market.

Soft lenses are based on synthetic hydrogels, that is covalently or ionically cross-linked hydrophilic polymers that have swollen in water to form a soft, elastic, gel-like material [178]. The existence of water in hydrogels deeply affects their mechanical and barrier properties, and puts them in a unique position in relation to their ability to interface with blood and living tissues [54].

The equilibrium water content (EWC) is the most fundamental property of a hydrogel [141]. It depends, of course, on the balance of hydrophilic and hydrophobic moieties of the polymer, and can be carefully tuned by copolymerization of different monomers. Oxygen permeability of SL is controlled by EWC, since oxygen from atmosphere can reach the cornea by diffusion through the aqueous phase of the hydrogel. For EWC above 30% (a condition that is satisfied by almost all SL) it has been found that the permeability of dissolved oxygen (Pd) at 34 °C can be predicted from the EWC (W%) by the following expression [141]:

$$Pd = 24 \times 10^{-10} e^{0.0443w} [cm^3 (STP) mm\, cm^{-2} s^{-1} cm^{-1} Hg] \qquad (12.3)$$

The relationship between water content and oxygen permeability prompted chemists to develop hydrogel based lenses characterized by very high EWC, besides, of course, adequate mechanical properties and dimensional stability. The original work of Wichterle and Lim was mainly based on HEMA and some of the homologous esters of the glycol methacrylate series, slightly cross-linked with a dimethacrylate glycol [179, 180]. The EWC of these materials is about 40%. Many other compositions have been later patented, and an interesting review of

Table 12.6 Some of the materials used in soft contact lenses

Trade name	Principal components*	Water content (%)
Accugel	HEMA, PVP, MA	46.6
Accusoft	HEMA, PVP	47
Comfort Flex	HEMA, BMA	43
CS-151	MMA, GM	41
Hydromarc	HEMA, MA	43
Flexol	HEMA, AMA, MA	72
Scanlens	Amido-amino copolymer	73.5
Snoflex 50	MMA, VP, MHPM	52.5

*HEMA = hydroxyethylmethacrylate, PVP = poly(vinyl pyrrolidone), MA = methacrylic acid, GM = glyceryl methacrylate, BMA = buthyl methacrylate, VP = n-vinyl pyrrolidone, AMA = alkyl methacrylate, MHPM = 3-methoxy-2-hydroxypropylmethacrylate.

the technical and patent literature of SL can be found in ref. [181]. Some of the compositions in use today are shown in Table 12.6, together with the trade name and the EWC. As suggested in the quoted review, it is interesting to note the more extensive evaluation and use of novel hydrogel structures in the contact lens field in comparison with the strong dependence upon PHEMA in other medical areas. This state of the art mainly arises because of the comparative ease of clinical investigations in the former case, coupled with the fact that biocompatibility problems in the eye are slight in comparison with those involving blood contact discussed in the first part of this chapter [181].

With respect to surface properties, hydrogel based lens, because of their high water content, are, in general, very wettable, but important exceptions to this rule exist. In fact, the aqueous environment has a plasticizing effect on chains [15], which are very mobile and can reorient in response to the interfacing medium [133]. As discussed in the chapter on contact angle measurement (Chapter 4) PHEMA surfaces expose preferentially the hydrophobic methyl groups at the polymer–air interface and the hydrophilic groups at the polymer–water interface [182]. Thus, in hydrogels containing both hydrophilic and hydrophobic groups, high water content does not necessarily mean high wettability.

Reorientation of mobile hydrogel surfaces can also be stimulated by biological macromolecules adsorbed from tears, and affect the hydrogel response to the tear components [133].

Another point linked to surface properties is that SL, especially those with a very high water content, are prone to adsorb proteins and lipoproteinaceous debris. The interaction of hydrogel materials with natural tears and tear-simulating solutions has been the subject of many studies. The problem is, of course, difficult to analyse, since tears are very complex fluids, with individual-related differences as to quantity and quality of components. From its side, the complex nature of hydrogel surfaces, as underlined, for instance, by the previously quoted reorientation behaviour, further complicates the problem [183].

Despite these difficulties a great body of knowledge on protein adsorption on SL has been accumulated [184–194]. The complete discussion of these findings is outside the scope of this text, but some interesting, SL surface-related aspects deserve some discussion. For instance, Sack and coworkers [192] described the strong effect of hydrogel chemistry on the lens-bound protein layer (LBPL). While in non-ionic, PHEMA based lenses the LBPL is thin and composed of highly denaturated tear proteins, in lenses containing methacrylic acid (MAA) moieties the LBPL consists primarily of a thick, loosely bound layer of lysozyme (which represents about 30% of tear proteins), much of which retains its structural integrity. Many other researchers found that the amount of adsorbed lysozyme is higher in MMA containing lenses. According to Castillo and coworkers, PHEMA/MAA (3%-MMA) lenses adsorb 30 times more lysozyme than PHEMA lenses [189] (probably, in this case, interpenetration of the polymer matrix by the protein occurs). The reason for this behaviour is probably the ionic interactions between carboxylate anions of the SL and lysozyme ionic groups; on the other hand, on non-ionic hydrogel surfaces, interactions likely occur through hydrophobic or non-ionic hydrophilic interaction, which favour protein denaturation by the exposure of non-ionic amino acids normally buried within the protein interior. These findings suggest that the LBPL should be considered as a highly selective, biologically active layer. As previously discussed for blood contacting devices, the characteristics of the material surfaces affect the overall behaviour through the adsorbed protein layer. From a practical point of view, this also means that HEMA intended for contact lens usage should be carefully purified, since MAA is a common contaminant of HEMA [195]. A very pure HEMA should ensure a low (but still significant) lysozyme adsorption. Another intriguing observation is that the fabrication process appears to induce different adsorption behaviour: lathe-cut lenses, produced by machining dehydrated PHEMA, adsorb twice the amount of protein compared with PHEMA spin-cast lenses. Clearly, fabrication defects induce surface imperfections which generate additional sites for the adsorption. Incidentally, this finding underlines the important role of all components of the hierarchy of surface structure as the result of interfacial events.

Despite the different nature of LBPL and the different kind of SL, protein adsorption is still an unresolved problem. Treatment with oxidizing solutions or enzymes is normally required to extend the useful life of lenses. Thus, it is interesting to note that, as in blood contacting devices, the unique properties of polyethyleneoxide (PEO) are exploited to produce protein resistant surfaces [98], as discussed in the first part of this chapter. In fact, the inclusion of significant amounts of PEO units in contact lenses is claimed to greatly reduce protein adsorption and allow comfortable extended wear [196]. In the quoted patent, dramatic decrease of *in vitro* lysozyme adsorption is observed in lenses containing several tens of per cent of methacrylate esters of polyethylene glycol copolymerized with conventional SL monomers. *In vivo* performance, with the contemporary action of different proteins, lipids and mucus is, at present, not known.

REFERENCES

[1] S. D. Bruck, *Biomater. Med. Devices Artif. Organs*, **5**, 97 (1977).
[2] B. Kasemo and J. Lausmaa, *CRC, Crit. Rev. Biocompatibility*, **2**, 335 (1986).
[3] L. L. Hench and E. C. Ethridge, *Biomaterials, an Interfacial Approach*, Academic Press, New York (1982).
[4] J. B. Park, *Biomaterials, Science and Engineering*, Plenum Press, New York (1984).
[5] *Biomedical Materials*, Sept. 1991, p. 8.
[6] *Biomedical Materials*, May 1990, p. 7.
[7] J. N. Mulvihill, J. P. Cazenave, A. Schmitt, P. Maisonneuve, and C. Pusineri, *Colloids Surfaces*, **14**, 317 (1985).
[8] S. Srinivasan and P. N. Sawyer, *J. Colloid Interface Sci.*, **32**, 456 (1970).
[9] P. N. Sawyer and S. Srinivasan, *Am. J. Surg.*, **114**, 42 (1967).
[10] P. N. Sawyer and S. Srinivasan, *J. Biomed. Mater. Res.*, **1**, 83 (1967).
[11] S. Srinivasan and P. N. Sawyer, *JAAMI*, **3**, 116 (1969).
[12] L. Vroman, *Bull. N.Y. Acad. Med.*, **48**, 302 (1972).
[13] R. Baier, *Polym. Sci. Tecnol.*, **8**, 139 (1975).
[14] I. Lundstrom, B. Ivarsson, U. Jonsson, and H. Elwing, in *Polymer Surfaces and Interfaces*, W. J. Feast and H. S. Munro (Eds), Wiley, Chichester (1987), p. 201.
[15] *Surface and Interfacial Aspects of Biomedical Polymers*, J. D. Andrade (Ed.), Plenum Press, New York (1985), Vol. 2.
[16] W. Norde, *Adv. Colloid Interface Sci.*, **25**, 267 (1986).
[17] T. Suzawa and H. Shirahama, *Adv. Colloid Interface Sci.*, **35**, 139 (1991).
[18] A. S. Hoffman, *ACS Organic Coatings and Applied Polymer Science Proceedings*, **48**, 28 (1983).
[19] J. E. Wilson, *Polym. Plast. Technol. Eng.*, **25**, 233 (1986).
[20] P. Didisheim, K. Hattori, and J. H. Lewis, *J. Lab. Clin. Med.*, **53**, 866 (1959).
[21] P. Fantl, *Aust. J. Exp. Biol.*, **39**, 403 (1961).
[22] E. F. Grabowski, P. Didisheim, J. C. Lewis, J. T. Franta, and J. Q. Stropp, *Trans. Am. Soc. Artif. Organs*, **23**, 141 (1977).
[23] P. Didisheim, J. Q. Stropp, J. H. Borowick, and E. F. Grabowski, *Am. Soc. Artif. Int. Organs*, **2**, 124 (1979).
[24] B. D. Ratner, A. Chilkoti, and G. P. Lopez, in *Plasma Deposition, Treatment, and Etching of Polymers*, Academic Press, New York (1990), p. 480.
[25] F. Silver and C. Doillon, *Biocompatibility*, VCH Publishers Inc., New York (1989), p. 235.
[26] P. N. Sawyer, *Ann. N.Y. Acad. Sci.*, **416**, 561 (1984).
[27] P. N. Sawyer and J. W. Pate, *Surgery*, **34**, 491 (1953).
[28] P. N. Sawyer and J. W. Pate, *Am. J. Physiol.*, **175**, 113 (1953).
[29] P. N. Sawyer and D. H. Harshaw, *Biophys. J.*, **6**, 653 (1965).
[30] R. I. Leininger, *CRC Critical Reviews in Bioengineering*, **1**, 333 (1972).
[31] J. M. Anderson and K. Koftke-Marchant, *CRC Critical Reviews in Biocomposition*, **1**, 111, 1985.

[32] M. C. Boffa, J. P. Farges, B. Dreyer, B. Conche, C. Pusineri, and G. Vantard, in *Biomaterials*, G. D. Winter, D. F. Gibbons and H. Plencks (Eds), Wiley, Chichester, (1980), pp. 399–408.
[33] S. D. Bruck, *J. Appl. Polym. Sci. Appl. Polym. Symp.*, **66**, 283 (1979).
[34] W. Drost-Hansen, *Fed. Proc. Fed. Am. Soc. Exp. Biol.*, **30**, 1539 (1971).
[35] S. D. Bruck, *Nature*, **243**, 416 (1973).
[36] S. D. Bruck, *Polymer*, **6**, 25 (1975).
[37] S. D. Bruck, *Polymer*, **5**, 435 (1964).
[38] S. D. Bruck, *Polymer*, **6**, 49 (1965).
[39] S. D. Bruck, *J. Polymer Sci.*, *Part C*, **17**, 169 (1967).
[40] E. Freund, *Med. Jahrb. Wien.*, **3**, 259 (1885).
[41] J. Bordet and O. Gengou, *Ann. de l'Inst. Pasteur*, **17**, 822 (1903).
[42] O. Neubauer and H. Lampert, *Muenchen Med. Wschr.*, **77**, 582 (1930).
[43] D. J. Lyman, W. M. Muir, and I. J. Lee, *Trans. Amer. Soc. Artif. Int. Organs*, **11**, 301 (1965).
[44] K. B. Bischoff, *J. Biomed. Mater. Res.*, **2**, 89 (1968).
[45] R. E. Baier, *Advan. Exp. Med. Biol.*, **7**, 235 (1970).
[46] R. E. Baier, *Bull. N.Y. Acad. Med.*, **48**, 257 (1972).
[47] R. E. Baier and R. C. Dutton, *J. Biomed. Mater. Res.*, **3**, 191 (1969).
[48] R. E. Baier, G. I. Loeb, and G. A. Wallace, *Fed. Proc.*, **30**, 1523 (1971).
[49] R. E. Baier, in *Proceedings of the Third International Congress on Marine Corrosion and Biofouling*, R. F. Acker, B. F. Brown, J. R. De Palma, and W. P. Iverson (Eds), Northwestern University Press, Evanston, Ilinois (1973), p. 633.
[50] S. C. Dexter, *J. Colloid Interface Sci.*, **70**, 346 (1979).
[51] M. E. Schrader, *J. Colloid Interface Sci.*, **88**, 296 (1982).
[52] J. D. Andrade, H. B. Lee, M. S. Jhon, S. W. Kim, and J. B. Hibbs Jr., *Trans. Amer. Soc. Artif. Int. Organs*, **19**, 1 (1973).
[53] J. D. Andrade, *J. Ass. Adv. Med. Inst.*, **7**, 110 (1973).
[54] *Hydrogels for Medical and Related Applications*, J. D. Andrade (Ed.), ACS Symposium Series N. 31, Amer. Chem. Soc., Washington, D.C. (1976).
[55] B. D. Ratner, A. S. Hoffman, S. R. Hanson, L. A. Harker, and J. D. Whiffen, *J. Polymer Sci. Polymer Symposium*, **66**, 363 (1979).
[56] D. L. Coleman, D. E. Gregonis, and J. D. Andrade, *J. Biomed. Mater. Res.*, **16**, 381 (1982).
[57] E. Nyilas, W. A. Morton, T. M. Lederman, T. H. Chiu, and R. D. Cumming, *Trans. Amer. Soc. Artif. Int. Organs*, **11**, 55 (1975).
[58] E. Nyilas and R. S. Ward Jr., *J. Biomed. Mater. Res. Symposium*, **8**, 69 (1977).
[59] D. H. Kaelble and J. Moacanin, *Polymer*, **18**, 475 (1977).
[60] D. H. Kaelble, *J. Adhes.*, **2**, 66 (1970).
[61] C. P. Sharma, *Biomaterials*, **2**, 57 (1981).
[62] L. Paul and C. P. Sharma, *J. Colloid Interface Sci.*, **84**, 546 (1981).
[63] C. P. Sharma and G. Kurian, *J. Colloid Interface Sci.*, **97**, 38 (1984).
[64] C. P. Sharma and P. V. Ashalatha, *Cellular Polymers*, **3**, 325 (1984).
[65] C. P. Sharma, *J. Colloid Interface Sci.*, **97**, 585 (1984).
[66] C. P. Sharma and A. K. Nair, *J. Colloid Interface Sci.*, **104**, 277 (1985).
[67] Y. Ikada, M. Suzuki, and Y. Tamada, *Polymer Preprints*, **24**, 1 (1983).
[68] Y. Ikada, *Adv. Polym. Sci.*, **57**, 103 (1984).
[69] M. Suzuki, Y. Tamada, H. Iwata, and Y. Ikada, in *Physicochemical Aspects of Polymer Surfaces*, K. L. Mittal (Ed.), Plenum Press, New York (1981), pp. 923–941.
[70] E. Ruckenstein and S. V. Gourisankar, *J. Colloid Interface Sci.*, **101**, 436 (1984).
[71] C. Jho, *J. Colloid Interface Sci.*, **109**, 588 (1986).

[72] S. V. Gourisanker and E. Ruckenstein, *J. Colloid Interface Sci.*, **109**, 591 (1986).

[73] C. P. Sharma, *J. Colloid Interface Sci.*, **110**, 292 (1986).

[74] E. Ruckenstein and S. V. Gourisankar, *J. Colloid Interface Sci.*, **110**, 293 (1986).

[75] E. Ruckenstein and D. B. Chung, U. S. Patent 4,929,510, to State University of New York, Albany, N.Y.

[76] V. L. Gott and A. Furuse, *Fed. Proc.*, **30**, 1679 (1971).

[77] H. Yasuda, M. O. Bumgarner, and L. G. Mason, in *Permeability of Plastic Films and Coatings to Gases, Vapor and Liquids*, Polymer Science and Technology Series, H. B. Hopfenberg (Ed), Plenum Press, New York (1974), Vol. 6, p. 453.

[78] A. S. Chawla, *Artif. Organs*, **3**, 92 (1979).

[79] A. S. Chawla, *Trans. Am. Soc. Artif. Int. Organs*, **25**, 287 (1979).

[80] A. S. Chawla, *Biomaterials*, **2**, 83 (1981).

[81] A. R. Ozdural, J. Hameed, M. Y. Boluk, and E. Piskin, *ASAIO J.*, **3**, 116 (1980).

[82] H. Yasuda, M. O. Bumgarner, and L. G. Mason, *Biomat. Med. Dev. Art. Org.*, **4**, 307 (1976).

[83] H. Yasuda and M. Gazicki, *Biomaterials*, **3**, 68 (1982).

[84] H. K. Yasuda, Y. Matsuzawa, S. R. Hanson, and L. A. Harker, *Trans. Soc. Biomat.*, **7**, 338 (1984).

[85] Y. S. Yeh, Y. Iriyama, Y. Matsuzawa, S. R. Hanson, and H. Yasuda, *J. Biomed. Mater. Res.*, **22**, 795 (1988).

[86] A. M. Garfinkle, A. S. Hoffman, B. D. Ratner, L. O. Reynolds, and S. R. Hanson, *Trans. Am. Soc. Artif. Int. Organs*, **30**, 432 (1984).

[87] A. S. Hoffman, B. D. Ratner, A. M. Garfinkle, L. O. Reynolds, T. A. Horbett, and S. R. Hanson, in *Polymers in Medicine II — Biomedical and Pharmaceutical Applications*, E. Chiellini, P. Giusti, C. Migliaresi, and L. Nicolais (Eds), Plenum Press, New York (1985), p. 157.

[88] D. Kiaei, A. S. Hoffman, B. D. Ratner, T. A. Horbett, and L. O. Reynolds, *A.C.S. Polym. Mater. Sci. Eng. Prepr.*, **56**, 710 (1987).

[89] D. Kiaei, A. S. Hoffman, B. D. Ratner, T. A. Horbett, and L. O. Reynolds, *J. Appl. Polym. Sci., Appl. Polym. Symp.*, **42**, 269 (1988).

[90] W. M. Reichert, F. E. Filisko, and S. A. Barenberg, in *Biomaterials: Interfacial Phenomena and Applications*, S. L. Cooper and N. A. Peppas (Eds), Adv. Chem. Ser., **199**, 177 (1982).

[91] E. W. Merrill and E. W. Salzman, *ASAIO J.*, **6**, 60 (1983).

[92] J. D. Andrade, S. Nagaoka, S. Cooper, T. Okano, and S. W. Kim, *ASAIO J.*, **10**, 75 (1987).

[93] D. J. Lyman, *Trans. Am. Soc. Artif. Int. Organs*, **10**, 17 (1964).

[94] D. J. Lyman and B. H. Loo, *J. Biomed. Mater. Res.*, **1**, 17 (1967).

[95] E. W. Merrill, E. W. Salzman, N. Mahmud, L. Kushner, and J. N. Lindon, *Trans. Amer. Soc. Artif. Int. Organs*, **28**, 482 (1982).

[96] Y. Mori and S. Nagaoka, *Trans. Amer. Soc. Artif. Int. Organs*, **28**, 459 (1982).

[97] S. Nagaoka, Y. Mori, H. Tanzawa, Y. Kikuchi, F. Inagaki, Y. Yokota, and Y. Noishiki, *ASAIO J.*, **10**, 76 (1987).

[98] J. H. Lee, J. Kopecek, and J. D. Andrade, *J. Biomed. Mater. Res.*, **23**, 351 (1989).

[99] R. Kijellander and E. Florin, *J. Chem. Soc. Faraday Trans.*, *1.*, **77**, 2053 (1981).

[100] S. I. Jeon, J. H. Lee, J. D. Andrade, and P. G. De Gennes, *J. Colloid Interface Sci.*, **142**, 149 (1991).

[101] S. I. Jeon and J. D. Andrade, *J. Colloid Interface Sci.*, **142**, 159 (1991).

[102] R. M. Pashley, P. M. McGuiggan, and B. W. Ninham, *Science*, **229**, 1088 (1985).

[103] S. Gogolewski, *Colloid Polym. Sci.*, **267**, 757 (1989).

[104] J. W. Boretos and W. S. Pierce, *Science*, **158**, 1481 (1967).

[105] J. B. Boretos, *Pure & Appl. Chem.*, **52**, 1851 (1980).

[106] *Polyurethanes in Biomedical Engineering*, H. Planck, G. Egbers, and R. Syre (Eds), Elsevier, Amsterdam (1984).

[107] S. C. Yoon and B. D. Ratner, in *Polymer Surface Dynamics*, J. D. Andrade (Ed.), Plenum Press, New York (1988).

[108] B. D. Ratner and R. W. Paynter, in *Polyurethanes in Biomedical Engineering*, H. Planck, G. Egbers, and R. Syre (Eds), Elsevier, Amsterdam (1984).

[109] S. L. Goodman, C. Li, J. B. Pawley, S. L. Cooper, and R. M. Albrecht, in *Surface Characterization of Biomaterials*, B. D. Ratner (Ed.), Elsevier, Amsterdam (1988), p. 281.

[110] S. L. Goodman, S. R. Simmons, S. L. Cooper, and R. M. Albrecht, *J. Colloid Interface Sci.*, **139**, 561 (1990).

[111] B. D. Ratner, in *Physicochemical Aspects of Polymer Surfaces*, K. L. Mittal (Ed.), Plenum Press, New York (1981), pp. 969–983.

[112] K. G. Tingey, J. D. Andrade, R. J. Zdrahala, K. K. Chittur, and R. M. Gendrau, in *Surface Characterization of Biomaterials*, B. D. Ratner (Ed.), Elsevier, Amsterdam (1988), p. 255.

[113] A. Takahara, N. J. Jo, K. Takamori, and T. Kajiyama, in *Progress in Biomedical Polymers*, C. G. Gebelein and R. L. Dunn (Eds), Plenum Press, New York (1990), p. 217.

[114] R. W. Paynter, in *Surface Characterization of Biomaterials*, B. D. Ratner (Ed.), Elsevier, Amsterdam (1988), p. 49.

[115] S. R. Hanson, L. A. Harker, B. D. Ratner, and A. S. Hoffman, *J. Lab. Clin. Med.*, **95**, 289 (1980).

[116] V. Sa Da Costa, D. Brier-Russell, E. W. Salzman, and E. W. Merrill, *J. Colloid Interface Sci.*, **80**, 445 (1981).

[117] A. Takahara, J. Tashita, T. Kajiyama, M. Takayanagi, and W. J. McKnight, *Polymer*, **26**, 987 (1985).

[118] S. J. Whicher and J. L. Brash, in *Physicochemical Aspects of Polymer Surfaces*, K. L. Mittal (Ed.), Plenum Press, New York (1981), pp. 985.

[119] N. Yui, T. Oomiyama, K. Sanui, and N. Ogata, *Makromol. Chem., Rapid. Commun.*, **5**, 805 (1984).

[120] N. Yui, J. Tanaka, K. Sanui, and N. Ogata, *Makromol. Chem.*, **185**, 2259 (1984).

[121] N. Yui, K. Kataoka, and Y. Sakurai, *Makromol. Chem.*, **187**, 943 (1986).

[122] N. Yui, K. Kataoka, and Y. Sakurai, in *Surface Characterization of Biomaterials*, B. D. Ratner (Ed.), Elsevier, Amsterdam (1988), p. 271.

[123] T. Okano, M. Katayama, and I. Shinohara, *J. Appl. Polym. Sci.*, **22**, 369 (1978).

[124] T. Okano, I. Kataoka, Y. Sakurai, M. Shimada, T. Akaika, and I. Shinohara, *Artif. Organs*, **5**, 468 (1981).

[125] T. Okano, *ASAIO J.*, **10**, 80 (1987).

[126] T. Okano, S. Nishiyama, I. Shinohara, T. Akaike, and Y. Sakurai, *Polymer Journal*, **10**, 223 (1978).

[127] F. Orang, F. Zenhausern, R. Emch, and P. Descouts, 7th International Conference of Surface and Colloid Science, Compiegne, France, Symposium C4, Poster no. 8.

[128] M. F. Refojo, in *Kirk-Othmer, Encyclopedia of Chemical Technology*, Wiley, New York (1984), Vol. 6, p. 720.

[129] W. B. Meyers, T. B. Harvey, III, and L. M. Bowman, in *Progress in Biomedical Polymers*, C. G. Gebelein and R. L. Dunn (Eds), Plenum Press, New York (1990), pp. 13–17.

[130] O. Wichterle and D. Lim, *Nature*, **185**, 117 (1960).

[131] L. M. Bowman, W. B. Meyers, and T. B. Harvey, III, in *Progress in Biomedical Polymers*, C. G. Gebelein and R. L. Dunn (Eds), Plenum Press, New York (1990), p. 7.

[132] F. J. Holly and M. A. Lemp, *Surv. Ophthalmol.*, **22**, 69 (1977).
[133] F. J. Holly, *Colloids and Surfaces*, **10**, 343 (1980).
[134] G. E. Brauninger, D. O. Shah, and H. E. Kaufman, *Am. J. Ophthalmol.*, **73**, 132 (1972).
[135] M. Norn, *Acta Ophthalmol.*, **57**, 766 (1979).
[136] B. V. Derjaguin, *Progress Colloid Polymer Sci.*, **74**, 17 (1987).
[137] J. N. Israelachvili, *Intermolecular and Surface Forces*, Academic Press, London (1985), p. 172.
[138] C. H. Dohlman, V. Kalevar, D. Yagoda, and E. Balazas, *Exp. Eye Res.*, **22**, 354 (1976).
[139] J. C. Moore and J. M. Tiffany, *Exp. Eye Res.*, **29**, 291 (1979).
[140] C. W. Chao, J. P. Vergnes, and S. I. Brown, *Exp. Eye Res.*, **36**, 139 (1983).
[141] B. J. Tighe, *Brit. Polym. J.*, **71**, Sept. 1976.
[142] H. D. Gesser and R. E. Warriner, Ger. Offen. 1,255,950.
[143] H. D. Gesser and R. E. Warriner, U. S. Pat. 3,925,178.
[144] du Pont de Nemour E. I. and Co., Neth. Appl. 75 04,289.
[145] P. Hoefer, and W. Kohl (to Titmus Eurocon Kontaktlinsen, K.G.), U. S. Pat. 4,214,014.
[146] Jpn. Kokai 57,182,326 (to Tokyo Contact Lens Research Inst.)
[147] T. Yuta and T. Takeuchi (to Tokai Rubber Industries, Ltd.), Jpn. Kokai 02 84,444.
[148] I. Tetsuo, M. Kimura, K. Yanagihara, and T. Suminoe, (to Japan Synthetic Rubber Co., Ltd.), Jpn. Kokai 61,166,841.
[149] H. Yasuda, *Plasma Polymerization*, Academic Press, Orlando (1985), p. 346.
[150] G. H. Butterfield and G. H. Butterfield, Jr. (to Butterfield Laboratories Inc.), Fr. Demande 2,181,065.
[151] G. H. Butterfield Jr. and G. H. Butterfield (to G. H. Butterfield and Son), U. S. Pat., 3,948,871.
[152] T. B. Harvey, III, W. B. Meyers, and L. M. Bowman, in *Progress in Biomedical Polymers*, C. G. Gebelein and R. L. Dunn (Eds), Plenum Press, New York (1990), p. 1.
[153] Jpn. Kokai 59,135,423 (to Barnes-Hind Hyrocurb, Inc.)
[154] N. G. Gaylord (to Polycon Laboratories), U. S. Pat. 3,808,178.
[155] K. Tanaka, M. Kanada, S. Kanome, and T. Nakajima (to Toyo Contact Lens Co., Ltd.), U. S. Pat. 4,139,513.
[156] K. Tanaka, K. Takahashi, M. Kanada, Y. K. Kasugai, and M. Ichihara (to Toyo Contact Lens Co., Ltd.), U. S. Pat. 4,139,692.
[157] A. Sugiyama, T. Nakajima, and Y. Taniyama (to Menicon Company Ltd.), U. S. Pat. 4,980,208.
[158] D. W. Fakes, J. M. Newton, J. F. Watts, and M. J. Edgel, *Surf. Interface Anal.*, **10**, 416 (1987).
[159] D. W. Fakes, M. C. Davies, A. Brown, and J. M. Newton, *Surf. Interface Anal.*, **13**, 233 (1988).
[160] D. W. Fakes, J. F. Watts, and J. M. Newton, in *Surface Characterization of Biomaterials*, B. D. Ratner (Ed.), Elsevier, Amsterdam (1988), p. 193.
[161] M. Morra, E. Occhiello, R. Marola, F. Garbassi, and D. Johnson, *J. Colloid Interface Sci.*, **137**, 11 (1990).
[162] M. Morra, E. Occhiello, F. Garbassi, M. Maestri, R. Bianchi, and A. Zonta, *Clin. Mat.*, **5**, 147 (1990).
[163] J. R. Hollahan and G. L. Carlson, *J. Appl. Polym. Sci.*, **14**, 2499 (1970).
[164] S. Wu, *Polymer Interface and Adhesion*, Marcel Dekker, New York (1982), p. 307.
[165] M. Morra, E. Occhiello, F. Garbassi, M. Maestri, R. Bianchi, and A. Zonta, unpublished results.

[166] S. Kanbe, and Y. Ikada (to Seiko Epson Corp.), Jpn. Kokai 02,220,024.

[167] H. K. Singer, E. C. Bellantoni, and A. R. LeBoeuf, in *Encyclopaedia of Polymer Science and Technology*, Wiley, New York (1986), Vol. 4, p. 164.

[168] B. Arkles, *Chemtech.*, **13**, 513 (1983).

[169] E. Masuhara, N. Tarumi, and T. Makoto (to Hoya Lens K. K.), Fr. Demande 2,407,232.

[170] F. J. Holly and M. J. Owen, in *Physicochemical Aspects of Polymer Surfaces*, K. L. Mittal (Ed.), Plenum Press, New York (1981), Vol. 2, p. 625.

[171] E. W. Merril, Fr. Demande FR 2,208,775.

[172] G. Torres (to Essilor International), Fr. Demande FR 2,646,672.

[173] R. A. Janssen, E. M. Freeman, and E. C. McCraw (to Ciba-Geigy, A.-G.), Eur. Pat. Appl. EP 378,511.

[174] H. Schaefer, G. Kossmehl, W. Neumann, and A. Fluthwedel (to Titmus Eurocon Kontaktlinsen, K.-G.), Ger. Offen DE 3,517,615.

[175] R. J. Robertson, K. C. Su, M. S. Goldenberg, and K. F. Mueller (to Ciba-Geigy, A. -G), Eur. Pat. Appl. EP 395,583.

[176] C. P. Ho and H. Yasuda, *Polym. Mater. Sci. Eng.*, **56**, 705 (1987).

[177] C. P. Ho and H. Yasuda, *J. Biomed. Mater. Res.*, **22**, 919 (1988).

[178] O. Wichterle, in *Encyclopedia of Polymer Science and Technology*, Interscience, (1971), Vol. 15, p. 274.

[179] O. Wichterle and D. Lim, U.S. Pat. 2,976,576.

[180] O. Wichterle and D. Lim, U.S. Pat. 3,220,960.

[181] D. G. Pedley, P. J. Skelly, and B. J. Tighe, *Brit. Polym. J.*, **12**, 99 (1980).

[182] F. J. Holly and M. F. Refojo, *J. Biomed. Mater. Res.*, **9**, 315 (1975).

[183] T. Bilbaut, A. M. Gachon, and B. Dastugue, *Exp. Eye Res.*, **43**, 153 (1986).

[184] F. J. Holly, *J. Polym. Sci., Polym. Symp.*, **66**, 409 (1979).

[185] M. F. Refojo and F. L. Leong, *J. Polym. Sci., Polym. Symp.*, **66**, 227 (1979).

[186] A. M. Gachon, T. Bilbaut, and B. Dastugue, *Exp. Eye Res.*, **40**, 105 (1985).

[187] T. Bilbaut, A. M. Gachon, and B. Dastugue, in *Biological and Biomechanical Performances and Biomaterials*, P. Christel and A. J. C. Lee (Eds), Elsevier, Amsterdam (1986), p. 171.

[188] E. J. Castillo, J. L. Koenig, J. M. Anderson, and J. Lo, *Biomaterials*, **5**, 319 (1984).

[189] E. J. Castillo, J. L. Koenig, J. M. Anderson, and J. Lo, *Biomaterials*, **6**, 339 (1985).

[190] E. J. Castillo, J. L. Koenig, J. M. Anderson, and N. *Jentoft Biomaterials*, **7**, 9 (1986).

[191] E. J. Castillo, J. L. Koenig, and J. M. Anderson, *Biomaterials*, **7**, 89 (1986).

[192] R. A. Sack, B. Jones, A. Antignani, R. Libow, and H. Harvey, *Invest. Ophthalmol. Vis. Sci.*, **28**, 842 (1987).

[193] J. L. Bohnert, T. A. Horbett, B. D. Ratner, and F. H. Royce, *Invest. Ophthalmol. Vis. Sci.*, **29**, 365 (1988).

[194] M. S. Goldenberg and A. C. Beekman, *Biomaterials*, **12**, 267 (1991).

[195] L. Pinchuk, E. C. Eckstein, and M. R. V. Mack, *J. Appl. Polym. Sci.*, **29**, 1749 (1984).

[196] M. Froix, U. S. Pat. 4,871,785.

Chapter 13

Friction and Wear

Although never in the mainstream of science, friction and wear phenomena are extremely important from a practical standpoint and attracted early attention (one of the first quantitative works was done by Leonardo da Vinci). Most of the work has been devoted to metal surfaces, although some literature has been produced for polymers as well. Most studies are quite old, going back to the 1950s and 1960s, owing to the importance of wear and friction for industries which were fast developing at that time, namely synthetic fibres, films and elastomeric goods [1–13].

Friction and wear are very complicated phenomena, which depend on both bulk and surface properties. They are bulk-dependent in that frictional forces depend on contact areas, which in turn are affected by the materials' mechanical properties and by its viscoelastic behaviour. They are surface dependent in that the changes in friction appear related to changes in surface energy [11, 12].

Friction is important in many applications, first of all polymer fibres, films and sheets need a reasonably low friction to be used in converting, for instance in packaging machines. Secondly in selected automotive applications, e.g. windshield blades, linings, low friction is required to lower the power of motors driving them. A pictorial representation of where friction and wear phenomena of polymeric materials are important in a car is presented in Figure 13.1.

Low friction is also required in magnetic storage media, e.g. floppies, to prevent sticking of the reading/writing head. Finally, in many medical applications, namely condoms and catheters, a low friction in a humid environment is mandatory. In other applications friction has to be enhanced, typical is the case of car tyres or of self-sustaining feminine socks.

Lowering friction may involve reducing the contact area (e.g. most linings for car windows) or increasing it (high performance tyres, socks). Chemically, lowering or increasing friction involves applying layers with a variable friction coefficient. This task can be pursued by bulk techniques such as coextrusion or coinjection, by the use of additives such as surface diffusing slip agents or by coating.

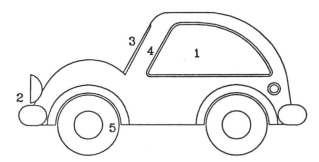

Figure 13.1 Pictorial representation of the importance of friction and wear of polymeric materials for applications in the automotive industry. Wear is important for transparent plastics in applications such as windows (1) and front-light (2). Low friction is required for wiper blades (3) and linings (4). Finally, good friction and low wear are mandatory for tyres (5).

Wear in turn is related to polymeric material removal upon contact with other bodies, it has typically to be prevented. Important applications involve first of all transparent objects, such as windshields, lenses, car lights. Further wear intensive uses are in gears, appliances, furniture.

Controlling wear involves either applying inherently wear resistant coatings, mimicking glasses, ceramics or metal, or working on wear coefficients by bulk methods, for instance, in the case of composite materials, fiddling with the chemical nature and morphological distribution of reinforcing phases.

With the exception of the quoted general references, most published information on wear and friction is to be found in patents and technical literature. Given the fact that good theoretical treatments can be found elsewhere ([1–13] and references therein), we chose to put more emphasis on the industrial aspect, i.e. on the practical application of the physical principles of the phenomenon.

13.1 COMPOUNDING

Friction is conveniently lowered by lubrication, which, in the case of metals, has to be external [3, 12]. On the other hand, for polymers it has long been known that it is possible to introduce lubricants within the polymer itself and to have it migrate slowly at the materials' surface. The relevant physics is the one underlying diffusion of low molecular weight species at the surface [12, 14]. Its most interesting characteristic is the fact that the amount worn away during friction is actually reintroduced at the surface from the bulk.

Going through the patent and trade literature, one finds a few general types of additives important for friction-related problems, namely internal lubricants, slip and antiblocking agents. Lists of products can be found in a few general references [15–16].

In the jargon of the plastic industry internal lubricant means a range of materials helping the movement of the polymer against itself or other materials or both. Further to the effect on the performance of manufactured goods, typical effects of lubricant addition to polymers include better flow and lower energy use during processing. Related effects can include the reduction of the temperature needed for processing and therefore the reduction of the tendency to thermal degradation.

These improvements are particularly critical for polymers such as PVC, whose processing cannot be improved by raising the temperature, since this could result in its degradation by dehydrochlorination [17–20]. The choice of the appropriate type and amount of internal lubricants is more important for single-screw systems than for twin-screw ones (Figure 13.2), because of the higher efficiency in transmitting shear of the latter, and for rigid than for flexible PVC, which is already compounded with plasticizers which have a lubricant effect themselves.

Slip agents are generally defined as those modifiers which are only meant to provide surface lubrication during and immediately after processing [15, 16]. They have a low compatibility with the polymer and therefore tend to exude at the materials' surface. Their action is essential for speeding up packaging operations using plastic films, such as in-line bagging and overwrapping. Poten-

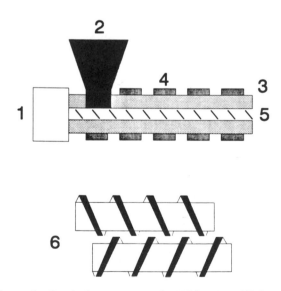

Figure 13.2 Schematic of a single-screw extruder, with motor (1), hopper (2), barrel (3), heating and cooling bands (4) and screw (5). For PVC often non-intermeshing counter-rotating twin screw extruders are used (6).

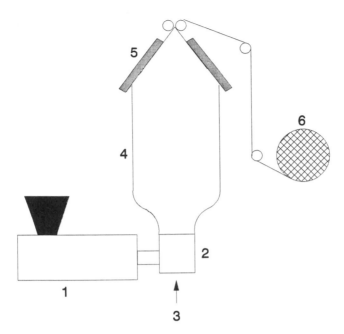

Figure 13.3 Schematic of a bubble-blowing process. The basic material, already formulated with additives, is fed into the extruder (1) and reaches the die (2), where air is introduced (3) forming the bubble (4). The film is then collected through a collapsing frame (5) and wound (6).

tial side-effects include improving antistatic properties as well. A schematic of a bubble-blowing process, used mainly for the production of polyolefin films, is presented in Figure 13.3.

Antiblocking agents have a somewhat similar definition, but their application is mainly aimed at avoiding the sticking of a plastic film on to itself. Beyond the previously quoted materials with a tendency to exude at the surface, antiblocking agents include inorganic particles such as silica, silicates and other fillers. These act by a different physics as compared to slip agents. Instead of lubricating, they introduce protrusions at the film surface, thereby reducing adhesion induced by the flatness (high contact area) of the film. The use of these compounds is particularly important for biaxially oriented PET; a schematic of a production line is presented in Figure 13.4.

The patent literature concerning lubricants, slip and antiblocking agents is rather extensive and is dominated by polymer producers and film makers. While it reflects the existence of general rules such as those related to the contact area (making microembossed films or adding inorganic particles), lubrication (similar to metal treatment) and exuding of additives (surface tension driven), empiricism is still very much a general rule. This is even more the case when combined

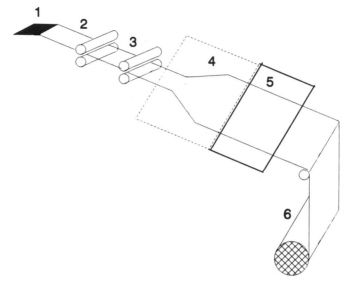

Figure 13.4 Schematic of the production of biaxially oriented PET film. The polymer, formulated with antiblocking additives in the polymerization phase, is fed to the extruder (1) and a sheet is cast (2). The latter is first stretched longitudinally using variable speed rolls (3), then undergoes transverse stretching at temperatures above. T_g (4), finally it is heat-set (5) and collected (6).

additive packages are designed to provide simultaneously more than one performance, e.g. antiblocking and antistatic.

A first important group of patents is devoted to the introduction of inorganic particles, which has been reported since the beginning of the film industry. The work on the size distribution and morphology of these particles has been particularly important for polymer-supported high density storage media, which require antiblocking performance, but also good evenness of the film surface. Reviews concerning these topics have been published in the past [21, 22].

More recently, for biaxially oriented PET films, patent emphasis shifted from conventional fillers (e.g. silica) [23], to different kinds of inorganic particles, such as silicates, ziroconyl compounds, barium sulphate and zeolites [24–28]. Furthermore, more emphasis than in the past is being devoted to films meant for the packaging industry, in particular biaxially oriented PP [29–31] and PE films [32–43]. Finally, some innovations have been reported for polymers meant for technical applications, namely polycarbonate [44–51], with interesting suggestions such as the use of glass microspheres.

The other prominent group of activities is related to various kinds of chemicals meant for getting at the polymer/environment interface and acting as a lubricant or a slip agent. Chemically, they can be classified in a few classes, namely fatty acids, esters, alcohols and amides; metallic stearates; paraffinic waxes; silicones and siloxanes [15–16].

Fatty acids, esters and alcohols have been primarily disclosed for PVC, although some applications have been reported for other polymers such as PS, PC and ABS [47, 52–57]. On the other hand, fatty amides (primary, secondary, alkanolamides) are important particularly for polyolefins, mainly PE and PP, for improving slip, antiblock and sometimes antistatic properties [32, 49, 58–64]. Fatty acid salts, in particular stearates, which can act in different ways, e.g. antiblocking, altering melt viscosity, stabilizing, etc., are widely used for polymers such as PVC, PS, thermosets [15, 16, 58, 65–72].

Siloxanes and silicones have attracted most patent coverage in recent years, mainly as additives for polyolefin films (PE and PP) meant for high speed packaging [29, 38, 41, 73–83]. Other patents concerned applications in diverse fields such as magnetic media, photo-imaging and heat transfer films [84–86].

To further evidence the fact that the chemical nature of the material and its mechanical properties are important for friction related properties, it is interesting to observe that in the field of packaging films improvements in antiblocking properties of polyolefins have been reported even by blending them with polyamides [40, 87, 88], polymethacrylates [89] and other polyolefins [90, 91]. Some patents disclosed polyolefins synthetized explicitly with the aim of improving friction-related properties [92, 93].

While in the case of thermoplastics friction is mainly adhesion driven, for rubbers both adhesion and deformation contributions are present [4, 5, 94]. The number of applications where rubber surfaces need reduced friction is rather lower than for thermoplastics, yet the methods to reduce friction by compounding are partly similar. In particular the additives most frequently claimed again include waxes or silicone fluids [95–106]. The aim of course is to have a surface migration of the relevant additive with increased surface slipperiness and reduced wear. In some cases, applications of silicone-grafted elastomers have also been reported [107, 108], the latter being suggested for biomedical applications.

Compounding is again of paramount importance for objects made in reinforced plastics and used for wear-intensive applications. In this case the solution lies in the use of fillers and the improvement is particularly good for situations when there is interfacial wear, e.g. when there is a constant sliding of plastics against smooth metals. Little benefit is obtained when wear is abrasive, i.e. derived by contact of the polymer surface with hard external irregularities [7, 11].

For applications involving interfacial wear, getting a low wear rate is not really a surface problem, since the surface itself is continuously reformed. Rather it is a matter of choosing the correct reinforcing phase, working on both size and orientation of the latter [109]. Of course, short fibres have not the same effect as long fibres and aramids do not behave the same as carbon fibres [110]. Finally the orientation of fibres is important and woven reinforcements prove particularly effective, as compared to unidirectional systems [7, 111].

The patent literature does exist, but it is much less prominent than for friction-related problems [112–123].

13.2   CHEMICAL COATING

As mentioned in the preceding section, chemical coating has been of widespread use for solving friction problems in metals [3, 12]. In the case of polymers, one of the first areas where coating has become of great industrial interest for improving friction and wear-related characteristics is synthetic fibres. All fibres (both textiles and those meant for other uses, such as reinforcement) pass over guide surfaces in spinning and winding operations (see Figure 13.5). Furthermore, they rub against each other in typical textile processing steps, such as weaving and knitting. Thus, since the early beginnings of the fibre industry, much attention has been devoted to the development of finishes aimed at minimizing friction and wear [13, 124–128].

In the quoted general references, in particular in ref. [13], the physics underlying the control of friction and wear in fibre media are outlined. Further to these sources, there is a large amount of patent literature, which concentrates in the formulation of spin finishes, aimed generally at obtaining a good compromise of various surface related properties, such as friction, wear, antistaticity and antimicrobial behaviour.

Recently, the activity on finishes is still partially in the hands of polyester producers, but the role of specialized sizing manufacturers is increasing. For textile fibres, the most frequent chemistry involves polyoxyalkylenes and the

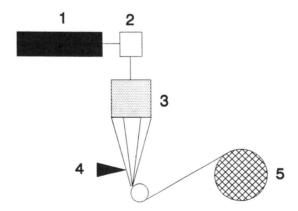

Figure 13.5   Schematic of a melt spinning process for PET fibres. The dry polymer (1) is fed into the extruder and, through a metering pump (2), directed into the spin pack (3), the quenched fibres (4) are coated with a sizing (5) and wound up for further processing.

most cited aim is at improving the heat and oxidation resistance of the sizing itself, which is particularly important, given the ever increasing spinning speeds [129–144]. Occasionally, other chemistries have been practised, such as the use of fluoro-derivatives [145, 146], sulphones [147], siloxanes [148] and epoxy resin reaction products [149, 150]. Partial cross-linking of the sizing has also been suggested to further improve antistaticity and soil release properties [151].

For polymer films, as an alternative to compounding, coating with slippery layers have been described (Figure 13.6). In recent years, claims related to packaging films included the application of sealing layers with slip additives, such as carnauba wax [152, 153], fluoropolymers [154] and alkali metal alkyl sulphates [155]. Other coatings included inherently low-friction or low-abrasion formulations, based on acrylics [156, 157] and siloxanes [158], with most applications for magnetic media.

The development of information technology led to an ever growing importance of storage media, such as tapes and disks. This generated a significant amount of literature relative to abrasion phenomena. In the academia this resulted in the attention to surface force phenomena, as reviewed in some recent papers [159–165]. On the industrial side the activity concentrated on the formulation of lubricants from different chemical classes, namely fatty acids [166, 167], hydrocarbons [168, 169], perfluoroalkylethers as such [170–172] or with reactive end groups [173] and silicones or siloxanes [174–179]. The latter family also included a number of chemical modifications, such as fluorination [180], alkylation [181, 182], hydroxyalkylation [183, 184] and mixing with nitrogen-containing compounds [185–187].

Elastomeric materials are themselves very good friction materials, therefore they can be used as coatings to increase friction, as claimed for instance in the case of turntables for disks [188]. Yet in many applications their tendency to friction and therefore abrasion is disadvantageous, thus an amount of patent literature has been devoted to friction-reducing and abrasion-resistant coatings.

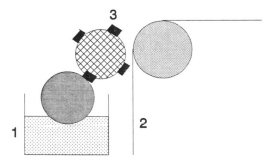

Figure 13.6 Schematic of a brush-coating machine. The coating (1) is transferred to the film (2) by brushes (3) and levelled by further mechanical action (4).

Typical applications where friction may be disadvantageous include transportation (window channels, wiper blades, etc.) and the biomedical industry (catheters, condoms, etc.). Solutions which have been suggested include the lamination of thermoplastic layers, e.g. polyethylene [189–191] and the formulation of polyurethane-based coatings and paints, for both automotive and biomedical applications [192–197], with alternatives including epoxy-based formulations [198]. More patent literature is related to coating with lubricants [199], vacuum deposited polymer layers [200] and fluorination [201, 202]. A further technique is grafting of acrylic layers, either from vapour solution, usually with the help of UV radiation and photosensitizers [203–206] or electron beams [207].

It should also be remembered that for high volume applications such as window channels for cars, the most widespread solution involves flocking, i.e. the application of a fur-like layer on top of the elastomer, to minimize contact area and friction (Figure 13.7).

Another area where an important patent literature has been generated is the protection of transparent wear-prone plastics, mainly polycarbonate. This activity has been driven by a few important applications, e.g. lenses, car windshields, front lights and transparent panels for glazing (Figure 13.8).

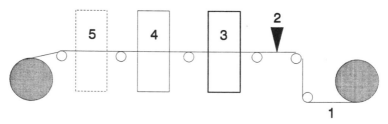

Figure 13.7 Schematic of a flocking line. The substrate (1) is coated with adhesive (2), passed through the flocking station (3) and then the adhesive is cured (4). Finally the coated material is brushed (5) and rewound.

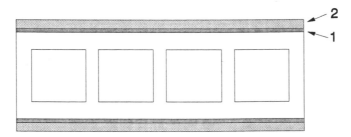

Figure 13.8 Section of a polycarbonate hollow sheet coated with a primer layer (1) and an abrasion resistant layer (2).

Most of the activity concentrated on the formulation of abrasion resistant coatings, on the processing of the coating [e.g. catalysts, UV curing), on side-effects such as additional UV protection for polycarbonate and on the adhesion of the top coat to the substrate [208–227].

Most of the typical chemistries of thermoset have been explored for the formulation of abrasion resistant coatings. Siloxane-based coatings proved particularly popular [221–253], but also acrylics [217, 218, 254–265], epoxies [215, 266, 267], polyurethanes [268], polyesters [269] and polyvinylalcohol [270] attracted research activity, which in turn produced patent literature. A number of the patents which have been quoted include the use of fine inorganic powders, most frequently silica, to improve wear resistance of the coating without damaging its transparency.

Further to formulation, many patents deal with the process of coating deposition and consolidation. The coatings which have been reported are typically quite hard and have dilation coefficients quite different from the substrate. Therefore adhesion cannot be taken for granted and a number of patents dealt with primer layers to avoid delamination of the coating [224, 232, 234, 238, 248, 271–273].

Another important step is the curing of the coating. A typical method for thermosets is heat curing, but for plastics such as polycarbonate this can induce dimensional variations leading either to adhesion problems or, more severely, to distortion of the whole object. Therefore, mainly in the case of acrylic coatings, alternatives such as the use of UV [217, 244, 245, 255, 257, 258, 266, 274, 275], radiochemical [232, 254] or glow discharge curing [276] have been suggested. Some of these patents are also involved in the introduction of the formulation of the appropriate sensitizers or catalysts [215, 229, 238, 241, 243, 258–259, 266, 275].

A further component of abrasion resistant coatings are compounds added to improve radiation stability [222, 225, 226, 228, 229, 231, 235, 239, 242, 250, 272, 275, 277, 278], which is not important for polycarbonate, and to alter characteristics such as reflectivity [279], which is important for lenses.

The alternative to thermoset coatings for improving the scratch resistance of polycarbonate is the production during processing of multilayer structures by coinjection [280] or coextrusion [281, 282]. Usually the internal layer is polycarbonate, which provides mechanical performance, while the external layer is polymethylmethacrylate, which is much less impact resistant but has a much better abrasion and damage resistance. This is best suited for injection-moulded parts, such as car front-lights, or extruded parts, such as panels for glazing.

13.3 HIGH ENERGY DENSITY TECHNOLOGIES

Given the scarce thermal resistance of polymer substrates, the application of high energy density technologies has been limited to those which do not induce sharp

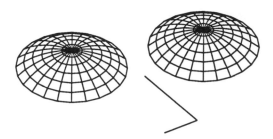

Figure 13.9   Lenses.

increases in substrate temperature [282–285]. Most published literature, both patent and academic, involves cold plasma treatments.

As reviewed by Yasuda [285–287], plasma polymers are very much cross-linked, and are therefore similar in nature to the thermoset coatings deposited by solution techniques. In the case of friction, the chemistries which have been suggested in most instances involve the use of fluorinated monomers [288–292], although acrylics have also been cited [293, 294]. The application of sputter etching for reducing friction in photographic supports has also been claimed [295].

More emphasis has been given to the deposition of abrasion-resistant layers. This utilized volatile silanes and siloxanes and started as an application in the 1970s [296–297]. The elimination of the organic part, which is critical for obtaining a very dense and hard coating, proved easier for silanes [298–301] than for siloxanes [302–305], although the latter have been preferred recently for their low toxicity and good chemical inertness.

Finally, some benefit for wear resistance is also given in the case of lenses (Figure 13.9) by the application of anti-reflection silicon oxide coatings by electron beam induced evaporation [306], a technique very much similar to that used for imparting barrier (see Chapter 11).

REFERENCES

[1]   B. J. Briscoe, in *Physicochemical Aspects of Polymer Surface*, K. L. Mittal (Ed.), Plenum, New York (1981).

[2]   L. H. Lee (Ed.), *Advances in Polymer Friction and Wear*, Plenum, New York (1974).

[3]   F. P. Bowden and D. Tabor, *Friction and Lubrication of Solids*, Oxford University Press, Oxford (1950).

[4]   D. Moore, *The Friction and Lubrication of Elastomers*, Pergamon, New York (1972).

[5]   G. M. Bartenev and V. V. Lavrentev, *Friction and Wear of Polymers*, Elsevier, Amsterdam (1981).

[6]   V. A. Belyi, A. I. Sviridyonok, M. I. Petrokovesto, and V. G. Savkin, *Friction and Wear in Polymer-based Materials*, Pergamon, New York (1981).

[7] K. Friedrich (Ed.), *Composite Materials Series, Vol. 1: Friction and Wear of Polymer Composites*, Elsevier, Amsterdam (1986).

[8] U. S. Tewari, S. K. Sharma, and P. Vasudevan, *J. Macromol. Sci.-Rev.*, **C29**, 1 (1989).

[9] H. E. Garey, *Polym. Plast. Technol. Eng.*, **28**, 73 (1989).

[10] K. L. Mittal, *J. Adhes. Sci. Technol.*, **1**, 247 (1987).

[11] J. K. Lancaster, in *Encyclopaedia of Polymer Science and Technology*, Wiley, New York (1989), Vol. 1, p. 1.

[12] O. J. Sweeting (Ed.), *The Science and Technology of Polymer Films*, Wiley, New York (1971).

[13] M. J. Schlick (Ed.), *Surface Characteristics of Fibers and Textiles*, Dekkar, New York (1975), Vol. I, p. 1.

[14] S. Wu, *Polymer Interface and Adhesion*, Dekker, New York (1982).

[15] *Encyclopedia of Modern Plastics*, Wiley, New York, updated every year.

[16] *Plastics Technology, Manufacturing Handbook and Buyers' Guide*, Plastics Technology, New York, updated every two years.

[17] I. I. Rubin (Ed.), *Handbook of Plastic Materials and Technology*, Wiley, New York (1990), pp. 525, 549.

[18] L. I. Nass (Ed.), *Encyclopedia of PVC*, Dekker, New York (1976).

[19] H. A. Sarvetnick, *Polyvinylchloride*, Reinhold, New York (1969).

[20] W. V. Titow, in *Developments in PVC Production and Processing*, Applied Science, Barking (1977), Vol. 1, p. 63.

[21] E. Werner, S. Janocha, M. J. Hopper, and K. J. Mackenzie, in *Encyclopedia of Polymer Science and Technology*, Wiley, New York (1989), Vol. 12, p. 193.

[22] C. J. Heffelfinger and K. L. Knox, ref. 12, p. 587.

[23] T. Ono, A. Kinji, T. Asai, and Y. Murakami, EP 415383 (1991).

[24] H. Kawai, J. Takase, and K. Yonezawa, US 4981897 (1991).

[25] K. Kuze, H. Hashimoto, T. Ohta, T. Akaishi, K. Takeuchi, and H. Kamatani, EP 83746 (1983).

[26] D. Schuhmann, H. Pfeiffer, U. Murschall, and G. Schloegl, DE 4923272 (1992).

[27] B. J. Sublett, EP 412029 (1991).

[28] T. Kagiyama, Y. Oguri, K. Kunugihara, E. Hattori, K. Endo, N. Shudo, C. Kawaguchi, T. Watanabe, and Y. Meguro, EP 257611 (1988).

[29] U. Murshall, A. Speith, and H. Pfeiffer, DE 4031784 (1992).

[30] H. Pfeiffer, U. Murshall, and G. Schloegl, DE 4037417 (1992).

[31] L. J. Gust, E. F. O'Sullivan, and R. C. Wood Jr., EP 124310 (1984).

[32] G. Lueers and R. Sobottka, DE 3337356 (1985).

[33] S. B. Ohlsson, P. De Cambourg, and W. J. J. Leyson, EP 477035 (1992).

[34] E. M. Mount III and K. P. Morgan, EP 461820 (1991).

[35] Y. Itaba, M. Izawa, K. Saito, and T. Kondo, EP 369705 (1990).

[36] H. G. Schirmer, EP 358461 (1990).

[37] S. Bekele, US 4909726 (1990).

[38] J. M. Sullivan, GB 2220372 (1990).

[39] A. Scheller, N. E. Luker, and G. A. Le Grange, EP 342822 (1989).

[40] D. V. Dobreski, J. J. Donaldson, and B. E. Nattinger, US 4786678 (1988).

[41] R. Belloni, J. K. Keung, D. A. Liestman, and A. M. Nahmias, EP 273680 (1988).

[42] A. R. Wolfe, US 4611024 (1986).

[43] I. Nagayasu, K. Wakita, GB 2219801 (1989).

[44] O. M. Boutni, US 4663391 (1987).

[45] O. M. Boutni, US 4749758 (1988).

[46] P. Y. Liu, US 4622359 (1986).

[47] M. E. Shepard, D. E. Galloway, and K. D. Lind, US 4877684 (1989).
[48] S. Miyata and S. Oishi, EP 301509 (1989).
[49] K. Yamada, K. Miyazaki, Y. Oowatari, Y. Egami, and T. Honma, EP 257803 (1988).
[50] H. Hagens and H. Dallmann, DE 4034869 (1992).
[51] S. C. Chu and K. A. Kirk, EP 381454 (1990).
[52] K. Vorschech, P. Wedl, E. U. Brand, and E. Fleischer, DE 3820065 (1989).
[53] E. L. White, ref. 48, ch. 13.
[54] R. A. Lindner, US 5134185 (1992).
[55] F. L. Heinrichs, G. Hohner, and J. P. Piesold, DE 4020483 (1992).
[56] G. M. Conroy and K. R. Wursthorn, US 4696754 (1987).
[57] K. Worschech, F. Loeffeloholz, P. Wedl, and B. Wegemund, DE 3420226 (1985).
[58] K. Klaar, M. Simm, and R. Schoepe, DE 4041453 (1992).
[59] G. Mueller, J. Pohrt, and G. Hatzmann, DE 4009065 (1991).
[60] Y. Ono, A. Nabara, and T. Ikeda, US 4476195 (1984).
[61] R. H. Doyen, US 4618527 (1986).
[62] P. Breant, WP 91 12291.
[63] F. Possemeyer and W. Schlarmann, DE 3724413 (1989).
[64] K. Masahiro and I. Seiichiro, EP 249342 (1987).
[65] G. Mueller, J. Pohrt, and G. Hatzmann, DE 4009065 (1990).
[66] D. Rudd and S. W. Rice, EP 323599 (1989).
[67] K. Worschech, P. Wedl, H. Kachel, and G. Schult, DE 3630778 (1988).
[68] L. G. Bourland, US 4665118 (1987).
[69] D. J. Bower, US 4554694 (1985).
[70] L. R. Brecker and C. Keeley, US 4500665 (1985).
[71] K. Worschech, G. Stoll, E. U. Brand, and P. Wedl, DE 4018293 (1991).
[72] J. Boussely, EP 439395 (1991).
[73] L. Bothe and G. Crass, DE 3509384 (1986).
[74] G. Crass, L. Bothe, and S. Janocha, DE 3444158 (1986).
[75] G. Crass, L. Bothe, W. Dietz, and H. Pfeiffer, DE 3502680 (1986).
[76] G. Crass, S. Janocha, and L. Bothe, DE 3444866 (1986).
[77] K. Kondo, N. Ishiguro, T. Wano, T. Tada, and T. Yoshida, EP 454420 (1991).
[78] R. Belloni, J. K. Keung, and L. E. Keller, EP 411968 (1991).
[79] R. Belloni and M. A. Nahmias, EP 378445 (1990).
[80] A. Pernot, EP 365442 (1990).
[81] L. Bothe, G. Crass, and G. Schloegl, DE 3637471 (1988).
[82] E. Fleury, L. Tabus, and L. Vovelle, WO 91 07456.
[83] H. C. Park and A. M. Nahmias, EP 181065 (1985).
[84] E. L. Grubb, US 4756991 (1988).
[85] K. Ohno, M. Yoshimoto, and Y. Ozaki, DE 3719342 (1987).
[86] H. Hamano, K. Hasegawa, Y. Noumi, and H. Katoh, EP 229670 (1987).
[87] J. Kersten and L. Lecomte, WO 87 (1990).
[88] C. M. Lulham and G. G. Toney, EP 487268 (1992).
[89] I. Schinkel, U. Reiners, and A. Krallmann, DE 4006402 (1991).
[90] P. P. Shirodkar, US 4957972 (1990).
[91] D. W. Dobreski, EP 262795 (1988).
[92] M. Kioka, M. Nakano, and A. Toyota, EP 360491 (1990).
[93] R. Bachl and H. Vogt, DE 3633131 (1988).
[94] Y. Uchiyama, *Intern. Polym. Sci. Technol.*, **19**, T98 (1992).
[95] W. H. Harrop, US 4020048 (1977).
[96] Y. Tatsukami, T. Futami, and Y. Oda, US 3864433 (1975).

[97] A. Nakamura, US. 4341675 (1982).
[98] T. J. Boran and H. C. Stevens, US 3953223 (1976).
[99] R. J. Eldred, US 3988227 (1976).
[100] Y. Ishida, M. Mitsuboshi, H. Inoue, I. Otsuka, and K. Iio, EP 343566 (1989).
[101] H. Nakahama, T. Kimura, and T. Misima, EP 358393 (1990).
[102] W. E. Peters, US 4962136 (1990).
[103] G. Bayan and A. S. Esposito, US 4978714 (1990).
[104] D. W. Carlson and W. D. Breach, US 5100950 (1992).
[105] R. Nakata and H. Hayashi, DE 4034541 (1991).
[106] Y. Ishida, M. Mitsuboshi, H. Inoue, I. Otsuka, and K. Ilio, EP 343566 (1989).
[107] G. Bayan and A. S. Esposito, US 4978714 (1990).
[108] U. Vogt and W. Wenneis, DE 3816830 (1989).
[109] A. C. M. Yang, J. E. Ayala, A. Bell, and J. C. Scott, *Wear*, **146**, 349 (1991).
[110] O. Jacobs, K. Friedrich, G. Marom. K. Schulte, and H. D. Wagner, *Wear Mater.*, **7**, 495 (1989).
[111] K. Friedrich and O. Jacobs, *Compos. Sci. Technol.*, **43**, 71 (1991).
[112] K. L. Price, US 5147918 (1992).
[113] W. Baun and W. Denneler, DE 4041534 (1992).
[114] P. Laflin, J. E. Kerwin, G. Colley, and D. R. Newton, WO 90 13592.
[115] K. H. Matucha, T. Steffens, H. P. Baureis, W. Bickle, and J. Braus, DE 3815265 (1989).
[116] (a) T. Minamisawa, M. Mitsuuchi, and H. Kitamura, EP 339910 (1989).
      (b) A. Buonaura and J. W. Moore, WO 89 03407.
[117] P. R. Doose, WO 86 00326.
[118] G. M. Lorenz, E. Gebauer, U. Schuster, M. Tschacher, and B. Schoenrogge, EP 450488 (1991).
[119] R. L. Brandon, A. R. Pokora, and W. L. Cyrus, EP 452100 (1991).
[120] M. Kobayashi and T. Nagata, GB 2241246 (1991).
[121] S. O. Friend, J. J. Barber, R. D. Creehan, and C. E. Snyder, WO 91 01621.
[122] K. Seki and N. Hashimoto, DE 3939481 (1990).
[123] E. Parker and B. Grele, US 4775705 (1988).
[124] W. E. Morton and J. W. S. Hearle, *Physical Properties of Textile Fibres*, Heinemann, London (1975).
[125] H. F. Mark, S. M. Atlas, and E. Cernia (Eds), *Man-Made Fibers—Science and Technology*, Interscience, New York (1968).
[126] G. W. Davis and J. R. Talbot, in *Encyclopedia of Polymer Science and Engineering*, Wiley, New York (1988), Vol. 12, p. 118.
[127] H. G. Howell, K. W. Miezskis, and D. Tabor, *Friction in Textiles*, Butterworths, London (1959).
[128] W. A. Zisman, *Adv. Chem. Ser.*, **43**, 1 (1964).
[129] R. B. Login and D. D. Newkirk, US 4165405 (1979).
[130] D. D. Newkirk, B. Thir, and R. B. Login, US 4179544 (1979).
[131] R. B. Login and D. D. Newkirk, US 4199647 (1980).
[132] R. B. Login and D. D. Newkirk, US 4217390 (1980).
[133] R. B. Login and D. D. Newkirk, US 4245004 (1981).
[134] M. Takahashi, K. Yoshida, S. Ohwaki, and H. Moriga, EP 24375 (1981).
[135] D. D. Newkirk and B. Thir, US 4259045 (1981).
[136] E. M. Dexheimer, US 4384104 (1983).
[137] B. Thir, S. E. Eisenstein, and E. M. Dexheimer, EP 90273 (1983).
[138] E. M. Dexheimer, M. J. Anchor, B. Thir, and S. E. Eisenstein, US 4426301 (1984).
[139] D. D. Newkirk, US 4460486 (1984).

[140] B. Thir and S. E. Eisenstein, US 4470914 (1984).
[141] R. L. Camp, E. M. Dexheimer, and B. Thir, US 4496632 (1985).
[142] R. B. Login and M. J. Anchor, EP 189804 (1986).
[143] E. M. Dexheimer, EP 197355 (1986).
[144] R. Kleber, S. Billenstein, L. Jaeckel, and I. Wimmer, DE 3701303 (1988).
[145] P. M. Murphy, EP 458356 (1991).
[146] E. A. Hosegood and L. E. Saufert, US 4565717 (1986).
[147] E. M. Dexheimer, US 4503212 (1985).
[148] P. D. Seemuth, US 4999120 (1991).
[149] J. Wickelhaus, S. Rebouillat, J. Andres, and W. Gruber, EP 392477 (1990).
[150] S. Makino, EP 358023 (1990).
[151] G. H. Greene, US 4463035 (1984).
[152] T. J. Prosser, EP 261337 (1988).
[153] S. C. Chu, P. D. Heilman, and K. A. Kirk, EP 376560 (1990).
[154] P. Corsi and S. Traversier, FR 2654674 (1991).
[155] S. C. Chu and K. A. Kirk, EP 434280 (1991).
[156] H. G. Fiechtmeier, EP 417901 (1991).
[157] E. C. Culbertson and J. M. Heberger, EP 436152 (1991).
[158] D. J. Steklenski, US 4404276 (1983).
[159] J. Israelachvili, *J. Vac. Sci. Technol.*, **A10**, 2961 (1992).
[160] G. McClelland and S. R. Cohen, *Springer Ser. Surf. Sci.*, **22**, 419 (1990).
[161] M. Yang, S. K. Ganapathi, R. D. Balanson, and F. E. Talke, *IEEE Trans. Magn.*, **27** (1991).
[162] A. B. Van Groenou, *J. Magn. Magn. Mater.*, **95**, 289 (1991).
[163] K. Tanaka, S. Ueda, and Y. Enoki, *Surf. Coat. Technol.*, **43–44**, 790 (1990).
[164] P. Hedenqvist, M. Olsson, S. Hogmark, and B. Bhushan, *Wear*, **153**, 65 (1992).
[165] S. Jahanmir, S. M. Hsu, and R. G. Munro, *ASTM Spech. Tech. Publ.*, **1017**, 340 (1989).
[166] M. Nishimatsu, O. Shinoura, S. Shimada, K. Tamazaki, and Y. Kubota, DE 3347528 (1984).
[167] K. Namikawa, T. Goshima, T. Hamaguchi, M. Nakamura, and A. Kuroda, DE 3414055 (1984).
[168] H. Zaitsu, T. Yamase, S. Yamashita, and K. Mizushima, US 4223361 (1980).
[169] P. Datta and E. S. Poliniak, US 4410748 (1983).
[170] G. Caporiccio and M. A. Scarati, EP 194465 (1986).
[171] M. Shoji, T. Nakakawaji, Y. Ito, S. Komatsuzaki, and A. Mukoh, EP 361346 (1990).
[172] G. Caporiccio, E. Strepparola, and M. A. Scarati, EP 165650 (1985).
[173] R. A. Janssen and G. D. Sorenson, US 4642246 (1987).
[174] T. Hamaguchi, K. Namikawa, M. Nakamura, A. Kuroda, and T. Akai, US 4646284 (1987).
[175] A. Kuroda, A. Hata, and N. Fujii, EP 181753 (1986).
[176] T. Hamaguchi, K. Namikawa, T. Goshima, M. Nakamura, A. Kuroda, and Y. Kudo, DE 3415029 (1984).
[177] P. Dutta and R. N. Friel, US 4416807 (1983).
[178] N. J. Probst, J. Iker, and J. Autin, US 4412941 (1983).
[179] D. A. Berry, J. R. Preston, and L. J. Hillenbrand, US 4342660, (1982).
[180] S. Kimura, M. Somezawa, Y. Hinoto, and H. Yoshioka, DE 3341613 (1984).
[181] P. Datta and R. N. Riel, US 4416807 (1983).
[182] C. C. Wang, L. Ekstrom, T. C. Lausman, and H. Wielicki, DE 2934282 (1980).
[183] C. C. Wang and R. F. Bates, DE 3202493 (1982).
[184] C. C. Wang and R. F. Bates, US 4414660 (1983).

[185] I. Shidlowsky and W. E. Harty, US 4416789 (1983).
[186] D. A. Berry, US 4351048 (1982).
[187] L. J. Hillebrand, J. R. Preston, and D. A. Berry, US 4340629 (1982).
[188] N. V. Desai, US 4524109 (1985).
[189] C. O. Edwards, EP 472268 (1992).
[190] V. R. Cross and C. B. Shulman, WO 91 17900.
[191] P. E. Jourdain and S. B. Ohlsson, EP 341068 (1989).
[192] T. K. Larsen, H. O. Larsen, and P. Wolff, WO 90 05162.
[193] R. A. Burns and J. W. Vanwyk, US 4694038 (1987).
[194] K. Y. Chihara and G. M. Lenke, US 5115007 (1992).
[195] R. K. Elton, EP 454293 (1991).
[196] K. Y. Chihara and G. M. Lenke, EP 430421 (1991).
[197] K. R. Strudwick, R. C. Wallis, and R. D. Bridges, WO 90 06598.
[198] K. Yonekura and H. Shimotsu, EP 482480 (1992).
[199] W. D. Best and E. T. Fisher, WO 88 02395.
[200] V. G. Romberg, P. H. Kiang, and W. T. Curry, WO 88 08012.
[201] H. Wollenberg and H. Schneider, DE 3415381 (1985).
[202] G. J. Tennernhouse, US 3423377 (1969).
[203] Quantum, Inc., GB 1120803 (1968).
[204] Quantum, Inc., GB 1120804 (1968).
[205] M. Morra, E. Occhiello, and F. Garbassi, Brev. It. MI 91 A 002528 (1991).
[206] Y. Uyama, H. Tadokoro, and Y. Ikada, *J. Appl. Polym. Sci.*, **39**, 489 (1990).
[207] W. J. Burlant, US 3632400 (1972).
[208] R. Payne and W. Goetz, EP 327983 (1989).
[209] R. A. Philips and T. A. Haddad, EP 102224 (1983).
[210] H. Steinberger, M. Schoenfelder, H.-H. Moretto, and C. Wegner, US 4315091 (1982).
[211] R. F. Hegel, US 4785064 (1988).
[212] R. H. Chung, US 4478876 (1984).
[213] H. Weber, US 4490495 (1984).
[214] L. T. Ashlock, H. Mukamal, and W. H. White, US 4540634 (1985).
[215] C. Bluestein and M. S. Cohen, US 4738899 (1988).
[216] C. B. Quinn, R. R. McClish, and J. E. Moore, US 4714657 (1987).
[217] Y. Kishima, US 4668588 (1987).
[218] B. D. Scott and J. W. Watkins, US 4617194 (1986).
[219] J. E. Moore, US 4557975 (1985).
[220] J. E. Moore, US 4552936 (1985).
[221] M. Schoenfelder, H. Steinberger, and H. Schmid, US 4456647 (1984).
[222] R. A. Sallavanti, J. L. Dalton, and S. M. Olsen, US 4800122 (1989).
[223] A. Factor and G. A. Patel, US 4680232 (1987).
[224] J. C. Goossens, US 4615947 (1986).
[225] D. R. Olson and T. W. O'Donnell, US 4533595 (1985).
[226] H. A. Vaughn, US 4436851 (1984).
[227] D. R. Olson, US 4439494 (1984).
[228] F. Sawaragi and M. Funaki, EP 475149 (1992).
[229] J. D. Basil, H. Franz, C. C. Lin, and R. M. Hunia, EP 465918 (1992).
[230] J. L. Cottington and A. Revis, EP 439293 (1991).
[231] C. C. Lin, J. D. Basil, and R. M. Hunia, US 5035745 (1991).
[232] J. L. Cottington and A. Revis, EP 439294 (1991).
[233] G. A. Patel, EP 339257 (1989).
[234] J. E. Doin and S. E. Hayes, US 4443579 (1984).

[235] C. C. Lin, B. E. Yoldas, R. M. Hunia, J. D. Basil, and C. A. Falleroni, EP 263428 (1988).
[236] G. Boccalon, A. Tintinelli, P. Carciofi, M. De Antoniis, and G. Mazzamurro, EP 242889 (1987).
[237] J. A. G. Gent, EP 215676 (1987).
[238] P. Ascarelli, G. Boccalon, and M. De Antoniis, EP 146995 (1985).
[239] R. B. Frye, US 4477519 (1984).
[240] D. R. Olson, US 4439494 (1984).
[241] J. P. Lampin and Y. Leclaire, EP 89279 (1983).
[242] D. R. Olson and O. V. Orkin, 4382109 (1983).
[243] J. R. January, US 4355135 (1982).
[244] R. H. Chung, US 4348462 (1982).
[245] R. H. Chung, WO 82 00295.
[246] W. D. Kray, US 4298655 (1981).
[247] J. S. Humphrey, US 4232088 (1980).
[248] J. T. Conroy, US 4311763 (1982).
[249] D. R. Olson, O. V. Orkin, and K. K. Webb, US 4308317 (1981).
[250] R. B. Frye, US 4299746 (1981).
[251] R. W. Ubersax, US 4177315 (1979).
[252] H. A. Clark, US 3986997 (1976).
[253] R. H. Baney and F. K. Chi, US 4275118 (1981).
[254] A. Revis and C. W. Evans, US 5075348 (1991).
[255] D. Katsamberis, EP 442305 (1991).
[256] J. L. Cottington and A. Revis, US 4973612 (1990).
[257] E. T. Crouch and R. F. Sieloff, EP 371413 (1990).
[258] S. Tayama and T. Ishii, EP 331087 (1989).
[259] T. A. Misev, EP 336474 (1989).
[260] S. R. Kerr, EP 323560 (1989).
[261] H. Belmares, EP 314979 (1989).
[262] E. T. Crouch, EP 228671 (1987).
[263] R. H. Chung, US 4348462 (1982).
[264] E. H. Kroeker and W. J. Croxall, US 244655 (1948).
[265] E. H. Kroeker, US 2997745 (1958).
[266] L. J. Cottington, EP 445791 (1991).
[267] A. Fujioka, K. Sakiyama, A. Takigawa, and M. Yoshida, US 4405679 (1983).
[268] C. R. Coleman, DE 3323684 (1984).
[269] F. Renzi, F. Rivetti, U. Romano, and C. Gagliardi, EP 241997 (1987).
[270] B. L. Laurin, US 4064308 (1977).
[271] G. A. Patel, US 5041313 (1991).
[272] J. C. Goossens, EP 227070 (1987).
[273] K. L. Benjamin, EP 119482 (1984).
[274] P. Hirshmann, DE 4025811 (1992).
[275] M. G. Tilley and P. M. Miranda, EP 445570 (1991).
[276] R. A. Philips and T. A. Haddad, EP 102224 (1984).
[277] D. R. Olson, US 4410594 (1983).
[278] D. R. Olson,US 4404257 (1983).
[279] Y. Leclaire, EP 429325 (1991).
[280] H. Vetter, W. Hoess, and W. Siol, DE 3642138 (1988).
[281] J. H. Im and W. E. Shrum, EP 138194 (1985)
[282] (a) J. Lehmann, H. Vetter, and W. Arnold, EP 65619 (1982).
[282] (b) J. R. Hollahan and A. T. Bell (Eds), *Techniques and Applications of Plasma Chemistry*, Wiley, New York (1974).

[283] J. L. Vossen and W. Kern (Eds), *Thin Film Processes*, Academic Press, New York (1978).

[284] H. V. Boenig, *Plasma Science and Technology*, Cornell University Press, London (1982).

[285] H. Yasuda, *Plasma Polymerization*, Academic Press, Orlando, (1985).

[286] D. L. Cho and H. Yasuda, *J. Appl. Polym. Sci., Appl. Polym. Symp.*, **42**, 139 (1988).

[287] D. L. Cho and H. Yasuda, *Polym. Mater. Sci. Eng.*, **56**, 420 (1987).

[288] M. Sakai, EP 264227 (1988).

[289] R. A. Auerbach, US 4188246 (1980).

[290] R. Chasset, G. Legeay, J. C. Touraine, and B. Arzur, *Eur. Polym J.*, **24**, 1049 (1988).

[291] Y. Momose, T. Takada, and S. Okazaki, *J. Appl. Polym. Sci., Appl. Polym. Symp.*, **42**, 79 (1988).

[292] K. Harada, *J. Appl. Polym. Sci.*, **26**, 3707 (1981).

[293] Y. Okada and Y. Ikada, *Makromol. Chem.*, **192**, 1705 (1991).

[294] R. A. Gorelik, E. A. Kukhovskoi, A. M. Kleiman, N. A. Kleimenov, A. M. Markevich, A. N. Ponomarev, A. A. Silin V. M. Skok, V. L. Talroze, A. V. Khomyakov, and A. Y. Lyapunov, US 4374180 (1983).

[295] Z. Bilkadi and W. A. Hendrickson, US 4568598 (1986).

[296] J. R. Hollahan, US Pat. Appl. 797217 (1977).

[297] T. Wydeven, *Appl. Opt.*, **16**, 717 (1977).

[298] C. W. Reed, S. J. Rzad, and J. C. Devins, US 4927704 (1990).

[299] C. W. Reed, S. J. Rzad, and J. C. Devins, WO 89 01957.

[300] R. C. Custer, A. Register, A. Johncock, S. J. Hudgens, R. Burckardt, and K. Dean, US 4783374 (1988).

[301] J. C. Devins, C. W. Reed, and S. J. Rzad, EP 285870 (1988).

[302] S. J. Rzad, D. J. Conley, and C. W. Reed, US 5156882 (1992).

[303] F. A. Sliemers, U. S. Nandi, P C. Behrer, and G. P. Nance, EP 252870 (1988).

[304] E. S. Lopata and J. T. Felts, EP 299754 (1989).

[305] K. Enke, DE 3838905 (1990).

[306] H. R. Dobler and R. Eichinger, US 3984581 (1976).

Index

Abrasion resistant coatings 444–445
Acid–base interactions 188–190, 194
Acrylonitrile–butadiene–styrene (ABS) 441
Activation energy, dynamics of polymer surfaces 310
Additives, surface diffusion 436–442
Adhesion,
 acid–base interactions 338–339
 aluminium-poly(ethylene) 337
 chemical interactions 337–339
 chemical treatments effect on 350–353
 crosslinking effects 355, 356, 362–363, 373
 crystallinity effects 353
 electrostatic attraction 336–337
 IETS analysis 139
 interdiffusion 336
 morphological effects 334–335
 surface crystallinity effects 362–363
 surface tension effects 348
 theories 332–333
 wetting effects on 339
 wood 334
Adhesion cross test 341
Adhesion tension, effective 172
Adhesion, work of 190, 194
Adsorption,
 free energy 194
 gas–solid 193–194
 on polymers 112
Ageing 372
 crosslinking effect 422

plasma treated polymers 62–64, 228, 307
Airopak^R process 383
Aluminium, as barrier layer 385–388
Angular resolved XPS 94
Antiblocking agents 439
Antifogging 317
Antistaticity 439
Aramid fibres 370–371
Atomic force microscope (AFM) 415
Attenuated total reflectance (ATR), see IRS
Auger electron spectroscopy (AES) 95

Beam damage 103
Benzophenone 265–266
Biaxially oriented film 385–388
Bioadhesion 404
Biocompatibility, plasma polymers 408–409
Biocompatible, intrinsically 411–414
Biomaterials, survey 396
Biomedical devices 395
Biopolymers 403
Blend surfaces 275–285
Blending, for surface modification 275–285
Blends, lamellar morphology 379, 385
Blends, polymer, see Polymer blends
Block copolymer surfaces 285–296
Block copolymers
 by neutron reflectivity 217
 IRS analysis 120

Block copolymers (*cont.*)
  SIMS analysis 84–85
  surface composition 103
  surface induced orientation 84–85
  XPS analysis 103
Block-copolymerization, for surface
    modification 285–296
Blood clothing 403
Blood compatibility, 396–416
  morphological effects 399, 410
Blood motion 398–406
Blood–material interaction, testing 400
Blow molding 382–385, 439
Bromination 386
Butt joint test 343

Capillarity 303
Captive bubble method 168
Carbon fibres 344
Cassie angle 173
Cassie equation 173
Cassie–Baxter equation 175
Cellulose acetate butyrate (CAB) 423
Chain length, PEO 413
Chain motions, PEO 413
Charging, electrostatic 76–77
Chemical shift, X-ray photoelectron
    spectroscopy (XPS) 95, 99
Chemical treatments, for surface
    modification 242–257
Chromic acid etching, effect on
    adhesion 351, 369
Coating 379–382, 442–445
Coating apparatus 380
Coextrusion 379, 385, 436
Coinjection 379
Cold plasma treatment, for surface
    modification 226–229
Collagen 403
Composite materials
  surface studies 113
  IRS analysis 120–121
  fibre-matrix interaction 344–347
Composite materials 369–370, 441
Composite surface 175
Compounding 437–442
Compressive microdebond test 346
Compton effect 263, 264
Conductive polymers 111–112
Conductometric titration 209

Contact angle 161–195, 309
  advancing 165
  captive bubble method 168
  definition 164–166
  heterogeneity contribution 173, 177
  receding 166
  roughness contribution 172
  sessile drop method 166–167
  shear stress contribution 169–171
  stage speed effect 181
  tilting plate method 168
  time dependency 178–182
  Wilhelmy plate technique 168–169,
    180
Contact angle hysteresis 165–166, 169,
    171–182, 320
  kinetic contribution 178–182
  measurement 166–171
  thermodynamic contribution 171–178
Contact lenses 416–430
Contact lenses, wettability 429, 428
Cooling 427
Cornea 419
Corona discharge 358
Corona treatment 369
  for surface modification 225–226
Coupling agents 370
Crazes
  and polymer surface dynamics 57–61
  healing of 57, 58
  TSC spectra of 59–61

Debye interaction 7
Debye length 20–23
  effect of electrolytes on 22, 23
Derivatization 101, 102
Derjaguin approximation 11
Diffuse reflectance spectroscopy (DRS),
  composite materials 125
  filler/matrix interaction 125–7, 112
  Kevlar^R fibres 128
  Kubelka-Munk model 123, 130
  physical basis 122–123
  polymer films 127
  UV-vis mode 123, 125
Diffuse reflectance spectroscopy
    122–128
Diffusion 383
Dipole–dipole interactions 190
Disjoining pressure 419

Dispersive forces, Lifshits–van der
    Waals, forces 320
DLVO theory 27–32
    and bacterial adsorption 29–31
    and protein adsorption 29–31
Double layer 15–27
    Gouy–Chapman theory of 17–22
    Gouy–Stern theory of 22–26
Double-layer potential 211
Double-layer structure 208
DRIFT *see also* DRS
    applications 125–128
    infrared mode 123–124
    instrumentation 124
Drop motion 302
Dynamics, of polymer surfaces 307–315

Electro-osmosis 211
Electrodynamic interactions 190
Electrokinetic phenomena 211, 215
Electron beam treatment, for surface
    modification 232–233
Electron induced vibrational
        spectroscopy
    see also HREELS 141–142
Electrons, inelastic mean free path 93
Electrophoresis 211
Electrostatic double layer 337
Electrostatic forces 207
    and Van der Waals forces, *see* DLVO
        theory 27–32
    in liquid medium 14–27
Electrostatic interaction measurement
    207–216
Emission spectroscopy,
    experimental set-up 136
    infrared 136–137
    physical basis 133–134
Endothelium 402
Energetic beams 360–361
Energy analysers, cylindrical mirror
        analyzer 71
Equilibrium water content (EWC) 427
Etching
    by plasma 228
    of fluoropolymers 243–246
    of polyolefins 246–251
Etching, chemical 350
Ethylene vinylalcohol–ethylene
        copolymers (EVOH) 379, 380
Evanescent wave, in IRS 114

Evaporation 385
External reflection spectroscopy (ERS)
    133–136
    instrumentation 134
    physical basis 133–134
    sensitivity 136
Extrusion 438

Fast atom bombardment mass
        spectrometry (FABMS) 75, 77, 89
Fibre–matrix interaction 344–347
Fibers, critical length 347
Fillers, in composite materials 112, 133,
    139
Flame treatments, effect on adhesion
    353–354
Flame treatment, for surface
        modification 223–224
Flexible lenses 423–427
Flory interaction parameter,
    effect of bulk miscibility 276
    effect of steric interaction 40, 42
    effect on surface enrichment of block
        copolymers 293–295
Fluorination 352, 384–385
Fluorocarbons, water repellents
    327–328
Fluoropolymers 359–362
    etching 243–246
Fluoropolymer films 410
Fluorosiloxane 423
Force–distance law 204–205
Fowkes theory 190, 194
Friction 436–446
Fuel tanks 382–385

Gamma ray treatment, for surface
        modification 232
Gasoline 380
Gibbs free energy 162
Glass fibres 344
Good–Girifalco equation 185
Good–Girifalco theory 404
Gore-tex^R
    water repellents 410–411
Grafting 257–269, 370, 426, 427
    by chemical reactions 260–261
    by glow discharge 267–269
    by ionizing radiation 261–264
    by UV light 264–268, 359
    for surface modification 257–269

Grafting (*cont.*)
  in bulk   359
  IRS analysis   120
  on polymers   110–111
Griffith surface energy   406

Hamaker constant   8–12, 320
  calculation of   11, 12
  effect of the medium on   12
Hard lenses   418, 421–423
Helmholtz free energy   162
Heparin   400
High resolution electron energy loss
    spectroscopy (HREE)   141–142
Holly–Refojo model   178, 181
Hot plasma treatment, for surface
    modification   229–231
Hydrocarbons water repellents   324
Hydrogels   419, 427–430
Hydrolysis,
  for surface modification   251–254
  of Kevlar^R 254
  of poly(ethylene terephtalate)
    251–254
  of polyimides 251–253
Hydrophilicity   317
Hydrophilization   305–307
Hydrophobic force   35
Hydrophobic recovery   307–315,
  372–373
Hydroxyethyl methacrylate (HEMA)
  429

Imaging,
  in NMR   147
  in SIMS   89
  in XPS   113
Implants   406
Impurities, surface active   181
In vitro experiments   398, 402
In vivo experiments   398
Inelastic electron tunneling spectroscopy
    (IETS)   137–141
  adhesion study   139
  experimental set-up   137
  physical basis   138–139
  sample preparation   139
  silane/substrate study   139
Interaction parameters   185
Interdiffusion coefficients, in polymer
  blends   85

Interfacial energetics   403–405, 408–406
Interfacial equilibrium   165
Interfacial free energy   411, 420–421
Interfacial tension, solid–liquid   187
Interlaminar shear strength   370
Internal lubricants   437
Internal reflection element (IRE)
  115–116
Internal reflection spectroscopy (IRS)
  113–122, 426
  CIRCLE mode   117–118
  evanescent wave   114
  fluorescence mode   116, 120
  infrared mode   116–117, 120
  istrumentation   115, 118
  internal reflection element (IRE)
    115–116
  physical basis   113–115
  Raman mode   116–117, 120
  sampling depth   114
  ultraviolet mode   116
Inverse gas chromatography (IGC)
  193–195
  physical basis   193–194
  retention volume   193
Ion beam treatment, for surface
    modification   233–234, 368
Ion bombardment   70
Ion bombardment damage   79–81
Ion bombardment, depletion effects
  368–369
Ion scattering spectroscopy (ISS)   69–75
  physical basis   71

Johnson–Dettre model   173

Kaelble theory   406
Keesom and Debye interactions   338,
  349
Keesom interaction   7
Kubelka–Munk model   123, 130

Lap shear test   343
Laplace equation   165, 321
Laser microprobe microanalysis
  (LAMMA)   91
Laser treatment, for surface
    modification   106, 231–232
Lifshitz theory   11
Lifshitz–van der Waals' components
  190

Lifshitz–van der Waals' forces 339
London interaction 8
Low energy ion scattering (LEIS), *see*
    Ion scattering spectroscopy
Lubricants 440, 443
Lubrication 439

Magnetic media 441, 443
Mechanical interlocking 333–335
Membranes 390
Metallization, for surface modification
    234–235, 362, 365, 367, 385
Methylmethacrylate/
    hydroxyethylmethacrylate
    copolymers 215
Microdebond test 371
Microdroplet pull-off test 346
Microembolization 410
Microphase separated surfaces 414–416
Misting 319
Modification, of surfaces 305

Neumann theory 186
Neutron reflectivity 216–217
Nuclear magnetic resonance (NMR)
    144–147, 413
    compatible polymer blends 144, 147
    CP-MASS set-up 145
    imaging 147
    silane/substrate study 144–146
    solid state 145
    theory 145
Nuclear reaction analysis (NRA) 74

Organic monolayers, dynamic of polymer
    surfaces 315

Packaging 379, 380, 438
Peeling 341–343
Permeability 379, 383, 387–388, 418,
    421–423, 426, 427
Photoacoustic spectroscopy (PAS)
    128–133
    depth profiling 131–132
    infrared mode 130
    instrumentation 130, 132
    physical basis 128–130
    polymer analysis 130–133
    quantitative analysis 131

sample form effect 131
    thermal diffusion length 130
    ultraviolet mode 133
Photoelectric effect 93, 263, 264
Photoinitiator 265
Photosensitizer 265
Photothermal spectroscopy 133
Plasma etching, of polymers 177
Plasma polymerization 389
Plasma polymers 422, 446
Plasma polymers, permeability 390
Plasma proteins 406–407
Plasma treated polymers, SIMS analysis
    88
Plasma treatment 309, 356, 362, 365,
    367, 369–370, 389, 425
    ageing 318–319
    of polymers 105
Plasma treatment, *see* Surface
    modification
Platelet adhesion 398, 400
Point adhesion 355
Poly(acrylates), XPS analysis 103
Poly(acrylic acid) 206
Poly(dimethyl siloxane) 178, 214, 426
Poly(etheretherketone), hydrophobic
    recovery 314
Poly(ethylene oxide) (PEO) 411–414
Poly(ethylene terephtalate) 363–366,
    386
    biaxially oriented film 440
    fibres 442–443
    hydrophobic recovery 314
Poly(ethylene) 383–385, 440–441
    fibres 369–370
    hydrophobic recovery 308, 311
    methods to increase adhesion
        350–359
Poly(hydroxyethyl acrylate) 180
Poly(hydroxyethyl methacrylate)
    (PHEMA) 178, 181, 429
    grafting 318
    hydrophilic surface 315
Poly(methacrylates),
    SIMS analysis 81–83
    XPS analysis 103
Poly(methyl methacrylate) 180, 214,
    318, 416–417, 421, 423, 445
Poly(propylene) 209, 440–441
    grafted 180
    hydrophobic recovery 308, 311

Poly(propylene) (*cont.*)
    methods to increase adhesion
        350–359
Poly(styrene) 214
    dynamics of polymer
        surfaces 312, 318
Poly(tetrafluoroethylene) 184, 360–363
    contact angle 176–177
    hydrophobic recovery 315
Poly(urethanes) 177, 216
Poly(vinyl chloride) 214, 438
Poly(vinylidene chloride) (PVDC) 379,
    380
Poly(2-chloroxylylene) 384–385
Polyamic acids 367
Polyamide, SIMS analysis 81
Polycarbonate hydrophobic recovery
    314, 440–441, 444–445
Polyimide 366–369, 403
Polymer blends,
    IRS analysis 120
    surface composition (by SIMS)
        84–85, 90
    surface composition (by XPS)
        106–110
Polymer blends, diffusion in by neutron
    reflectivity 217
Polymer surfaces, dynamics 307–315
Polymerization from plasma 229
Polymers surfaces, electrical properties
    400–403
Polyolefins 350–359
    etching 246–251
    flame treatment of 224
    oxidation and etching of 246–251
    SIMS analysis 77, 83
Polysaccharides 382
Potentiometric titration 209
Primers 351
Priming 372
Protein adsorption 411–414, 427, 428,
    429
Pull-out test 340, 342, 371

Raman scattering, surface enhanced 142
Reduction, surface of fluoropolymers
    360–361
Reflection-absorption spectroscopy, *see*
    ERS
Rigid gas permeable (RGP) 423–427

Rigid gas-permeable lenses 418
Rutherford backscattering (RBS) 73

Secondary ion mass spectrometry
    (SIMS) 75–92, 309–310, 354, 426
    dynamic 75, 85
    imaging 89
    instrumentation 76–77, 89–91
    interdiffusion coefficients 85
    isotopic studies 85
    physical basis 75–76
    sample charging 76–77
    static 75
    Tandem spectrometer 91
Segmented polyuretanes 414–416
Sessile drop method 166–167
Short range interactions 43–45
Silica 440
Silicon dioxide (SiO2) 389–390
Silicon oxides, as barrier layer 387–388
Silicone rubber 419, 441
Silicones water repellents 325–327
Siloxane–acrylate lenses 426
Siloxanes 389–390
Single fibre pull-out test 346
Sizings 370, 442–443
Slip agents 441
Soft lenses 427–430
Solvation forces 34, 35
Spreading pressure 165
Sputtering, for surface modification
    235–237
Sputtering process 75–76
Steric interactions 35–43
    HVO theory of 39–42
    Mackor model of 37–39
Stern potential 211
Streaming current, *see* Streaming
    potential
Streaming potential 211, 213, 215
    experimental set-up 212, 214
Structural interactions 32–43
Sulphonation 383
Surface analysis techniques 69–219
    acronyms 70
    analysis of polymers 72
    intrumentation 72
Surface charge measurement 208, 215
Surface charges 402
Surface dynamics,
    and ageing 62–64

effect of the interfacing medium on 55–57, 61–63
experimental evidence of 57–65
of polymer 49–66
simulation of 52–55
Surface electromagnetic wave spectroscopy (SEWS) 142
Surface energetics 161–195
Surface energy 436
definition 161–163
Surface energy, polar and dispersive components 406
Surface enhanced Raman scattering (SERS) 142
Surface enrichment,
and Flory interaction parameter 293–295
in block copolymers 285–296
in polymer blends 109, 277–285
spectral simulation (XPS) 110
Surface etching 362
Surface force,
apparatus 201–206
application to polymers 206
measurement 200–207
Surface forces 3–48
Surface functionalization 254–257
Surface ion bombardment 70
Surface modification 223–296
by blending 275–285
by block-copolymerization 285–296
by chemical treatments 242–257
by cold plasma treatment 226–229
by corona treatment 225–226
by electron beam treatment 232–233
by flame treatment 223–224
by gamma ray treatment 232
by grafting 257–269
by hot plasma treatment 229–231
by hydrolysis 251–254
by ion beam treatment 233–234
by laser treatment 231–232
by metallization 234–235
by sputtering 235–237
by UV treatment 230–231
by X-ray treatment 232
of polymers 305
Surface modification techniques, of polymers 105–106
Surface oxidation 350–352
Surface oxides 365

Surface picosecond Raman gain spectroscopy 142
Surface spectroscopies, comparison 148–150
Surface tension,
components 187–192
critical 183–184
definition 161–163
measurement 164, 182–192
of polymers 304
Swelling 180

Tagging, *see* Derivatization
Tear film 419, 420
Textiles 323
Three-phase boundary 164
Thromboembolic cascade 397–399, 402
Thrombogenicity 397, 400–402, 404
Tilting plate method 168
Triboelectrification 318
Tris(trimethylsiloxy)silylpropyl methacrylate 418, 423, 424, 425

Ultraviolet photoelectron spectroscopy (UPS) 95
UV treatment, for surface modification 230–231

Valence band spectra, in XPS 95
Van der Waals forces 6–14
between macroscopic bodies and electrostatic forces 27–32
between microscopic bodies 6–8
Vascular grafts 396–416
Vibrational spectroscopies 133
Vibrational spectroscopy, transmission 143
compatible polymer blends 143
silane/substrate study 143

Water repellents 323–328
Waterproofing 322–323
Waveguide Raman spectroscopy 142–143
Weak boundary layer (WBL) 351–352, 355, 357, 360–361
Wear 436–446
Wenzel angle 172
Wenzel equation, roughness factor 172

Wettability,
  hard lenses   421–423
  porous structures   321–322
  roughness effect   320
Wetting   301–303
Wicking   175
Wilhelmy plate technique   58, 168–169,
    180
  instrumentation   168
Work of adhesion   190, 194, 338, 407

X-ray irradiation effect in XPS   103
X-ray photoelectron spectroscopy
    (XPS)   92–113, 313, 353, 426
  angular resolved   94
  chemical shift   95, 99
  imaging   113
  instrumentation   94

physical basis   93
shake-up peaks   95
valence band spectra   95
X-ray treatment for surface modification
    232

Young angle   164, 172, 182
Young equation   164–165, 183, 185, 188,
    192
Young–Good–Girifalco–Fowkes
    equation
  geometric mean method   188
  harmonic mean method   188

Zeta potential   211–216
Zisman critical surface tension   404
Zisman plot   183–184, 186